e^+e^- Annihilation: New Quarks and Leptons

e^+e^- Annihilation: New Quarks and Leptons

Robert N. Cahn, *Editor*

Lawrence Berkeley Laboratory
University of California

A volume in the Annual Reviews Special Collections Program

The Benjamin/Cummings Publishing Company, Inc.
Menlo Park, California • Reading, Massachusetts • Wokingham, U.K.
Don Mills, Ontario • Amsterdam • Sydney • Singapore
Tokyo • Mexico City • Bogota • Santiago • San Juan

Front Cover: Redrawn from a computer reconstruction of a ψ (3684) $\rightarrow \pi^+ \pi^- \psi$ (3095) $\rightarrow \pi^+ \pi^-$ $e^+ e^-$ event in the LBL-SLAC detector seen in projection on the plane perpendicular to the intersecting beams. [Redrawn from G.S. Abrams *et al. Phys. Rev. Letters* **34,** 1181 (1975).]

Cover art by Rebecca Felt

ISBN 0-8053-1610-8

ABCDEFGHIJ-MA-8987654

The Benjamin/Cummings Publishing Company, Inc.
2727 Sand Hill Road
Menlo Park, California 94025

Annual Reviews Sources

Schwitters, R.F. and K. Strauch from *Ann. Rev. Nucl. Sci.* 1976. 26:89-149. E. Segre, J.R. Grover and H.P. Noyes (Eds.).

Chinowsky, W. from *Ann. Rev. Nucl. Sci.* 1977. 27:393-464. E. Segre, J.R. Grover and H.P. Noyes (Eds.).

Appelquist, T., R.M. Barnett and K. Lane from *Ann. Rev. Nucl. Part. Sci.* 1978. 28:387-499. J.D. Jackson, H.E. Gove and R.F. Schwitters (Eds.).

Goldhaber, G. and J.E. Wiss from *Ann. Rev. Nucl. Part. Sci.* 1980. 30:337-381. J.D. Jackson, H.E. Gove and R.F. Schwitters (Eds.).

Perl, M.L. from *Ann. Rev. Nucl. Part. Sci.* 1980. 30:229-335. J.D. Jackson, H.E. Gove and R.F. Schwitters (Eds.).

Bloom, E.D. and C.W. Peck from *Ann. Rev. Nucl. Part. Sci.* 1983. 33:143-197. J.D. Jackson, H.E. Gove and R.F. Schwitters (Eds.).

Franzini, P. and J.L. Franzini from *Ann. Rev. Nucl. Part. Sci.* 1983. 33:1-29. J.D. Jackson, H.E. Gove and R.F. Schwitters (Eds.).

Kohaupt, R.D. and G.-A. Voss from *Ann. Rev. Nucl. Part. Sci.* 1983. 33:67-104. J.D. Jackson, H.E. Gove and R.F. Schwitters (Eds.).

Publisher's Foreword

The proliferation of scientific information in recent years has been so rapid that carefully written and well referenced reviews are of greater importance to scientists than ever before. Such reviews are critical resources for students just entering a research field and for researchers whose interests are broadening, as well as essential references for many specialists. Recognizing the broad utility of review coverage at this level, Benjamin/Cummings has joined with Annual Reviews Inc. to provide access to its outstanding scientific reviews in new formats: The Benjamin/Cummings—Annual Reviews Special Collections Program. Each volume in this program is dedicated to a single topic of current scientific interest and consists of articles taken from one or more of the Annual Review series. Compiled and introduced by an eminent scientist, the articles in each volume provide review coverage and exhaustive referencing of the original literature in the area discussed. By bringing together the rigorous scholarly standards of Annual Reviews articles and Benjamin/Cummings' worldwide resources and commitment to educational publishing in science, we believe this unique program will be of real utility to those active in science today as well as those who will be active tomorrow.

James W. Behnke
Editor-in-Chief
The Benjamin/Cummings Publishing Company, Inc.
Menlo Park, California
November 1984

Preface

This special collection of articles from the *Annual Review of Nuclear and Particle Science* provides a self-contained introduction to e^+e^- physics and a comprehensive summary of its remarkable achievements. The articles are accessible to students of particle physics and nonspecialists, but at the same time are definitive reviews providing the researcher with an extensive reference work. The articles are written by distinguished participants in the revolutionary discoveries reviewed here: The ψ resonances, the charmed mesons, the τ lepton, and the Υ resonances. The excitement of the discoveries and the painstaking efforts which produced them are clearly conveyed to the reader.

The astonishing announcement on Monday, November 11, 1974 that a very narrow resonance with a mass of about 3.1 GeV had been observed both at the Stanford Linear Accelerator Center and at Brookhaven National Laboratory began a new era in particle physics. The "November Revolution" followed a series of important experimental and theoretical advances establishing that the proton could be understood as being composed of point-like constituents or partons, eventually identified as quarks and gluons. Feynman compared the strongly interacting particles (hadrons) to watches whose constituents could be studied by smashing two watches together, or by looking at one watch with "light," that is, with the virtual photons produced in scattering by electrons. Experiments done at the Stanford Linear Accelerator Center indeed showed clear evidence in support of the parton model.

Annihilation of electrons and positrons permits probing of the structure of matter in an even clearer way than electron-proton scattering. The electromagnetic energy produced in the annihilation is turned into matter without bias towards those constituents found in the proton. Quarks of all kinds can be produced in e^+e^- annihilation so long as the available energy is great enough to create their mass. Without a doubt, the J/ψ resonance at 3.1 GeV was matter of a new sort.

The great discoveries of 1974 were built on the pioneering efforts at Frascati, Orsay, Novosibirsk, and Cambridge, Massachusetts. This early history is reviewed by Schwitters and Strauch, writing not long after the discovery of the J/ψ. The fundamental principles of e^+e^- collisions are presented in the first article in the collection. The brief description of the charmed quark–anti-quark system anticipates future discoveries and includes a discussion of the then experimentally puzzling C-even states. The imposter X(2.8) identified as the 1^1S_0 state is already on the scene.

The second article, by Chinowsky, describes the wealth of data then available on decays of the ψ (3096), i.e., J/ψ, and the ψ (3684), i.e. ψ', including the decays within the $c\bar{c}$ system: $\psi' \rightarrow \psi\eta$, $\psi' \rightarrow \psi\pi\pi$, and $\psi' \rightarrow \chi\gamma$. The branching ratios for the radiative decays to the p-wave states were already near the presently accepted values, but the confusion of the X(2.8) was compounded by the appearance of a *C*-even state at 3.455 GeV with properties in conflict with theoretical expectations. Were these the 1^1S_0 and 2^1S_0 states? Chinowsky found the evidence for these states "mushy." In contradistinction, the existence of charmed mesons was judged "conclusive."

These experimental achievements had been, to a significant extent, anticipated by a number of theorists. The charmed quark was proposed by Bjorken and Glashow, made compelling by Glashow, Iliopoulos, and Maiani, and had its mass estimated by Gaillard and Lee, all before the discovery of the ψ. Indeed, the phenomenology of charmed particles was set out in a review (!) and the existence of a series of $c\bar{c}$ bound states was predicted before November 11, 1974. These and later theoretical developments are reviewed and compared with available data in the article by Appelquist, Barnett, and Lane. The extensive success of the theoretical predictions was clouded only by the continuing confusion over the states at 2.830 GeV and 3.455 GeV.

To confirm that the ψ truly contained the predicted charmed quark required identifying the charmed mesons and establishing that their weak decays conformed to the extended Cabibbo model. The discovery of charmed mesons and the elucidation of their decay modes is reviewed by Goldhaber and Wiss. The ψ (3770) resonance provides a "factory" for D-mesons, without contamination from the vector D^* states. Studies at the ψ (3770) indicated a significant difference between the lifetimes of the charged and neutral D-mesons. The decay patterns of the D^* mesons are especially intricate because the vector-pseudoscalar splitting is nearly equal to the π meson mass, so not all $D^* \rightarrow D\pi$ decays are possible.

The discovery of the τ-lepton did not come overnight as did that of the ψ. Persistent work on isolating events with only an electron and a muon as charged particles in the final state enabled Perl and co-workers finally to establish this unanticipated addition to the list of fundamental fermions. As Perl describes in his article, every test showed the τ-lepton to be an exact analogue of the electron and muon, with its own neutrino and pure V-A weak couplings.

The continuing problem with the *C*-even $c\bar{c}$ states was finally overcome with the Crystal Ball Detector, designed to provide high efficiency, high-resolution detection of photons. Bloom and Peck show how the earlier findings on the p-wave states were confirmed by the Crystal Ball, while the states and 2.830 and 3.455 GeV were discredited. Instead, convincing candidates for the 1^1S_0 and 2^1S_0 states were found at 2.984 and 3.592 GeV, in accord with theoretical expectations. Radiative decays of the ψ revealed new particles outside the $c\bar{c}$ system, potential candidates for gluonic mesons.

The remarkable developments in the $c\bar{c}$ system are now being paralleled in the $b\bar{b}$ system. Franzini and Lee-Franzini review the understanding of the Υ resonances obtained at Cornell and Hamburg. Despite much lower event rates, many states and decays have already been measured. Because an extra level each of p-wave and s-wave bound states exists, the spectroscopy is much more complex than for the ψ system.

The breakthroughs in particle physics arising from e^+e^- annihilation studies were dependent on advances in accelerator physics. The future of the subject is dependent on adequate luminosities being attained at the next two facilities: SLC at SLAC and LEP at CERN. The knowledge obtained from the circular machines, SPEAR, DORIS, PEP, PETRA, and CESR is reviewed by Kohaupt and Voss in the final article.

This collection of articles shows how continued efforts and detector improvements have together produced increasingly refined results. Today, these fields are very active, and new results are appearing in journals and at major conferences. To provide a partial update of the measurements discussed in the reviews, a short appendix is included from the 1984 edition of the *Review of Particle Properties*. Not even these values can be considered complete or final because the τ, c-quark, and b-quark systems continue to be primary topics for experimental work in particle physics and will provide new insights into the fundamental nature of matter in the coming years.

Robert N. Cahn
Berkeley, California
November 1984

CONTENTS

Ann. Rev. Nucl. Sci. 1976. 26: 89–149

1

THE PHYSICS OF e^+e^- COLLISIONS[1]

Roy F. Schwitters
Stanford Linear Accelerator Center, Stanford, California 94305

Karl Strauch
Department of Physics, Harvard University, Cambridge, Massachusetts 02138

CONTENTS

[1] Work supported by Energy Research and Development Administration.

1 INTRODUCTION

Within a ten-year period, experiments investigating the products of collisions between positrons and electrons have grown from eclectic curiosities to become major research programs profoundly contributing to our knowledge of elementary particle physics. This metamorphosis has been paced by significant developments in accelerator design that have allowed the promise of e^+e^- physics to be realized in the laboratory. Today, colliding-beam devices and their associated particle detectors are among the most sought-after research tools in high-energy physics.

The importance of e^+e^- collisions to elementary particle physics is derived from the simplicity provided by the e^+e^- initial state. Discrete additive quantum numbers of this state, such as electric charge, baryon number, lepton number, and strangeness are all zero. Thus the total energy of the electron and positron are available to produce any final-state configuration of particles that satisfies energy-momentum conservation and has zero net charge, baryon number, lepton number, strangeness, etc. As is discussed below, the dynamics of the initial-state electron and positron are well described by the theory of quantum electrodynamics, thereby allowing unknown structures and forces of final-state particles to be probed by the known electromagnetic force. At the high energies of interest here, the dominant state formed by the positron and electron has the quantum numbers of a single photon, namely total spin one, negative parity, and negative charge conjugation. Thus e^+e^- collisions provide a unique opportunity to study high-energy systems of particles in a nearly pure quantum state having the quantum numbers of the photon.

In this article we review what has been learned to date from the study of e^+e^- collisions at energies above the threshold for hadron production. In this field, experimental observations are usually on much firmer ground than their theoretical interpretation, so we concentrate almost exclusively on the experiments. After an introduction to nomenclature and kinematics of e^+e^- collisions, we present a brief description of colliding-beam devices and the history of their development. In Section 2 the underlying theoretical principles necessary to the interpretation of experimental results are presented. The e^+e^- system provides important experimental testing grounds for the theory of quantum electrodynamics at high energies and large values of momentum transfer. Experimental tests of QED are discussed in Section 3. In recent years there has been great interest in the study of hadron production in e^+e^- collisions. We report on the current status of this rapidly changing field in Section 4. Other topics, such as photon-photon processes, new particle searches, and weak interaction effects are included where appropriate within these sections.

1.1 *Notation and Kinematics*

The system of units where $\hbar = c = 1$ is used throughout this article, so that many quantities are expressed in units of energy. The electron rest mass is often ignored in what follows because of the ultrarelativistic nature of electrons in these experiments. The following symbols are frequently used in describing e^+e^- colliding-beam physics:

W = center-of-mass (c.m.) energy of the e^+e^- system.
E = single-beam energy ($= W/2$).
s = square of c.m. energy.
L = luminosity. The conventional units for L are $cm^{-2}\,sec^{-1}$.
θ = production angle of any final-state particle with respect to the incident positron direction.
ϕ = azimuthal angle of any final-state particle measured from the plane of the colliding-beam orbit.
R = ratio of the total cross section for hadron production in e^+e^- annihilation to the total cross section for muon pair production.

That colliding-beam systems permit more efficient use of accelerator energy to obtain a high center-of-mass energy is a simple consequence of elementary mechanics that has been taught to generations of students. Consider the collision of two particles, oppositely directed, of total energies E_1 and E_2, and rest masses m_1 and m_2, respectively. The square of the c.m. energy, s, is easily calculated to be

$$s = 2E_1E_2\{1+\sqrt{[1-(m_1/E_1)^2][1-(m_2/E_2)^2]}\}+m_1^2+m_2^2. \quad 1.1.$$

The most important aspect of Equation 1.1 for high-energy physics is the product E_1E_2; accelerating the "target" particle as well as the "projectile" increases s by essentially the ratio of the target energy to its rest mass, which can be a considerable factor for high-energy beams of light particles such as electrons. In the limit $E_1 = E_2 = E \gg m_1, m_2$, the c.m. energy is simply:

$$W = 2E. \quad 1.2.$$

However, if a projectile of energy $E \gg m_1, m_2$ strikes a stationary target of mass $m_2 = m$, W is given by:

$$W = \sqrt{2mE}. \quad 1.3.$$

For instance, Equations 1.2 and 1.3 show that e^+e^- interactions at $W = 10$ GeV can be studied with head-on collisions of two 5-GeV beams or with a 100,000-GeV e^+ beam hitting a stationary target. Obviously, colliding beams provide the only practical means for studying high-energy e^+e^- collisions.

The kinematics of the final state are also extremely simple in colliding beams because the laboratory and c.m. reference systems are the same (or nearly the same if the beams cross at a small angle).

1.2 *Colliding-Beam Devices*

The advantages in kinematics provided by colliding beams are gained at the expense of rate of interactions. The luminosity L, defined as the interaction rate per unit cross section, is considerably smaller in colliding beams compared to conventional targets because of the extremely low density of particles in a beam. (The interaction rate N for a given process is related to the cross section σ for that process by $\dot{N} = \sigma L$.) The task of obtaining sufficiently high luminosities for performing practical experiments is an epic story in itself, performed by a small number of accelerator physicists working with particle physicists. An introduction to the science of colliding-beam devices has been presented by Sands (1); a more advanced discussion is contained in the review article by Pellegrini (2).

All present colliding-beam devices make use of one or two storage rings into which beams of positrons and electrons are introduced. Storage rings confine the beams to periodic orbits within an ultrahigh vacuum so that the beams continue to circulate for periods up to several hours. At several points around the storage ring, the beams intersect and collisions can take place. There are typically two to eight intersection regions in a colliding-beam device, and at these regions the experimental apparatus is mounted. Electromagnetic forces between the particles of a beam and the storage ring structure, the other beam, and particles within the same beam all limit the maximum attainable luminosity. In general, L may be a strong function of energy of a given storage ring and may vary greatly among different colliding-beam devices of comparable energy. For example, L varies as E^4 for $E < 3.5$ GeV at the SPEAR e^+e^- colliding-beam facility. Transverse beam dimensions, typically 1 mm or less, and the spread of energies within the beams [typically $\delta E/E \simeq 0.05\% \times E$ (GeV)] are governed by quantum fluctuations in the synchrotron radiation process.

The birth of experimental high-energy colliding-beam physics occurred at the end of 1963 inside AdA (Anello di Accumulazione) when Bernadini et al (3) observed single bremsstrahlung produced in the collision between 250-MeV positrons and 250-MeV electrons. This pioneering single-ring device (4), designed and built at the Frascati Laboratory in Italy and tested at the Orsay Laboratory in France, achieved its goal of demonstrating that counter-rotating positron and electron beams of sufficient intensity could be built up, stored, and made to collide so that interesting experiments can be performed.

The first two high-energy physics experiments with colliding beams were performed in 1965; they tested the predictions of quantum electrodynamics (QED) for e^-e^- scattering at previously unattained high values of momentum transfer. The Princeton-Stanford group used intersecting double rings (5) located at Stanford initially at $W = 600$ MeV (6), and later on at $W = 1.12$ GeV (7). The Novosibirsk group used a similar though somewhat smaller device, VEPP-1 (8), up to W values of 320 MeV (9). The information obtained and the lessons learned from the building, debugging, and operation of the two double-ring devices (no longer in operation), were most important for the subsequent development of higher-energy colliding-beam systems.

Definite results on hadron production in e^+e^- collisions were first reported at the 1967 Electron-Photon Symposium by the Novosibirsk group (10) and by the Orsay group (11). The Soviet group presented a beautiful excitation curve exhibiting resonance structure for the reaction $e^+e^- \rightarrow \pi^+\pi^-$ at values of W corresponding to the ρ mass; these results were obtained with the single-ring system VEPP-2 (12). The French group, also using a single ring called ACO (13), reported the very large cross section at the energy corresponding to the peak of the ρ mass distribution. VEPP-2 is now part of the higher luminosity system VEPP-2M (14). ACO has continued operation.

With the beginning of operation for experiments in 1971 of the single-ring ADONE (15) at the Frascati Laboratory, the available energy region was expanded to $W = 2.5$ GeV (later to 3.1 GeV), and the luminosity increased substantially. The predictions of QED for several reactions have been verified, and hadron production has and is being studied extensively by several groups. A particularly important result was the observation that the total cross section for hadron production was substantially higher than had been expected for values of W well above the resonance region.

During its brief period of operation (1972–1973), the Cambridge Electron Accelerator BYPASS system (16) demonstrated that "low-beta" sections could be used to increase the luminosity and showed that the hadron production cross section exceeded expectations even more at $W = 4$ and 5 GeV than had been observed with ADONE, suggesting the existence of some new hadronic phenomena. Predictions of QED for several reactions were also verified at these two energies.

e^+e^- colliding beam physics entered a new regime of energy, data rates, and detail of experimental information in 1973 with the start of operation of SPEAR and its magnetic detector (17). SPEAR is a single-ring device located at the Stanford Linear Accelerator Center (SLAC); it operates at c.m. energies between 2.5 and 8 GeV with luminosities in the range from 10^{29} to 10^{31} cm^{-2} sec^{-1} One of its two interaction regions is surrounded by the SLAC/Lawrence Berkeley Laboratory (LBL) magnetic detector, which is able to cover approximately two thirds of 4π solid angle with momentum analysis and particle identifications. Certainly, the zenith of e^+e^- physics to date occurred in November 1974 with the simultaneous announcement of the discovery at SPEAR by the SLAC/LBL group (18) and at the Brookhaven National Laboratory (BNL) by the MIT/BNL group (19) of the very narrow J/ψ resonance, and the subsequent discovery of a second narrow resonance, the ψ', found just ten days after the first at SPEAR (20).

SPEAR was joined in 1974 by DORIS (21) at the German electron-synchrotron laboratory DESY in Hamburg, Germany. DORIS consists of two intersecting storage rings in which the beams collide at an angle of 1.4°. This construction permits future experiments with e^-e^- or e^+e^+ collisions in addition to e^+e^- collisions. DORIS is designed to cover the same energy range as SPEAR with comparable luminosity.

Two new projects are well under way at the end of 1975. The Orsay group is building a novel four-beam device known as DCI (22) that is designed to operate up to $W = 3.6$ GeV with higher luminosity than other systems in this energy region. The Novosibirsk group is working toward the operation of VEPP-4, which should provide e^+e^- collisions up to energies of $W = 14$ GeV (23).

The next generation of colliding-beam devices is planned to commence operation in 1979 to 1980. Two projects, PEP (24) at SLAC and PETRA at DESY (25), are in the final design and initial construction phases. Both are designed to open a new energy regime up to $W \simeq 35$ GeV. A Cornell group (26) has proposed construction of a device to operate up to $W = 18$ GeV.

The operating regions of these e^+e^- colliding beam devices are summarized in Figure 1.

2 THEORETICAL INTRODUCTION

When positrons and electrons collide, they may scatter elastically, annihilate into states containing no e^+e^-, or produce final states containing e^+e^- in addition to other particles. At presently available energies $W \lesssim 10$ GeV, e^+e^- interact primarily through the electromagnetic force and, except for very small and as yet unobserved weak interaction corrections, the e^+e^- system can be described by QED.

2.1 *One-Photon Exchange*

At lowest order in electromagnetic coupling, e^+e^- annihilation to inelastic final states proceeds through the one-photon [1γ] channel represented by Figure 2. The pioneering work by Tsai (27), Cabibbo & Gatto (28) and a recent paper by Tsai (29) are useful references to the physics of the 1γ channel.

The amplitude for any 1γ process is proportional to

$$A \sim \frac{e^2}{s} j_\mu J^\mu, \qquad 2.1.$$

where

$$j_\mu = \bar{v}\gamma_\mu u \qquad 2.2.$$

is the e^+e^- current. $\bar{v}$ and u are the usual Dirac spinors and γ_μ are Dirac matrices.

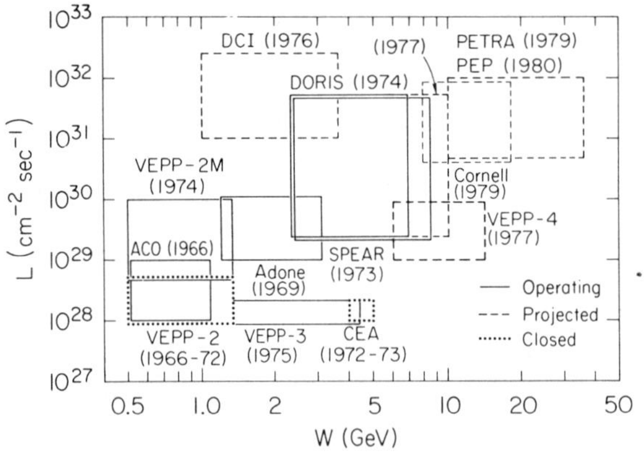

Figure 1 Operating regions of various e^+e^- colliding-beam devices. The indicated luminosities are approximate ranges of peak luminosity and do not represent the actual energy dependence.

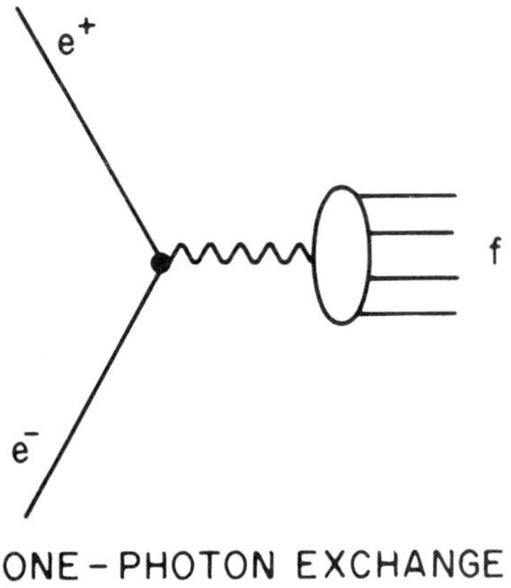

Figure 2 Feynman diagram for one-photon exchange processes.

The $1/s$ factor comes from the photon propagator and e is the electron charge. The "physics" of the final state is contained in the current J^μ; the object of many experiments is to measure matrix elements of various J^μ.

j_μ can be computed by standard techniques (30). In the c.m. frame, j_0 is zero (a consequence of current conservation) and the space components of j_μ are

$$\mathbf{j} = \frac{1}{m}\phi^+\left[E\boldsymbol{\sigma} - \frac{\boldsymbol{\sigma}\cdot\mathbf{p}_+}{E+m}\mathbf{p}_+\right]\chi, \qquad 2.3.$$

where χ and ϕ are two-component Pauli spinors for the e^+e^- spins in their respective rest frames. $\boldsymbol{\sigma}$ are the Pauli spin matrices, m is the electron mass and $\mathbf{p}_+$ is the e^+ 3-momentum. Direct calculation of Equation 2.3 shows that the longitudinal component of $\mathbf{j}$—the component parallel to $\mathbf{p}_+$—is the order of m/E smaller than the transverse component, and therefore negligible in high-energy processes. The transversality of $\mathbf{j}$ is equivalent to the fact that the e^+e^- annihilate only through states of net helicity one.

The beams will naturally polarize in high-energy storage rings due to their interaction with synchrotron radiation (31). A complete discussion of this phenomenon is given by Baier (32), including calculations of various depolarizing effects. A more pedagogical treatment and an excellent list of references to both the theoretical and experimental literature are contained in the review article by Jackson (33). In the absence of depolarizing effects, the polarization P of each beam builds up in time according to

$$P(t) = P_0[1-\exp(-t/T)], \qquad 2.4.$$

$$T \simeq 98 \text{ sec} \times \frac{R_0^3(m)}{E^5\,(\text{GeV})} \times \frac{R_{\text{avg}}}{R_0}, \qquad 2.5.$$

$$P_0 = \frac{8\sqrt{3}}{15} \approx 0.924, \qquad 2.6.$$

where R_0 is the bending radius and R_{avg} is the average radius of the storage ring. The positrons (electrons) are polarized parallel (antiparallel) to the guide magnetic field that we choose to lie along the $-y$ axis. Our coordinate system is defined in Figure 3. At low energies P can generally be neglected, but at high energies the characteristic time T is less than typical storage times, and polarization effects become important. With the e^+e^- completely polarized in opposite directions, the only nonvanishing component of j^μ will lie along the mutual polarization direction, in this case the y axis. If this mutual polarization direction were to be aligned with the beam direction, j^μ, and hence 1γ, exchange processes would vanish.

Symmetry principles and gauge invariance restrict the possible forms for the final-state current J^μ. Following Tsai (29), the most general form for single-particle inclusive cross sections in the 1γ channel can be written

$$E_f \frac{d\sigma}{d^3\mathbf{p}_f} = \frac{\alpha^2}{2s^2}[(W_1+W_0)+(W_1-W_0)(\cos^2\theta+P^2\sin^2\theta\cos 2\phi)], \tag{2.7.}$$

where $\mathbf{p}_f$ and E_f are the single-particle momentum and energy. P^2 is the product of the e^+, e^- transverse polarizations, which are assumed to be oppositely directed in the y direction. The structure functions W_0 and W_1 are functions of s, E_f, and the type of particle produced. They are defined by

$$W_0 = \sum_{\substack{\text{all final}\\ \text{states}\\ \text{except } \mathbf{p}_f}} (2\pi)^3\,\delta^4(P_i-P_f)\,|\langle f|J_{z'}|0\rangle|^2, \tag{2.8.}$$

$$W_1 = \sum_{\substack{\text{all final}\\ \text{states}\\ \text{except } \mathbf{p}_f}} (2\pi)^3\,\delta^4(P_i-P_f)\,|\langle f|J_{t'}|0\rangle|^2, \tag{2.9.}$$

where $\langle f|J_{z'}|0\rangle$ is the matrix element of the component of J^μ parallel to $\mathbf{p}_f$ that gives rise to the state f; $\langle f|J_{t'}|0\rangle$ is the matrix element of a component of J^μ perpendicular to $\mathbf{p}_f$; P_i and P_f are the net four-momenta of the initial and final states, respectively; δ is the Dirac delta-function; and W_0 represents the probability that the final state has zero net helicity along $\mathbf{p}_f$, while W_1 gives the probability that the final state has net helicity one along $\mathbf{p}_f$.

Bjorken (34) has argued that at high energies hadron production by 1γ annihilation should exhibit scaling, such that W_0 and W_1 become functions of only one dimensionless quantity x, the ratio of E_f to the beam energy E. If scaling holds, the inclusive cross section can be written

$$\frac{d\sigma}{d\Omega\,dx} = \frac{\alpha^2}{8s}\beta x\{[W_1(x)+W_0(x)]+[W_1(x)-W_0(x)](\cos^2\theta+P^2\sin^2\theta\cos 2\phi)\}, \tag{2.10.}$$

where $\beta = p_f/E_f$ is the particle velocity and $x = E_f/E$. The total cross section for a process that exhibits scaling falls with energy as s^{-1}.

Some specific examples of 1γ annihilation processes follow.

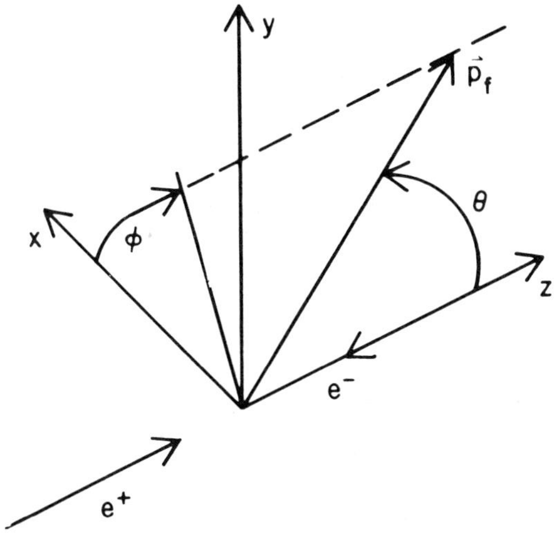

Figure 3 Coordinate system. The incident e^+ direction is the z axis and the guide magnetic field points in the $-y$ direction. Any final-state particle is described by the momentum vector $\mathbf{p}_f$ and polar and azimuthal angles θ and ϕ.

2.1.1 $e^+e^- \to \mu^+\mu^-$ This is the standard of reference for all 1γ processes. The structure functions for a pair of pointlike spin-1/2 particles can be calculated from QED and are

$$W_0 = 2\delta(E - E_f)m_l^2/E_f, \qquad W_1 = 2\delta(E - E_f)E_f, \tag{2.11.}$$

where m_l is the muon rest mass.

The differential cross section is given by

$$\frac{d\sigma_\mu}{d\Omega} = \frac{\alpha^2}{4s}\beta[(2-\beta^2) + \beta^2(\cos^2\theta + P^2\sin^2\theta\cos 2\phi)]. \tag{2.12.}$$

As usual, β is the muon velocity.

The total cross section for muon-pair production is

$$\sigma_\mu = \frac{2\pi\alpha^2}{3s}\beta(3-\beta^2). \tag{2.13.}$$

At high energies where $\beta \approx 1$, σ_μ is numerically

$$\sigma_\mu \simeq \frac{21.7\,(\text{nb})}{E^2\,(\text{GeV})} = \frac{86.8\,(\text{nb})}{s\,(\text{GeV})^2}. \tag{2.14.}$$

2.1.2 $e^+e^- \to \pi^+\pi^-$ The current for two spinless particles of unknown charge structure contains one unknown function of s, the form factor $F(s)$. The cross section for pion-pair production is

$$\frac{d\sigma}{d\Omega} = \frac{\alpha^2}{8s} |F_\pi(s)|^2 \beta^3 \sin^2\theta (1 - P^2 \cos 2\phi). \qquad 2.15.$$

2.1.3 $e^+e^- \rightarrow p\bar{p}$ In this case, when a pair of spin-1/2 particles with internal structure are produced, the current will contain two independent form factors. One choice of form factors is to multiply W_0 and W_1 for the μ-pair case (Equation 2.11) by $|G_E(s)|^2$ and $|G_M(s)|^2$ respectively. The cross section for $p\bar{p}$ production can then be written:

$$\frac{d\sigma}{d\Omega} = \frac{\alpha^2}{4s} \beta [|G_M|^2 (1 + \cos^2\theta + P^2 \sin^2\theta \cos 2\phi)$$

$$+ (1 - \beta^2)|G_E|^2 \sin^2\theta (1 - P^2 \cos 2\phi)]. \qquad 2.16.$$

2.2 *Other QED Processes of Order* α^2

In addition to μ-pair production, two other purely electromagnetic processes of order α^2 occur in e^+e^- interactions. The elastic scattering of e^+e^-, known as Bhabha scattering, may be represented by the Feynman diagrams in Figure 4. In addition to the one-photon exchange contribution discussed above, there is an important t-channel amplitude. The cross section for Bhabha scattering is (29)

$$\frac{d\sigma}{d\Omega} = \frac{\alpha^2}{8s} \frac{1}{(1 - \cos\theta)^2} [(3 + \cos^2\theta)^2 - P^2 \sin^4\theta \cos 2\phi]. \qquad 2.17.$$

Small-angle Bhabha scattering is useful for monitoring luminosity because of the relatively large cross section; it is studied at large angles as a test of QED for large spacelike and timelike values of the virtual photon mass.

The second process is e^+e^- annihilation into two photons. Lowest-order diagrams for this process are given in Figure 5 and the cross section is

$$\frac{d\sigma}{d\Omega} = \frac{\alpha^2}{s} \left[\frac{1 + \cos^2\theta}{\sin^2\theta} + P^2 \cos 2\phi \right]. \qquad 2.18.$$

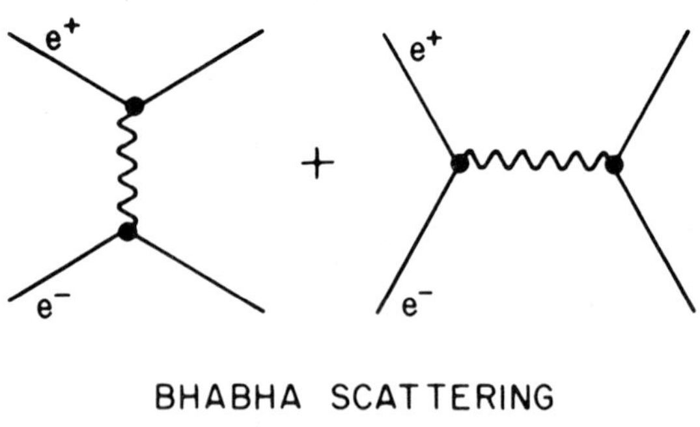

Figure 4 Lowest-order Feynman diagrams for Bhabha scattering.

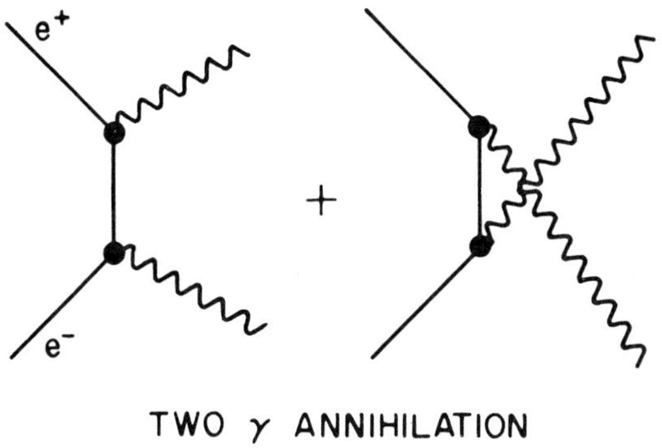

Figure 5 Lowest-order Feynman diagrams for e^+e^- annihilation into two photons.

2.3 *Photon-Photon Processes*

It has been pointed out by a number of authors (35–37) that the cross section for the production of systems of invariant mass, which is small compared to W, together with forward-peaked e^+e^-, will compete with one-photon exchange processes at high energies. These reactions, shown schematically in Figure 6, can be thought of as the collision of two nearly real photons emanating from the initial e^+ and e^-, and, unlike 1γ annihilation, they result in states of positive charge conjugation. There is a vast theoretical literature on this subject; it has been reviewed by Terazawa (38).

The essential physics of photon-photon (2γ) processes can be derived from the equivalent photon approximation (38), where the flux of nearly real photons of energy k associated with a high-energy electron is approximately

$$N(k)\,\mathrm{d}k \simeq \frac{2\alpha}{\pi}\ln\left(\frac{E}{m}\right)(1-k/E+k^2/2E^2)\frac{\mathrm{d}k}{k}, \qquad 2.19.$$

where m is the electron mass.

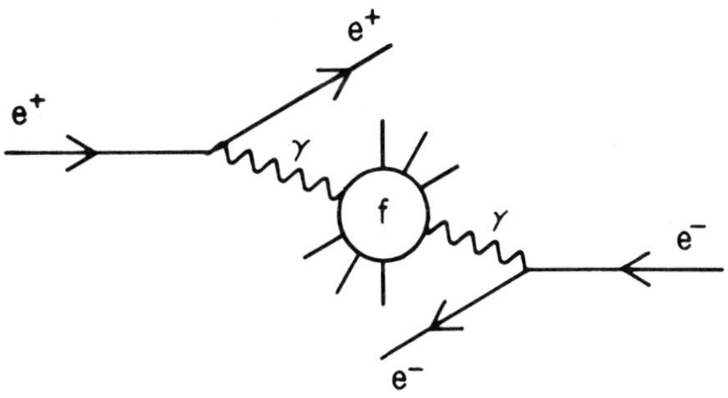

Figure 6 Schematic diagram of photon-photon process leading to the final state e^+e^-f.

The cross section $d\sigma_{ee \to eex}$ for producing the state x through the 2γ process is related to the cross section $d\sigma_{\gamma\gamma \to x}$ for producing x in a collision of two photons by

$$d\sigma_{ee \to eex} = \int dk_1\, dk_2 N(k_1) N(k_2)\, d\sigma_{\gamma\gamma \to x}(k_1, k_2). \qquad 2.20.$$

Even though 2γ processes are higher order in α than 1γ processes, the $[\ln(E/m)]^2$ terms augment the cross section so that at high energies they may exceed 1γ cross sections that are falling as E^{-2}. As opposed to 1γ reactions, 2γ reactions will also occur in e^-e^- or e^+e^+ collisions.

Due to differences in the photon energies, the state x is generally moving relative to the laboratory frame. The spectrum of photons is peaked at very low energies, so it is likely that states of low invariant mass are formed. It is very unlikely to have more than one half of the total e^+e^- energy appear in the state x. The final-state e^+, e^- are strongly peaked about their initial direction, leading to a characteristic signature for these processes.

2.4 *Radiative Corrections*

Emission of additional photons, both virtual and real, over those considered in lowest-order results in modifications to lowest-order cross sections and angular distributions. The application of radiative corrections to obtain bare cross sections from experiment is a subtle problem far beyond the scope of this article. The paper by Tsai (39) is a classic in this field. Bonneau & Martin (40) have calculated the lowest-order radiative corrections to one-photon exchange shown in Figure 7.

Briefly, radiative corrections can be categorized into single hard-photon emission, multiple soft-photon emission, and vertex corrections and vacuum polarization. For hard-photon emission the incident beams can be considered as being accompanied by a flux of photons similar to Equation 2.19. The e^+e^- then annihilate at an energy less than W, where the cross section may be different. The treatment of multiple soft-photon emission dates back to the work of Bloch & Nordsieck (41). In the present context the hard-photon spectrum is usually modified by a factor $(k/E)^t$ to account for multiple soft photons. The effective radiator thickness t is given by

$$t = \frac{2\alpha}{\pi}[\ln(s/m^2) - 1]. \qquad 2.21.$$

Vertex corrections and vacuum polarization effectively modify the flux of the single virtual photons involved in one-photon exchange.

The net radiative correction to 1γ exchange taken from Bonneau & Martin (40) and modified for soft-photon emission is

$$\sigma_{\text{exp}}(s) = \sigma_0(s)\left[1 + \frac{2\alpha}{\pi}\left(\frac{\pi^2}{6} - \frac{17}{36}\right) + \frac{13}{12}t\right]$$
$$+ t\int_0^E \frac{dk}{k}\left(\frac{k}{E}\right)^t (1 - k/E + k^2/2E^2)[\sigma_0(s - 4Ek) - \sigma_0(s)], \qquad 2.22.$$

where $\sigma_{\text{exp}}(s)$ is the experimentally measured cross section and $\sigma_0(s)$ is the one-photon exchange cross section. The first term in Equation 2.22 contains the vertex modification and vacuum polarization corrections; the second term is for external photon emission. Equation 2.22 is displayed here to illustrate the types of radiative corrections involved in e^+e^- collisions. It does not represent a universal formulation to be applied to all cases.

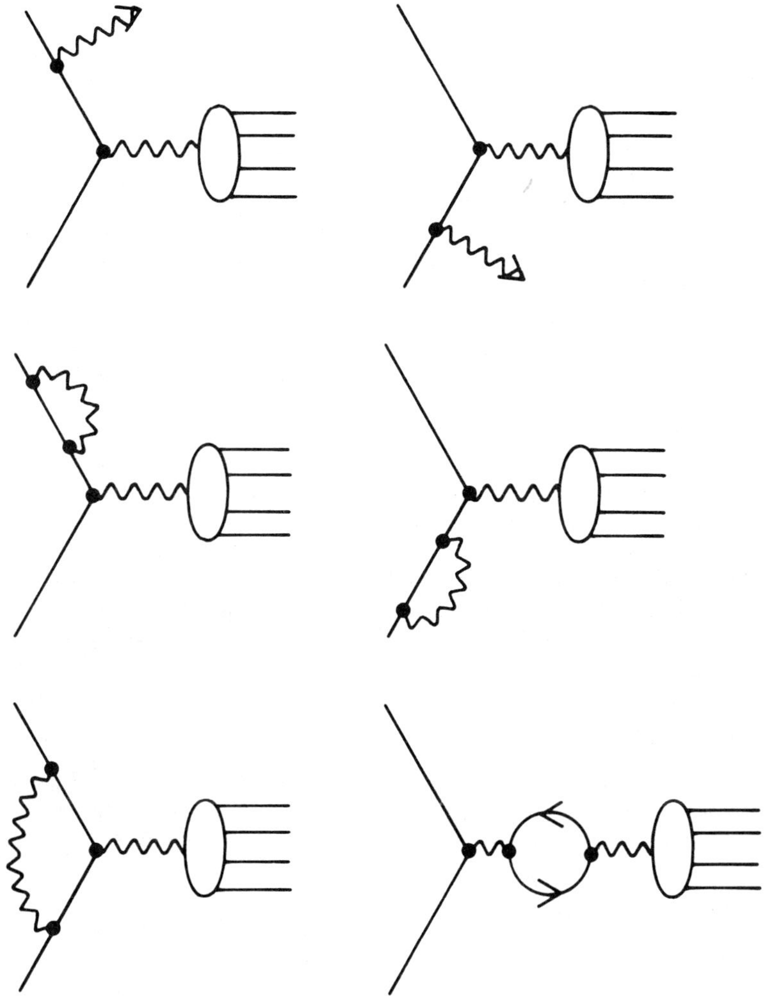

Figure 7 Feynman diagrams showing lowest order radiative corrections to one-photon exchange processes that lead to nonelectrodynamic final states f.

3 ELECTRODYNAMIC FINAL STATES

In this section we review the experimental information on final states of e^+e^- collisions where only leptons and photons are observed. For the most part, the reactions under consideration can be described by the theory of quantum electrodynamics, and therefore these experiments constitute tests of QED at high energies. The relevance of such tests to the theory has been discussed by Brodsky & Drell (42). The basic elements of the theory being tested are the assumptions that both electrons and muons may be treated as pointlike Dirac particles and that the Coulomb field satisfies the $1/r^2$ law. Deviations between experiment and theory would imply one or more of the following: (*a*) a breakdown of the fundamental assumptions of QED; (*b*) interference with nonelectromagnetic sources of electrodynamic final states, such as the strong or weak interactions; and (*c*) the presence of entirely new interactions and particles.

To date, there is no evidence for a failure of the basic assumptions of QED; indeed, it is the most successful theory in physics. There are several examples of hadronic contributions to electrodynamic final states, there is much theoretical anticipation concerning the study of weak interaction effects at the next generation of storage rings (43) but no experimental data, and there is a hint that new electrodynamic particles may be produced in e^+e^- collisions.

3.1 *Breakdown Parameters*

In experimental tests of QED it is conventional to compare experimental yields with the "predictions of QED" in which the theory, including explicit radiative corrections, is calculated for the particular experimental conditions of angular acceptance, momentum and angle resolution, and kinematic cuts. Deviations from QED are usually parameterized by "breakdown" parameters Λ having dimensions of energy. The scale of distance corresponding to a given Λ is $\hbar c/\Lambda$. There is no unique way to modify QED with breakdown parameters; it has become standard practice (44, 45) to replace the usual photon propagator by

$$\frac{g_{\mu\nu}}{q^2} \to \frac{g_{\mu\nu}}{q^2} \mp \frac{g_{\mu\nu}}{q^2 - \Lambda_\mp^2}, \qquad 3.1.$$

where $g_{\mu\nu}$ is the metric tensor and q^2 is the invariant mass of the virtual photon. The choice of sign corresponds to "negative" or "positive" metric corrections. Such a modification would change one-photon exchange cross sections by

$$\sigma_{\text{modified}} \simeq \sigma_{1\gamma}\left(1 \pm \frac{2s}{\Lambda_\mp^2}\right). \qquad 3.2.$$

An alternative parameterization, which leads to modifications similar to Equation 3.2, is to assign form factors to the electron and muon. The charge radius of e or μ would then be proportional to Λ^{-1}.

Based on the work of Kroll (46), reactions involving a lepton propagator, such as $e^+e^- \to \gamma\gamma$ annihilation, are parameterized by

$$\frac{\sigma_{\text{modified}}}{\sigma_{\text{QED}}} = 1 \pm \frac{2q^4}{\Lambda_\pm^4}, \qquad 3.3.$$

where q^2 is the invariant mass of the virtual lepton.

We wish to stress that breakdown parameters should be used only as a guide in interpreting the data. Only detailed examination of individual experiments will disclose how stringent is the particular test of QED.

3.2 $e^-e^- \to e^-e^-$

The e^-e^- final state was the first to be studied experimentally with colliding beams at the Stanford storage rings (6, 7) and at Novosibirsk (9). This process, known as Møller scattering, is represented in lowest order by the diagram of Figure 8. The Stanford apparatus consisted of scintillation counters and optical spark chambers covering approximately $\pm 30°$ in ϕ and the polar angle range $35° < \theta < 115°$. There was no magnetic analysis or independent luminosity monitor. The measured angular distribution at $W = 1112$ MeV is presented in Figure 9, along with QED predictions normalized to the same total number of events and corrected for detector acceptance. The data are in excellent agreement with theory and establish lower limits for breakdown parameters in the range 2–3 GeV (95% confidence level).

3.3 $e^+e^- \to e^+e^-$

Bhabha scattering has been widely studied in e^+e^- colliding-beam experiments. This final state is characterized by a collinear pair of showering particles and it tests the photon propagator for both timelike and spacelike values of momentum transfer. The most readily available theoretical calculation, including detailed radiative corrections, that is suitable for comparison with experiment is given by Berends, Gaemers & Gastmans (47).

The first absolute measurement of Bhabha scattering with colliding beams was performed at ACO (48). A nonmagnetic detector with scintillation counters and

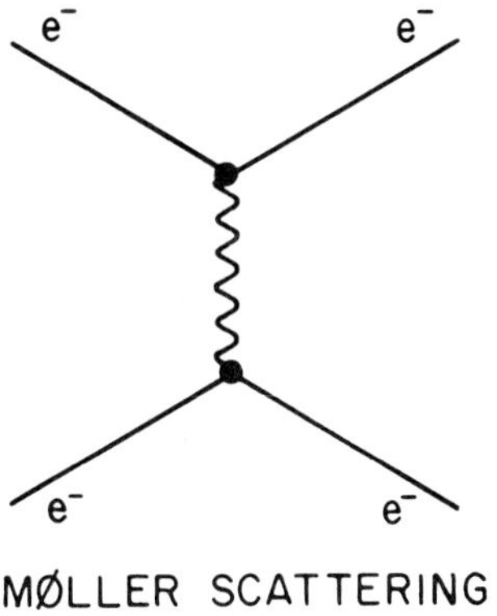

Figure 8 Feynman diagram for Møller scattering.

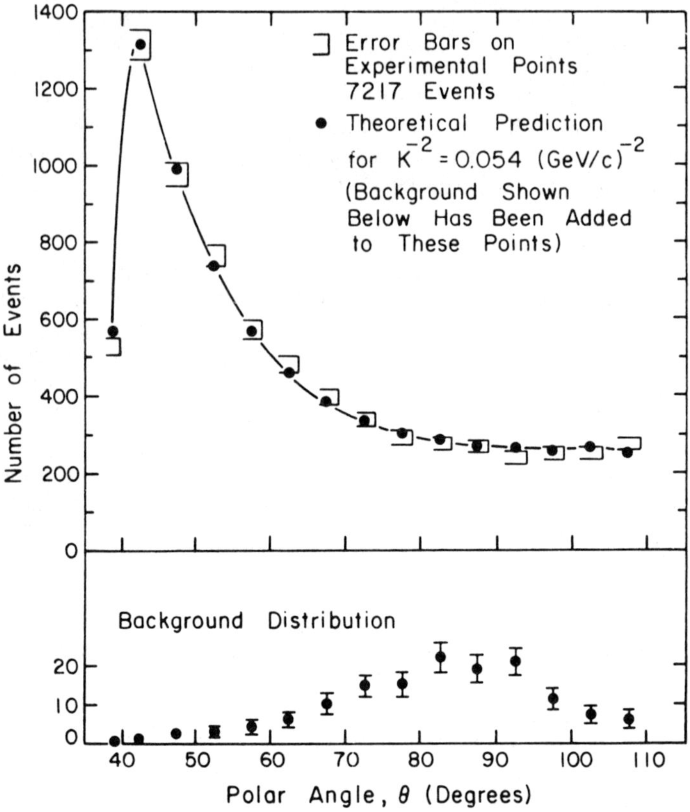

Figure 9 Experimental angular distribution for Møller scattering at $W = 1112$ MeV from (7). The theoretically expected angular distribution, normalized to the total number of events and corrected for detector acceptance, is also shown.

optical spark chambers was used to measure this process at $W = 1020$ MeV. Luminosity was monitored by detecting small-angle gamma rays from double bremsstrahlung ($e^+e^- \rightarrow e^+e^-\gamma\gamma$). This experiment also set breakdown parameter limits in the 2-GeV to 3-GeV range.

At ADONE, several groups used optical spark chambers and scintillation counters to detect large-angle Bhabha scattering; luminosity was monitored through small-angle Bhabha scattering. The early experiments (49, 50) verified the validity of QED in absolute normalization and energy dependence to a level of 5% to 10% over the energy range 1.6 GeV $\leqq W \leqq 2.0$ GeV. The effect of radiative corrections on the collinearity of events was discussed by the "BCF" group (50, 51). The ratio of experimental yields for Bhabha scattering to those theoretically expected from more recent measurements performed by the "boson" group (52, 53) and the "$\mu\pi$" group (54) are plotted in Figure 10 along with results at other energies.

The boson group (53) quoted an average ratio $\bar{R}$ of measured-to-theoretical yield over the energy range 1.4 GeV $\leqq W \leqq$ 2.4 GeV of

$$\bar{R} = 1.05 \pm 0.04 \; (\pm 0.065 \text{ systematic}).$$

The $\mu\pi$ group (54) were able to set the limit

$$\Lambda_+ > 6 \text{ GeV } (95\% \text{ confidence})$$

with data over the same energy range. The BCF group (55) described the energy dependence of the Bhabha scattering yield in their detector over the energy range 1.2 GeV $\leqq W \leqq$ 3.0 GeV by the form:

$$\text{Yield} = As^n$$

and found $A_{\text{exp}}/A_{\text{QED}} = 1.00 \pm 0.02$ and $n = -(0.99 \pm 0.02)$. They also observed acoplanar e^+e^- events that result from radiative corrections (56).

Experiments performed with the nonmagnetic BOLD detector at CEA measured wide-angle Bhabha scattering at $W = 4$ GeV (57) and 5 GeV (58). These measurements, which were normalized to double bremsstrahlung, agreed with QED and

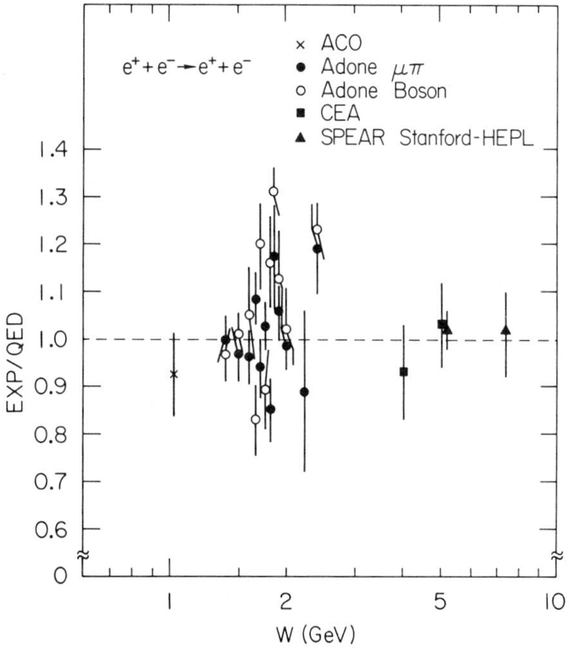

Figure 10 Ratio of experimental yield to that expected from QED for Bhabha scattering. Detector acceptances and normalization procedures differ between experiments. Data points shown are from the following references: ACO (48), Adone $\mu\pi$ (54), Adone Boson (53), CEA (57, 58), and SPEAR (59, 60).

pushed lower limits on breakdown parameters to nearly 10 GeV. These results are summarized in Figure 10. The coplanarity distribution observed in the CEA experiments (58) is in good agreement with theoretical radiative corrections (47).

Two SPEAR groups have reported tests of QED through Bhabha scattering. Using a novel detector consisting of two large NaI crystals, a Stanford-HEPL group (59) measured 90° Bhabha scattering at 5.2 GeV and, more recently (60), have presented preliminary results at 7.4 GeV. These results, plotted in Figure 10, agree with QED and raise the lower limit on breakdown parameters to approximately 20 GeV, an order of magnitude higher than the earliest results discussed above. The SLAC/LBL collaboration (61) measured Bhabha scattering with a magnetic detector at 3.0, 3.8, and 4.8 GeV, and for the first time were able to distinguish the charge of the outgoing particles. The angular distribution obtained in this experiment is presented in Figure 11*a* and is seen to be in excellent agreement with QED. Their data were fitted to several sets of breakdown parameters; the results are given in Table 1. The lower limits are generally the order of 20 GeV, demonstrating the validity of QED in this process to scales of distance less than 10^{-15} cm.

The beam polarization term in Bhabha scattering (see Equation 2.17) has been

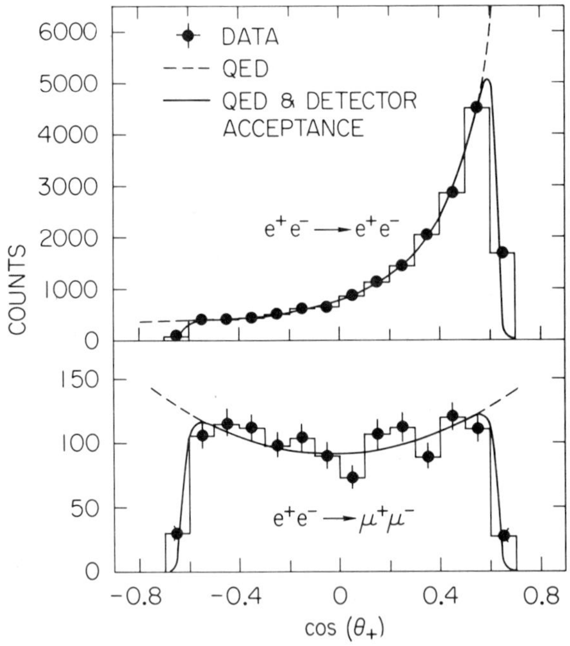

Figure 11 Angular distribution for Bhabha scattering and muon-pair production at $W =$ 4.8 GeV, measured by the SPEAR magnetic detector (61). Predictions of QED normalized to the total number of Bhabha scattering events and corrected for detector acceptance are indicated.

Table 1 Breakdown parameters determined in (61) by fitting to Bhabha scattering and muon-pair production data at three energies, $W = 3.0$, 3.8, and 4.8 GeV[a]

Data used	Model	Fitted parameters (Λ in GeV)	Λ at 95% CL (GeV) positive metric	negative metric
ee only				
	a	$1/\Lambda_S^2 = 0.0008 \pm 0.0022$ $1/\Lambda_T^2 = 0.0013 \pm 0.0031$ correlation coefficient = 0.82	$\Lambda_{S+} > 15$ $\Lambda_{T+} > 13$	$\Lambda_{S-} > 19$ $\Lambda_{T-} > 16$
	b	$1/\Lambda^2 = 0.0007 \pm 0.0022$	$\Lambda_+ > 15$	$\Lambda_- > 19$
$\mu\mu$ and *ee*				
	a	$1/\Lambda_S^2 = 0.0003 \pm 0.0013$ $1/\Lambda_T^2 = 0.0001 \pm 0.0005$ correlation coefficient = 0.23	$\Lambda_{S+} > 21$ $\Lambda_{T+} > 33$	$\Lambda_{S-} > 23$ $\Lambda_{T-} > 36$
	b	$1/\Lambda^2 = 0.0002 \pm 0.0004$	$\Lambda_+ > 35$	$\Lambda_- > 47$
	c	$1/\Lambda_e^2 = 0.0004 \pm 0.0011$ $1/\Lambda_\mu^2 = 0.0014 \pm 0.0021$ correlation coefficient = −0.97	$\Lambda_{e+} > 21$ $\Lambda_{\mu+} > 27$	$\Lambda_{e-} > 19$ $\Lambda_{\mu-} > 16$

[a] The various cases are (*a*) separate breakdown parameters for spacelike (*S*) and timelike (*T*) photons; (*b*) same breakdown parameters for *S*, *T*; and (*c*) separate form factors $ee\gamma$ and $\mu\mu\gamma$ vertices.

verified by a Wisconsin-Pennsylvania group (62) using data from the SLAC/LBL magnetic detector at SPEAR. Their experiment pointed out a sign error in the polarization term that had existed in the theoretical literature until the experiment was performed!

3.4 $e^+e^- \to \mu^+\mu^-$

The study of muon pair production in e^+e^- collisions is a classic test of QED for one-photon exchange processes and is sensitive to breakdowns in the formulation of the photon propagator for timelike values of momentum transfer or to possible μ-e differences. The complete α^3 cross section has been computed by Berends, Gaemers & Gastmans (63). The lowest-order cross section is given in Equation 2.12. Experiments on μ-pair production can be more demanding than Bhabha scattering measurements because of the lower counting rates and the lack of a characteristic electromagnetic shower to aid in particle identification. Cosmic rays can present serious backgrounds.

Balakin et al (64) measured muon pair production with VEPP-2 at four c.m. energies between 1020 and 1340 MeV. The apparatus included both optical and ferrite-core readout spark chambers, scintillation counters, and water-filled Cerenkov counters. Bhabha scattering events observed in the same appartus served to normalize the data. Their data, presented in Figure 12, agree with QED; the authors set a limit on possible breakdown parameters of $\Lambda_\pm > 3.1$ GeV. These results are not sensitive enough to reflect vacuum polarization effects induced by the ϕ meson; this topic is discussed below.

The μ-π group at ADONE (65) measured μ-pair production between 1.5 GeV and 2.1 GeV c.m. energy. They normalized to wide-angle e^+e^- events; the results, summarized in Figure 12, agree with QED within the estimated systematic errors of $\pm 6.5\%$. An early experiment by the BCF group (66) found that the ratio of $\mu\mu$-to-ee yields agreed with QED within $\pm 8\%$ error. Later results from this group (67) are plotted in Figure 12. They have also observed radiative effects (68) that agree with the theoretical calculations of (63).

At SPEAR, the Stanford-HEPL group has published results at $W = 5.2$ GeV (59) and presented preliminary results at $W = 7.4$ GeV (60). Their data, normalized to small-angle Bhabha scattering, are presented in Figure 12, as are results from the SLAC/LBL group (61), which compared $\mu^+\mu^-$ to e^+e^- yields in their magnetic detector at the three c.m. energies 3.0, 3.8, and 4.8 GeV. The polar angle distribution for muon pairs, obtained in the SLAC/LBL experiment, is shown in Figure 11*b* and is symmetric about 90° as expected for a one-photon exchange process. Various breakdown parameters, summarized in Table 1, were fitted to the SLAC/LBL

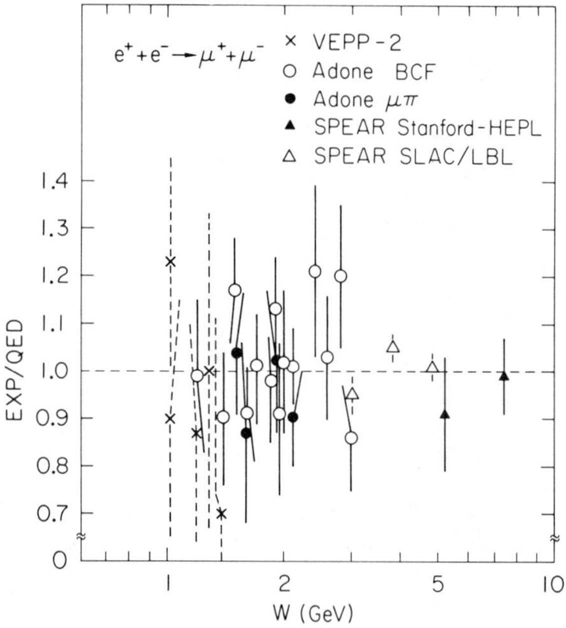

Figure 12 Ratio of experimental yield to that expected from QED for muon-pair production. Detector acceptance and normalization procedures differ between experiments. Data points shown are from VEPP-2 (64), Adone BCF (67), Adone $\mu\pi$ (65), SPEAR Stanford-HEPL (59, 60), and SPEAR SLAC/LBL (61).

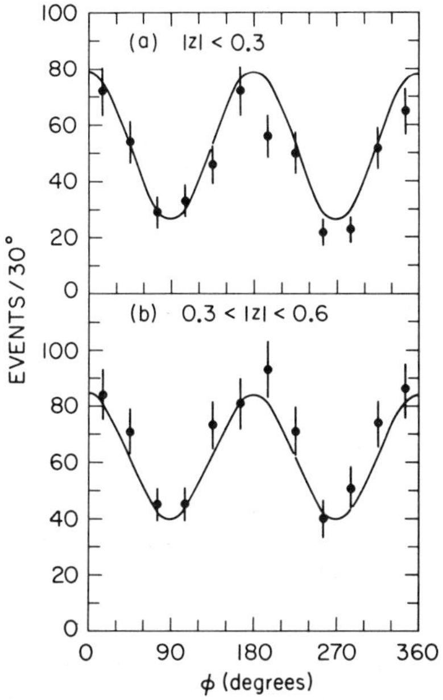

Figure 13 Azimuthal distributions for muon pair production with transversely polarized beams at $W = 7.4$ GeV from (62). Upper data are with the cut $z = |\cos\theta| < 0.3$. Lower data are for $0.3 < |\cos\theta| < 0.6$. The curves are predictions of QED fitted for the observed beam polarization.

data; both SPEAR groups have set lower limits in the vicinity of 30 GeV for possible modifications of the timelike photon propagator.

An azimuthal asymmetry in μ-pair production due to polarized incident beams (see Equation 2.12) has been observed at SPEAR (62) and at VEPP-2M (69). The azimuthal angle distribution from the SPEAR experiment is given in Figure 13, where the strong $\cos 2\phi$ term is evident.

3.5 $e^+e^- \rightarrow \gamma\gamma$

e^+e^- annihilation into two photons provides a test of the off-shell lepton propagator for spacelike values of invariant mass. A detailed cross-sectional calculation has been performed by Berends & Gastmans (70); the lowest-order cross section is given by Equation 2.18. Because of the technical problems of photon detection and the small cross section, $e^+e^- \rightarrow \gamma\gamma$ is difficult to study experimentally.

In Figure 14 we plot the ratio of experiment to theory, as a function of W for experiments that have measured this reaction (59, 71–74). This ratio has quite different meanings for the various experiments. Balakin et al (71) at VEPP-2 compared 2γ yields to e^+e^- over the same angular range. At ADONE, Bacci et al (72) compared their experimental rate over two regions of polar angle, 20°–45° and

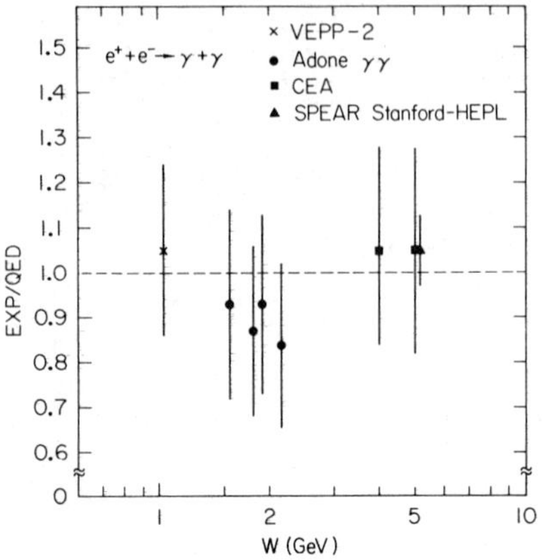

Figure 14 Ratio of experimental yield to that expected from QED for e^+e^- annihilation into two photons. Detector acceptance and normalization procedures differ between experiments. Data points shown are from VEPP-2 (71), Adone $\gamma\gamma$ (72), CEA (73, 74), and SPEAR (59).

70°–110°. The CEA measurements at 4 GeV (73) and 5 GeV (74) observed photons over the range $50° < \theta < 130°$ and normalized to double bremsstrahlung. The Stanford-HEPL group (59) measured gamma rays with their NaI crystals near 90° and normalized to small-angle Bhabha scattering.

Within the rather large errors, all measurements of the reaction $e^+e^- \to \gamma\gamma$ agree with the appropriate QED prediction. The most sensitive test of a possible breakdown of the lepton propagator comes from the SPEAR experiment (59) which sets the limits $\Lambda_+ > 6.2$ GeV, $\Lambda_- > 6.9$ GeV (95% confidence).

3.6 $e^+e^- \to e^+e^-\gamma$

Experiments sensitive to the lepton propagator at timelike values of invariant mass have been reported from ACO (75) and ADONE (76). The process studied gives

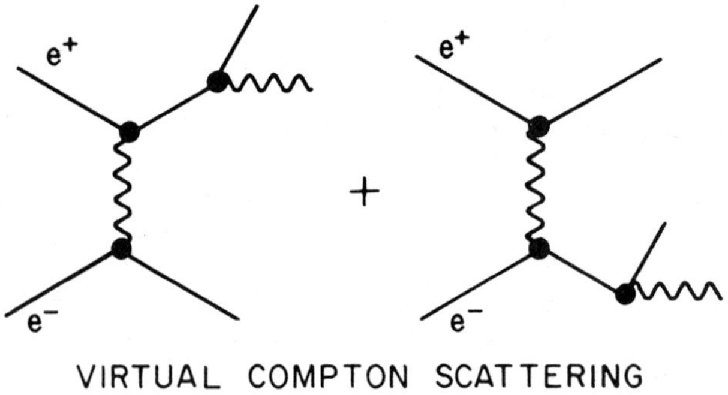

Figure 15 Feynman diagrams for virtual Compton scattering.

final states of e^+, e^-, and a hard photon at large angles from the charged particles; it is known as virtual Compton scattering and can be represented by the diagram in Figure 15. Both experiments agree with theoretical calculations. Bacci et al (76) present their results in terms of the $e^\pm, \gamma$ invariant mass spectrum, here represented in Figure 16. These data are evidence against any $e\gamma$ excited state in the mass range 0.6 to 2.4 GeV/c^2.

3.7 $e^+e^- \rightarrow e^+e^-e^+e^-$, $e^+e^-\mu^+\mu^-$

The first experimental observation of a photon-photon process was reported by Balakin et al (77). They observed approximately 100 events at $W = 1020$ MeV that contained acollinear, coplanar pairs of particles that could not be understood in terms of the usual final states found near the ϕ meson mass. These events were interpreted as coming from the reaction $e^+e^- \rightarrow e^+e^-e^+e^-$ and are adequately described by the equivalent photon approximation (38).

Bacci et al (78) observed similar events near $W = 2$ GeV. They demanded a small-angle particle to be tagged in special counters placed near the incident beam line and at least two particles to be detected in their large-angle apparatus. Because of trigger biases, most of their 29 events were interpreted not as coming from the classic photon-photon process, but rather from a similar process where one of the photons is highly virtual. This points out a potentially serious background for experiments intending to exploit the photon-photon process to study nearly real gamma-ray collisions. Parisi (79) has estimated some of these background processes.

Barbiellini et al (80) observed $\mu^+\mu^-$ and e^+e^- production through the photon-photon process. They tagged and momentum-analyzed (using the storage ring magnets) one or both of the forward-going e^+, e^- and detected the other two particles at large angles. Shower criteria were used to separate electrons and muons; kinematic constraints identified the muons. These results are in good agree-

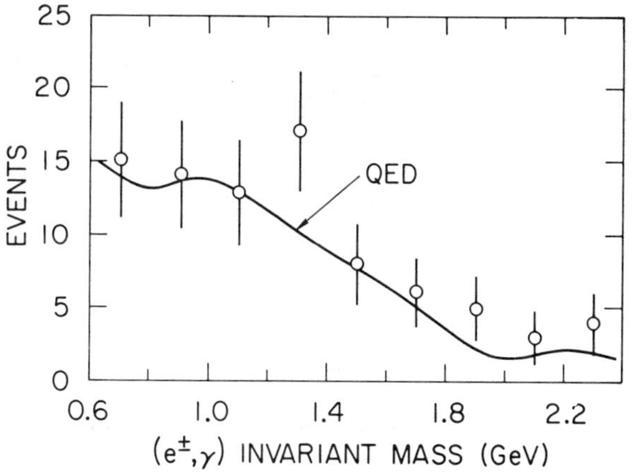

Figure 16 Experimental yield of virtual Compton scattering events from (76) compared to predictions of QED.

ment with the equivalent photon approximation and demonstrate the feasibility of studying $\gamma\gamma$ collisions in e^+e^- colliding beams.

3.8 *Hadronic Contributions to e^+e^-, $\mu^+\mu^-$ Final States*

There are now several examples (18, 81–88) where the usual formulations of QED fail to describe the cross sections for e^+e^- and $\mu^+\mu^-$ production at certain c.m. energies. These energies correspond to resonance production of various final states through the one-photon channel with the result that interference effects between resonance production and QED can be observed. For example, consider the amplitude for μ-pair production shown schematically in Figure 17. The QED amplitude is real with a value of minus-one unit. The resonance amplitude is described by a Breit-Wigner formula. Unitarity, μ-e universality, and causality demand that the Breit-Wigner amplitude have relative strength $3B_{ee}/\alpha$ (B_{ee} is the resonance branching ratio to lepton pairs) and be superimposed on the QED amplitude as indicated in Figure 17. The resonance amplitude is represented by the circle in Figure 17; passing-through resonance with increasing c.m. energy corresponds to counterclockwise motion of the net amplitude A about the circle. Thus a characteristic interference pattern emerges: below resonance there is destructive interference; then the cross section peaks before returning to the QED value well above the resonant energy. The observed μ-pair cross section σ_{exp} is related to the QED cross section σ_{QED} (Equation 2.13) by

$$\frac{\sigma_{\text{exp}}}{\sigma_{\text{QED}}} = \left| -1 + \frac{3B_{ee}}{\alpha} \frac{\Gamma/2}{M - W - i\Gamma/2} \right|^2, \qquad 3.4.$$

where M, Γ are the mass and full width of the resonance. Of course, Equation 3.4 will be modified by radiative corrections and the effect of energy spread within the beams (89).

The pioneering experimental work on this subject was done by the Orsay group

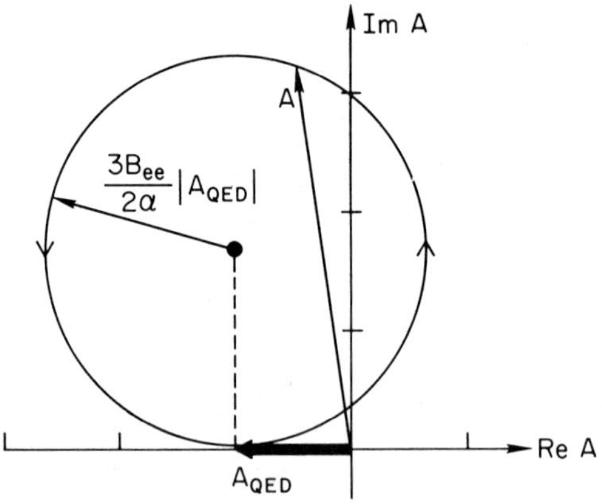

Figure 17 Schematic diagram of the amplitude A for muon-pair production in the vicinity of a vector-meson resonance with electron pair branching ratio B_{ee}.

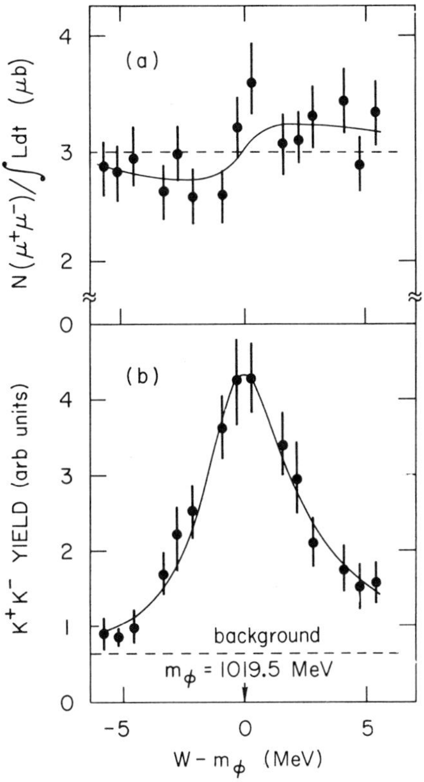

Figure 18 Experimental yield of muon pairs in the vicinity of ϕ meson (indicated by K^+K^- yield of lower figure) from (81).

(81), who studied μ-pair production in the vicinity of the ϕ meson and found evidence for the interference effect. The results from this experiment are presented in Figure 18. The most spectacular demonstrations of hadronic interference effects are the J/ψ (18, 82–86, 88) and ψ' (87) resonances. The SLAC/LBL group has observed the interference term in Equation 3.4 for both ψ particles (86, 87) through the ratio of $\mu^+\mu^-$ to e^+e^- production. Their results are given in Figure 19.

In these experiments we have very clear deviations from the simple predictions of QED which we interpret not as a breakdown of QED, but rather as a method for determining the branching ratios of vector particles to lepton pairs!

e^+e^- annihilation into two photons, which does not proceed via the one-photon channel, shows no interference effects at the J/ψ (88).

3.9 $e^+e^- \to e^\pm\mu^\mp, e^\pm\mu^\mp X$

Processes where an electron and a muon are the only charged particles in the final state of an e^+e^- collision are expected to be extremely rare in all known interactions of e^+e^-. States of $e\mu$ are therefore a unique signal for new interactions or new particle production. One possibility leading to such final states is the production of pairs of particles that subsequently decay to electrons or muons plus undetected neutrals. A heavy charged lepton (90) that decays to either electrons or muons plus neutrinos is an example.

$e\mu$ final states have been searched for without success at ADONE (91–93). Orito et al (93) studied their sensitivity to various types of heavy charged leptons and concluded that no additional charged leptons exist with masses less than 1.15 GeV.

There is now evidence from the SLAC/LBL group reported by Perl et al (94) that the reaction

$$e^+ + e^- \rightarrow e^\pm + \mu^\mp + \geqq 2 \text{ undetected particles}$$

does exist at c.m. energies above 4 GeV. In a sample of two-prong, net charge zero, acoplanar events, where both momenta are greater than 650 MeV/c, they observe an excess of events identified $e^+\mu^-$ or $e^-\mu^+$ with no simultaneous detection of gamma rays. Particle identification relies on Pb-scintillator shower counters and a 20-cm-thick iron hadron absorber. Consequently, there are rather

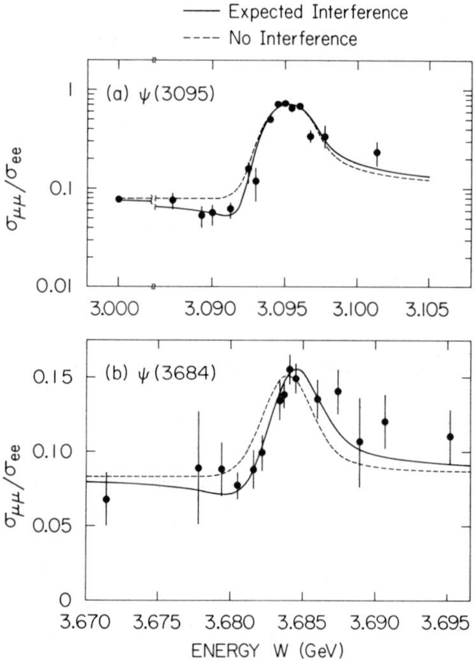

Figure 19 Ratio of muon-pair yields to Bhabha scattering in the vicinity of the $J/\psi(3095)$ and $\psi'(3684)$ resonances showing interference effect expected. The additional t-channel diagram for Bhabha scattering (see Figure 4) gives it a different energy dependence than muon-pair production. Data are from (86) and (87).

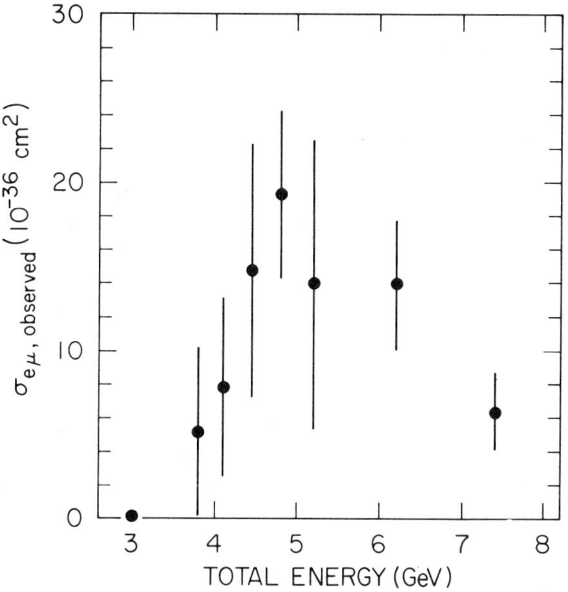

Figure 20 Experimental cross section for anomalous $e\mu$ events reported in (94).

large misidentification probabilities (which are measured with multihadronic final states) for hadrons to simulate electrons and muons. Misidentification results in a 25% contamination of the $e\mu$ signal events by normal processes, but still leaves 64 events for which there is no conventional explanation. The photon-photon process is not expected to contribute significantly; the fact that there is no like-charge signal confirms this. The experimental cross section for $e\mu$ events, given in Figure 20, shows evidence for a threshold near 4 GeV. Kinematic properties of the charged particle lead the authors to state that two or more missing particles are associated with these events.

At present there is no unique hypothesis for the origin of these $e\mu$ events. They certainly indicate the presence of some new physical process occurring above 4 GeV. If this turns out to be another member of the lepton family, it is a major discovery indeed.

4 HADRONIC FINAL STATES

As discussed above, e^+e^- collisions provide a unique tool with which hadrons can be studied in nearly pure quantum states having the quantum numbers of the photon. For purposes of orientation, we present in Figure 21 a summary of the current knowledge of R, the ratio of the total cross section for hadron production by one-photon annihilation to the total μ-pair cross section (Equation 2.13). Data are now available at c.m. energies from approximately 0.5 GeV to nearly 8 GeV.

Many groups at several laboratories have made valuable contributions to the results shown in Figure 21; we examine individual contributions below.

Figure 21 shows that hadron production is comparable to or greater than μ-pair production over the entire range of W studied to date. At certain energies R is punctuated by large peaks that correspond to the direct coupling of vector particles to the virtual annihilation photon. The large cross sections at resonant energies have permitted detailed studies of hadrons produced in decays of these states. These studies lead to determinations of quantum numbers, resonance parameters, and decay-branching ratios for the vector mesons. The very small cross sections away from resonances have limited us to a more qualitative picture of hadron production at these energies. We study these two general regions of cross section in sequence.

4.1 *Resonance Production*

The resonances observed to date in hadron production by e^+e^- annihilation correspond to vector mesons that are also observed in many other processes. Near a resonance, the cross section σ_f for producing some particular final state f can be described by a Breit-Wigner formula

$$\sigma_f = \frac{12\pi}{s} B_{ee} B_f \frac{M^2\Gamma^2}{(M^2-s)^2+M^2\Gamma^2}. \qquad 4.1.$$

The resonance is assumed to have spin 1. B_{ee} and B_f are branching ratios to electron pairs and the state f, respectively; M and Γ are the mass and full width of

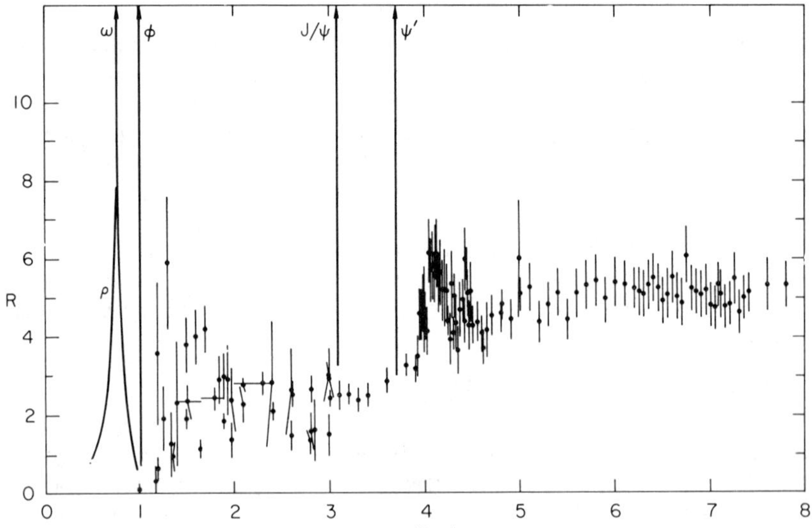

Figure 21 A summary of data for R, the ratio of the cross section for hadron production by e^+e^- annihilation to the muon pair cross section, versus center-of-mass energy W.

the resonance. The observed cross section will be a convolution of the Breit-Wigner shape with the energy spectrum of the incident e^+e^- beams and the effective energy spectrum of the initial state that arises from radiative corrections. The effect of beam energy spread and radiative corrections on resonance line shapes is discussed in (89). In general, radiative corrections give a high-energy "tail" to the observed cross section, the classic example of which is the J/ψ resonance shown in Figure 22. After correction for radiative effects, the value of σ_f at the peak of the resonance can be used to determine the product $B_{ee}B_f$

$$\sigma_f(s = M^2) = \frac{12\pi}{M^2} B_{ee} B_f. \tag{4.2.}$$

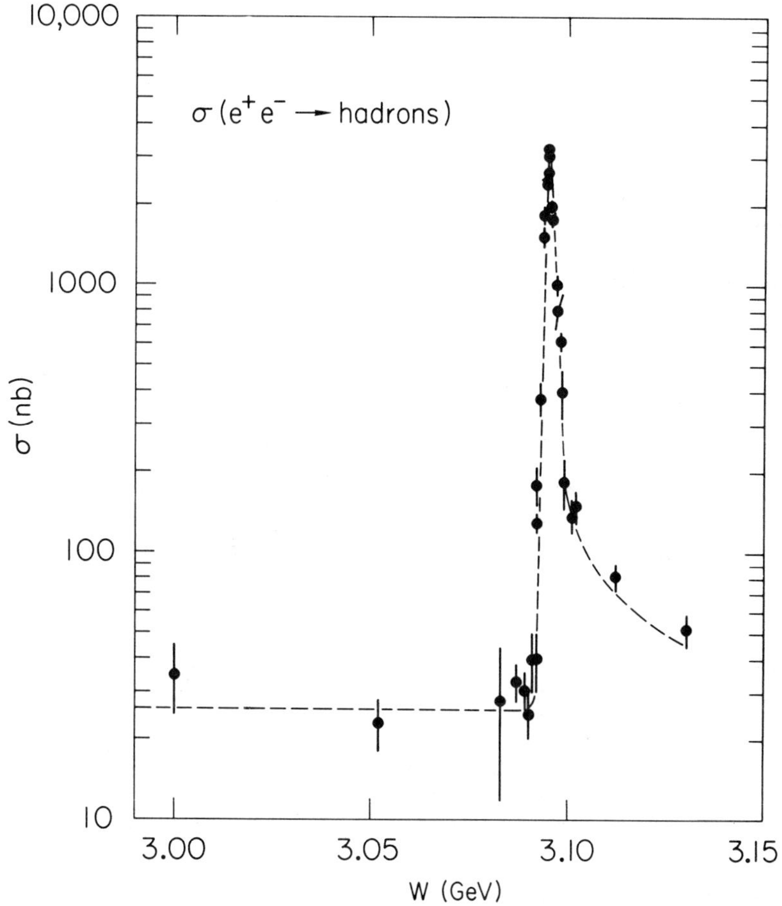

Figure 22 Total hadron production cross section in the vicinity of the J/ψ resonance showing "radiative tail." Data are from (86).

With separate knowledge of B_f, for example, Equation 4.2 gives a determination of B_{ee}. When the full width of a resonance is narrower than the energy spread of the colliding beams, the line shape cannot be resolved, yet the area under the resonance can be measured; this area is related to the resonance parameters by

$$\int \sigma_f \, dW = \frac{6\pi^2}{M^2} B_{ee} B_f \Gamma. \qquad 4.3.$$

The partial width of a resonance decaying to lepton pairs Γ_{ee} ($\Gamma_x \equiv B_x \Gamma$) is proportional to the area under the total production cross section. Γ and B_{ee} can be determined separately from the areas under both the total and elastic resonance cross sections, even in cases where the resonance line shape cannot be resolved.

The coupling of vector mesons to virtual photons is often parameterized by a coupling constant g_V that is related to Γ_{ee} by

$$\frac{g_V^2}{4\pi} = \tfrac{1}{3}\alpha^2 \frac{M_V}{\Gamma_{Vee}}, \qquad 4.4.$$

where V refers to the particular vector meson and M_V is its mass.

4.1.1 THE ρ, ω, ϕ RESONANCES The first hadronic state to be studied in e^+e^- annihilation was $\pi^+\pi^-$ in the vicinity of the ρ meson (95–97). The most recent results from the Novosibirsk group (98) and Orsay group (99) are shown in Figure 23. In this figure, the square of the pion form factor $|F_\pi|^2$, defined by Equation 2.15, is plotted.

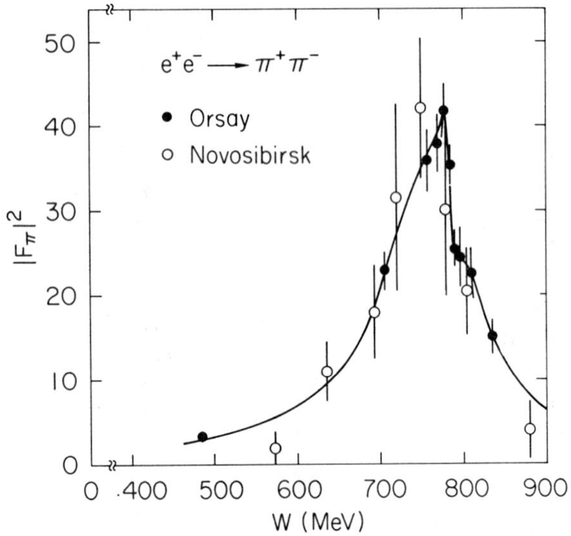

Figure 23 The pion form factor squared in the vicinity of the ρ and ω resonances. Data are from Orsay (99, 102) and Novosibirsk (98). The curve is a fit to the Gounaris-Sakurai formula (100), including ρ-ω interference.

Table 2 Parameters of the vector mesons ρ, ω, ϕ, J/ψ, and ψ' as measured in e^+e^- colliding-beam experiments

V	ρ (99)[a]	ω (104)[a]	ϕ (Ref.)	J/ψ (86)[a]	ψ' (87)[a]
m (MeV)	772.3 ± 5.9	no precision value reported	1019.4 ± 0.3 (111)	3095 ± 4	3684 ± 5
Γ (MeV)	135.8 ± 15.1	9.1 ± 0.8	4.09 ± 0.29 (109) 4.67 ± 0.42 (107) 3.81 ± 0.37 (108)	0.069 ± 0.015	0.228 ± 0.056
Γ_{ee} (keV)	5.8 ± 0.5	0.76 ± 0.08	1.41 ± 0.12 (109) 1.31 ± 0.12 (107) 1.27 ± 0.11 (110)	4.8 ± 0.6	2.1 ± 0.3
$B_{ee} = \Gamma_{ee}/\Gamma$	$(4.2 \pm 0.4) \times 10^{-5}$	$(0.83 \pm 0.10) \times 10^{-4}$	$(3.45 \pm 0.27) \times 10^{-4}$ (109) $(2.81 \pm 0.25) \times 10^{-4}$ (107) $(3.3 \pm 0.3) \times 10^{-4}$ (110)	0.069 ± 0.009	0.0093 ± 0.0016
$g_V^2/4\pi$	2.38 ± 0.18	18.4 ± 1.8	12.9 ± 1.1 (109) 13.8 ± 1.3 (107) 14.3 ± 1.2 (110)	11.4 ± 1.4	31.1 ± 4.5

[a] Numbers in parentheses are references.

The data do not follow a simple Breit-Wigner curve. There are two reasons for this. First, in studying the analytic properties of F_π, Gounaris & Sakurai (100) showed that a more complicated formulation than an s-wave Breit-Wigner amplitude is appropriate for the ρ because of its large width and the fact that the $\pi^+\pi^-$ are in a relative P wave. Second, the ω meson, to be discussed below, has a G parity–violating decay mode to $\pi^+\pi^-$ that interferes with the ρ amplitude and gives the sharp high-energy edge to $|F_\pi|^2$. The Orsay group (99) have fitted both the Novosibirsk and Orsay data to the Gounaris-Sakurai form factor F_{GS}, plus an interference term from ω decay

$$|F_\pi|^2 = \left| F_{GS} + Ae^{i\phi} \frac{M_\omega^2}{M_\omega^2 - s - iM_\omega\Gamma_\omega} \right|^2. \qquad 4.5.$$

A represents the strength of the ω coupling to e^+e^- and its subsequent G parity–violating decay, ϕ is the phase of the $\rho\omega$ interference, and M_ω, Γ_ω are the mass and full decay width of the ω meson. Resonance parameters describing the ρ and ω mesons determined from the data of Figure 23 are summarized in Table 2. ϕ was determined to be $88.3 \pm 15.8°$ (99), so the coefficient of the ω contribution to the $\pi^+\pi^-$ amplitude is almost purely imaginary. If we consider the G parity–violating transition as the ω coupling to a virtual ρ, which then decays to $\pi^+\pi^-$, then the $\omega\rho$ coupling is nearly real and the ρ propagator evaluated near M_ρ^2 gives the G parity–violating amplitude its large imaginary part (101). There is a recent point, also plotted in Figure 23, that was measured with the new Orsay magnetic detector at the c.m. energy $W = 484$ MeV (102). This point agrees with the Gounaris-Sakurai formula.

The $\pi^+\pi^-\pi^0$ decay mode of the ω meson has been studied with two different experimental setups at Orsay (103, 104). The later apparatus (105), shown schematically in Figure 24, subtended a large solid angle ($0.6 \times 4\pi$ sr) with cylindrical

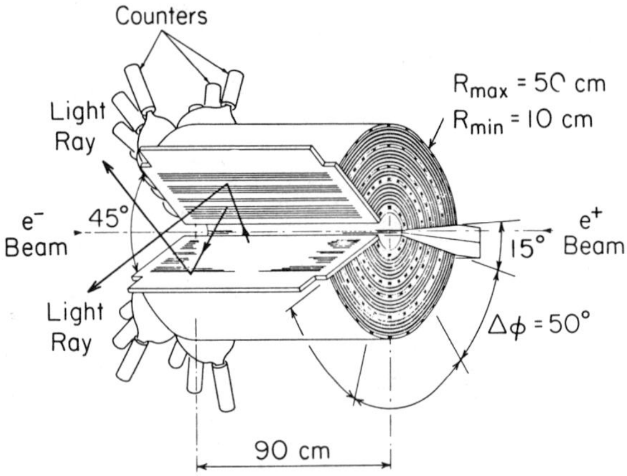

Figure 24 Schematic diagram of the Orsay detector described in (105).

spark chambers and scintillation counters. The inner spark chambers were used to track charged particles and the outer chambers were interspersed with Pb sheets to identify electron- and photon-induced showers. This apparatus was used in many of the Orsay experiments and has been one of the most successful detectors in e^+e^- physics. The cross section for $\pi^+\pi^-\pi^0$ production in the ω region is shown in Figure 25; resonance parameters derived from these data (104) are given in Table 2.

The ϕ meson has been studied extensively in e^+e^- annihilation. Excitation curves for the final states $K_S^0K_L^0$ (106–109) and K^+K^- (106, 107, 110) have been measured at Novosibirsk and Orsay. Resonance parameters determined from these experiments are given in Table 2. The most precise determination of the ϕ decay width comes from the experiment of Bizot et al (109). In a very interesting experiment performed at Novosibirsk, Bukin et al (111) were able to accurately calibrate the VEPP-2M energy by measuring the electron spin precession frequency of particles stored in the machine. This calibration yields a very precise determination of the ϕ mass.

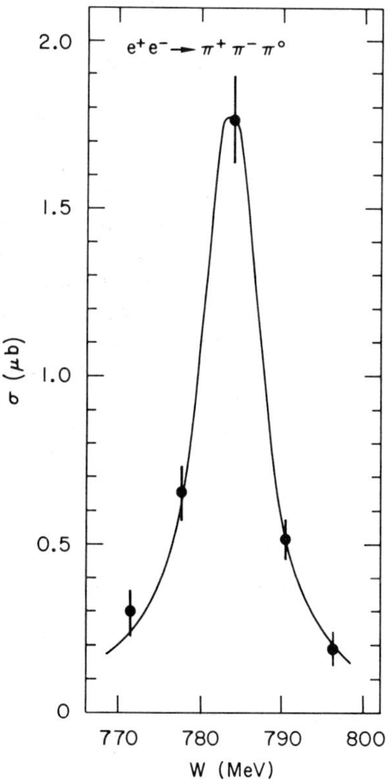

Figure 25 Cross section for the reaction $e^+e^- \rightarrow \pi^+\pi^-\pi^0$ in the vicinity of the ω meson from (104).

Table 3 Branching ratios for radiative decays of ω and ϕ determined from e^+e^- colliding-beam experiments

Decay mode	Ratio (%)	Ref.
$\dfrac{\Gamma(\omega \to \pi^0\gamma)}{\Gamma(\omega \to \pi^+\pi^-\pi^0)}$	10.9 ± 2.5	(112)
$\dfrac{\Gamma(\omega \to \eta\gamma)}{\Gamma(\omega \to \pi^0\gamma)}$	<27 (90% CL)	(112)
$\dfrac{\Gamma(\omega \to \pi^0\pi^0\gamma)}{\Gamma(\omega \to \pi^0\gamma)}$	<15 (90% CL)	(112)
$\dfrac{\Gamma(\phi \to \eta\gamma)}{\Gamma(\phi \to \text{all})}$	1.5 ± 0.4	(102)
$\dfrac{\Gamma(\phi \to \pi^0\gamma)}{\Gamma(\phi \to \text{all})}$	0.14 ± 0.05	(102)
$\dfrac{\Gamma(\phi \to \pi^+\pi^-\gamma)}{\Gamma(\phi \to \text{all})}$	<0.7 (90% CL)	(110)

Radiative decays of the ω and ϕ mesons leading to $\pi^0\gamma$ or $\eta\gamma$ states were studied by Benaksas et al (112). They detected events with three coplanar γ rays. By measuring the angles between the γ rays, the energies of all three γ rays can be calculated, assuming there were no missing particles in the event. The three energies can then be plotted on the triangular Dalitz plot shown in Figure 26*a*. Events where two γ rays have a definite invariant mass are confined to three straight lines on the Dalitz plot. If the γ energies are ordered, only the triangular sector of the Dalitz plot shown in Figure 26*b* will be populated. An example of one of these Dalitz plots for 3 γ events at the ϕ mass (112) is given in Figure 27. The $\pi^0\gamma$ and $\eta\gamma$ bands are broadened by experimental resolution, yet they clearly show an accumulation of events. Various branching ratios for radiative decays of ω and ϕ are given in Table 3. The ϕ results are from some very recent work performed by the Orsay group (102).

In a remarkable tour de force reported in (102), the Orsay group of Parrour et al have evidence for an interference between the ω and ϕ in the $\pi^+\pi^-\pi^0$ channel. In spite of their narrow widths and wide spacing, the ω and ϕ seem to interfere as if the phase were simply given by the ω propagator evaluated near the ϕ mass, but with coupling constants of opposite sign (112a).

4.1.2 THE NARROW ψ RESONANCES In the brief period of time since their discovery (18–20), the J/ψ and ψ' resonances have received an extraordinary amount of attention from both an experimental and theoretical point of view. It is much too soon to present a complete review of the properties of these states; we cover only a small fraction of the experimental work performed to date.

Two pieces of experimental apparatus have made major contributions to our knowledge of the new particles. The SLAC/LBL magnetic detector of SPEAR (61), shown schematically in Figure 28, consists of a large solenoid magnet with magnetostrictive spark chambers for tracking charged particles and measuring their momenta. Time-of-flight counters, shower counters, and range chambers are used for particle identification. The full momentum-analysis and particle-identification capabilities of this device extend over 0.65 of 4π sr solid angle. Two charged particles are necessary to trigger the device. The double-arm spectrometer DASP (113) at DESY consists of two symmetric, high-resolution, magnetic spectrometers, each covering approximately 0.45 sr with excellent particle identification capability. Surrounding 0.7 of 4π sr solid angle is a nonmagnetic inner detector, consisting of scintillation counters and proportional chambers, which is well suited for tracking

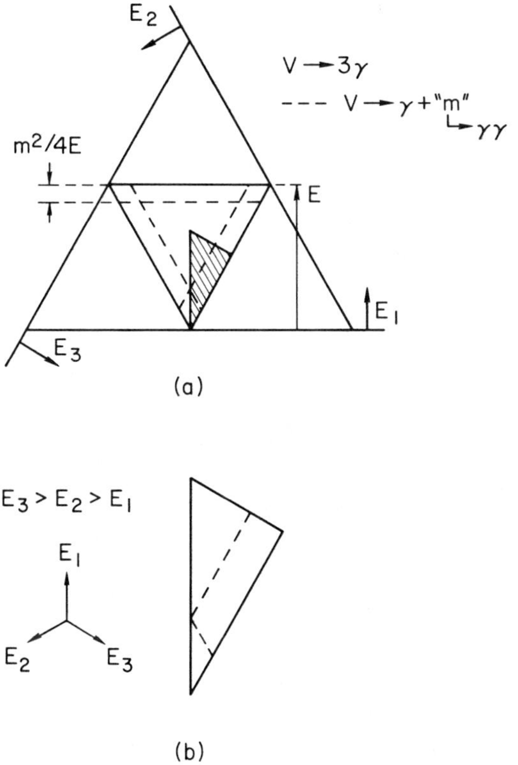

Figure 26 (*a*) Dalitz plot for 3γ final states of e^+e^- annihilation. Events populate the interior triangle. Dashed lines represent regions populated by events where two of the gamma rays are decay products of a state of mass m. (*b*) Same as *a*, except only the populated region where the gamma rays are ordered by energy is shown. This corresponds to the shaded region of *a*.

charged particles and γ rays. A schematic drawing of the DASP detector is given in Figure 29.

The SLAC/LBL collaboration measured the total hadronic, e^+e^-, and $\mu^+\mu^-$ cross sections at the J/ψ (86) and ψ' (87) and from these derived the resonance parameters given in Table 2. From the angular distributions of e^+e^- and $\mu^+\mu^-$ final states and the interference with QED in the $\mu^+\mu^-$ channel, noted in Section 3, the SLAC/LBL group conclusively demonstrated that the new resonances are vector states.

The most remarkable properties of the new particles are their exceedingly narrow decay widths. Even though they are over three times as massive as the ρ, ω, and ϕ mesons, their decay widths are two to three orders of magnitude smaller. A detailed theoretical understanding of the decay rates of the new particles is one of the challenging problems in elementary particle physics today. Below, we briefly discuss some of the theoretical ideas concerning the new particles.

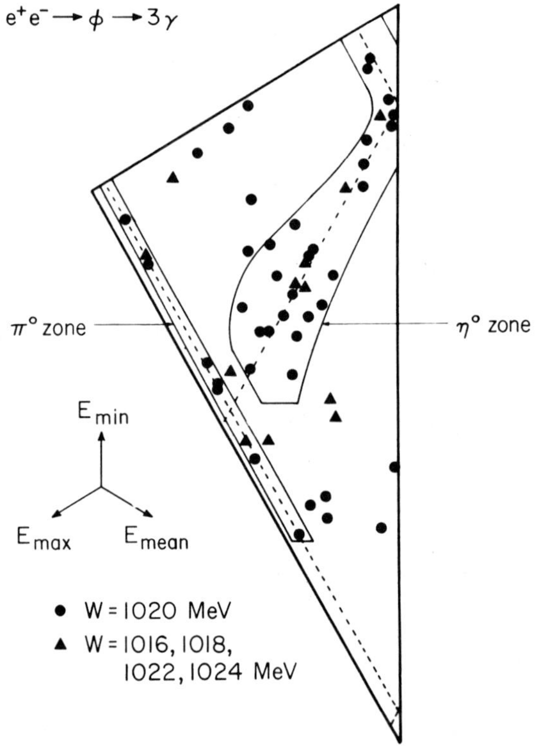

Figure 27 Dalitz plot for 3γ final states in the vicinity of the ϕ meson from (112). The π^0 and η^0 bands are indicated.

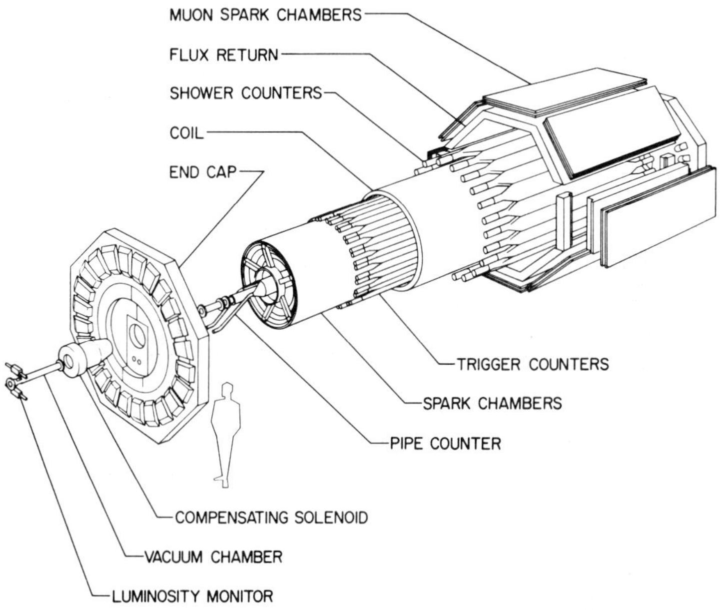

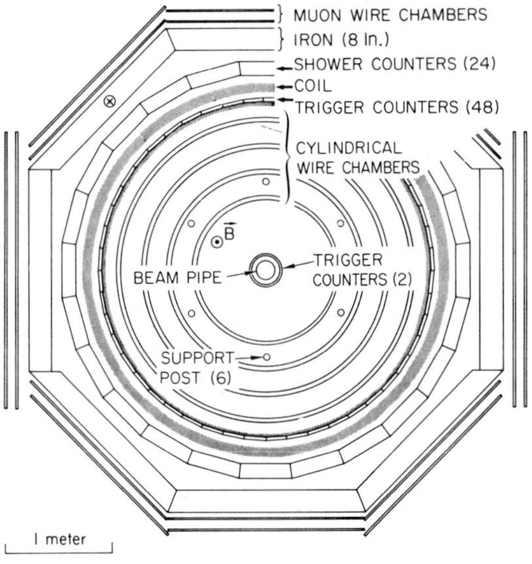

Figure 28 Schematic diagrams of the SLAC/LBL SPEAR magnetic detector.

There are two general decay modes for these resonances: direct decays and second-order electromagnetic decays. These are shown schematically in Figure 30. Direct decays display the intrinsic properties of the resonance, while second-order electromagnetic decays display the same characteristics as one-photon exchange

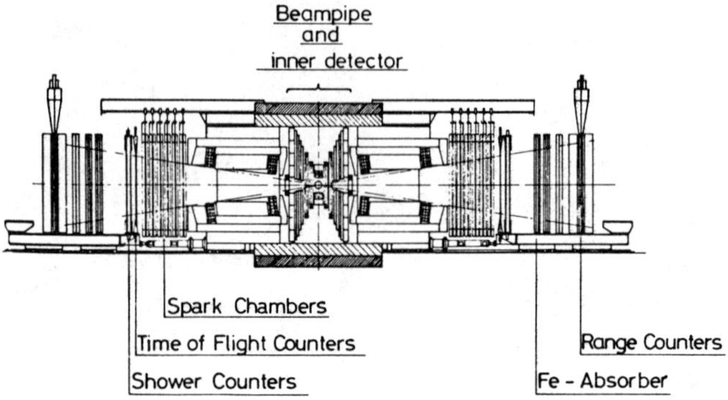

Figure 29 Schematic diagram of the DASP double arm spectrometer at DESY.

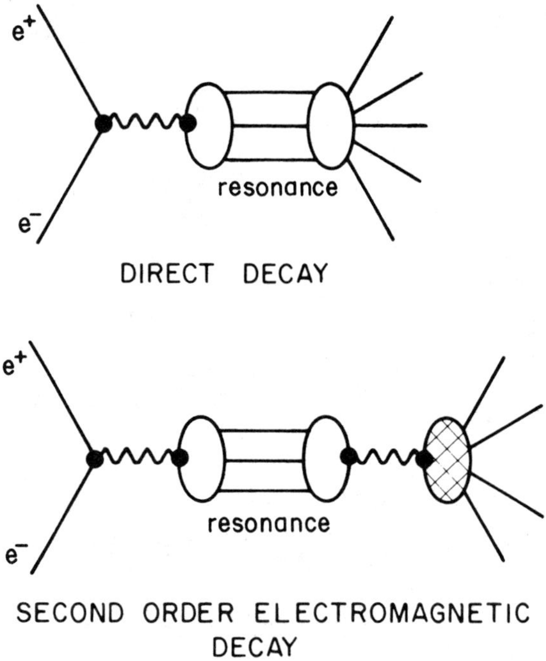

Figure 30 Schematic diagram showing the difference between "direct" and "second-order electromagnetic" decays of a resonance in e^+e^- annihilation.

processes at nearby nonresonant energies. The μ-pair final state is an excellent monitor of second-order electromagnetic processes, so that comparison of particular decay rates to the μ-pair rate, both on and off resonance, will differentiate between the two modes.

Jean-Marie et al (114) compared ratios of multipion final states to μ-pair production for data taken at the J/ψ energy and $W = 3.0$ GeV. Both sets of data were taken with the SLAC/LBL magnetic detector. The results, given in Figure 31, show states with an even number of pions to be second-order electromagnetic decays, while states with an odd number of pions come from direct decays of the J/ψ. Thus G parity is a good quantum number in J/ψ decays to pion states with the value -1. In the same paper Jean-Marie et al studied isotopic ratios for the states $\rho\pi$ and showed that the isotopic spin of the J/ψ is zero. Corroborating evidence for this assignment comes from observations of decays to $\Lambda\bar{\Lambda}$ (115), $p\bar{p}$ (115, 116), and the stringent upper limits placed on $\pi^+\pi^-$ and K^+K^- final states by the DASP group (116). Many hadronic decays of the J/ψ have been identified by the SLAC/LBL group; a summary of these is given in (115).

The ψ' decays predominantly to states containing a J/ψ (117, 117a). The presence of J/ψ among decay products of the ψ' is most easily established through the $\mu^+\mu^-$ decay mode of the J/ψ. The invariant mass spectrum for μ-pair candidates in

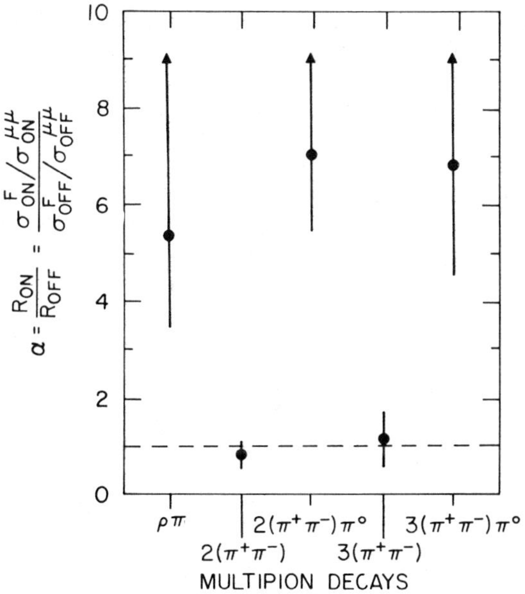

Figure 31 Comparison of multipion decays of J/ψ resonance to an adjacent nonresonance energy to find direct decay modes of J/ψ. ON and OFF refer to data taken on the resonance and just below the resonant energy, respectively. Values of α greater than one indicate direct decays of the J/ψ. Data taken from (114).

the SLAC/LBL magnetic detector (117), shown in Figure 32, has two peaks; one corresponds to μ-pair production at the ψ' mass, the other corresponds to J/ψ decays. At present, the following cascade decays have been identified:

$$\psi' \to J/\psi + \pi^+\pi^- \quad (113, 117), \qquad 4.6a.$$

$$\psi' \to J/\psi + \pi^0\pi^0 \quad (113, 179), \qquad 4.6b.$$

$$\psi' \to J/\psi + \gamma\gamma \quad (118, 119), \qquad 4.6c.$$

$$\psi' \to J/\psi + \eta \quad (113, 120). \qquad 4.6d.$$

Branching ratios for these decays are given in Table 4.

These important cascade decays point out the close relationship between the J/ψ and ψ'. The relative rates for the dipion decays and the η decay mode indicate that the isotopic spin of the ψ' is the same as J/ψ, namely zero.

4.1.3 THE NEW PARTICLE SPECTROSCOPY In a study of reaction 4.6c, the DASP collaboration (118) reported evidence that the γ-ray cascade proceeds through an intermediate state in the sequence:

$$\psi' \to \gamma + P_c \hookrightarrow \gamma + J/\psi, \qquad 4.7.$$

where the P_c had a unique mass either near 3500 or 3300 MeV. The SLAC/LBL group (119) also observed this decay scheme by detecting the $\mu^+\mu^-$ decay of the J/ψ and one or both γ rays. In seven of their events, one of the γ rays converts to an e^+e^- pair in the beam vacuum chamber so that its energy can be measured accurately from the e^+e^- momenta. From this measurement and more recent DASP results (113), the observed width of the P_c state is compatible with the experimental resolution of approximately ± 10 MeV. In principle, the present ambiguity in the mass of the P_c can be resolved by observing Doppler broadening of the second γ ray; as yet, this has not been done.

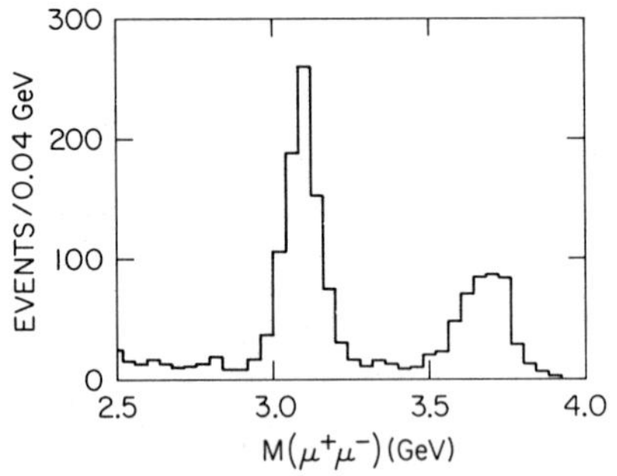

Figure 32 Invariant-mass spectrum of muon-pair candidates in decays of the ψ' from (117). The peak near 3.7 GeV corresponds to muon-pair production; the peak near 3.1 GeV indicates the presence of J/ψ in decays of the ψ'.

Table 4 Branching ratios for cascade decays of $\psi' \to J/\psi$

Decay mode	Ratio (%)	Ref.
$\frac{\Gamma(\psi' \to J/\psi + \text{anything})}{\Gamma(\psi' \to \text{all})}$	57 ± 8	(117)
$\frac{\Gamma(\psi' \to J/\psi + \pi^+\pi^-)}{\Gamma(\psi' \to \text{all})}$	31 ± 4 36 ± 6	(117) (113)
$\frac{\Gamma(\psi' \to J/\psi + \pi^0\pi^0)}{\Gamma(\psi' \to \text{all})}$	18 ± 6	(113)
$\frac{\Gamma(\psi' \to J/\psi + \eta)}{\Gamma(\psi' \to \text{all})}$	4.3 ± 0.8 3.7 ± 1.5	(120) (113)
$\frac{\Gamma(\psi' \to J/\psi + \gamma\gamma)}{\Gamma(\psi' \to \text{all})}$	3.6 ± 0.7 4 ± 2	(119) (113)
$\frac{\Gamma(\psi' \to J/\psi + \text{neutrals})}{\Gamma(\psi' \to J/\psi + \text{anything})}$	44 ± 3	(117)
$\frac{\Gamma(\psi' \to J/\psi + \pi^0\pi^0)}{\Gamma(\psi' \to J/\psi + \pi^+\pi^-)}$	64 ± 15	(179)

Evidence for radiative decays of the ψ' to hadronic states has been reported by the SLAC/LBL group (121). They observed charged hadrons and inferred the presence of single γ rays from the energy-momentum balance of the charged particles. These results indicate the decay scheme:

$$\begin{array}{l}\psi' \to \gamma + \chi \\ \qquad\quad \hookrightarrow 4\pi^{\pm}, 6\pi^{\pm}, \pi^{\pm}K^{\pm}, \pi^+\pi^-, K^+K^-.\end{array} \qquad 4.8.$$

At least two different χ states are present, a narrow state at 3410 ± 10 MeV and a broader state at 3530 ± 20 MeV. The state at 3410 MeV decays to $\pi^+\pi^-$ and K^+K^-, therefore its spin and parity must be even. It is tempting to associate the P_c with the $\chi(3530)$, but the broader width of the $\chi(3530)$ implies either that they are not the same, or that there is more than one state near 3500 MeV in mass.

Finally, two DESY groups (113, 122) have examined coplanar 3γ-ray final states of the J/ψ to look for radiative decays to new states that subsequently decay to 2γ rays. Using a Dalitz plot analysis technique similar to that discussed above in connection with ω and ϕ radiative decays, they have evidence for a new state near 2800 MeV that has been designated $X(2800)$.

We have summarized the current state of knowledge of the new particle spectroscopy in the energy-level diagram of Figure 33. The higher-mass solution was assigned to the P_c. It should be noted that the information on many of these new states is very preliminary. All of the events observed to date at the P_c, χ, and X states number only a few hundred, compared to the hundreds of thousands of J/ψ

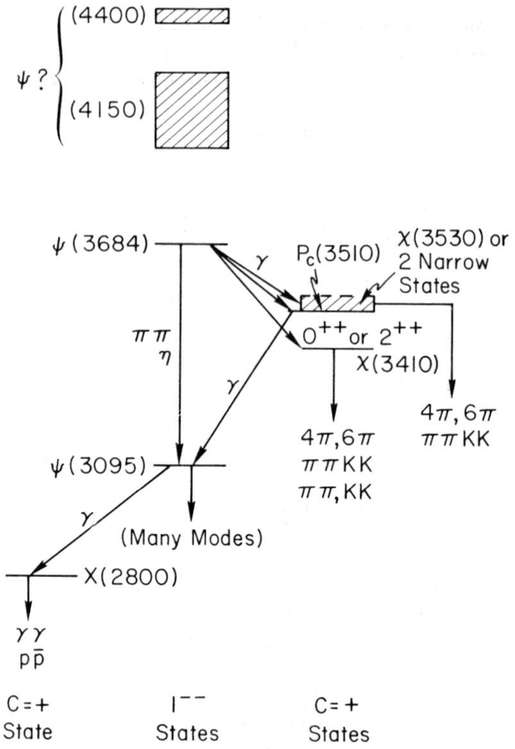

Figure 33 Energy-level diagram summarizing the new heavy-particle spectroscopy.

and ψ' decays that have been studied. The existence of a new heavy-particle spectroscopy is well established; the job of determining the masses, quantum numbers, and decay rates of the new particles lies before us.

4.1.4 OTHER RESONANCES IN e^+e^- ANNIHILATION The first resonance beyond the ρ, ω, and ϕ to be studied in e^+e^- colliding beams was the $\rho'(1600)$. It was first observed in e^+e^- collisions as a broad enhancement in the $2\pi^+\pi^-$ final state at ADONE (123). There is a complete discussion in (124) of the apparatus and analysis methods used by the $\mu\pi$ group that has performed much of the work on the $\rho'(1600)$ discussed here. This state has also been observed in photoproduction (125). The most recent data (126) indicate that the mass of the $\rho'(1600)$ is (1550 ± 60) MeV and its total decay width is (360 ± 100) MeV. Ceradini et al (127) studied 23 events from the ρ' energy region with four charged prongs in a nonmagnetic detector and concluded that most of the decays proceeded through the quasi–two-body final state $\rho^0\varepsilon^0$. Given the small sample of events, the broad ε^0 width used in the analysis, and the small available phase space, the notion of an ε^0 in decays of the $\rho'(1600)$ seems dubious. However, ρ^0 production does seem to be important at this energy. Bernardini et al (128) and Alles-Borelli et al (129) showed that the 4π enhancement at 1.6 GeV may simply be a consequence of the opening of four-pion phase space coupled with a general decrease in production amplitude that could give a peak near 1.6 GeV.

Conversi et al (126) speculate on the existence of a resonance at 1250 MeV in the $\pi^+\pi^-\pi^0\pi^0$ final state with limited statistics. The analysis depended on a calculation of the decay $\rho'(1600) \to \pi^+\pi^-\pi^0\pi^0$ that was subtracted from the measured yields. Statistically marginal departures of the measured pion form factor (129, 130) from the Gounaris-Sakurai formula (100) is also taken as evidence for a $\rho'(1250)$. More experimental information is required before a $\rho'(1250)$ is established.

The SLAC/LBL group (131) reported a broad structure in R at 4.15 GeV. The area under this peak is comparable to the narrow ψ resonances. Preliminary data from a detailed scan of the 4-GeV region (132) shown in Figure 34 confirmed the structure at 4.1 GeV and indicated that there may be another, narrower resonance near 4.4 GeV and that the 4.1-GeV peak may contain a substructure.

The 4-GeV region in e^+e^- annihilation appears to contain a vast richness of structure that we are only beginning to resolve at present. Note also that this is the same energy as the apparent threshold for μe events discussed in Section 3.

4.1.5 SEARCHES FOR NEW RESONANCES e^+e^- colliding-beam devices are ideally suited for high-resolution scanning for new vector states. After the discovery of the J/ψ, several groups (20, 132–136) performed scans where short data runs were taken at fine c.m. energy intervals. Typical c.m. energy spacings were of order 2 MeV. By far the most successful of these scans was the first, which discovered the ψ' (20). Since then, no additional narrow resonances have been uncovered. The c.m. energy ranges scanned to date are 770 to 1340 MeV (133); 1910 to 2545 MeV and 2970 to 3090 MeV (134, 135); and 3.2 to 7.7 GeV (132, 136). The sensitivity of these

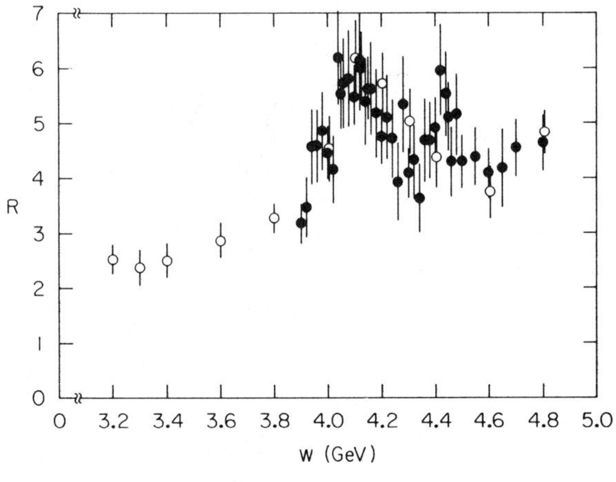

Figure 34 R vs W in the 4-GeV region. Open points are from (131); closed points are preliminary measurements reported in (132). These data indicate there are at least two and possibly more resonances in the 4-GeV region.

searches varies greatly, but narrow ($\Gamma \lesssim 10$ MeV) resonances decaying to lepton pairs at partial widths of order 20% or larger of the J/ψ value should have been observed. Undoubtedly, more sensitive searches will continue to be made until new resonances are found or the experimentalists become exhausted.

4.2 *Nonresonant Hadron Production*

4.2.1 TWO-BODY FINAL STATES Measurements of hadronic pair production at nonresonant energies in e^+e^- annihilation provide information on electromagnetic form factors of the hadrons for timelike values of momentum transfer. In general, these measurements are difficult to perform because of the very small cross sections involved and the large competing reactions such as Bhabha scattering and μ-pair production, which also give two final-state particles.

The pion form factor has been studied at energies above the ρ mass in experiments at Novosibirsk (130) and Frascati (137, 138). The Novosibirsk group used threshold water Cerenkov counters to separate π and K mesons. The Frascati $\mu\pi$ group (137) used sufficient absorber so that at $W = 1.25$ and 1.52 GeV, only pions could be detected; at higher energies no separation of π and K mesons was

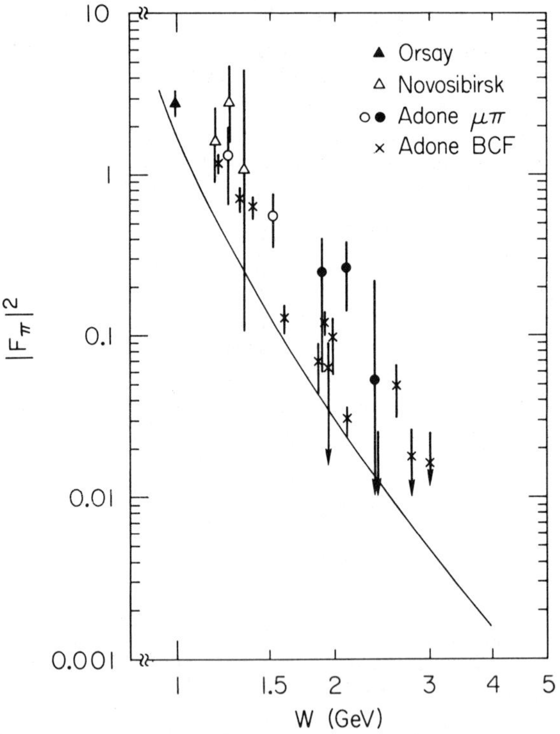

Figure 35 $|F_\pi|^2$ vs W above the ρ mass region. Data are from Orsay (99), Novosibirsk (130), Adone $\mu\pi$ (137) (full circles represent upper limits), and Adone BCF (138).

possible and the corresponding results represent upper limits to the pion form factor. For $W \leqq 1.40$ GeV the BCF group (138–140) detected only pions; between 1.50 and 1.70 GeV, π and K mesons were separated by a range method; above $W = 1.85$ GeV no such separation was possible and a theoretical ratio of $(\pi^+\pi^-)$ and (K^+K^-) pairs based on SU(3) was applied to the data to calculate the pion form factor. These results are shown in Figure 35, where we also plot an extrapolation of the Gounaris-Sakurai formula (100) for these energies. Particularly between 1.0 and 1.5 GeV, the data are consistently higher than the theoretical extrapolation. As noted previously, this has been taken as evidence for the existence of additional resonances.

The information on the K^+ form factor obtained from experiments of the Novosibirsk group (130) and the BCF group (139, 140) suggests that the kaon and pion form factors are comparable within the rather large errors.

The $p\bar{p}$ final state has been studied by the Naples group (141) at ADONE. Their result for the $p\bar{p}$ cross section is $\sigma(e^+e^- \to p\bar{p}) = (0.91 \pm 0.22)$ nb at the energy $W =$ 2.1 GeV. The $p\bar{p}$ were identified through ionization in a nonmagnetic detector. It is impossible to separately extract the form factors $G_{\rm E}$ and $G_{\rm M}$ (see Equation 2.16) for this measurement. Assuming the magnitudes of the two to be equal, they find

$$|G_{\rm M}| = |G_{\rm E}| = 0.27 \pm 0.04.$$

4.2.2 TOTAL HADRONIC CROSS SECTION The preliminary results on multihadron production from groups working at ADONE were presented at the Kiev Conference (142), causing great excitement because of the "large" total hadronic cross section $\sigma_{\rm T}$. Since then we have come to realize the fundamental role of $\sigma_{\rm T}$ in the study of hadron structure. In studying the Bjorken scaling (34) properties of inclusive proton production by e^+e^- annihilation and its connection to deep inelastic electron scattering, Drell, Levy & Yan (143) predicted a $1/s$ energy behavior for $\sigma_{\rm T}$, using parton model arguments. The parton model (144) is a very convenient framework in which to organize the experimental data, and, as we shall see, provides a simple intuitive picture of many of the phenomena actually observed. In this picture hadrons are built out of constituents, the partons, which have pointlike couplings to the electromagnetic current. For example, spin-$\frac{1}{2}$ partons would couple to the virtual-annihilation photon in the same way as muons, except for differences in net electric charge. As indicated schematically in Figure 36, hadron production is viewed as

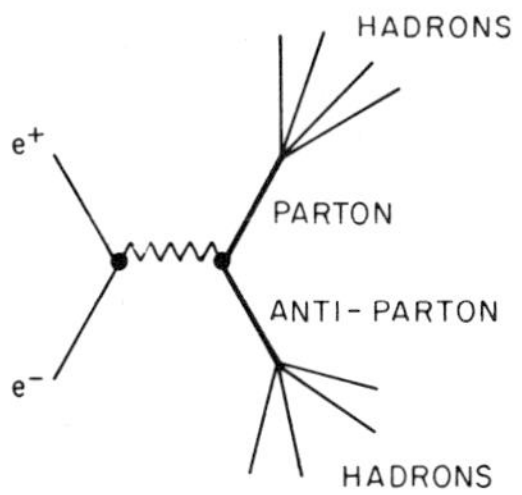

Figure 36 Schematic diagram of hadron production in the parton model.

the production of pairs of parton-antipartons that subsequently decay to hadrons. At sufficiently high energies, where hadron masses can be neglected, the total hadronic cross section is equal to the sum of the individual parton cross sections that are proportional to σ_μ, the ratio σ_T/σ_μ, denoted R, therefore depends only on the sum of squares of parton charges (145, 146):

$$R = \sum_{J=\frac{1}{2}} Q_i^2 + \tfrac{1}{4} \sum_{J=0} Q_i^2. \qquad 4.9.$$

The indicated summations are over spin-$\frac{1}{2}$ and spin-0 partons, respectively. The Q_i are parton charges in units of e. R is a convenient experimental quantity as well, because experimenters usually normalize their hadronic yields to the yield from some electrodynamic process that would normally be proportional to σ_μ. If R is a constant over some range of energies, then hadron production is said to exhibit "scaling" over that energy range. The actual value of R gives information about the number and properties of partons.

The important factors limiting the precision to which R can be measured are (*a*) systematic errors in hadronic event detection efficiency, (*b*) systematic errors in luminosity monitoring, and (*c*) statistical errors of the event sample. In virtually all measurements of R, the detection efficiency is estimated by an "unfold" procedure. The number of events N_q observed in some configuration q (e.g. the number of observed charged prongs) is related to the number of events $\tilde{N}_p$ produced in the configuration p by

$$N_q = \sum_p \varepsilon_{qp} \tilde{N}_p, \qquad 4.10.$$

where ε_{qp} is the appropriate detection efficiency for observing an event in configuration q when it was produced in configuration p. The ε_{qp} are usually computed by Monte Carlo techniques. Known properties of the apparatus and a model of representative final states are necessary ingredients in the calculation. Parameters of the model are varied to obtain agreement with observed quantities such as angle, multiplicity, and momenta spectra. Such calculations are discussed in reference 147. Equation 4.10 is then inverted, usually by maximum likelihood methods, subject to the constraint $\tilde{N}_p \geqq 0$. The average detection efficiency $\bar{\varepsilon}$ is defined by

$$\bar{\varepsilon} = \frac{\sum N_q}{\sum \tilde{N}_p}, \qquad 4.11.$$

and the total cross section is computed from:

$$\sigma_t = \frac{\sum N_q}{\bar{\varepsilon} \int L \, dt}, \qquad 4.12.$$

where $\int L\,dt$ is the time-integrated luminosity. Usually, no corrections are made for all-neutral final states; an all-π^0 state is excluded by charge conjugation invariance. The advantage of the unfold procedure is that maximum use is made of experimental data; model dependence does not directly enter into the calculation of $\bar{\varepsilon}$, but enters in an average way through the ε_{qp}. Generally, σ_t and the mean charged-particle multiplicity $\langle n_{\text{ch}} \rangle$ depend only weakly on a particular choice of model. However, individual partial cross sections usually are not well determined by these methods.

Several experimental groups at Frascati have measured σ_t. The first group to publish multiparticle results was the boson group (148, 149); they concluded that σ_t was greater than 30 nb for the c.m. energy range 1.4–2.4 GeV. A final report of this experiment is contained in (53). The $\gamma\gamma$ group (150, 151) measured σ_t between 1.4 and 3.0 GeV and presented positive arguments that these multiparticle events were hadronic. A detailed description of the apparatus and data analysis methods used by the $\mu\pi$ group can be found in (124). Their most recent cross section results were reported by Ceradini et al (152). The BCF group calibrated their apparatus in beams of hadrons and leptons and verified the hadronic nature of the multiparticle events. Their most recent results on σ_t are given in (153). In this paper two different methods were used to compute $\bar{\varepsilon}$ and they did not agree at all energies.

The most recent of the ADONE results on R are plotted in Figures 21 and 37, along with lower-energy data from Novosibirsk (154) and Orsay (155). In spite of the large errors and disagreements between various groups, it is clear that multihadron production is significant at energies above $W = 1.2$ GeV. By far the most serious problems with these pioneering measurements of R come from the relatively small

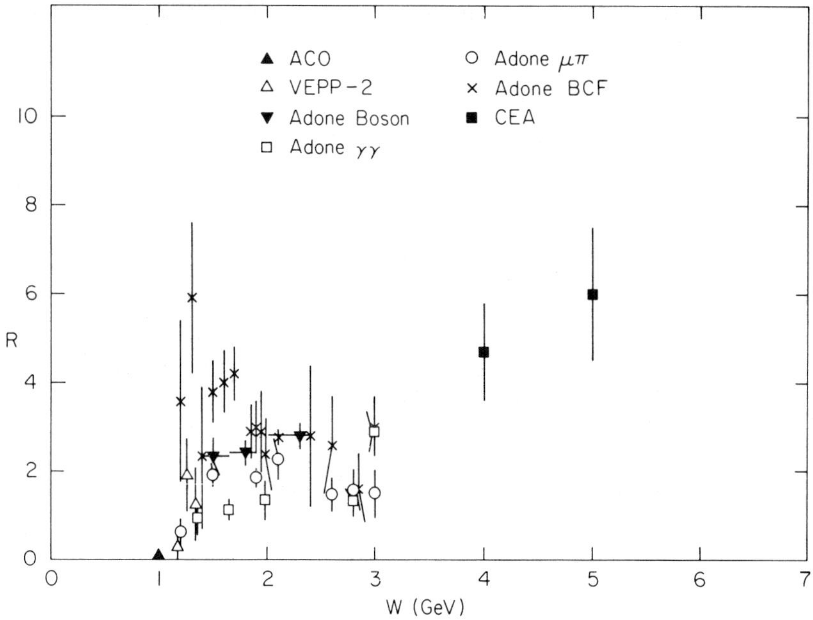

Figure 37 Results on R vs W from ACO (155), Novosibirsk (154), Adone Boson (53), Adone $\gamma\gamma$ (151), Adone $\mu\pi$ (124, 152), Adone BCF (153), and CEA (156, 157).

solid angle of the detectors. Typically, the early Frascati detectors covered only about 20% of 4π sr solid angle, required at least two charged particles to trigger, and had interaction-region volumes that were nearly as long as the detectors. Thus correction factors as large as 10 to 50 were applied to the experimental cross sections to obtain R. The model dependence and other systematic errors in $\bar{\varepsilon}$ cannot be reliably controlled with such small acceptance. These detectors were simply not prepared for the new world of multihadron production that they discovered.

The next round of σ_t information came from the nonmagnetic BOLD detector at CEA (156, 157). The larger solid angle of this device was much better suited for determining R, but low counting rates limited measurements to the two energies $W = 4$ and 5 GeV. Nevertheless, these results, plotted in Figure 37, were a forerunner of the "new physics," shortly to be discovered.

The SLAC/LBL group (131) published measurements of R at several c.m. energies in the range from 2.4 to 5.0 GeV. In Figures 21 and 38, the published SPEAR results and recently presented values for R up to $W = 7.8$ GeV (132) are shown. The large solid-angle and momentum-analysis capability of the SPEAR magnetic detector have substantially reduced the correction factors needed to determine R. The SLAC/LBL group estimate the systematic uncertainty in the absolute value of R to be $\pm 10\%$. A further $\pm 15\%$ smooth variation could occur between lowest and highest energies covered.

Aside from the narrow resonances J/ψ and ψ', the most conspicuous features of our present knowledge of R at energies above 1 GeV are the two regions where R is roughly independent of c.m. energy and the transition region between these. As shown in Figures 21 and 38, between 2.4 and 3.5 GeV the data are consistent with the value $R = 2.5$. The data between 1 and 2.4 GeV could also have this same value for R, but the large experimental uncertainties prevent us from drawing any

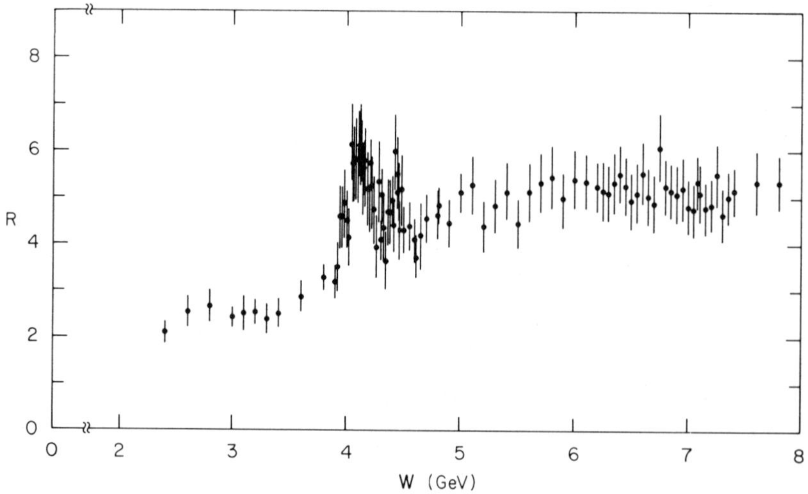

Figure 38 R vs W from SPEAR (131, 132).

conclusions except that there probably exists at least one broad resonance here, the $\rho'(1600)$. Above 5.0 GeV R is again nearly constant at a level approximately twice that of the lower-energy region. Between 3.5 and 5.0 GeV there is a complicated transition region with several possible resonances.

From the constancy of R we see that hadron production exhibits scaling in two different regions of c.m. energy. In the context of the parton model, it follows from Equation 4.9 that new partons must be coming into play in the transition region of R to effect an increase in R. The lower-energy scaling region, where R is approximately 2.5, is compatible with parton models where the partons are nine Gell-Mann/Zweig quarks (158, 159) arranged in three "flavors" and three "colors." This quark arrangement, so very successful in explaining the spectroscopy of "old" hadrons, predicts $R = 2$. The increase in R in going to the higher-energy scaling region indicates that new processes are adding 2 to 3 units of R to the "old physics" of the lower-energy scaling region. The fact that this rise in R occurs at energies just above the masses of the J/ψ and ψ' suggests that the "new physics" represented by the increase of R is related to the new particles. The rich structure observed in the transition region also hints at such a relationship; it may reflect the breaking of the selection rules responsible for the very long lifetimes of the new particles. The anomalous $e\mu$ events (94) may also be related to an increase in R near $W = 4$ GeV if the parent particles have significant decays to hadrons (160).

The notions of "new" and "old" physics are made explicit in the theoretical models that attempt to explain the properties of the new particles (161). The charmed-quark model (162–164) proposes that the J/ψ and ψ' are bound states of a fourth type of quark and its antiquark, which possess a new quantum number called charm. Above the threshold for production of particles having this new quantum number (charmed particles), charmed-quark models predict that R should approach $3\frac{1}{3}$; below threshold, R is equal to 2. The additional $1\frac{1}{3}$ units of R come from charmed particle production. Another feature of the charm models (163, 164) is the atomlike spectroscopy of the bound charmed quark-antiquark system. In addition to the J/ψ and ψ', a host of intermediate states having various quantum numbers and broad states above the ψ' were predicted. Indeed, these predictions motivated much of the experimental effort that ultimately led to the new particle spectroscopy summarized in Figure 32. While the data on the new particle spectroscopy (113, 165) and R disagree in detail with present theoretical calculations, the charmed-quark hypothesis provides an impressive framework in which the energy dependence of R and the new particle spectroscopy can be understood in qualitative terms. A search for charmed particles was carried out in the "new" physics region at $W = 4.8$ GeV by the SLAC/LBL group (166) with no success. As pointed out by Einhorn & Quigg (167) and others, the null results of the SPEAR experiment are unpleasant but not catastrophic for the charm model.

In another class of models, which follow from the quark model of Han & Nambu (168), this general picture of isolated new particle states connected to an increase in R recurs. In these models the new physics of so-called colored states adds 2 units of R to the low-energy value, which is also equal to 2.

4.2.3 PHOTON-PHOTON PROCESSES An early criticism of experiments measuring R was that the photon-photon process could be responsible for a substantial fraction of the large multihadron cross section. Most of the groups that have published the results summarized in Figure 21 discussed possible contamination from photon-photon processes. Experiments (53, 151, 156) calculated possible contaminations and concluded that they were negligible. Experiments described in (131, 152, 157) used small-angle tagging counters to detect the forward electrons in coincidence with multihadron events; they experimentally verified that the contamination of R by photon-photon processes can be neglected at present energies.

Hadronic final states produced through photon-photon processes have been observed, albeit with very low rates. Orito et al (169) observed two events consistent with the reaction

$$e^+e^- \rightarrow e^+e^-\pi^+\pi^-. \qquad 4.13.$$

Paoluzi et al (170) reported evidence for the production of acoplanar pairs of pions

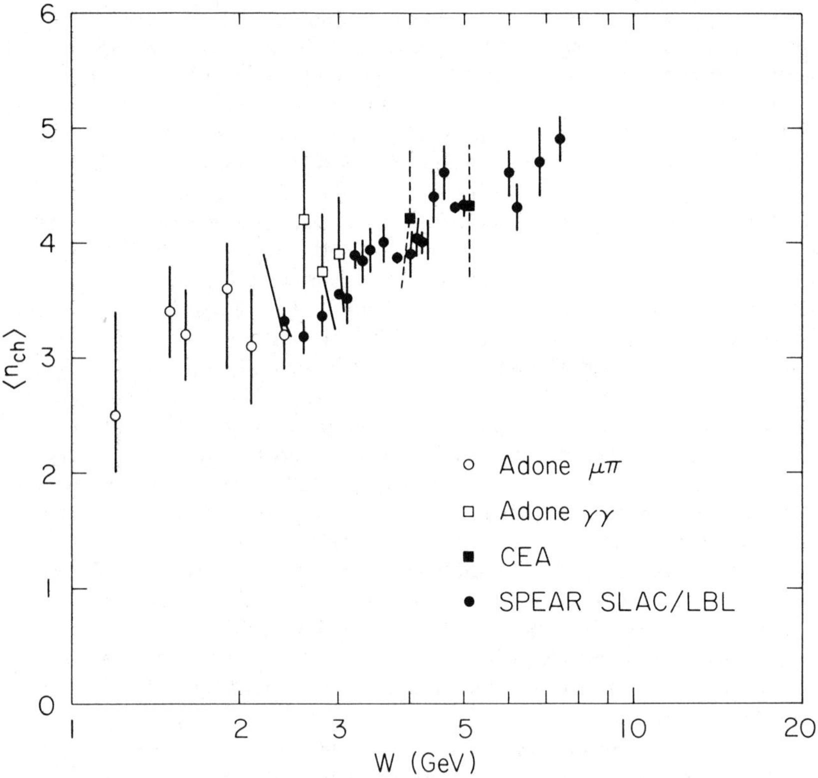

Figure 39 Mean charged multiplicity vs W for hadron production in e^+e^- annihilation. Data are from Adone $\mu\pi$ (171), Adone $\gamma\gamma$ (172), CEA (156, 157), and SPEAR (131, 132).

that would indicate the presence of additional neutral particles in the final state. Again, two events were detected. Both experiments used the same apparatus and demanded that the small-angle e^+ and e^- be tagged and momentum-analyzed.

4.2.4 AVERAGE PROPERTIES OF MULTIHADRON FINAL STATES The mean charged-particle multiplicity $\langle n_{ch} \rangle$ can be computed from the unfold procedure used to determine R. The data from experiments at ADONE (124, 152, 171, 172), CEA (156, 157), and SPEAR (131, 132) are presented in Figure 39. The energy dependence of these data is consistent with the logarithmic growth

$$\langle n_{ch} \rangle = a + b \ln s, \quad 4.14.$$

where (165) $a = 1.93$, $b = 0.75$, and s is in units of GeV^2. Of course, some other slow growth with energy cannot be ruled out. This behavior is reminiscent of the multiplicity growth in many other hadronic processes at comparable energies (173). There is no evidence for abrupt changes in $\langle n_{ch} \rangle$ in the transition region of R, although the experimental uncertainties are rather large and could obscure important effects. Bjorken & Brodsky (174) have discussed the energy dependence of $\langle n_{ch} \rangle$ in terms of either parton or statistical models of final states. The observed slow growth of $\langle n_{ch} \rangle$ favors the parton picture.

The mean fraction of c.m. energy appearing in charged particles was measured with the SLAC/LBL magnetic detector (132) and is shown in Figure 40. In the analysis, all particles were assumed to be pions and corrections were applied for expected geometrical losses of particles. An imporant feature of Figure 40 is the

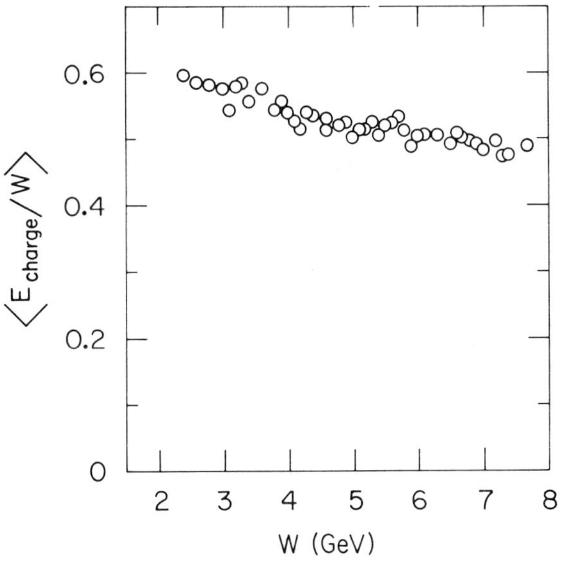

Figure 40 Average fraction of c.m. energy appearing in charged hadrons from the SPEAR experiments (132).

fact that the charged-energy fraction is less than the naive expectation of $\frac{2}{3}$ at all energies. That it is equal to 0.6 at low energies is probably a consequence of the production of heavy particles such as nucleons, kaons, and etas, in addition to pions. Why it should fall with increasing c.m. energy is not understood.

At present there is little experimental information available on the relative abundance of different types of hadrons in multihadron events. The most detailed information has been obtained by the DASP group (113) on the mass of the J/ψ. Preliminary nonresonant data were presented in (172) by the SLAC/LBL group. The Princeton Pavia Maryland group (175) published results of a single-arm spectrometer measurement at nonresonant energies. In the broadest terms, these measurements have shown that pions dominate the population of final-state particles, while roughly 10% of charged particles are kaons and 5% or so are protons.

4.2.5 INCLUSIVE MOMENTUM SPECTRA The scaling of R immediately suggests (143) that single-particle momentum spectra may show Bjorken scaling (34). The SLAC/LBL group (132) has measured inclusive momentum spectra at several c.m. energies. Their results for the quantity $s\, d\sigma/dx$ at three c.m. energies are plotted versus x in Figure 41. In this case, the scaling variable x is computed from the particle momentum rather than total energy because no hadron identification was available.

The spectra for all three energies shown in Figure 41 rise sharply at small values of x, peak at relatively low x, then fall with increasing x. Above $x = 0.4$, the spectra are equal within experimental error and thus are consistent with Bjorken scaling. To study scaling more critically, the SLAC/LBL results for $s\, d\sigma/dx$ are plotted versus W for several x intervals in Figure 42. Bjorken scaling implies

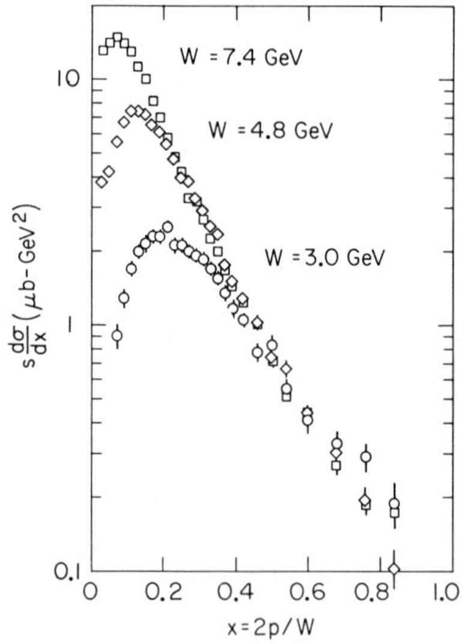

Figure 41 $s(d\sigma/dx)$ vs x for three values of W from (132).

(see Equation 2.10) that $s\, d\sigma/dx$ should not change with W at fixed values of x. For the lowest x interval near $x = 0.1$, scaling is badly broken. By $x = 0.2$, the data are roughly independent of W for W greater than 4 GeV. For $x \geqq 0.4$, the data exhibit Bjorken scaling to the 20% level over the entire energy range studied.

The scaling observed for x values greater than 0.4 is quite remarkable in light of the doubling of R over this same energy range, and suggests that the "new" physics is confined to relatively small values of x. However, the interplay between the changes in R, the mean charged-particle energy, and inclusive momentum spectra as a function of W have made it impossible to isolate two components of the data relating to "old" and "new" physics.

4.2.6 ANGULAR DISTRIBUTIONS The most general single-particle inclusive angular distribution for hadrons produced through one-photon exchange has been given in Equation 2.7. The coefficient of the $\cos^2\theta$ term α is defined by:

$$\alpha = \frac{W_1 - W_0}{W_1 + W_0} \qquad 4.15.$$

where W_1 and W_0 are the structure functions defined in Section 2. α is bounded between -1 and 1. Experiments measuring only polar angle distributions have been

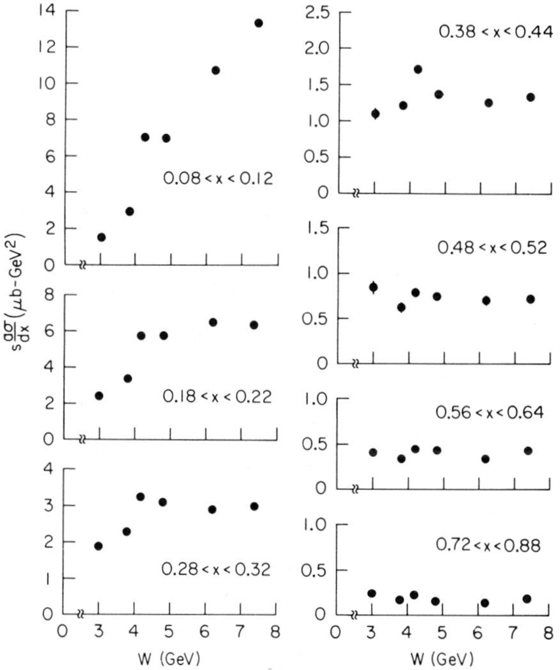

Figure 42 $s(d\sigma/dx)$ vs W for various regions of x from (132). Bjorken scaling implies $s(d\sigma/dx)$ should be independent of W for fixed values of x.

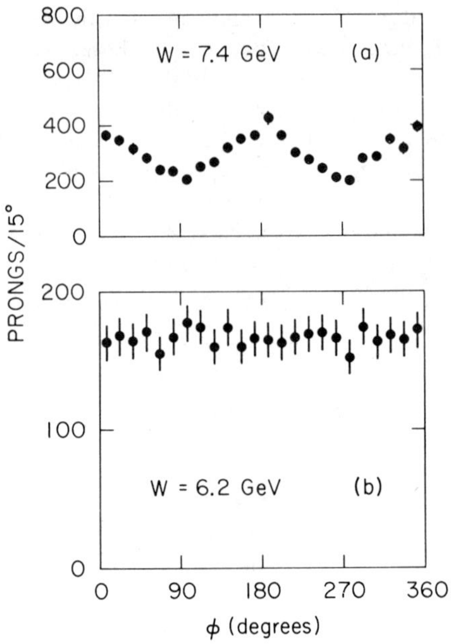

Figure 43 Azimuthal distributions for hadrons with $x > 0.3$ at two values of W from (177). At $W = 7.4$ GeV the e^+e^- beams were polarized, while at $W = 6.2$ GeV they were not.

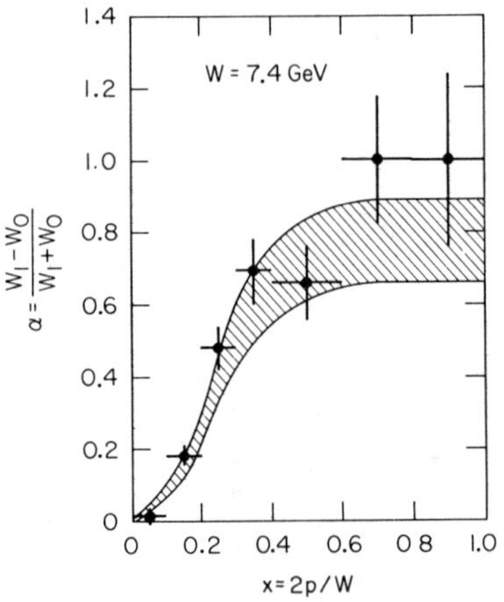

Figure 44 α vs x at $W = 7.4$ GeV determined from the θ and ϕ dependence of inclusive hadron production in the case of polarized e^+e^- beams (177). The shaded region is the prediction of the "jet" model of (178) where $\alpha_{jet} = 0.78 \pm 0.12$.

reported by the BCF group at ADONE (176) and the SLAC/LBL group (172). The ADONE group covered the c.m. energy range from 1.2 to 3.0 GeV; the SPEAR experiment covered 3.0 to 4.8 GeV. In both cases the observed angular distributions were essentially isotropic, but α was poorly determined because of the relatively small range of $\cos^2\theta$ covered in these experiments. The SLAC/LBL group (177) have measured α as a function of the scaling variable x at $W = 7.4$ GeV, where the beam polarization is significant. In this experiment α was determined from the azimuthal, as well as the polar angle, distribution of hadrons where complete coverage was possible. Two examples of their azimuthal distributions are given in Figure 43. At $W = 6.2$ GeV there is no azimuthal dependence because the beams are depolarized due to a spin-precession resonance (32). At $W = 7.4$ GeV the polarization was large and the strong $\cos 2\phi$ term evident in Figure 43 could be used to determine α. The values for α as a function of x are given in Figure 44. It is seen that particles with low x are produced isotropically, while hadrons of large x are

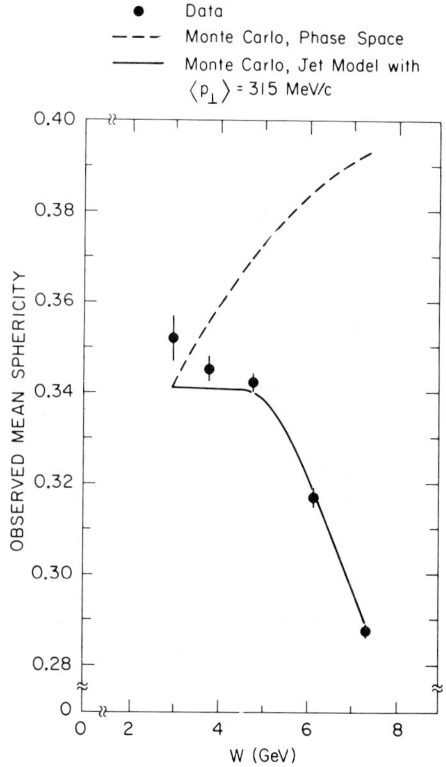

Figure 45 Mean sphericity vs W from (178). Smaller values of sphericity indicate events are more jetlike. The predictions of two models of final states are shown.

produced predominantly through a coupling to the virtual photon that is transverse to their direction of motion.

4.2.7 JET STRUCTURE An important prediction of the parton model (144) is that jets of hadrons moving in opposite directions should be produced in e^+e^- collisions. The schematic drawing of hadron production by partons given in Figure 36 provides an intuitive picture of jets—a pair of partons are produced and they subsequently radiate hadrons along their common direction of motion with limited transverse momentum relative to that axis. An operational definition of jets is any multiparticle correlation that leads to some preferred direction in a system of hadrons about which the transverse momentum is limited.

The SLAC/LBL group (178) has searched for jet structure in multihadron events by finding that axis in each event that minimizes the sum of squares of charged particle momenta perpendicular to it. For each event, a parameter called the sphericity, S, was computed. S is a measure of the jetlike character of events and is defined by:

$$S = \frac{3\sum_i p_{\perp i}^2}{2\sum_i p_i^2}, \qquad 4.16.$$

where summations run over all charged particles observed in the event, p_i is the particle momentum, and $p_{\perp i}$ is the momentum perpendicular to the jet axis. S is bounded between 0 and 1; events with small S are jetlike, events with large S are not.

The mean sphericity reported by (178) is plotted versus c.m. energy in Figure 45; it decreases substantially with increasing c.m. energy, suggesting the presence of jets in the high-energy data. Two models of final states, an invariant-phase space model, and a jet model that modified phase space to give limited transverse momentum about a jet axis, were compared with the data. The mean transverse momentum of particles in the jet model was 315 MeV/c at all c.m. energies. In Figure 45 the jet model is seen to agree with the data at all c.m. energies, while the invariant-phase space model fails to describe the data. The authors argue that the observed small values of S are not simple kinematic reflections, but represent multiparticle correlations that cannot be explained by energy-momentum conservation alone.

The complete (θ and ϕ) angular distribution for the jet axis at $W = 7.4$ GeV displays the usual one-photon exchange form with $\alpha = 0.78 \pm 0.12$, where α is defined by Equation 4.15. This is very close to the μ-pair distribution where $\alpha = 1$, and strongly implies that if partons are responsible for jets, then the parton spin is $\frac{1}{2}$.

The jet model of (178) is able to reproduce well the inclusive angular distribution of hadrons at $W = 7.4$ GeV. The predictions of this model are given by the shaded area in Figure 44. The simple jet picture provided by the parton model with spin-$\frac{1}{2}$ partons gives a remarkably accurate description of the sphericity and angular distributions of hadrons produced in e^+e^- annihilations at high energy.

5 CONCLUSIONS

In this article we have reviewed experimental results from the study of e^+e^- interactions at high energies. This field is in a period of exceptionally rapid growth at present, so that we have only been able to outline current areas of research.

In many examples we have seen the importance of the simple initial state provided by the e^+e^- system. Throughout the studies of quantum electrodynamics and hadron production, the one-photon exchange channel has played a dominant role. Its associated simple angular distribution has been observed in electrodynamic processes as well as in hadron production. Radiative corrections are necessary for a quantitative understanding of all processes, but have not introduced any essential difficulties to understanding experimental results.

The photon-photon process has played a minor role to date, but has been observed in electrodynamic and hadronic final states and promises to be more important in the future.

Final states involving only electrodynamic particles are well described by the theory of quantum electrodynamics when calculable hadronic vacuum polarization corrections are included. The only exception to this rule is the sample of anomalous $e\mu$ events discovered at SPEAR. A great deal more experimental information is required in order to establish the origin of these very significant events.

Predictions of QED have been experimentally verified to the 10% to 20% level in many reactions over a broad range of energies. Lower limits on possible breakdown parameters are in the vicinity of 30 GeV. Equivalently, these experiments have checked the validity of QED down to scales of distance less than 10^{-15} cm. Hadronic corrections to QED have been observed in the vicinity of the narrow resonances ϕ, J/ψ, and ψ'.

e^+e^- experiments have played a major role in the discovery and determination of properties of the vector mesons. The spectrum of vector mesons now includes the $\rho, \omega, \phi, J/\psi, \psi', \rho'(1600)$, and several broad states with masses near 4 GeV. Because of their extraordinary properties, the J/ψ and ψ' appear to represent new hadronic degrees of freedom not present in ordinary hadrons. A new heavy-particle spectroscopy related to the narrow resonances is presently being explored. e^+e^- colliding-beam experiments are uniquely responsible for our present depth of understanding of the new particles.

The total hadronic cross section is scaling in two different energy regions. For c.m. energies between 2.4 and 3.5 GeV, R is approximately 2.5, while between 5.0 and 7.8 GeV, R is roughly 5. A complicated transition region connects these scaling regions. The mean charged-particle multiplicity slowly increases with energy, while the mean fraction of c.m. energy appearing in final-state charged hadrons decreases as the c.m. energy increases. Single-particle inclusive momentum spectra exhibit Bjorken scaling for $x > 0.4$ over the energy range 3.0 GeV $\leq W \leq$ 7.4 GeV. There is jet structure in multihadronic final states at high energies. The average transverse momentum of hadrons with respect to the jet axis is approximately

315 MeV/c. Jets couple to the virtual photons mainly throu, a transverse coupling similar to muon pairs.

Qualitative features of hadron production by e^+e^- collidi g beams are remarkably well described by the quark-parton model. The constancy of R, scaling of inclusive momentum spectra, and jets are natural consequences of the parton model. The azimuthal dependence of hadron production with polarized beams is strong evidence for the spin-$\frac{1}{2}$ nature of the partons.

Specific quark models are able to describe qualitative features of the data, but fail quantitatively. The new resonances J/ψ and ψ' and their associated heavy-particle spectroscopy look very much like bound states of a new charmed quark, but details of the spectrum and decay rates disagree with current estimates. A direct connection between the "new" physics represented by the increase in R between 3.5 and 5.0 GeV and the new particles has not yet been established.

Much experimental data is still needed in order to understand hadron production in e^+e^- collisions. Specifically, the c.m. energy region 1.0 to 2.5 GeV needs to be explored with new detectors more suited to studying multihadron final states. The approach to scaling and spectrum of vector mesons in this energy range is of crucial importance to our understanding of the structure of the "old" physics. Likewise, detailed information on R in the 4-GeV transition region and knowledge of the spectrum of states, quantum numbers, and decay rates for the new particle spectroscopy are essential for understanding the "new" physics.

With the new storage rings now being built, a vast, uncharted sea of e^+e^- physics awaits us. Our guides, the theory of QED and the parton ideas, will again be tested. We can expect to encounter new physics involving the weak interactions. From past experience we have every reason to believe that the next generation of experiments will uncover new phenomena as rich and important to our understanding of fundamental processes as those we have outlined here.

Literature Cited

1. Sands, M. 1971. *Proc. Int. Sch. Phys. "Enrico Fermi," Course 46*, pp. 257–411. New York/London: Academic
2. Pellegrini, C. 1972. *Ann. Rev. Nucl. Sci.* 22:1
3. Bernardini, C. et al 1964. *Nuovo Cimento* 34:1473
4. Bernardini, C., Corazza, G. F., Ghigo, G., Touscheck, B. 1960. *Nuovo Cimento* 18:1293
5. O'Neill, G. K. 1959. *Proc. Int. Conf. High Energy Accel. Instrum., Geneva*, p. 125
6. Barber, W. C., Gittelman, B., O'Neill, G. K., Richter, B. 1966. *Phys. Rev. Lett.* 15:1127
7. Barber, W. C., Gittelman, B., O'Neill, G. K., Richter, B. 1971. *Phys. Rev. D* 3:2796
8. Budker, G. I. et al 1965. *Proc. Int. Conf. High Energy Accel., 5th, Frascati*, p. 390
9. Budker, G. I. et al 1965. *At. Energ.* 19: 497
10. Auslander, V. L. et al 1967. *Proc. Int. Symp. Electron Photon Interactions High Energies, Stanford*, pp. 385, 601
11. Augustin, J.-E. et al 1967. See Ref. 10, pp. 385, 600
12. Auslander, V. L. et al 1965. See Ref. 8, p. 394
13. Orsay Storage Ring Group 1965. See Ref. 8, p. 271
14. Budker, G. I. et al 1972. *Sov. Conf. Charged Part. Accel.* Moscow, 1:318
15. Amman, F. et al 1963. *Proc. Int. Conf. High Energy Accel., Dubna*, p. 249
16. Hofmann, A. et al 1967. *Proc. Int. Conf. High Energy Accel., 6th, Cambridge*, p. 112
17. SPEAR Storage Ring Group 1971. *Proc. Int. Conf. High Energy Accel., 7th,*

Geneva, p. 145
18. Augustin, J.-E. et al 1974. *Phys. Rev. Lett.* 33:1406
19. Aubert, J. J. et al 1974. *Phys. Rev. Lett.* 33:1404
20. Abrams, G. S. et al 1974. *Phys. Rev. Lett.* 33:1453
21. DESY Storage Ring Group 1974. *Proc. Int. Conf. High Energy Accel., 9th, Stanford*, p. 43
22. Marin, P. 1974. See Ref. 21, p. 49
23. Sidorov, V. A. 1975. Private communication
24. LBL-SLAC PEP Study Group 1974. LBL Rep. No. 2688, SLAC Rep. No. SLAC-171
25. PETRA Storage Ring Group 1974. DESY Proposal
26. Cornell Storage Ring Group 1975. Cornell Rep. CLNS-301
27. Tsai, Y. S. 1960. *Phys. Rev.* 120:269
28. Cabibbo, N., Gatto, R. 1961. *Phys. Rev.* 124:1577
29. Tsai, Y. S. 1976. *Phys. Rev. D* 12:3533
30. Bjorken, J. D., Drell, S. D. 1964. *Relativistic Quantum Mechanics*. New York: McGraw-Hill
31. Sokolov, A. A., Ternov, I. M. 1964. *Sov. Phys. Dokl.* 8:1203
32. Baier, V. N. 1971. See Ref. 1, pp. 1–49
33. Jackson, J. D. 1976. *Rev. Mod. Phys.* 48:417
34. Bjorken, J. D. 1969. *Phys. Rev.* 179:1547
35. Brodsky, S. J., Kinoshita, T., Terazawa, H. 1970. *Phys. Rev. Lett.* 25:972
36. Arteaga-Romero, N., Jaccarini, A., Kessler, P., Parisi, J. 1971. *Phys. Rev. D* 3:1569
37. Baier, V. N., Fadin, V. S. 1971. *Nuovo Cimento Lett.* 1:481
38. Terazawa, H. 1973. *Rev. Mod. Phys.* 45:615
39. Tsai, Y. S. 1960. *Phys. Rev.* 120:269
40. Bonneau, G., Martin, F. 1971. *Nucl. Phys. B* 27:381
41. Bloch, F., Nordsieck, A. 1937. *Phys. Rev.* 52:54
42. Brodsky, S. J., Drell, S. D. 1970. *Ann. Rev. Nucl. Sci.* 20:147
43. Godine, J., Hankey, A. 1972. *Phys. Rev. D* 6:3301; Cung, V. K., Mann, A. K., Paschos, E. A. 1972. *Phys. Lett. B* 41:355
44. Drell, S. D. 1958. *Ann. Phys.* 4:75
45. Lee, T. D., Wick, G. C. 1969. *Nucl. Phys. B* 9:209
46. Kroll, N. M. 1966. *Nuovo Cimento A* 45:65
47. Berends, F. A., Gaemers, K. J. F., Gastmans, R. 1974. *Nucl. Phys. B* 68:541
48. Augustin, J.-E. et al 1970. *Phys. Lett. B* 31:673
49. Bartoli, B. et al 1970. *Nuovo Cimento A* 70:603
50. Alles-Borelli, V. et al 1972. *Nuovo Cimento A* 7:345
51. Alles-Borelli, V. et al 1971. *Phys. Lett. B* 36:149
52. Bartoli, B. et al 1971. *Phys. Lett. B* 36:593
53. Bartoli, B. et al 1972. *Phys. Rev. D* 6:2374
54. Borgia, B. et al 1971. *Phys. Lett. B* 35:340
55. Bernardini, M. et al 1973. *Phys. Lett. B* 45:510
56. Bernardini, M. et al 1973. *Phys. Lett. B* 45:169
57. Madaras, R. et al 1973. *Phys. Rev. Lett.* 30:507
58. Newman, H. et al 1974. *Phys. Rev. Lett.* 32:483
59. Beron, B. L. et al 1974. *Phys. Rev. Lett.* 33:663
60. Hofstadter, R. 1975. *Proc. 1975 Int. Symp. Lepton Photon Interactions High Energies, Stanford*, p. 869
61. Augustin, J.-E. et al 1975. *Phys. Rev. Lett.* 34:233
62. Learned, J. G., Resvanis, L. K., Spencer, C. M. 1975. *Phys. Rev. Lett.* 35:1688
63. Berends, F. A., Gaemers, K. J. F., Gastmans, R. 1972. *Nucl. Phys. B* 57:381; 75:546
64. Balakin, V. E. et al 1971. *Phys. Lett. B* 37:435
65. Borgia, B. et al 1972. *Nuovo Cimento Lett.* 3:115
66. Alles-Borelli, V. et al 1972. *Nuovo Cimento A* 7:330
67. Alles-Borelli, V. et al 1975. *Phys. Lett. B* 59:201
68. Bollini, D. et al 1975. *Nuovo Cimento Lett.* 13:380
69. Kurdadze, L. M. et al 1975. Novosibirsk Prepr. 75-66
70. Berends, F. A., Gastmans, R. 1973. *Nucl. Phys. B* 61:414
71. Balakin, V. E. et al 1971. *Phys. Lett. B* 34:99
72. Bacci, C. et al 1971. *Nuovo Cimento Lett.* 2:73
73. Hanson, G. et al 1973. *Nuovo Cimento Lett.* 7:587
74. Law, M. E. et al 1974. *Nuovo Cimento Lett.* 11:5
75. Cosme, G. et al 1973. *Nuovo Cimento Lett.* 8:509
76. Bacci, C. et al 1973. *Phys. Lett. B* 44:530
77. Balakin, V. E., Bukin, A. D., Pakhtusova, E. V., Sidorov, V. A., Khabakhpashev,

A. G. 1971. *Phys. Lett. B* 34:663
78. Bacci, C. et al 1972. *Nuovo Cimento Lett.* 3:709
79. Parisi, J. 1974. *J. Phys. Paris Colloq. C2* 35:51
80. Barbiellini, G. et al 1974. *Phys. Rev. Lett.* 32:385
81. Augustin, J.-.E. et al 1973. *Phys. Rev. Lett.* 30:462
82. Barbiellini, G. et al 1974. *Nuovo Cimento Lett.* 11:718
83. Baldini–Celio, R. et al 1974. *Nuovo Cimento Lett.* 11:711
84. Braunschweig, W. et al 1974. *Phys. Lett. B* 53:393
85. Braunschweig, W. et al 1975. *Phys. Lett. B* 56:491
86. Boyarski, A. M. et al 1975. *Phys. Rev. Lett.* 34:1357
87. Lüth, V. et al 1975. *Phys. Rev. Lett.* 35:1124
88. Ford, R. L. et al 1975. *Phys. Rev. Lett.* 34:604
89. Jackson, J. D., Scharre, D. L. 1975. *Nucl. Instrum. Methods* 128:13
90. Tsai, Y. S. 1971. *Phys. Rev. D* 4:2821
91. Alles-Borelli, V. et al 1970. *Nuovo Cimento Lett.* 4:1151
92. Alles-Borelli, V. et al 1970. *Nuovo Cimento Lett.* 4:1156
93. Orito, S. et al 1974. *Phys. Lett. B* 48:165
94. Perl, M. L. et al 1975. *Phys. Rev. Lett.* 35:1489
95. Auslander, V. L. et al 1967. *Phys. Lett. B* 25:433
96. Augustin, J.-E. et al 1968. *Phys. Rev. Lett.* 20:126
97. Augustin, J.-E. et al 1969. *Phys. Lett. B* 28:508
98. Auslander, V. L. et al 1969. *Sov. J. Nucl. Phys.* 9:114
99. Benaksas, D. et al 1972. *Phys. Lett. B* 39:289
100. Gounaris, G. J., Sakurai, J. J. 1968. *Phys. Rev. Lett.* 21:244
101. Gourdin, M., Stodolsky, L., Renard, F. M. 1969. *Phys. Lett. B* 30:347
102. Bemporad, C. 1975. See Ref. 60, p. 113
103. Augustin, J.-E. et al 1969. *Phys. Lett. B* 28:513
104. Benaksas, D. et al 1972. *Phys. Lett. B* 42:507
105. Cosme, G., Jean-Marie, B., Jullian, S., Lefrançois, J. 1972. *Nucl. Instrum. Methods* 99:599
106. Augustin, J.-E. et al 1969. *Phys. Lett. B* 28:517
107. Balakin, V. E. et al 1971. *Phys. Lett. B* 34:328
108. Cosme, G. et al 1974. *Phys. Lett. B* 48:159
109. Bizot, J. C. et al 1970. *Phys. Lett. B* 32:416
110. Cosme, G. et al 1974. *Phys. Lett. B* 48:155
111. Bukin, A. D. et al 1975. Novosibirsk Prepr. 75-64
112. Benaksas, D. et al 1972. *Phys. Lett. B* 42:511
112a. Renard, F. M. 1974. *Nucl. Phys. B* 82:1
113. Wiik, B. H. 1975. See Ref. 60, p. 69
114. Jean-Marie, B. et al 1976. *Phys. Rev. Lett.* 36:291
115. Abrams, G. S. 1975. See Ref. 60, p. 25
116. Braunschweig, W. et al 1975. *Phys. Lett. B* 57:297
117. Abrams, G. S. et al 1975. *Phys. Rev. Lett.* 34:1181
117a. Hilger, E. et al 1975. *Phys. Rev. Lett.* 35:625
118. Braunschweig, W. et al 1975. *Phys. Lett. B* 57:407
119. Tanenbaum, W. et al 1975. *Phys. Rev. Lett.* 35:1323
120. Tanenbaum, W. et al 1976. *Phys. Rev. Lett.* 36:402
121. Feldman, G. J. et al 1975. *Phys. Rev. Lett.* 35:821
122. Heintze, J. 1975. See Ref. 60, p. 97
123. Barbarino, G. et al 1972. *Nuovo Cimento Lett.* 3:689
124. Grilli, M. et al 1973. *Nuovo Cimento A* 13:593
125. Bingham, H. H. et al 1972. *Phys. Lett. B* 41:635
126. Conversi, M. et al 1974. *Phys. Lett. B* 52:493
127. Ceradini, F. et al 1973. *Phys. Lett. B* 43:341
128. Bernardini, M. et al 1974. *Phys. Lett. B* 53:384
129. Alles-Borelli, V. et al 1975. *Nuovo Cimento A* 30:136
130. Balakin, V. E. et al 1972. *Phys. Lett. B* 41:205
131. Augustin, J.-E. et al 1975. *Phys. Rev. Lett.* 34:764
132. Schwitters, R. F. 1975. See Ref. 60, p. 5
133. Aulchenko, V. M. et al 1975. Novosibirsk Prepr. 75-65
134. Bacci, C. et al 1975. *Phys. Lett. B* 58:481
135. Esposito, B. et al 1975. *Phys. Lett. B* 58:478
136. Boyarski, A. M. et al 1975. *Phys. Rev. Lett.* 34:762
137. Barbiellini, G. et al 1973. *Nuovo Cimento Lett.* 6:557
138. Bollini, D. et al 1975. *Nuovo Cimento Lett.* 14:418
139. Bernardini, M. et al 1973. *Phys. Lett. B* 44:393

140. Bernardini, M. et al 1973. *Phys. Lett. B* 46:261
141. Castellano, M. et al 1973. *Nuovo Cimento A* 14:1
142. Wilson, R. 1972. *Proc. Int. Conf. High Energy Phys., 15th,* Kiev, p. 219
143. Drell, S. D., Levy, D., Yan, T. M. 1969. *Phys. Rev.* 187:2159
144. Feynman, R. P. 1972. *Photon-Hadron Interactions.* Reading, Mass: Benjamin
145. Cabibbo, N., Parisi, G., Testa, M. 1970. *Nuovo Cimento Lett.* 4:35
146. Ferrara, S., Greco, M., Grillo, A. F. 1970. *Nuovo Cimento Lett.* 4:1
147. Feldman, G. J., Perl, M. L. 1975. *Phys. Rep. C* 19:235
148. Bartoli, B. et al 1970. *Nuovo Cimento A* 70:615
149. Bartoli, B. et al 1971. *Phys. Lett. B* 36:598
150. Bacci, C. et al 1972. *Phys. Lett. B* 38:551
151. Bacci, C. et al 1973. *Phys. Lett. B* 44:533
152. Ceradini, F. et al 1973. *Phys. Lett. B* 47:80
153. Bernardini, M. et al 1974. *Phys. Lett. B* 51:200
154. Kurdadze, L. M., Onuchin, A. P., Serednyakov, S. I., Sidorov, V. A., Eidelman, S. I. 1972. *Phys. Lett. B* 42:515
155. Cosme, G. et al 1972. *Phys. Lett. B* 40:685
156. Litke, A. et al 1973. *Phys. Rev. Lett.* 30:1189
157. Tarnopolsky, G. et al 1974. *Phys. Rev. Lett.* 32:432
158. Fritzsch, H., Gell-Mann, M. 1972. *Proc. Int. Conf. High Energy Phys., 16th,* Batavia, Ill.
159. Bjorken, J. D. 1973. *Proc. Int. Symp. Electron Photon Interactions High Energies, 6th,* Bonn, p. 25
160. Gilman, F. J. 1975. See Ref. 10, p. 131
161. Harari, H. 1975. *Proc. 1975 Int. Symp. Lepton Photon Interactions High Energies,* Stanford, p. 317
162. Gaillard, M. K., Lee, B. W., Rosner, J. L. 1975. *Rev. Mod. Phys.* 47:277
163. Appelquist, T., De Rújula, A., Politzer, H. D., Glashow, S. L. 1975. *Phys. Rev. Lett.* 34:365
164. Eichten, E. et al 1975. *Phys. Rev. Lett.* 34:369
165. Feldman, G. J. 1975. See Ref. 60, p. 39
166. Boyarski, A. M. et al 1975. *Phys. Rev. Lett.* 35:196
167. Einhorn, M. B., Quigg, C. 1975. *Phys. Rev. Lett.* 35:1114
168. Han, M. Y., Nambu, Y. 1963. *Phys. Rev. B* 139:1006
169. Orito, S., Ferrer, M. L., Paoluzi, L., Santonico, R. 1974. *Phys. Lett. B* 48:380
170. Paoluzi, L. et al 1974. *Nuovo Cimento Lett.* 10:435
171. Ceradini, F. et al 1972. *Phys. Lett. B* 42:501
172. Richter, B. 1974. *Proc. Int. Conf. High Energy Phys., 17th, London,* pp. IV–37
173. Whitmore, J. 1974. *Phys. Rep. C* 10:273
174. Bjorken, J. D., Brodsky, S. J. 1970. *Phys. Rev. D* 1:1416
175. Atwood, T. L. et al 1975. *Phys. Rev. Lett.* 35:704
176. Bernardini, M. et al 1975. *Nuovo Cimento A* 26:163
177. Schwitters, R. F. et al 1975. *Phys. Rev. Lett.* 35:1320
178. Hanson, G. et al 1975. *Phys. Rev. Lett.* 35:1609
179. Hilger, E. et al 1975. *Phys. Rev. Lett.* 35:625

Ann. Rev. Nucl. Sci. 1977. 27: 393–464

2

PSIONIC MATTER[1]

W. Chinowsky

Department of Physics and Lawrence Berkeley Laboratory, University of California, Berkeley, California 94720

CONTENTS

[1] Work supported by the Energy Research and Development Administration, under the auspices of the Division of Physical Research.

INTRODUCTION

Before November of 1974, heavy hadrons that decay to hadrons were neatly classified into two groups, distinguished by their decay rates. Those are, very roughly, either of the order of 10^{-9} sec or 10^{-23} sec (1). Strangeness conservation distinguishes the longer-lived group from the shorter-lived one. The strikingly different decay rate of the J/ψ,[2] orders of magnitude different from either of these values, indicates the operation of a quite different kind of dynamics. With further developments, particularly the observation in e^+e^- annihilation of heavier, directly formed vector states as well as others produced in e^+e^- annihilations to multiparticle states, interpretation has narrowed. Arguments in favor of a new property of matter and a new conservation law for strong interactions have become very convincing, if not yet entirely compelling. Indeed, all observations now fit very comfortably into a theoretical framework originating with the introduction of a fourth quark precisely of the character proposed earlier, and for quite different reasons, by Glashow and co-workers (2, 3). In this article, I review the phenomenology of psion production and decay, and also charmed-hadron production and decay, as relevant to a broader conception of the subject. The simple model of the ψ states as composites of a charmed quark and its antiparticle dominates the interpretive discussion (4, 5).

1.1 *Old Vector Mesons*

It is inappropriate to give an extensive review of properties of the older hadronic resonant states, but still useful to summarize some characteristics of production and decay to contrast with the ψ-family characteristics. Only those vector mesons that couple directly to the photon, i.e. ρ^0, ω^0, and ϕ^0, are discussed. Table 1 shows selected decay parameters of these vector mesons (1). Values of full width, Γ, and partial width, Γ_{ee}, for decay to e^+e^- pairs, have been extracted from measurements of cross sections for e^+e^- annihilation to particular hadronic final states, $\pi^+\pi^-$ for ρ^0, $\pi^+\pi^-\pi^0$ for ω^0, and $K\bar{K}$ for ϕ (6). The annihilation cross section in the vicinity of a resonance varies with center of mass (c.m.) energy E, according to the Breit-Wigner expression

$$\sigma_f(E) = \frac{(2J+1)\pi}{E^2} \frac{\Gamma_{ee}\Gamma_f}{(E-M)^2+\Gamma^2/4},$$

Table 1 Parameters of vector mesons ρ^0, ω^0, and ϕ^0

Meson	Mass (MeV/c^2)	Γ(MeV)	Γ_{ee}(keV)	$f_v^2/4\pi$
ρ^0	773 ± 3	152 ± 3	6.5 ± 0.7	2.1 ± 0.2
ω^0	782.7 ± 0.3	10.0 ± 0.4	0.76 ± 0.17	18.3 ± 4.0
ϕ^0	1019.7 ± 0.3	4.1 ± 0.2	1.31 ± 0.08	13.8 ± 0.8

[2] From now on, I eschew the typographically awkward "J/ψ" and use ψ or ψ (3095) for the lightest psion.

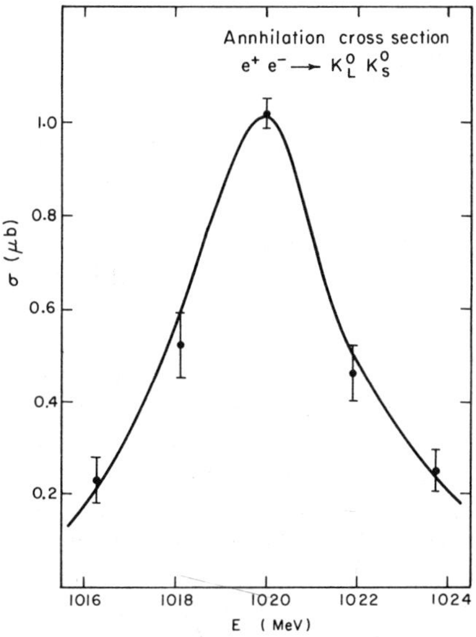

Figure 1 Cross section for e^+e^- annihilation to $K_S^0K_L^0$ as a function of c.m. energy, showing ϕ^0-meson resonance excitation; data from (7).

where Γ_f and Γ_{ee} are the partial widths for decay to the final states f and e^+e^- respectively, and Γ is the full width. Suitable modification of this simple behavior must be made to take account of initial-state radiation in order to obtain the resonance widths. An exemplary excitation curve, for $K_L^0K_S^0$ production (7), is shown in Figure 1. The full width and the product $\Gamma_{ee}\Gamma_f$ are determined from the annihilation cross sections near the resonance. Values of Γ_f are obtained from independent sources, thus allowing determination of electron-pair partial widths. A useful parameterization is in terms of the coupling of the vector meson to the virtual photon, indicated in the vector-dominance (8) diagram of Figure 2. The coupling constant f_v, defined by

$$G_v \equiv M_v^2/f_v,$$

is related to Γ_{ee} by (9)

$$f_v^2/4\pi = \frac{1}{3}\,\frac{\alpha^2 M_v}{\Gamma_{ee}}. \tag{1.1}$$

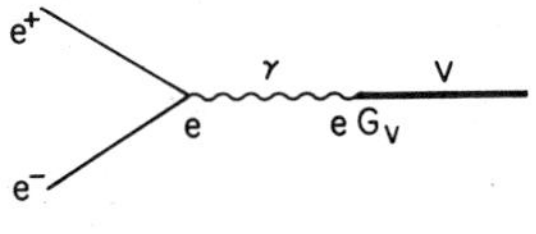

Figure 2 Direct coupling of a vector meson to a virtual photon.

Values of Γ in Table 1 are averages of values obtained from measurements of resonance production in hadron-hadron interactions, as well as the e^+e^- annihilation results. As has been remarked often, it is noteworthy that the width of the ϕ^0 is only one half that of the ω^0 and more remarkable still that the partial width for ϕ^0 decay to $\pi^+\pi^-\pi^0$ is just 0.67 ± 0.07 MeV, while the partial width is 9.0 ± 0.4 MeV for decay of the lighter ω^0 to $\pi^+\pi^-\pi^0$. Indeed the principal decay mode of the ϕ^0 meson is $K\bar{K}$, with 82% branching fraction, in spite of the relatively small phase space that is available to that state.

Meson resonances of the old style are produced in hadron-hadron collisions, restricted only by requirements of the strong-interaction conservation laws. Typical exclusive reaction cross sections, e.g. for $\pi p \to \rho p$, are in the range 0.1–10 mb (10), values typical of strong interactions. Data on inclusive vector-meson production in hadron-hadron interactions are sparse and suffer poor statistical accuracy. Inclusive ρ^0 production rates in π-p and p-p collisions have been determined from observations of interactions in bubble chambers (11). Cross sections increase smoothly from ~ 2 mb at 4.9 GeV c.m. energy to ~ 10 mb at 19 GeV. Inclusive production of ϕ^0 appears to be smaller by about a factor of ten, judging from measured values of 0.16 ± 0.04 mb at 6.8 GeV (12) and 0.6 ± 0.2 mb at 16.8 GeV c.m. energy (13). The first result is again from a pp bubble-chamber exposure. The other is from an experiment that measured rates for muon pair production in π^+ and proton collisions with Be nuclei. It is described in some detail below.

In the context of vector-meson dominance (8), vector-meson photoproduction (14) and scattering are related according to (15)

$$\frac{d\sigma}{dt}(\gamma p \to Vp) = \frac{4\pi}{f_v^2}\,\alpha\,\frac{d\sigma}{dt}(Vp \to Vp) \qquad 1.2$$

where t is the momentum transfer. With the optical theorem, neglecting the real part of the forward scattering amplitude,

$$\frac{d\sigma}{dt}(\gamma p \to Vp) = \frac{\alpha}{16\pi}\,\frac{4\pi}{f_v^2}\,\sigma_{\mathrm{T}}^2(Vp \to Vp) \qquad 1.3$$

at $t = 0$. The value of the total vector-meson-nucleon interaction cross section, σ_{T}, is then obtained with f_v determined from the e^+e^- partial width Γ_{ee}.

Values of $d\sigma/dt$ at $t = 0$ for ρ^0 photoproduction on protons are, to within $\sim 15\%$, constant at 115 μb/GeV^2 for 3.6 GeV $< (s)^{1/2} <$ 5.9 GeV (16). Using the vector-

Table 2 Properties of quarks

Flavor	Spin	I	I_3	Q/e	B	S	Y	C
u	1/2	1/2	1/2	2/3	1/3	0	1/3	0
d	1/2	1/2	−1/2	−1/3	1/3	0	1/3	0
s	1/2	0	0	−1/3	1/3	−1	−2/3	0
c	1/2	0	0	2/3	1/3	0	0	1

dominance equation (1.3) with $f_\rho^2/4\pi$ determined from a weighted average of measured $e^\pm$ and $\mu^\pm$ partial widths (1), we obtain the total cross section $\sigma(\rho\text{-nucleon}) = 24$ mb in that range of c.m. energy. Results of measurements of ϕ^0 photoproduction (17) at various energies in the range 2.2 GeV $< (s)^{1/2} <$ 4.2 GeV are compatible with a constant value for the differential cross section at $t = 0$, $d\sigma/dt = 2.49 \pm 0.15$ μb GeV^{-2}. Vector dominance then yields $\sigma(\phi\text{-nucleon}) = 8.7 \pm 0.5$ mb (17). Those values for ρ^0 and for ϕ^0 interaction cross sections are typical of strong interactions.

1.2 *Quark-Model Classification*

Hadron states group into multiplets specified by spin, parity, and baryon numbers indicative of the exact SU(2) symmetry and approximate SU(3) symmetry of the strong interactions. Baryons populate SU(3) multiplets of dimensionality ten and eight. Mesons fit into octet and singlet representations. The observed states can be considered composite structures whose basic constituents are the three spin-$\frac{1}{2}$ quarks of the 3 representation of SU(3). Their properties are listed in Table 2. In this model, baryons are systems of three quarks, and mesons are quark-antiquark composites. Properties of meson states are given in terms of the orbital and spin angular momentum of the $q\bar{q}$ system and its isospin, namely spin $J = |\mathbf{L}+\mathbf{S}|$, parity $P = (-1)^{L+1}$, charge conjugation $C = (-1)^{L+S}$ and G-parity $G = C(-1)^I$. Members of the lowest-lying meson multiplets are s states of the $q\bar{q}$ system. The quark contents assigned to the "old" vector mesons are

$$\rho^0 = (2)^{-1/2}[u\bar{u} - d\bar{d}]$$

$$\omega^0 = (2)^{-1/2}[u\bar{u} + d\bar{d}]$$

$$\phi^0 = s\bar{s}.$$

An essentially ad hoc rule based on this composition and those of the pseudoscalar mesons, the Okubo (18)-Zweig (19)-Iizuka (20) (OZI) rule, has been introduced to explain the small width and dominant $K\bar{K}$ decay mode of the ϕ. This rule proclaims that a process is inhibited if its duality diagram contains disconnected quark lines. Such lines can be isolated in a duality diagram by drawing a line that crosses no quark lines. Allowed and forbidden decays of the ϕ-meson are illustrated in Figure 3. Some attempts have been made to put the OZI rule on a firmer dynamical foundation based on dual models (21) or asymptotically free gauge theories (22), but only semiquantitative success has yet been achieved.

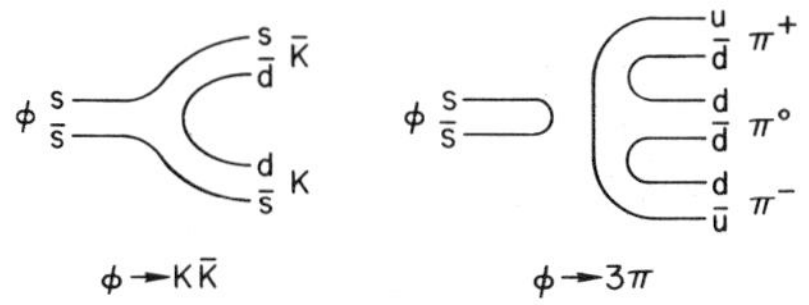

Figure 3 Duality (quark-line) diagrams for $K\bar{K}$, OZI-allowed; and 3π, OZI-forbidden, decays of the ϕ meson.

2 PRODUCTION OF $\psi(3095)$ AND $\psi(3684)$

The existence of the narrow state $\psi(3095)$ was first established independently, with quite different experimental techniques, by Aubert et al (23) and Augustin et al (24).

The Brookhaven National Laboratory-Massachussets Institute of Technology (BNL-MIT) (23) group measured the effective mass spectrum of e^+e^- pairs produced in the interaction of 28.5 GeV/c protons with beryllium, $p + \text{Be} \rightarrow e^+ + e^- + X$. In this experiment, a double-arm magnetic spectrometer, designed to have maximum detection efficiency for heavy, unstable particles produced at rest in the center of mass, produced the mass spectrum shown in Figure 4. A narrow resonance was definitively established at mass 3.112 GeV/c^2 and width less than the 5-MeV/c^2 experimental resolution of the apparatus. More details of this experiment and subsequent hadronic-production experiments are given below.

The Lawrence Berkeley Laboratory-Stanford Linear Accelerator Center (LBL-SLAC) (24) group measured the total cross section for e^+e^- annihilation as a function of energy. Measurements were made in small energy intervals near 3100 MeV with a

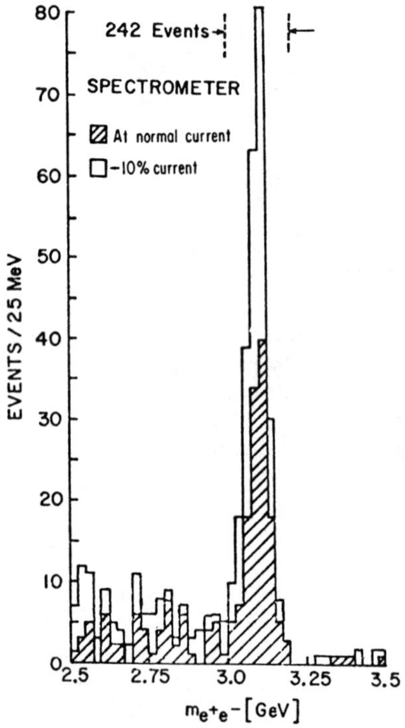

Figure 4 Invariant-mass spectrum of e^+e^- pairs produced in collisions of 28-GeV/c protons with beryllium showing J-meson production; from (23).

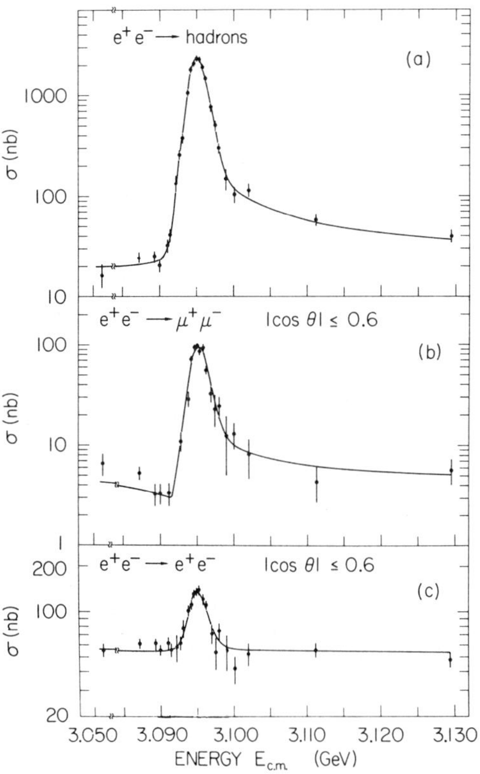

Figure 5 Cross sections for e^+e^- annihilation to (*a*) hadrons, (*b*) muon pairs, and (*c*) electron pairs, showing ψ(3095)-resonance excitation; from (43).

cylindrical detector array in an axially directed solenoidal magnetic field. Yields were measured for annihilation to e^+e^- pairs, $\mu^+\mu^-$ pairs, and multiparticle hadronic states. Results, shown in Figure 5, indicate a resonance at 3095 MeV, coupled to e^+e^-, $\mu^+\mu^-$, and hadronic states. Its width is less than 2 MeV. With the same technique, the LBL-SLAC group observed a second resonance at 3684 MeV, as indicated by the data of Figure 6 (25). Again the hadronic and lepton-pair decay modes appear, with the latter less prominent than at 3095 MeV.

The ψ resonance peak has been observed in e^+e^- annihilations at the storage ring ADONE (26); both ψ and ψ' have been observed at the higher-energy e^+e^- storage rings DORIS (27). Those results confirmed the existence of the two states and generally agree with the SLAC-LBL measurements of energy and decay widths of the resonances.

Further analyses of the shapes of these yield curves, particularly by the LBL-SLAC group, taking account of broadening due to finite beam-energy spread and initial-state radiation, led to determination of the full widths 69 ± 15 keV and

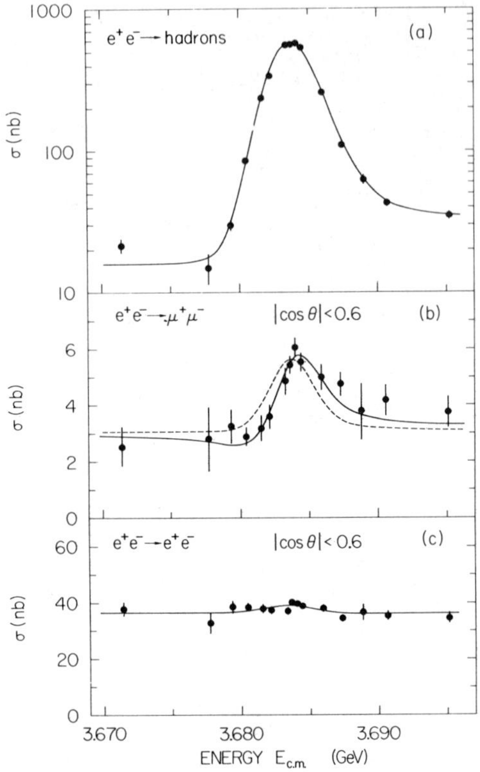

Figure 6 Cross sections for e^+e^- annihilation to (*a*) hadrons, (*b*) muon pairs, and (*c*) electron pairs, showing $\psi(3684)$ resonance excitation; from (51).

228 ± 56 keV for the lower- and higher-mass resonances, respectively. Compared with previously established resonances at considerably smaller energies, these widths are startingly small, indicating the effect of some new mechanism inhibiting their decays.

2.1 *Production of ψ and ψ' in Hadron-Hadron Collisions*

All such experiments have until now searched for narrow peaks in mass distributions of pairs of oppositely charged leptons as a signature of decays of heavy particles produced in the primary hadronic reactions. Severe experimental difficulties in detecting decay products in heavy backgrounds of hadronic debris have been overcome with sophisticated, complex experimental arrangements with special capabilities (28, 29). These include $e^{\pm}$ or $\mu^{\pm}$ identification and rejection of $\pi^{\pm}$, $K^{\pm}$, $p^{\pm}$; momentum- and angle-measurement precision sufficient to yield mass resolution of order 10–100 MeV/c^2; multicoincident counting of particles to reduce accidental rates to acceptable levels; good multitrack efficiency and position resolution; rejection of photon and neutron backgrounds associated with incident beams and from secondary, diffuse sources; large angle and large mass acceptance; and, since rates are very low, reliability over long periods of experimental operation. Instrumental limitations restrict the detected regions of production and decay phase space, and further introduce mass dependence in the acceptance of the apparatus. To obtain absolute yields, corrections must be made for these geometric deficiencies, as well as detection inefficiencies. Calculations are made with Monte Carlo simulations of detector response with an assumed dependence on production and decay kinematic variables.

2.1.1 28.5-GeV/c AND 20-GeV/c PROTONS ON BERYLLIUM Aubert et al (BNL-MIT) (30) used an arrangement designed for maximum acceptance of heavy particles produced at rest in the center of mass in the reaction $p + \text{Be} \rightarrow e^+e^- + X$ (X signifies all other particles in the final state). The experiment was made with incident protons of 28.5 GeV/c momentum. Schematic views of the detector are shown in Figure 7. To extract a ψ production cross section from the observed peak in the mass distribution, they assumed the differential cross section

$$\frac{d^3\sigma}{dP_{\parallel}^* \, dP_{\perp}^{*2}} \propto \frac{\exp(-6P_{\perp}^*)}{E^*}$$

for production of a ψ with c.m. longitudinal momentum $P_{\parallel}^*$ and transverse momentum $P_{\perp}^*$ and allowed the ψ decay to be isotropic at rest. They obtained a

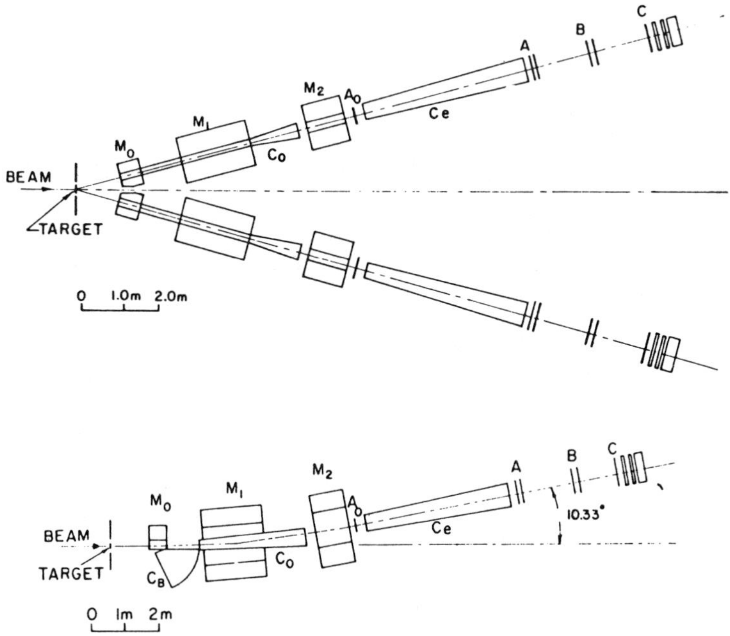

Figure 7 Spectrometer detection apparatus of the MIT-BNL group (23, 28–30) at Brookhaven National Laboratory.

production cross section $\sigma \sim 10^{-34}$ cm^2/nucleon for e^+e^- pairs with mass in the peak. With 20-GeV/c incident protons, the cross section is ten times smaller. No production of heavier particles was observed, again at 28.5 GeV/c, yielding an estimated upper limit $\sigma \sim 10^{-36}$ cm^2, with 95% confidence, for e^+e^- pairs of mass 3.2 GeV/c^2 < M < 4.0 GeV/c^2.

2.1.2 300-GeV NEUTRONS ON NUCLEI Knapp et al (31) and later, Binkley et al (32, 33), measured the yield of muon pairs in interactions of neutrons with Be, Al, Cu, and Pb nuclei. A primary objective was to determine the A dependence of resonance production so that cross sections for single nucleon interaction could be determined by extrapolation. Neutrons were produced by 300-GeV/c protons in the earlier experiments and by 400-GeV/c protons in the later runs. The resultant neutron-energy spectra peak at ~250 GeV and ~300 GeV for the lower and higher proton energy, respectively. Binkley et al (32) made fits to the yields of dileptons of momentum $p < 75$ GeV/c with the power law A^γ, obtaining the best-fit value $\gamma = 0.93 \pm 0.04$ for $\psi(3095)$ production. In contrast, they find $\gamma = 0.62 \pm 0.03$ for ρ^0 and ω^0. Fits were made to the observed distributions in the dimuons' $P_\perp$ and $P_\parallel$ (Figure 8), taking account of acceptance limitations and resolution, using assumed production dependences on x ($\equiv P_\parallel^*/P_{\text{max}}^*$) and $P_\perp$ of the form

$$E\frac{d^3\sigma}{dP^3} = C(1-x)^\alpha \exp(-bP_\perp) \qquad \text{for} \quad x_0 < x < 1,$$

and

$$E\frac{d^3\sigma}{dP^3} = C(1-x_0)^\alpha \exp(-bP_\perp) \qquad \text{for} \quad 0 < x < x_0.$$

For dimuon decay products of $\psi(3095)$, they obtain values $\alpha = 5.2 \pm 0.5$, $b = (1.6 \pm 0.2)$ c GeV^{-1} with $x_0 = 0.3$. With these, they calculated an integrated yield of ψ-

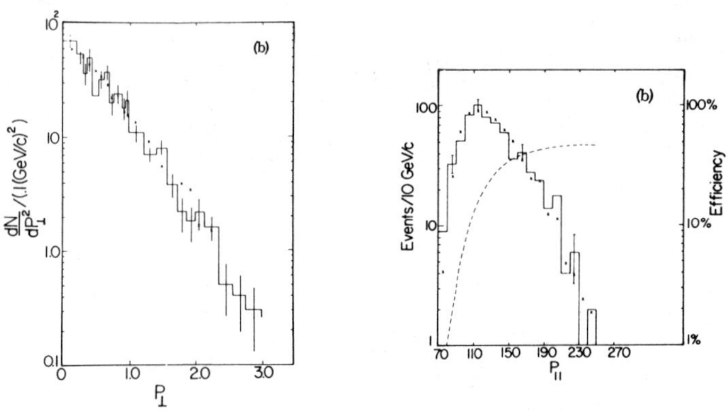

Figure 8 Differential yield of ψ mass dimuons as functions of transverse ($P_\perp$) and longitudinal ($P_\parallel$) momentum from interactions of ~300-GeV neutrons with nuclei; ×'s are results of a Monte Carlo calculation; from (33).

mass $\mu^{\pm}$ pairs

$$\int_{0.24}^{1.0} B_{\mu\mu} \frac{d\sigma}{dx} dx \simeq 3.5 \times 10^{-33} \text{cm}^2/\text{nucleon}.$$

With the further assumption $d\sigma/dy = \text{constant}$ ($y \equiv$ c.m. rapidity) in the region $0 < x < 0.3$, extrapolation into the unobservable region $x < 0.24$ yields the estimate for product of branching ratio, $B_{\mu\mu}$, and total inclusive cross section, $B_{\mu\mu}\sigma = 22 \times 10^{-33}$ cm^2/nucleon for 300-GeV neutrons. With no evidence for ψ' production, they conclude

$$B_{\mu\mu}(\psi')\sigma(\psi')/B_{\mu\mu}(\psi)\sigma(\psi) < 0.015.$$

2.1.3 150-GeV/c π^+ AND PROTONS ON BERYLLIUM An extensive series of measurements of inclusive $\mu^{\pm}$ production by 150-GeV/c protons and positive pions on Be has been carried out by Anderson et al (34, 35). Muon pairs with mass 0.456 GeV/$c^2 < M < 3.5$ GeV/c^2 were detected efficiently, so that ρ^0 (or ω^0) and ϕ^0 decays were observed, as well as ψ decays. Their apparatus is shown in Figure 9. Muons are specified by the requirement that they penetrate a total of 4.7-m iron absorber. The observed dependence of dimuon yield on x and transverse momentum $P_{\perp}$ of the dimuon was fit with a parameterization of the invariant differential cross section, also of the form

$$E\frac{d^3\sigma}{dP^3} = C(1-x)^{\alpha} \exp(-bP_{\perp})$$

after appropriate corrections for detection efficiency. Events with $x > 0.15$ and $P_{\perp} > 200$ MeV/c were included in the fits. For the dependence of $\psi(3095)$ yield on x, they find $\alpha = 2.9 \pm 0.3$ for production by protons and $\alpha = 1.7 \pm 0.4$ for production by pions. The steeper x dependence with incident protons than with pions agrees with earlier results of Blanar et al (36) for π^- and proton interactions with iron. The parameters of the $P_{\perp}$ dependence obtained are $b = 2.1 \pm 0.3$ c GeV^{-1} and $b = 2.6 \pm 0.4$ c GeV^{-1} for incident protons and pions, respectively. It is worth noting that the falloff with $P_{\perp}$ is significantly slower than that observed for ρ^0 (or ω^0) or ϕ^0 production, for which the corresponding parameter $b \sim 4$ for protons or pions. Integrated inclusive cross sections for production on single nucleons,

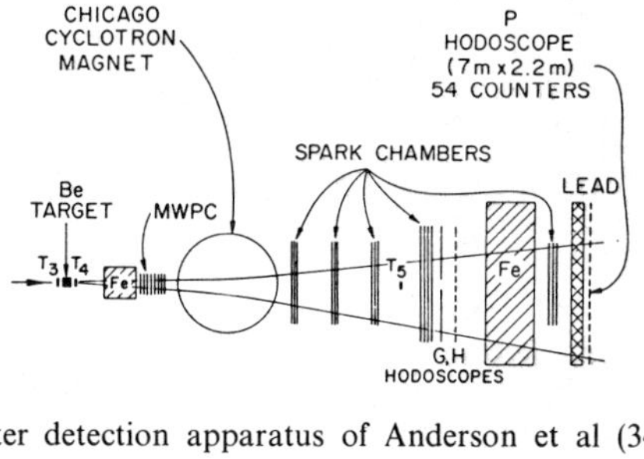

Figure 9 Spectrometer detection apparatus of Anderson et al (34) at Fermi National Accelerator Laboratory.

obtained by extrapolating the fit of $d\sigma/dx$ to $x = 0$ and using the dependence on A of (32), are $\sigma_p B_{\mu\mu} = (3.3 \pm 1.1) \times 10^{-33}$ cm^2/nucleon for incident protons and $\sigma_\pi B_{\mu\mu} = (6.5 \pm 2.2) \times 10^{-33}$ cm^2/nucleon for incident pions, for forward, $x > 0$, psions.

2.1.4 70-GeV/c PROTONS ON BERYLLIUM An arrangement similar to the above was used by Antipov et al (37) to measure $\mu^\pm$ pair yields from 70 GeV/c protons on beryllium. They fit the x distribution for ψ production with $\exp[-(6.0 \pm 1.2)x]$ for $x > 0.3$ and constant at smaller x. The $P_\perp$ dependence has the form

$$d\sigma/dP_\perp^2 \sim \exp[-(1.8 \pm 0.03)P_\perp^2].$$

For the inclusive $\psi(3095)$ production yield, they obtained

$$\sigma B_{\mu\mu} = (9.5 \pm 2.5) \times 10^{-33} \text{ cm}^2/\text{nucleus},$$

and, with the $A^{0.93}$ dependence of Binkley et al,

$$\sigma B_{\mu\mu} = (1.2 \pm 0.3) \times 10^{-33} \text{ cm}^2/\text{nucleon},$$

2.1.5 400-GeV PROTONS ON Be Hom et al (38, 40) and Snyder et al (39) have measured both $e^\pm$ and $\mu^\pm$ pair yields from interactions of 400-GeV protons with beryllium. Their spectrometer acceptance limits the detectable phase space to the small x interval $-0.06 < x < 0.08$ and to $P_\perp < 2$ GeV/c. They observed a transverse-momentum dependence well represented by either

$$E\frac{d^3\sigma}{dP^3} = (1.7 \pm 0.4) \times 10^{-32} \exp[-(1.1 \pm 0.4)P_\perp^2] \text{ cm}^2/\text{nucleus}$$

or

$$E\frac{d^3\sigma}{dP^3} = (2.5 \pm 0.6) \times 10^{-32} \exp[-(1.6 \pm 0.4)P_\perp] \text{ cm}^2/\text{nucleus}.$$

That result, extrapolated away from $x \simeq 0$ with the functional form

$$E\frac{d^3\sigma}{dP^3} = C(1-x)^{4.3} \exp(-1.6\, P_\perp),$$

yields an inclusive cross-section branching-ratio product

$$\sigma B = (1.0 \pm 0.3) \times 10^{-31} \text{ cm}^2/\text{nucleus},$$

which gives, with $A^{0.93}$ dependence,

$$\sigma B = (1.3 \pm 0.3) \times 10^{-32} \text{ cm}^2/\text{nucleon}.$$

A small peak at a mass consistent with ψ' yields an estimate of the ratio of production cross sections $\sigma(\psi')/\sigma(\psi) = (10 \pm 3)\%$.

2.1.6 26-GeV PROTONS ON 26-GeV PROTONS In p-p collisions at the CERN Intersecting Storage Rings (ISR), production of ψ has been observed, with ψ identified by e^+e^- (41) and $\mu^+\mu^-$ (42) decays. From the $\mu^+\mu^-$ yield at c.m. energy $(s)^{1/2} = 52$

GeV, there results an estimate of total cross section $\sigma B = (4.2 \pm 1.9) \times 10^{-32}$ cm^2, based on detection of eleven dimuon events above background.

2.1.7 SUMMARY Dividing the measured values of dilepton yields by $B_{\mu\mu} = 0.069$, the branching ratio for $\psi(3095)$ decay to lepton pairs obtained by the LBL-SLAC group (43), one obtains the various ψ-production total cross-section estimates plotted as a function of c.m. energy in Figure 10. This shows a rise of more than two orders of magnitude from the lowest energy, $(s)^{1/2} = 7.4$ GeV, to the highest, $(s)^{1/2} = 52$ GeV. Cross sections for ϕ production are smaller than those for ρ by about a factor of ten. Production of the ψ's is more grossly inhibited; their rates are a factor of 1000 smaller still. Average transverse momenta of the ψ's are about 1 GeV/c, roughly twice as large as those of ρ^0 and ϕ^0.

2.2 *Photoproduction of ψ and ψ'*

Measurement of the ψ differential photoproduction cross section, extrapolated to $t = 0$, with application of the vector-meson dominance-model prescription, yields

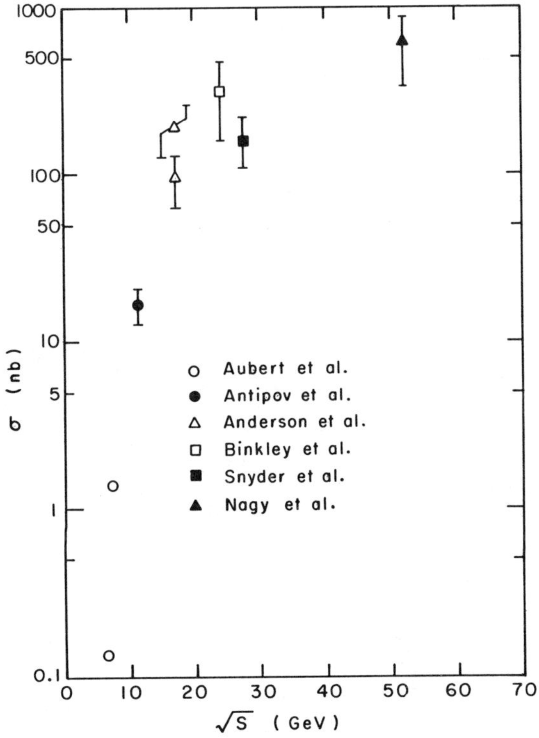

Figure 10 Compendium of results on the energy dependence of the cross section for inclusive $\psi(3095)$ production in hadron-hadron collisions; data from (30, 32–39, 42).

estimates of the ψ-nucleon total cross section (see Section 1.1) and so is important for determination of the strength of ψ coupling to hadrons.

Knapp et al (44) first observed ψ photoproduction and its diffractive character. Photons were produced by 300-GeV protons interacting with beryllium and had a continuous energy distribution to a maximum of ~200 GeV. The photon beam bombarded a beryllium target, and produced muon pairs were detected. Camerini et al (45) observed production of electron pairs and muon pairs, made by bremsstrahlung photons of various energies from 13 GeV to 21 GeV, interacting with deuterium. Gittelman et al (46) measured ψ production by 11.8-GeV bremsstrahlung photons on a beryllium target, detecting electron-pair decays. Nash et al (47) measured electron-pair yields from bombardment of deuterium by photons of mean energy 55 GeV, in a range from 31–80 GeV. A typical yield curve is in Figure 11, showing the electron-pair effective mass distribution produced in the experiment of Gittelman et al. Figure 12, the momentum-transfer distribution of dimuon pairs with mass near 3100 MeV/c^2, the results of Knapp et al, shows the characteristic

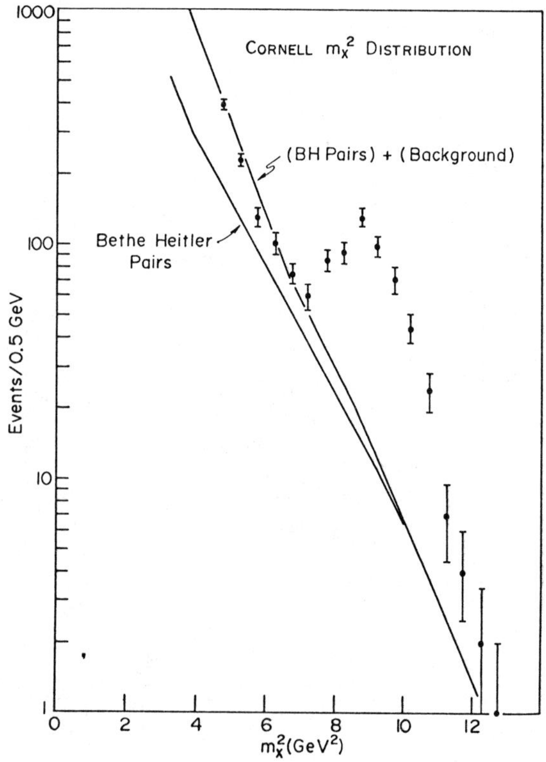

Figure 11 Invariant-mass distribution of dielectrons photoproduced on beryllium at 11 GeV, showing ψ(3095) yield; from (46).

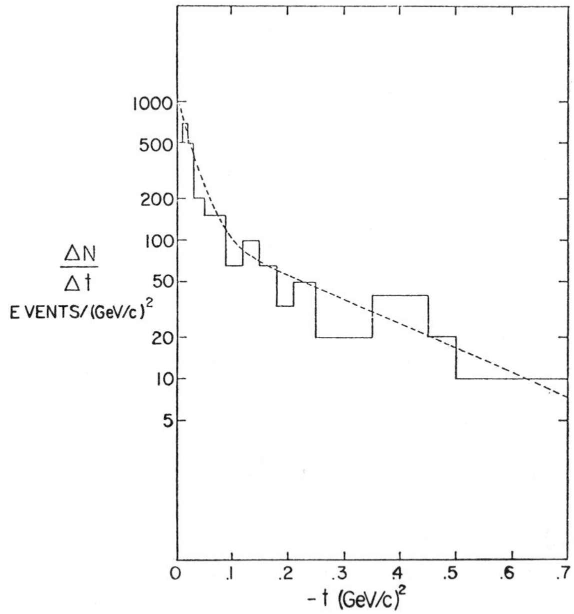

Figure 12 Differential yield of $\psi(3095)$ photoproduction on beryllium as a function of momentum transfer. The superimposed curve is the calculated t distribution of the form $A^2 \exp(40t) + A \exp(4t)$ after correction for experimental resolution and detector acceptance; from (44).

exponential behavior near zero momentum transfer. In a subsequent experiment at SLAC, Camerini et al (45) detected single-muon decay products of $\psi(3095)$ produced by photons on beryllium and tantalum. They provide a measurement of photoproduction rates for, among others, 9.5-GeV incident energy, the lowest energy at which $\psi(3095)$ photoproduction has been observed and find, in fact, the smallest differential cross section, 0.18 ± 0.08 nb $(\text{GeV}/c)^{-2}$ at the minimum momentum transfer. We have applied the vector-dominance relation (Equation 1.3) to calculate ψ-nucleon total cross sections from reported values of $d\sigma/dt$ at $t = 0$, neglecting the real part of the forward-scattering amplitude and using the lepton-pair partial width $\Gamma_{ee} = 4.8$ keV (43). These values are shown in Figure 13 as a function of c.m. energy. The experiment of Camerini et al (45) has provided a direct measurement of the ψ-nucleon cross section, inferred from the ratio of yields in Be and Ta. From results with 20-GeV incident photons, they find $\sigma(\psi N) = 2.8 \pm 0.9$ mb, where only statistical errors are included in the uncertainty. The systematic error was estimated at $\sim \pm 0.5$ mb. The first directly measured ψ-interaction cross section agrees with values extracted from the photoproduction cross sections with vector-dominance model.

It appears that the ψN total cross section approaches a value $\sigma \simeq 1$ mb, small, but within the range typical of strong-interaction processes. It is noteworthy that

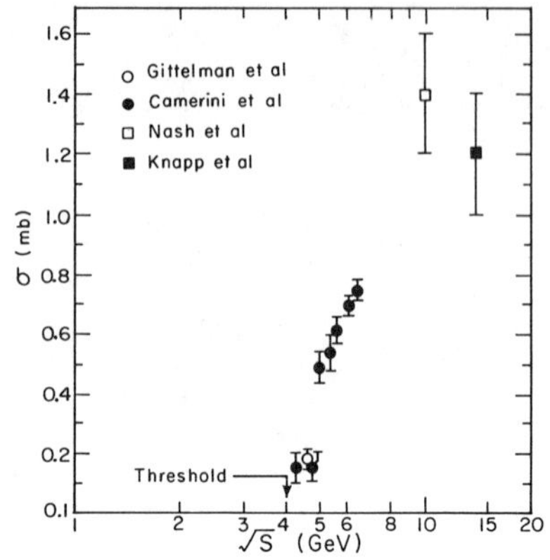

Figure 13 Compendium of results on the energy dependence of the $\psi(3095)$-nucleon total-interaction cross section obtained from photoproduction data of (44–48) and the vector-dominance model.

this ψ-nucleon cross section is smaller by a factor of ten than the ϕ-nucleon total cross section. Also the energy dependence has a shape somewhat suggestive of a threshold for an inelastic ψ-nucleon reaction at an energy near 5 GeV in the ψ-nucleon c.m. system. Candidates for such reactions are the processes yielding pairs of charmed mesons, implicit in the four-quark view of hadron microstructure.

The yield of ψ' in photoproduction is considerably smaller than that of ψ. Only Camerini et al (45) report a positive result. With 21-GeV photons, they measure $d\sigma/dt = (2.1 \pm 0.8) \times 10^{-33}$ cm^2 (GeV/c)$^{-2}$ at minimum momentum transfer, just about one seventh of the value for ψ photoproduction at the same energy and its minimum momentum transfer.

3 PROPERTIES OF $\psi(3095)$ AND $\psi(3684)$

Essentially all information on the characteristics of the resonant states has come from analysis of e^+e^- annihilations and mostly from the SLAC-LBL group at SPEAR and the Double Arm Spectrometer (DASP) group at DORIS (27). The LBL-SLAC magnetic detector (49) is a cylindrical arrangement of counters and spark chambers arrayed around the direction of the colliding beams as the axis, all in an axial magnetic field of $\sim$4000 G uniform to a few percent over its $\sim$3-m-diameter-by-$\sim$3-m-length volume. Charged particles and photons are detected over about 65% of the full solid angle. A schematic view of a vertical cross section through the detector axis is shown in Figure 14. The DASP apparatus (50), Figure 15, is a symmetric pair of magnetic spectrometers providing precise momentum and time-of-flight measurements of charged particles produced in a solid angle of $\sim 0.1 \times 4\pi$ sr. The nonmagnetic inner detector covers $\sim 0.7 \times 4\pi$ sr. Both the SLAC-LBL and DASP apparatuses detect photons, DASP with better efficiency, energy, and position resolution.

3.1 *Masses and Widths*

The resonant-state energies, full widths, and partial widths were determined from data obtained at SPEAR on the energy dependence of the annihilation cross section shown in Figures 5 and 6 (43, 51). For these data the incident flux was obtained from measurement of Bhabha scattering at ~25 mrad. Hadron yields were corrected for the solid-angle and momentum acceptance of the apparatus, using a Monte Carlo simulation of the detector response and a phase-space production model (52). That calculation yields a net detection efficiency $\varepsilon = 0.4$ with $\pm 15\%$ estimated uncertainty. Lepton-pair cross sections refer only to the angular range $|\cos\theta| \leqq 0.6$, where θ is the angle between the final-state leptons' and initial

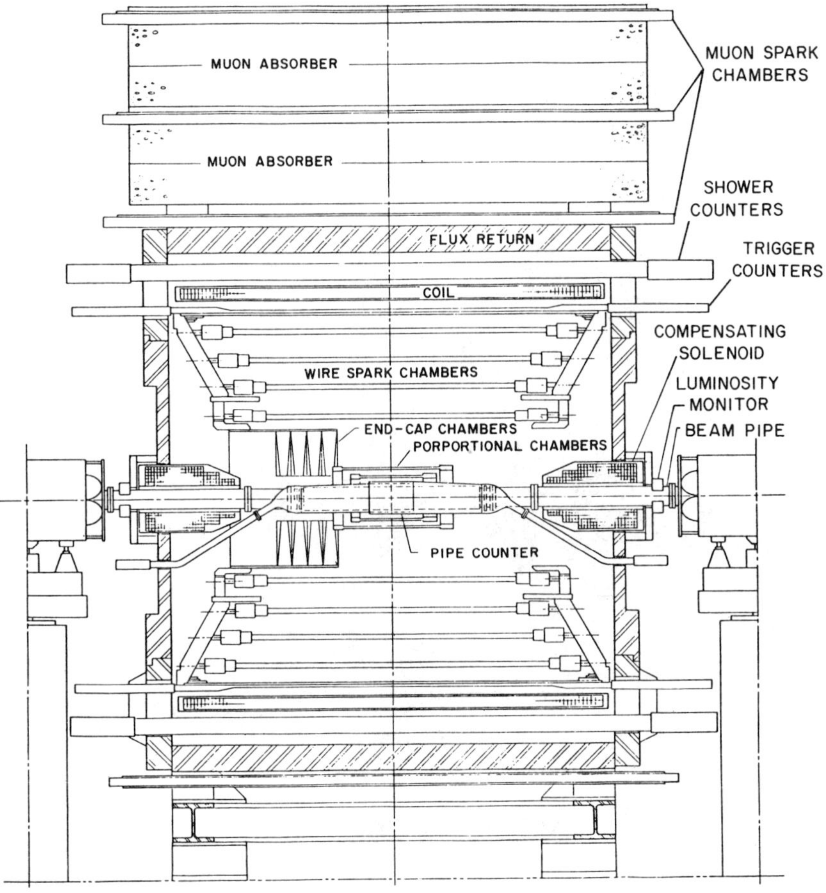

Figure 14 The SLAC-LBL magnetic detector at the e^+e^- storage ring SPEAR; a view in the vertical plane through the intersecting beams.

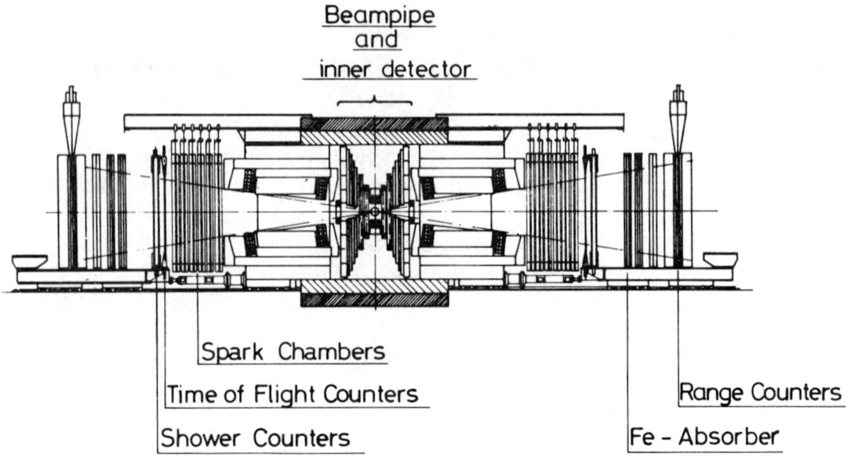

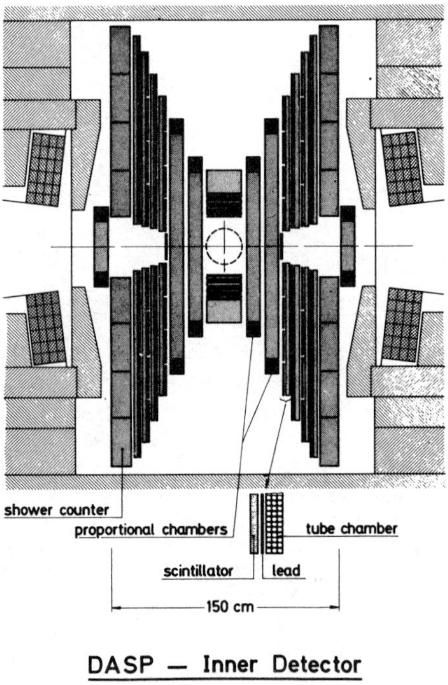

Figure 15 The Double Arm Spectrometer detector (DASP) at the e^+e^- storage rings DORIS.

Table 3 Parameters of ψ particles

Particle	Mass (MeV/c^2)	Γ (MeV)	Γ_{ee} (keV)	$\Gamma_{\mu\mu}$ (keV)	Γ_h (keV)	$\Gamma_{\gamma h}$ (keV)	$f_v^2/4\pi$
$\psi(3095)$[a]	3095 ± 4	0.069 ± 0.015	4.8 ± 0.6	4.8 ± 0.6	59 ± 15	12 ± 2	11.5 ± 1.4
$\psi(3684)$[b]	3684 ± 5	0.228 ± 0.056	2.1 ± 0.3	2.1 ± 0.3	224 ± 56	7 ± 1	31.1 ± 4.5
$\psi(4414)$[c]	4414 ± 7	33 ± 10.0	0.44 ± 0.14				178 ± 57

[a] LBL-SLAC values (43).
[b] LBL-SLAC values (51).
[c] Reference 101.

beams' directions. The observed excitation curves were each fit to a function consisting of a Breit-Wigner line shape for a resonance at energy M of full width Γ,

$$\sigma_\beta(W) = \frac{(2J+1)\pi}{W^2} \frac{\Gamma_{ee}\Gamma_\beta}{(W-M)^2 + \Gamma^2/4}, \tag{3.1}$$

convoluted with the energy-resolution function of the colliding beams. The parameters Γ_{ee} and Γ_β are the partial widths for decays to electron pairs and final state β, respectively. The observed cross section at nominal interaction energy W_0 (the storage-ring energy setting) is then

$$\sigma_\beta(W_0) = \int G_R(W_0 - W)\sigma_\beta(W)\,dW, \tag{3.2}$$

where initial-state radiation has been taken into account by replacing the resolution function $G(W_0 - W)$ by a radiatively corrected form (53)

$$G_R(W_0 - W) = t \int_0^{W_0/2} \left(\frac{2k}{W_0}\right)^t G(W_0 - W - k)\frac{dk}{k}, \tag{3.3}$$

where $t = 2(\alpha/\pi)[\ln(W_0/M_e)^2 - 1]$ and k is the energy of a radiated photon. Radiation causes a "tail" on the high-energy side of a resonance, decreasing the maximum observed cross section, but does not appreciably change the position of the maximum. Best-fit values obtained for M, Γ, Γ_{ee}, $\Gamma_{\mu\mu}$, and Γ_h, the hadron width, are listed in Table 3. It was assumed that no decay modes escaped detection, i.e. $\Gamma = \Gamma_h + \Gamma_{\mu\mu} + \Gamma_{ee}$. It is of some value to note here the more transparent relation between the resonance parameters and the area under the resonance peak

$$\iint \sigma_\beta(W_0)G(W_0 - W)\,dW\,dW_0 = \frac{2\pi^2}{M^2}(2J+1)\frac{\Gamma_{ee}\Gamma_\beta}{\Gamma}, \tag{3.4}$$

independent of the shape of the resolution function. Assuming no modes escape detection, the sum of the areas under the three resonance peaks shown determines the electron-pair partial width directly.

Included in Γ_h is a contribution $\Gamma_{\gamma h}$, from resonance decays via a second-order electromagnetic process, indicated in Figure 16*b*. With the usual assumption that

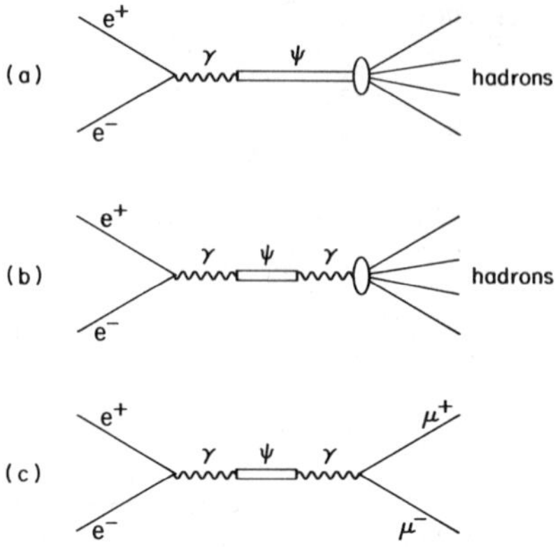

Figure 16 Diagrams of resonance production in lowest order in e^+e^- annihilation with decays to (*a*) hadrons, (*b*) hadrons via second-order electromagnetic interaction, and (*c*) muon pairs.

lepton pairs couple to the resonant state via an intermediate photon, it follows

$$\Gamma_{\gamma h} = \left[\frac{\sigma(e^+e^- \to \text{hadrons})}{\sigma(e^+e^- \to \mu^+\mu^-)}\right]\Gamma_{\mu\mu},$$

where the quantity in brackets is evaluated at an energy just below resonance. Values of the cross-section ratio measured by the SLAC-LBL group (52) were used to calculate the second-order electromagnetic widths given in the table. The fraction of indirect hadronic decays, $\Gamma_{\gamma h}/\Gamma$, is 0.17 ± 0.03 for $\psi(3095)$ and 0.031 ± 0.009 for $\psi(3684)$.

3.2 *Spin, Parity, and Charge Conjugation*

These quantum numbers are directly established to have values $J = 1$, $P = \text{odd}$, and $C = \text{odd}$ from the observation of interference between single-photon–exchange and ψ-exchange amplitudes in the process $e^+e^- \to$ leptons near resonance, Figure 16*c* (43, 51). If the intermediate meson has the same quantum numbers as the photon, the muon-pair-annihilation differential cross section is

$$\frac{d\sigma}{d\Omega} = \frac{9}{16W^2}(1+\cos^2\theta)\left|\frac{-2\alpha}{3} + \frac{\Gamma_{\mu\mu}}{M - W - i\Gamma/2}\right|^2,$$

showing destructive interference below the resonant energy M. To minimize the effects of uncertainties in incident-flux determinations, the SLAC-LBL group examined the ratio of electron-pair to muon-pair cross sections, integrated over the angular interval $-0.6 < \cos\theta < 0.6$ covered by the detector. That ratio is shown as a function of energy in Figure 17, together with calculated curves, due account having been taken of radiative corrections. Both $\psi(3095)$ and $\psi(3684)$ data are consistent with expectations for spin-1 and exclude spin-0. Interference observations directly establish that both psions and the photon have the same value, -1,

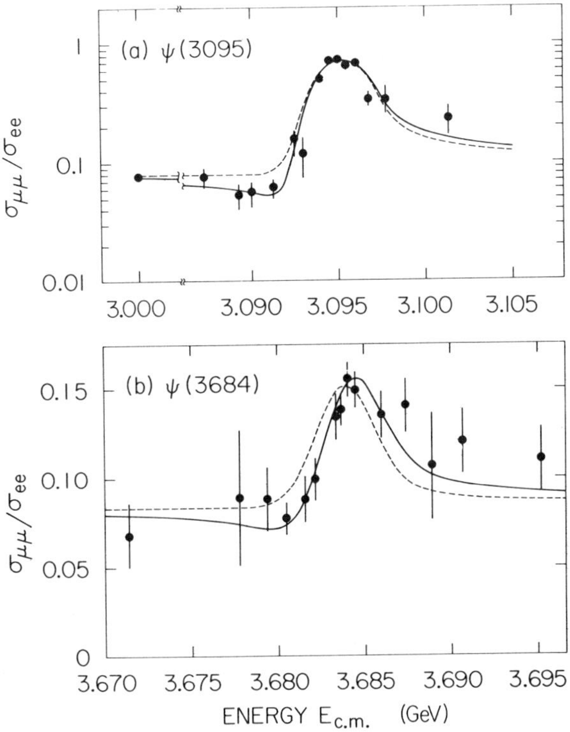

Figure 17 Ratio of cross sections for e^+e^- annihilation to muon pairs to that for annihilation to electron pairs at energies near (*a*) 3095 MeV and (*b*) 3684 MeV. The full and dashed curves show maximal and zero interference, respectively, between one photon and resonance amplitudes; from LBL-SLAC (43, 51).

of parity and charge-conjugation quantum numbers. Spin greater than one is not excluded a fortiori by an interference observation because the unit helicity projection of the intermediate ψ will produce interference effects in the energy variation of that fraction of the total cross section included in the limited angular range of detection. Since the QED amplitude is nonzero only for transitions between lepton-pair states of helicity ± 1, the interference term in the cross section is proportional to the overlap integral of the spin-rotation functions $d^{J_\psi}_{11}(\theta)$ and $d^1_{11}(\theta)$. It follows from the properties of the $d^J_{11}(\theta)$ functions that ψ-spin values two and three give rise to constructive interference at energies below resonance. Higher spins produce negligible interference. Spins greater than, as well as less than, one are thus excluded by the LBL-SLAC data.

3.3 *Hadronic Decays of ψ(3095); Isospin and G-Parity; SU(3) Classification*

Table 4 presents information on hadronic-decay modes of $\psi(3095)$.

Table 4 Branching fractions for $\psi(3095)$ nonleptonic decays

Mode	Fraction (%)[a]	Ref.	Footnote(s)
$\pi^+\pi^-$	0.01 ± 0.007	54	b
$\pi^+\pi^-\pi^0$	1.6 ± 0.6	55	—
$\rho\pi$	1.3 ± 0.2	54, 55	c
$2\pi^+2\pi^-$	0.4 ± 0.1	55	d
$\pi^\pm A_2^\mp$	<0.4	54	d
$2\pi^+2\pi^-\pi^0$	4 ± 1	55	—
$\omega\pi^+\pi^-$	0.7 ± 0.2	57	e
ωf	0.19 ± 0.08	57	e
ρA_2	0.8 ± 0.5	57	e
$\rho\pi\pi$	1.2 ± 0.4	55	e
$3\pi^+3\pi^-$	0.4 ± 0.2	55	d
$3\pi^+3\pi^-\pi^0$	2.9 ± 0.7	55	—
$\omega 2\pi^+2\pi^-$	0.9 ± 0.3	57	f
$4\pi^+4\pi^-\pi^0$	0.9 ± 0.3	55	—
K^+K^-	0.02 ± 0.02	57	g
$K_S^0 K_L^0$	<0.008	57	g
$K_S^0 K^\pm \pi^\mp$	0.26 ± 0.07	57	—
$K^0 \bar{K}^{*0}$	0.27 ± 0.06	57	h, i, j
$K^\pm K^{*\mp}$	0.32 ± 0.06	57	h, j
$K^0 \bar{K}^{**0}$	<0.2	57	h, i, j
$K^\pm K^{**\mp}$	<0.15	57	h, j
$K^+K^-\pi^+\pi^-$	0.7 ± 0.2	57	—
$K^{*0}\bar{K}^{*0}$	<0.05	57	g, h, k
$K^{*0}\bar{K}^{**0}$	0.7 ± 0.3	57	h, i, k
$K^{**0}\bar{K}^{**0}$	<0.3	57	g, h, k
$\phi\pi^+\pi^-$	0.14 ± 0.06	57	k
ϕf	<0.04	57	—
$K^+K^-K^+K^-$	0.07 ± 0.03	57	—
ϕK^+K^-	0.09 ± 0.04	57	l
$\phi f'$	0.08 ± 0.05	57	l
$K^+K^-\pi^+\pi^-\pi^0$	1.2 ± 0.3	57	—
ωK^+K^-	0.08 ± 0.05	57	m
$\omega f'$	<0.02	57	m
$\phi\eta$	0.10 ± 0.06	57	m
$\phi\eta'$	<0.1	57	—
$K^+K^-2\pi^+2\pi^-$	0.3 ± 0.1	57	n
$\phi 2\pi^+2\pi^-$	<0.15	57	—
$\bar{p}p$	0.22 ± 0.02	54, 56	o
$\bar{N}N\pi$	0.4 ± 0.2	56	p
$\Lambda\bar{\Lambda}$	0.16 ± 0.08	56	o
$\gamma\gamma$	<0.05	74	q
$\gamma\pi^0$	<0.055	74	—
$\gamma\eta$	0.14 ± 0.04	50, 73	—
$\gamma\eta'$	<0.7	50, 73	—
$\gamma X(2800)$	<1.7	68	—
$\gamma X(\rightarrow 3\gamma)$	~ 0.015	50, 73	—

Table 4 Continued

[a] All upper limits refer to the 90% confidence level.
[b] Two events detected with background estimated at 0.24 events. An isospin-nonconserving, second-order electromagnetic mode.
[c] $\rho^0\pi^0 + \rho^{\pm}\pi^{\mp}$. Included in $\pi^+\pi^-\pi^0$ grouping.
[d] Proceeds via second-order electromagnetic interaction.
[e] One-π^0 mode included in $2\pi^+2\pi^-\pi^0$.
[f] One-π^0 mode included in $3\pi^+3\pi^-\pi^0$.
[g] Forbidden for SU(3) singlet.
[h] $K^* \equiv K^*(892)$, $K^{**} \equiv K^*(1420)$.
[i] Includes charge-conjugate state.
[j] $K_S^0K^{\pm}\pi^{\mp}$ fraction included in $K_S^0K^{\pm}\pi^{\mp}$ grouping.
[k] $K^+K^-\pi^+\pi^-$ fraction included in $K^+K^-\pi^+\pi^-$ grouping.
[l] $K^+K^-K^+K^-$ fraction included in $K^+K^-K^+K^-$ grouping.
[m] $K^+K^-\pi^+\pi^-\pi^0$ fraction included in $K^+K^-\pi^+\pi^-\pi^0$ grouping.
[n] $K^+K^-2\pi^+2\pi^-$ fraction included in $K^+K^-2\pi^+2\pi^-$ grouping.
[o] Angular distribution $1+\cos^2\theta$ assumed, consistent with data.
[p] Includes $p\bar{p}\pi^0$, $p\bar{n}\pi^-$, and $n\bar{p}\pi^+$.
[q] Forbidden for angular momentum $J = 1$.

Two-body branching ratios have been determined by Braunschweig et al (54) from data obtained with the double arm spectrometer DASP. Two-body final states were selected, using criteria of collinearity, momentum, energy loss, and range appropriate to the particular pair. Evidence for the modes $\pi^{\pm}\ \rho^{\mp}$ and $K^{\pm}\ K^{*\mp}$ is presented in the missing-mass spectra of Figure 18, showing the mass recoiling against single pions and kaons.

Multibody decay rates were measured by the SLAC-LBL group (55–57) using kinematical fits of multiprong events in which the detected prongs have zero total charge. Events with no missing particles were selected by requiring that the missing momentum be less than 100 MeV/c. As an example, the distribution in total energy of the detected prongs in four-prong events is shown in Figure 19. Four-particle

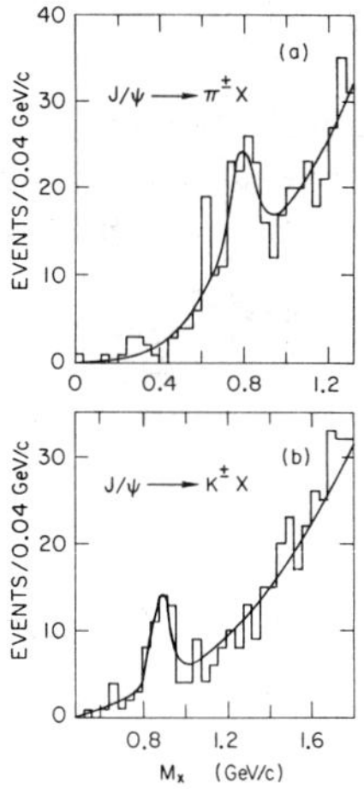

Figure 18 Distribution of mass recoiling against (*a*) single pions and (*b*) single kaons in multihadron decays of $\psi(3095)$ produced in e^+e^- annihilations; from DASP (54).

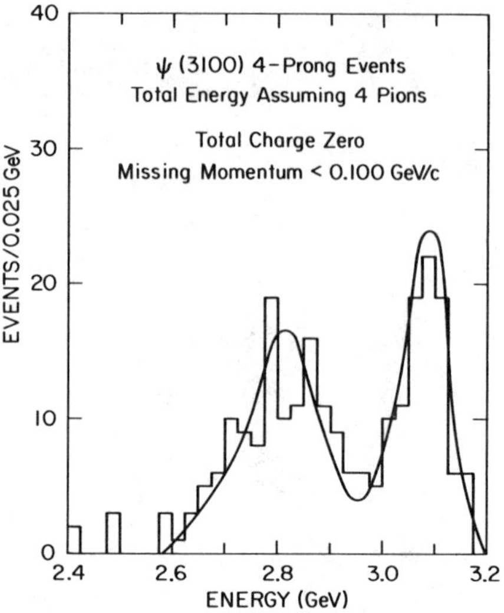

Figure 19 Total energy in four-prong $\psi(3095)$ decay events with missing momentum less than 100 MeV/c. The solid curve is a Monte Carlo fit to the data; from SLAC-LBL (55, 56).

states, presumed to be $\pi^+\pi^-\pi^+\pi^-$, are indicated by the presence of the peak near 3100 MeV in total detected energy. Events with momentum unbalance greater than 200 MeV/c exhibit the missing-mass spectrum of Figure 20 with clear evidence for the $2\pi^+2\pi^-\pi^0$ decay mode of the ψ and, for contrast, its relative absence at the nonresonant energy 3.0 GeV. Similar analysis techniques were used by the LBL-SLAC group to identify samples of the various final states of Table 4 and, after correction for losses due to detection inefficiency, to determine the branching ratios shown.

Application of these results to the determination of strong-interaction properties of the ψ requires isolation of the direct-decay contribution (Figure 16*a*) from that of the second-order electromagnetic decays (Figure 16*b*). That is done by comparing the relative branching ratio $\sigma_\beta/\sigma_{\mu\mu}$ for a final state β at resonance to the same quantity at 3.0 GeV, where the resonance contribution is negligible. Jean-Marie et al (55) show that this ratio, measured at resonance, equals the nonresonant value for states with even numbers of pions. States with odd numbers of pions have significantly larger cross-section ratios at resonance than below resonance, leading to the conclusion that ψ couples directly to states with odd numbers of pions and not to states with even numbers of pions. It follows that the ψ is a hadronic state of odd G-parity. Applying the relation between isospin and G-parity for a state of arbitrary number of pions,

$$G = C(-1)^I = (-1)^{I+1}$$

it follows that the ψ is a state of even isospin and that direct decays of the ψ follow the hadronic rule of isospin conservation.

The $\pi^+\pi^-\pi^0$ event class of Jean-Marie et al (55) is dominated by the quasi-two-body $\rho\pi$ state, as indicated in the Dalitz plot of Figure 21. From the distribution of events, they concluded that for ψ decay,

$$\frac{\Gamma_{\rho^0\pi^0}}{\Gamma_{\rho^+\pi^-}+\Gamma_{\rho^-\pi^+}} = 0.59 \pm 0.17.$$

Comparison with the values 0.5 and 2.0, appropriate to isospin-conserving decays of objects of isospin zero and two, respectively, uniquely establishes $I = 0$ for the ψ. Confirmation of this result came with observation of the $\bar{p}p$ mode by the DASP and LBL-SLAC groups. As pointed out by Braunschweig et al (54), the measured ratio $\Gamma_{\bar{p}p}/\Gamma_{\mu\mu}$ is much larger than would be obtained with any plausible extrapolation of the nucleon form factor, so that the $\bar{p}p$ mode must represent a direct decay, rather than a second-order electromagnetic process. That state can occur only with $I = 0$ or $I = 1$, and the latter is excluded by inference from the existence of decays to odd numbers of pions. Lastly, the LBL-SLAC team has reported a

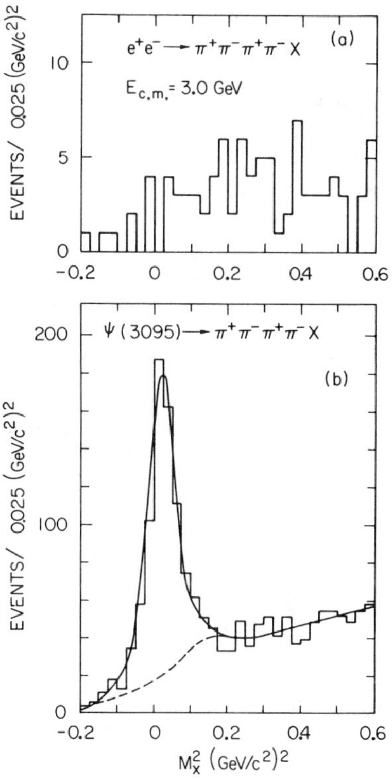

Figure 20 Missing mass-squared in four-prong events with missing momentum greater than 200 MeV/c produced in e^+e^- annihilations at (*a*) 3.0 GeV and (*b*) ψ(3095) resonance. The solid curve is a Monte Carlo fit to the data; from LBL-SLAC (55, 56).

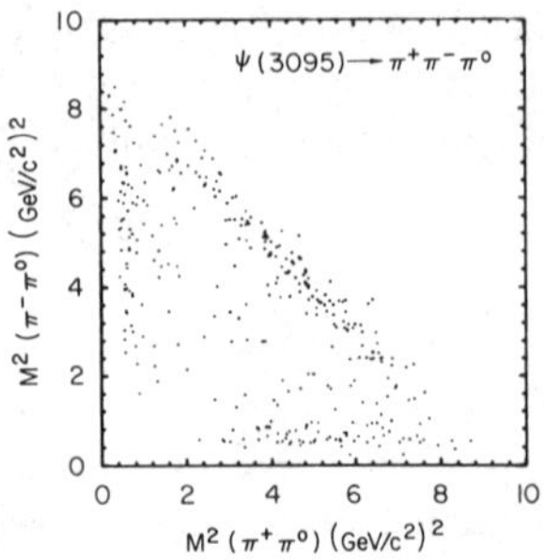

Figure 21 Dalitz plot for $\psi(3095)$ decays to $\pi^+\pi^-\pi^0$; from SLAC-LBL (55, 56).

measurement of the $\Lambda\bar{\Lambda}$ branching ratio, $\Gamma_{\Lambda\bar{\Lambda}}/\Gamma = 0.0016 \pm 0.0008$, based on twenty well-identified examples (56). None of those is consistent with primary $\Sigma^0\bar{\Lambda}^0$ ($\bar{\Sigma}^0\Lambda^0$) events. Since second-order electromagnetic decays populate $\Sigma\bar{\Lambda}$ states six times more frequently than $\Lambda\bar{\Lambda}$ states (58), it follows that the $\Lambda\bar{\Lambda}$ pairs are direct decay products. This observation provides another argument for assigning $I = 0$ to the $\psi(3095)$.

Two-body-decay branching ratios are especially useful in considering the applicability to ψ-decay interactions of the broader SU(3) strong-interaction symmetry (58, 59). Decay of an SU(3) singlet is constrained severely because only one amplitude determines the rates for all decays $\psi \to m_a + m_b$, where m_a and m_b represent any two members of meson octets. If m_a and m_b are the charged members of an $I = 1$ submultiplet, $m_a + m_b$ has even G-parity and is not an allowed final state for $\psi(3095)$ decay. By extension, decay to any two members of the same octet is forbidden. If m_a and m_b are the $I = 1$, $I_3 = 0$, $Y = 0$ members of different octets, a and b, and the charge-conjugation quantum numbers of m_a and m_b are equal, then m_a and m_b have the same G-parity, so the state $m_a + m_b$ also has even G-parity. Decay of an odd G-parity, SU(3) singlet to that state is again forbidden. Symmetry under SU(3) then forbids $\psi(3095)$ decay to a state of any two members of such octets a and b. Allowed decay modes are, for example, $\rho\pi$, $\eta\phi$, and $\bar{K}K^*(890)$, while K^+K^-, πA_2, $\bar{K}K^*(1420)$ are forbidden. Inspection of the branching fractions listed in Table 4 shows that the SU(3)-singlet-allowed modes $K\bar{K}^*(890)$ and $K^*(890)\bar{K}^*(1420)$ occur, and with branching fractions at least an order of magnitude larger than those for $\pi^+\pi^-$ and K^+K^-. Other forbidden modes, $\bar{K}K^*(1420)$, $K^*\bar{K}^*$, etc have not appeared, but present upper limits on their relative rates are not very convincingly small.

Exact SU(3) symmetry leads to the further prediction of equal branching ratios for allowed singlet decays to any two members, m_a and m_b, of any pair of octets a and b, e.g. to $\pi^0\rho^0$, $\pi^+\rho^-$, $\pi^-\rho^+$, $K^-K^{*+}(892)$, $K^+K^{*-}(892)$ etc. The relevant experimental ratios are

$$[\Gamma(\pi^+\rho^-)+\Gamma(\pi^-\rho^+)+\Gamma(\pi^0\rho^0)]/[\Gamma(K^+K^{*-})+\Gamma(K^-K^{*+})]/[\Gamma(K^0\bar{K}^{*0}) + \Gamma(\bar{K}^0K^{*0})] = (1.3 \pm 0.2)/(0.34 \pm 0.06)/(0.27 \pm 0.05),$$

not quite in agreement with the predicted 3/2/2, indicating some SU(3) breaking in the decay interaction.

In summary, ψ decays appear to have properties in agreement with the strictures of SU(3) symmetry appropriate to a singlet state, although there are some discrepancies.

3.4 *Hadronic Decays of $\psi(3684)$; Isospin and G-Parity*

The dominant hadronic decay mode of $\psi(3684)$ is the cascade decay to $\psi(3095)$, $\psi' \rightarrow \psi\pi\pi$. The modes $\psi\pi^+\pi^-$, $\psi\pi^0\pi^0$, together with $\psi \rightarrow \psi\eta$ and $\psi' \rightarrow \psi\gamma\gamma$, account for just over one-half of all decays. Table 5 lists branching ratios for all identified hadronic modes. Data from the DASP (50) and LBL-SLAC (60, 61) groups were obtained using the detection apparatus and analysis techniques described above. Hilger et al (62), at SPEAR, used two identical collinear spectrometers, each equipped with a large NaI crystal for photon and electron detection, an array of multiwire proportional chambers for track-trajectory sampling, and a magnetized iron block

Table 5 Branching fractions for $\psi(3684)$ nonleptonic decays

Mode	Fraction (%)[a]	Ref.	Footnote(s)
$\psi(3095)\pi^+\pi^-$	33 ± 3	50, 60–62	—
$\psi\pi^0\pi^0$	17 ± 3	50, 60–62	—
$\psi\eta$	4.2 ± 0.7	50, 61	—
$\psi\gamma + \psi\pi^0$	<0.15	61	b, c
$\pi^+\pi^-$	<0.04	54	d
$2\pi^+2\pi^-\pi^0$	0.35 ± 0.15	56	—
$\rho^0\pi^0$	<0.1	56	—
K^+K^-	<0.14	54	e
$p\bar{p}$	<0.05	54	f
$\gamma\gamma$	<0.5	69	g
$\gamma\pi^0$	<0.7	69	—
$\gamma\eta$	<0.13	50	—
$\gamma X(2800)$	<1	65, 68	—
$\gamma X(2800) \rightarrow 3\gamma$	<0.04	50	—
$\gamma\chi(3415)$	7 ± 2	65, 68	f
$\gamma\chi(3510)$	7 ± 2	68	h
$\gamma\chi(3550)$	7 ± 2	68	h
$\gamma\chi(3455)$	<2.5	68	—
$\gamma\chi(3455) \rightarrow \gamma\gamma\psi$	0.8 ± 0.4	65	—

[a] All upper limits refer to the 90% confidence level.
[b] $\psi\gamma$ forbidden by c invariance.
[c] $\psi\pi^0$ forbidden by isospin invariance.
[d] Isospin-nonconserving, second-order electromagnetic decay.
[e] Forbidden for an SU(3) singlet.
[f] Angular distribution $1 + \cos^2\theta$ assumed.
[g] Forbidden for angular momentum 1.
[h] Isotropic angular distribution assumed. With $1 + \cos^2\theta$ distribution, these branching fractions increase by a factor of 1.3.

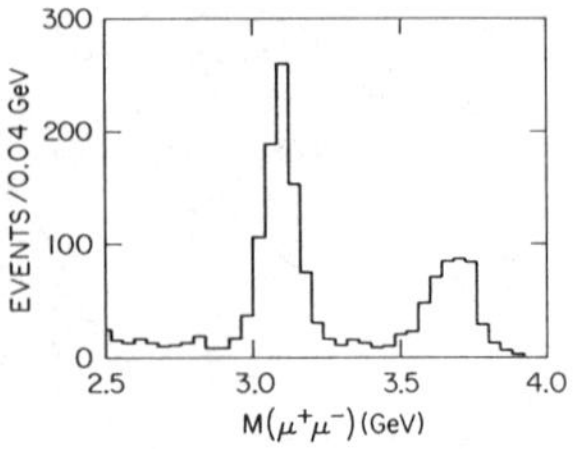

Figure 22 Inclusive distribution of muon-pair invariant masses from $\psi(3684)$ decays showing evidence of cascade decays to $\psi(3095)$; from LBL-SLAC (60).

for charged-particle momentum determination. The spectrometers were perpendicular to the direction of the colliding beams.

The LBL-SLAC evidence for the cascade decay $\psi(3684) \rightarrow \psi(3095) + \text{anything}$ (60) is presented in Figure 22. This shows the distribution in invariant mass of the two highest-momentum, oppositely charged prongs in the $\psi(3684)$ decays, calculated with the muon mass assumed for those two tracks. A peak at the ψ mass is apparent, as is the peak at 3684 MeV/c^2, due to direct decays of $\psi(3684)$ to muon pairs. Figure 23*a* shows the mass recoiling against all $\pi^+\pi^-$ pairs in all events, exhibiting again a clear peak at 3095 MeV/c^2. Figure 23*b* shows the recoil mass against those pion pairs in 4-prong events with zero total charge and, within experimental uncertainties, zero missing momentum, consistent with kinematics of the decay $\psi(3684) \rightarrow \mu^+\mu^-\pi^+\pi^-$. The peak results from that subset of $\psi(3684)$ cascade decays in which $\psi(3095)$ subsequently decays via its muon-pair mode. An example of a computer reconstruction of such an event is shown in Figure 24. The SLAC-LBL group identified $\psi' \rightarrow \psi\eta$ decays among the class of $\pi^+\pi^-\mu^+\mu^-$ events with missing momentum (61). The distribution in the square of the mass recoiling against the $\mu^+\mu^-$ pairs in Figure 25 shows a peak at the $\eta(\text{mass})^2$ whose

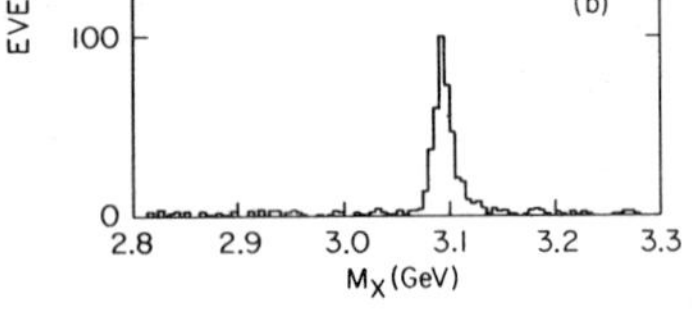

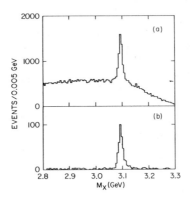

Figure 23 (*a*) Distribution of missing mass recoiling against all pairs of oppositely charged particles in $\psi(3684)$ decays to multihadron states. (*b*) same as (*a*), but for events consistent with zero missing energy and momentum; from SLAC-LBL (60).

width is consistent with that expected from experimental resolution folded with the natural width of the η meson. The $\psi\eta$ mode is also revealed in the LBL-SLAC group's mass spectrum of all-neutral states accompanying ψ decays via the $\mu^+\mu^-$

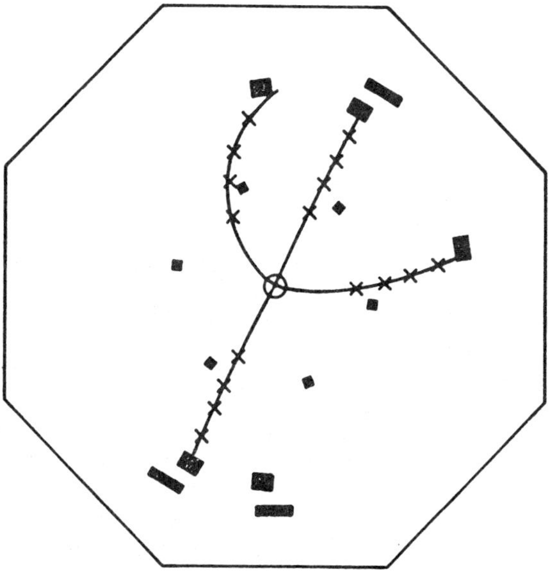

Figure 24 A computer reconstruction of a $\psi(3684) \to \pi^+\pi^-\psi(3095) \to \pi^+\pi^-e^+e^-$ event in the LBL-SLAC detector seen in projection on the plane perpendicular to the intersecting beams. The $\times$'s mark spark-chamber samplings. Black rectangles show counters. The straightest two tracks are the electron and positron; the others are the low-momentum pions; from SLAC-LBL (60).

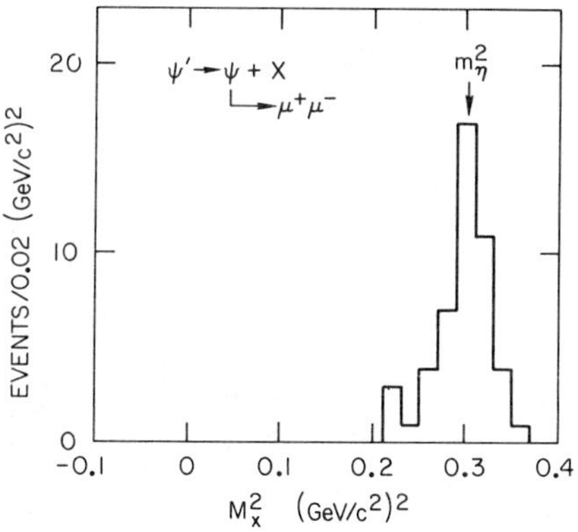

Figure 25 Distribution of square of recoil mass accompanying $\psi(3095)$-decay muons in $\psi' \to \psi\pi^+\pi^- \to \mu^+\mu^-\pi^+\pi^-$ events not consistent with kinematics of $\psi(3684) \to \psi(3095)\pi^+\pi^-$ decay; from LBL-SLAC (61).

mode. Figure 26 shows the spectrum of missing mass-squared in events of the type $\psi' \to \mu^+\mu^- +$neutrals, with a clear η peak. This is particularly prominent after subtracting a calculated contribution to the data expected from $\psi' \to \pi^0\pi^0\psi \to \pi^0\pi^0\mu^+\mu^-$, using the isospin-conservation prediction $\pi^0\pi^0/\pi^+\pi^- = \frac{1}{2}$ for a $\pi\pi$ system of zero isospin and the measured rate for $\psi' \to \psi\pi^+\pi^-$. Subsequently, direct observation was made by the DASP group of the decay $\psi' \to \psi\pi^0\pi^0$ (50), examples of which were reconstructed from events with ψ-decay muon pairs and detected γ rays. They measured a branching ratio

$$\Gamma(\psi' \to \psi\pi^0\pi^0)/\Gamma = 0.18 \pm 0.06.$$

The yield of events with muon pairs and charged particles provided data for their determination of the branching ratio

$$\Gamma(\psi' \to \psi\pi^+\pi^-)/\Gamma = 0.36 \pm 0.06,$$

in good agreement with the LBL-SLAC value 0.32 ± 0.04. The LBL-SLAC results yielded

$$\Gamma(\psi' \to \psi\pi^0\pi^0)/\Gamma = 0.17 \pm 0.04,$$

upon subtraction of the $\psi\eta$ (61) and $\psi\gamma\gamma$ (63) branching fractions from the directly measured value

$$\Gamma(\psi' \to \psi + \text{neutrals})/\Gamma = 0.25 \pm 0.04.$$

The LBL-SLAC and DASP values are in good agreement. Weighted averages are listed in Table 5, which summarizes the available data.

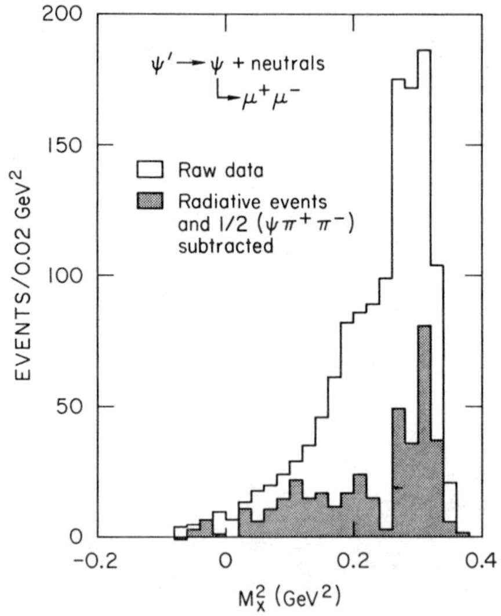

Figure 26 Distribution of square of missing mass accompanying ψ(3095)-decay muons in events containing $\mu^+\mu^-$ and undetected neutral particles; from SLAC-LBL (61).

Branching ratios to other exclusive hadronic states are relatively small. Comparison of the rates for ψ decay in Table 4 with those for ψ' in Table 5 reveals strong differences in the various branching ratios. In particular, the $\pi^+\pi^-\pi^+\pi^-\pi^0$ mode, prominent in $\psi(3095)$ decay, has a branching ratio smaller by an order of magnitude in $\psi(3684)$ decay.

Assignment of odd G-parity to $\psi(3684)$ follows directly from the existence of the $\psi\pi\pi$ mode and the value $G = -1$ for $\psi(3095)$ previously established.

Isospin is determined from the ratio of partial widths

$$\Gamma(\psi' \to \psi\pi^0\pi^0)/\Gamma(\psi' \to \psi\pi^+\pi^-)$$

which, corrected for the difference between the $\pi^+\pi^-$ and $\pi^0\pi^0$ phase space, has the predicted value 0.52 for $I = 0$, zero for $I = 1$, and 2.1 for $I = 2$. The measured value is

$$\Gamma(\psi' \to \psi\pi^0\pi^0)/\Gamma(\psi' \to \psi\pi^+\pi^-) = 0.5 \pm 0.1,$$

based on the combined results of the LBL-SLAC and DASP experiments. That conclusion appears to be consistent with the measurement of Hilger et al (62) of

$$\Gamma(\psi' \to \psi + \text{neutrals})/\Gamma(\psi' \to \psi\pi^+\pi^-) = 0.64 \pm 0.15,$$

considering that there is some unevaluated contribution from the $\psi\eta$ and $\psi\gamma\gamma$ modes to the $\psi +$ neutrals inclusive rate. Isospin conservation and the isosinglet character of ψ', as well as ψ, is established. This assignment of $I = 0$ is confirmed by the existence of the $\psi' \to \psi\eta$ decay mode.

3.5 *Hadronic Properties of ψ and ψ'*

In summary, the hadronic decays of the psions are characterized by strong-interaction conservation laws for isospin and G-parity. Both $\psi(3095)$ and $\psi(3684)$ have $I = 0$ and $G = -1$. SU(3) appears to be a symmetry of the decay of $\psi(3095)$, and that state has been provisionally classified as a singlet. Cascade to $\psi(3095) +$ pions dominates $\psi(3684)$ decays, with less suppression than other exclusive hadronic modes. These features of the decay dynamics, together with the estimates of $\psi(3095)$ interaction cross sections inferred from photoproduction, establish that the two lightest ψ particles are hadrons.

4 INTERMEDIATE STATES

No resonance peak at any energy between 3095 and 3684 MeV has been found in a series of measurements of e^+e^--annihilation cross sections at small energy intervals (25). Rather, such states have been observed in radiative decay, $\psi' \to \gamma\chi$, of the heavier ψ particle. Those states must thus have even charge-conjugation quantum numbers. The presence of such states has been revealed by the use of three distinct experimental techniques and analysis methods. First, the full cascade-decay chain $\psi' \to \gamma\chi \to \gamma\gamma\psi$ has been observed by detecting the lepton-pair decay products of the ψ with one or two coincident photons. Second, hadronic decay products of χ have been detected. Third, monochromatic photon lines have been observed in the inclusive energy distribution of photons produced in $\psi(3684)$ decays. For clarity of presentation, we use the generally agreed-upon symbol χ to denote the intermediate states of the psionic system.

4.1 *Cascade Radiative Decays*

Evidence for the existence of intermediate states was first obtained from the study of the $\psi' \to \psi\gamma\gamma$ decays by the DASP group (64) and later corroborated by the LBL-SLAC group (63). Both groups used the $\mu^+\mu^-$ decay mode as a signature of a ψ produced in ψ' decay. In one method of χ-mass determination, used by both groups, two photons were required to have been detected in shower counters. Since the energy resolution is poor, use was made only of the measured photon directions. Those, together with momentum and direction of the ψ, inferred from the $\mu^+\mu^-$ momentum, provided input for a two-constraint fit to the kinematics of the process $\psi' \to \psi\gamma\gamma$. It was estimated that the background from $\psi\pi^0\pi^0$ events is less than 10% of the number of events selected on the basis of quality of the kinematic fit. The DASP group found two clusters of points on a scatter plot of one photon energy versus the other. Since it is not known which of the two photons is the decay product of the χ, each group of photons yields two possible values of χ mass. Using central values of the energies of the two observed groups, they determined the masses to be (3.507 ± 0.007) or (3.258 ± 0.007) GeV/c^2 and (3.407 ± 0.008) or (3.351 ± 0.008) GeV/c^2, with decay widths consistent with zero.

To obtain better resolution in their χ-mass measurements, the LBL-SLAC group used the magnetic detector as a pair spectrometer to measure the energy of photons that had converted in the various pieces of the apparatus preceding the tracking

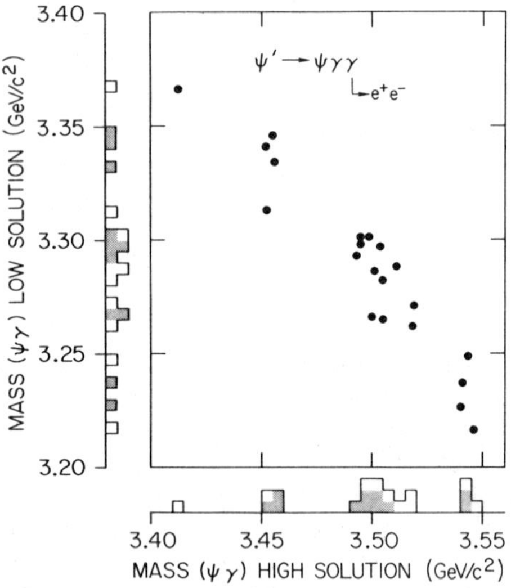

Figure 27 Scatter plot of $\psi\gamma_1$ invariant mass against $\psi\gamma_2$ invariant mass for $\psi(3684)$ decays to $\psi(3095)\gamma_1\gamma_2$; from LBL-SLAC (65).

chambers. The total conversion material has a thickness of some 5% of a radiation length. From an initial sample of 54 events with one converted photon, a subset of $\psi\gamma\gamma$ candidates was selected on the basis of consistency with zero missing mass and indication of the presence of a second photon, revealed by a shower-counter signal. With the elimination of events in which the inferred mass of the photon pair was consistent with the mass of the η, there remain 21 examples of $\psi' \rightarrow \psi\gamma\gamma$. One-constraint fits were made to the kinematics of $\psi' \rightarrow \psi\gamma\gamma$, in order to obtain values of both photon energies and, for each event, again two values of the mass of the parent of a $\psi\gamma$ pair. These masses are shown plotted against each other in Figure 27 (65). Three clusters are evident, at masses (3543 ± 10), (3504 ± 10), and (3454 ± 10) MeV/c^2. These, rather than the lower-mass alternatives, were chosen as χ masses because the spread of mass around the lower central values is in all cases consistent with that expected from Doppler broadening and the resolution of the apparatus. The one event with $M_\chi \simeq 3415$ MeV/c^2 appears to correspond to the two-event cluster in the DASP results, and the large clumps at $M_\chi = 3504$ MeV/c^2, it is presumed, are to be identified with the DASP events at $M_\chi = 3507$ MeV/c^2.

4.2 *Hadronic Decays of Intermediate States*

Hadronic decays of intermediate states were first identified by Feldman et al (66), who used the LBL-SLAC magnetic detector. Events were restricted to those con-

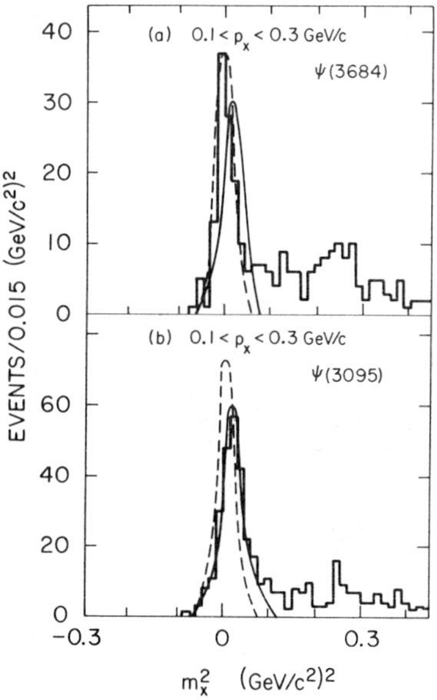

Figure 28 Spectrum of the square of missing mass in four-prong, zero-charge events with missing momentum between 100 and 300 MeV/c for (a) ψ(3095) and (b) ψ(3684) decays. The solid and dashed smooth curves show calculated resolution functions for a missing π^0 and missing photon, respectively; from SLAC-LBL (66).

taining even numbers of detected particles with total charge zero and momentum unbalance between 100 and 300 MeV/c. This restriction to small missing momentum allows a good separation of events with undetected π^0 from those with undetected photons, based on the observed distribution of missing mass. An example is the missing-mass distribution for the selected four-track events shown in Figure 28, which contrasts the results obtained for the $\psi(3095)$ and $\psi(3684)$ decays. The low-mass peak in the $\psi(3684)$ event sample is consistent with that expected for missing photons and inconsistent with that resulting from undetected neutral pions. Just the opposite conclusion is drawn from the $\psi(3095)$ sample. Adjusted values of particle momenta were obtained from one-constraint fits to the kinematics of $\psi' \to \gamma + \pi^+\pi^-\pi^+\pi^-$ and $\psi' \to \gamma + \pi^+\pi^- K^+K^-$. In subsequent analyses of a larger sample, the SLAC-LBL group used time-of-flight measurements, as well as the quality of the kinematic fit, as criteria for separating events with K^+K^- and $\bar{p}p$ pairs from those

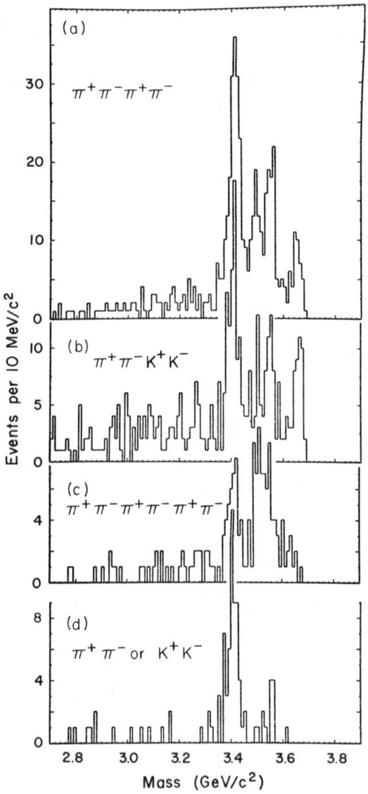

Figure 29 Spectra of invariant masses of states χ fit to the kinematics of the radiative decay $\psi(3684) \to \gamma\chi$ for χ constituent particles identified as indicated in the legends; from LBL-SLAC (67).

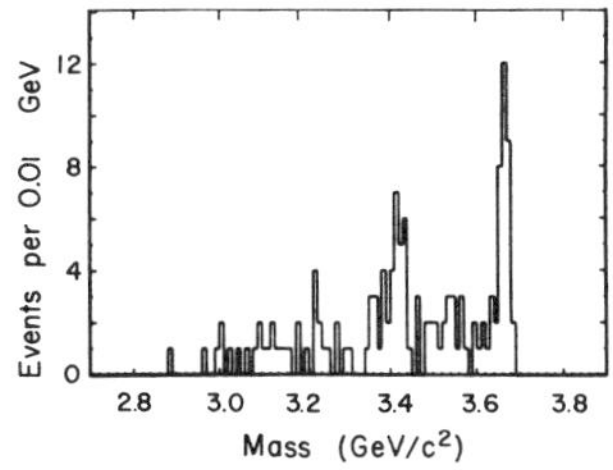

Figure 30 Invariant-mass spectrum of $\pi^+\pi^- p\bar{p}$ products of decays $\psi(3684) \rightarrow \gamma\pi^+\pi^- p\bar{p}$ identified with the aid of kinematic fits; from LBL-SLAC (67).

with pions only (67). Similar methods were applied to the analysis of two-prong and six-prong events. Results indicating the presence of decays of distinct states with small decay widths are shown in the mass distributions of Figures 29 and 30. All show prominent peaks at (3415 ± 10) MeV/c^2. Other peaks present at various levels of significance among the spectra are at (3500 ± 10) and (3550 ± 10) MeV/c^2. Noticeably lacking are two-body decays of the state at 3500 MeV/c^2. There is no indication of hadronic decay modes of the proposed state near 3450 MeV/c^2 inferred from results of analysis of the $\psi(3684) \rightarrow \psi(3095)\gamma\gamma$ events.

4.3 *Monochromatic Photons*

Certainly, the conceptually simplest and most direct evidence for distinct intermediate states would be the presence of monoenergetic lines in the spectrum of photon-decay products of $\psi(3684)$. Measurements of the yield of photons in the radiative decays of $\psi(3684)$ to the various χ states are essential for the determination of the $\psi' \rightarrow \gamma\chi$ and χ decay branching ratios.

The LBL-SLAC group measured the photon-energy spectra using only those photons that had converted to detected electron-positron pairs and whose momenta were then determined from measurements of reconstructed tracks sampled in the track chambers of the detector (65). Energy spectra of photons from $\psi(3684)$ decay and from $\psi(3095)$ are shown in Figure 31. The yield of photons from $\psi(3095)$ decays varies smoothly with photon energy. The shape of the $\psi(3684)$ distribution is similar, but for the evident peak at an energy determined to be (261 ± 10) MeV after correction for energy lost by the electron and positron by ionization. The observed width of the peak is consistent with a true width of zero, broadened by resolution. The inferred mass of the χ state accompanying the photon is (3413 ± 11) MeV/c^2. With appropriate corrections for detection inefficiencies and the assumption of a $1 + \cos^2\theta$ dependence of the photon angular distribution, the LBL-SLAC group finds a branching fraction of 0.075 ± 0.026 for radiative decay to this state. It is reasonably certain that this state is to be identified with the state at (3415 ± 10) MeV/c^2 whose hadronic decays have been observed.

Biddick et al (68) used an arrangement of proportional counters and NaI crystals arrayed about the other interaction region at SPEAR. Superior photon efficiency and energy resolution were achieved with that apparatus. Their results are shown in

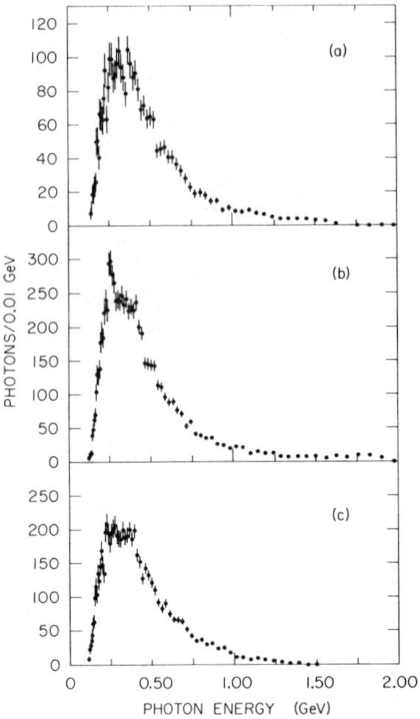

Figure 31 Inclusive energy spectra of photons from decays of (*a*) $\psi(3095)$ and (*b*) $\psi(3684)$; (*c*) a Monte-Carlo-simulated spectrum of π^0 decay photons in multihadron decays of $\psi(3095)$; from SLAC-LBL (67).

Figure 32, containing the spectrum of photons from $\psi(3684)$ decays and, for contrast, that from $\psi(3095)$ decays. In addition to the line at (260.6 ± 2.9) MeV, peaks above the continuum are evident at energies (169.2 ± 1.4) and (120.9 ± 1.3) MeV. The corresponding values of M_χ, 3413, 3511, and 3561 MeV/c^2, may be identified with the states at 3415, 3500, and 3550 MeV/c^2 discussed above, all consistent with the uncertainties in the various mass determinations. Branching ratios obtained by Biddick et al from these yields, corrected for detector acceptance, are

$$\Gamma[\psi' \to \gamma\chi(3415)]/\Gamma(\psi') = 0.072 \pm 0.023,$$

$$\Gamma[\psi' \to \gamma\chi(3510)]/\Gamma(\psi') = 0.071 \pm 0.019,$$

and

$$\Gamma[\psi' \to \gamma\chi(3550)]/\Gamma(\psi') = 0.070 \pm 0.020.$$

The first agrees well with the LBL-SLAC result, to within the substantial errors.

With values of the radiative-decay branching ratios and the branching-ratio products $\Gamma(\psi' \to \gamma\chi)/\Gamma(\psi') \times \Gamma(\chi \to f)/\Gamma(\chi)$ determined from the measured yields corrected for detection inefficiency, we obtain branching fractions for the various

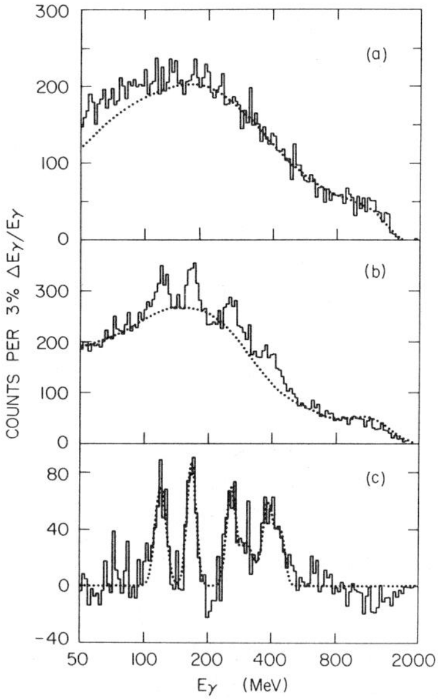

Figure 32 Inclusive energy spectra of photons from (*a*) $\psi(3095)$ and (*b*) $\psi(3684)$; (*c*) continuum subtracted from data of (*b*); from Biddick et al (68) at SPEAR.

detected decay modes of the χ states. The values in Tables 6–8 are to be considered estimates, accurate to approximately a factor of two.

Table 6 Branching fractions for $\chi(3415)$ decays

Mode	Branching fraction (χ) × branching fraction ($\psi' \to \gamma\chi$) (%)	Branching fraction (χ) (%)[a]	Ref.
$\pi^+\pi^-$	0.07 ± 0.02	1.0	67
K^+K^-	0.07 ± 0.02	1.0	67
$\pi^+\pi^-\pi^+\pi^-$	0.32 ± 0.06	4.6	67
$\pi^+\pi^-K^+K^-$	0.27 ± 0.07	3.9	67
$\pi^+\pi^-\bar{p}p$	0.04 ± 0.013	0.6	67
$\pi^+\pi^-\pi^+\pi^-\pi^+\pi^-$	0.14 ± 0.05	2.0	67
$\gamma\psi$	~ 1[b]	14.0	50
	0.2 ± 0.2[c]	3.0	65
	3.3 ± 1.7[d]	47.0	68

[a] Calculated from the branching-ratio product and 0.07 for the $\psi' \to \gamma\chi$ fraction. Estimates, with large uncertainties.
[b] Based on two events.
[c] Based on one event.
[d] Based on a fit to the inclusive energy spectrum of photon-decay products of $\psi(3684)$.

Table 7 Branching fractions for $\chi(3500)$ decay

Mode	Branching fraction (χ) × branching fraction ($\psi' \to \gamma\chi$) (%)	Branching fraction (χ) (%)[a]	Ref.
$\pi^+\pi^-$ and K^+K^-	<0.015	<0.2	67
$\pi^+\pi^-\pi^+\pi^-$	0.11 ± 0.04	1.6	67
$\pi^+\pi^-K^+K^-$	0.06 ± 0.03	0.9	67
$\pi^+\pi^-\bar{p}p$	0.01 ± 0.008	0.1	67
$\pi^+\pi^-\pi^+\pi^-\pi^+\pi^-$	0.17 ± 0.06	2.4	67
$\gamma\psi$	4 ± 2	57	50
	2.4 ± 0.8	34	65
	5.0 ± 1.5[b]	71	68
$\gamma\gamma$[c]	<0.0013	<0.02	50

[a] Calculated from the branching-ratio products and 0.07 for the $\psi' \to \gamma\chi$ fraction. Estimates, with large uncertainties.
[b] Based on a fit to the inclusive energy spectrum of photon decay products of $\psi(3684)$.
[c] Forbidden for spin-1.

Table 8 Branching fractions for $\chi(3550)$ decays

Mode	Branching fraction (χ) × branching fraction ($\psi' \to \gamma\chi$) (%)	Branching fraction (χ) (%)[a]	Ref.
$\pi^+\pi^-$ and K^+K^-	0.02 ± 0.01	0.3	67
$\pi^+\pi^-\pi^+\pi^-$	0.16 ± 0.04	2.3	67
$\pi^+\pi^-K^+K^-$	0.14 ± 0.04	2.0	67
$\pi^+\pi^-\bar{p}p$	0.02 ± 0.01	0.3	67
$\pi^+\pi^-\pi^+\pi^-\pi^+\pi^-$	0.08 ± 0.05	1.1	67
$\gamma\psi$	1.0 ± 0.6	14	65
	2.2 ± 1.0[b]	28	68

[a] Calculated from the branching ratio products and 0.07 for the $\psi' \to \gamma\chi$ fraction. Estimates, with large uncertainties.
[b] Based on a fit to the inclusive energy spectrum of photon decay products of $\psi(3684)$.

4.4 *Properties of the Intermediate States. J, P, C, G*

It follows directly from the observation of the radiative decays $\psi' \to \gamma\chi$ that the various χ states have even charge-conjugation quantum numbers. Information needed for assignment of parity and *G*-parity is available only for those three intermediate states whose decays to hadrons have been observed. Each decays to one or another state of even numbers of pions and thus has even *G*-parity. The states $\chi(3415)$ and $\chi(3550)$ decay to $\pi^+\pi^-$ pairs and so must be of natural spin and parity 0^+, 1^-, 2^+, etc. Of these, only the even spin, even parity assignments are compatible with the even charge conjugation of the χ states.

Angular correlations (70–72) among the various particles in the cascade-decay chains $e^+e^- \to \psi' \to \gamma\chi$, $\chi \to \gamma\psi$ or $\chi \to$ hadrons depend upon the spin of the state

χ, but are uniquely determined only in the case of zero spin. For instance, photons are produced in lowest-order e^+e^- annihilation with an angular distribution of the form

$$W(\theta_{\gamma_1}) = 1 + \alpha \cos^2 \theta_{\gamma_1},$$

where θ_{γ_1} is the angle between the photon and the initial e^+ (or e^-) direction. In the case that the χ spin is zero, $\alpha = 1$. For other spins, α cannot be specified further than by the general limit $|\alpha| < 1$ without knowledge of the strengths of the various multipole contributions to the transition amplitude. Although, in contrast to the situation with nuclear transitions, there are no strong arguments justifying neglect of higher multipoles, should it happen that $E1$ amplitudes are dominant, then the magnitude of the coefficient in the angular distribution is fixed by the χ spin, i.e. $\alpha = -1/3$ for $S_\chi = 1$ and $\alpha = +1/13$ for $S_\chi = 2$.

Figures 33 and 34 show the LBL-SLAC data on angular distributions (67) of the photons produced with those χ states that have substantial branching fractions to hadronic final states. Values of the inferred anisotropy coefficients for the three χ states are $\alpha = 0.3 \pm 0.4$ for $\chi(3550)$, $\alpha = 0.1 \pm 0.4$ for $\chi(3500)$, and $\alpha = 1.4 \pm 0.4$ for $\chi(3415)$. These results indicate spin-zero for $\chi(3415)$, and at best only weakly exclude spin-zero for the other two.

For those χ states decaying to two particles, as in, for example, $\psi' \to \gamma\chi \to \gamma\pi^+\pi^-$, the distribution in angle θ' between the photon and a decay particle, in the χ rest frame, is described by a polynomial in $\cos^2 \theta'$. For the decay into two pseudoscalars, this angular correlation function is isotropic for spin-zero and contains terms in $\cos^2 \theta'$ and $\cos^4 \theta'$ for spin-two and becomes increasingly complex for higher spins. The data of the SLAC-LBL group are shown in Figure 34*a* for $\chi(3415)$ decays to pion and kaon pairs. Poor statistical accuracy precludes the

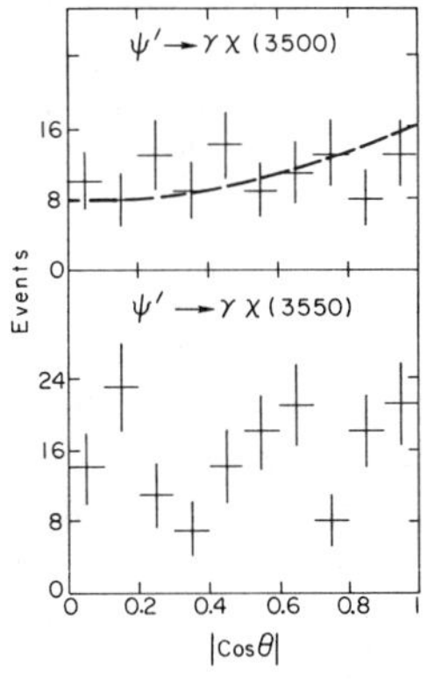

Figure 33 Distributions in the angle between the initial-state colliding beams' direction and the photons produced in $e^+e^- \to \psi(3684) \to \gamma\chi$. The χ masses are indicated in the legends. The smooth dashed curve gives the distribution for χ's of zero spin; from LBL-SLAC (67).

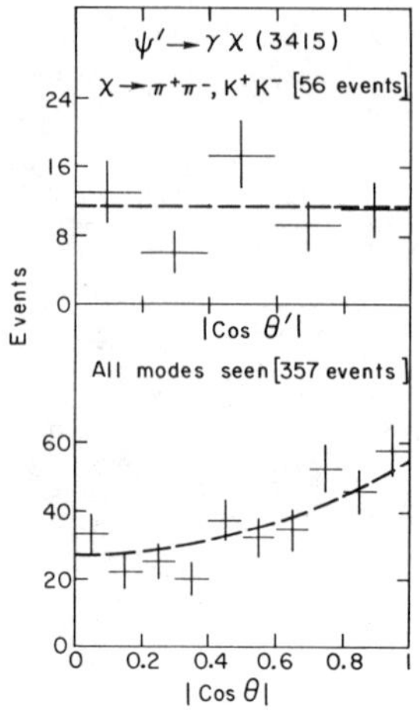

Figure 34 Angular correlations in radiative transitions $\psi(3684) \rightarrow \gamma\chi(3415)$ of psions created in e^+e^- annihilations. θ' is the angle between the photon and the dimeson direction in $\pi^+\pi^-$ or K^+K^- decays. θ is the angle between the photon and the colliding beams. The dashed curves are distributions for χ's of zero spin; from SLAC-LBL (67).

possibility of making definitive conclusions, but no deviation from isotropy is indicated, consistent with the expected distribution of the decay products of a zero-spin object.

As yet, there have been no reported measurements of angular correlations between the two photons in the cascade process $\psi' \rightarrow \gamma_1\chi \rightarrow \gamma_1\gamma_2\psi$. In this case, decays of states with angular momentum greater than zero may produce isotropic angular distributions. Lack of γ_1-γ_2 correlation is not a specific signature of zero spin. Further information about spin is contained in the correlation between γ_2 and lepton decay products of ψ in the processes $\psi' \rightarrow \gamma_1\gamma_2\psi \rightarrow \gamma_1\gamma_2 l^+l^-$. For zero spin, that correlation function has the simple form

$$W(\theta_1, \theta_2, \phi_2, \theta_l, \phi_l) = (1+\cos^2\theta_1)(1+\cos^2\theta_l),$$

where θ_l and ϕ_l are the polar and azimuthal angle of a decay lepton measured from an axis defined by the γ_2 direction in the rest frame of the ψ. Expressions for decay of states with greater spin are complicated, and their forms depend again on the multipolarity of the χ transitions.

In sum, the study of decay angular distributions has yielded one conclusive result, $S_\chi = 0$ for $\chi(3415)$ and tentative suggestions that $\chi(3500)$ and $\chi(3550)$ have greater spin. Nothing is known of the spin of $\chi(3455)$, observed so far only via its $\psi\gamma$ decay mode.

4.5 *Lower-Mass States*

Searches have not uncovered odd-*C* states formed in e^+e^- annihilations in the energy range 1.9–3.1 GeV (see Section 5). Evidence has been presented for existence of a state with mass near 2800 MeV/c^2, revealed in three-photon decays of $\psi(3095)$ produced in e^+e^- annihilations. Both the DASP group (50) and the Deutsches Electronen Synchrotron DESY-Heidelberg group (73, 74) measured only production angles of the three photons and determined their energies from the kinematics of the process $\psi(3095) \rightarrow 3\gamma$. The DASP data have been presented as the plot, reproduced in Figure 35, of the smallest against the largest opening angle of any pair of photons in each event. Decays of an intermediate state of unique mass in the sequence

$$\psi(3095) \rightarrow \gamma_1 X \rightarrow \gamma_1\gamma_2\gamma_3$$

produce points common to a locus determined by the mass of X. Figure 35 shows such curves for π^0, η, η' and $M_X = 2800$ MeV/c^2. An η signal is clear. A cluster of points is also shown in the region of the plot populated by events due to two-photon decays of a particle of mass $\sim$2800 MeV/c^2 produced in radiative decay of $\psi(3095)$. Calculations of background from three-photon events produced according to electrodynamics indicate smooth variation of density with mass in this region of the plot. It was concluded that a new narrow resonance had been observed with production-decay branching-ratio product

$$B[\psi \rightarrow \gamma X(2800)] \times B(X \rightarrow 2\gamma) \simeq 1.5 \times 10^{-4}.$$

Similar conclusions have been reached by the DESY-Heidelberg group, who detected three-photon events with the apparatus shown in Figure 36. The large array of NaI and lead-glass counters provides good photon-detection efficiency over $\sim$60% of the full-production solid angle. Their results are shown in Figure 37, a Dalitz plot of the lowest vs the highest γ-γ pair mass in each event. Data from

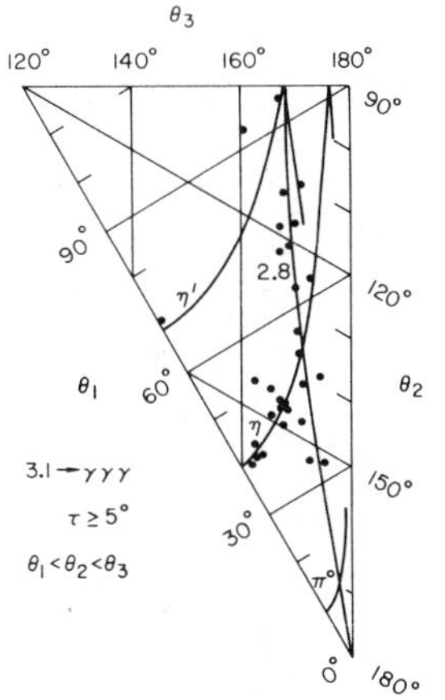

Figure 35 Scatter plot of the largest opening angle, θ_3, versus the smallest opening angle, θ_1, between any two photons in decays of $\psi(3095)$ to three photons. Loci of constant $\gamma\gamma\gamma$ invariant mass for η and η' and 2.8 GeV/c^2 are shown. Results from DASP (50) at DORIS.

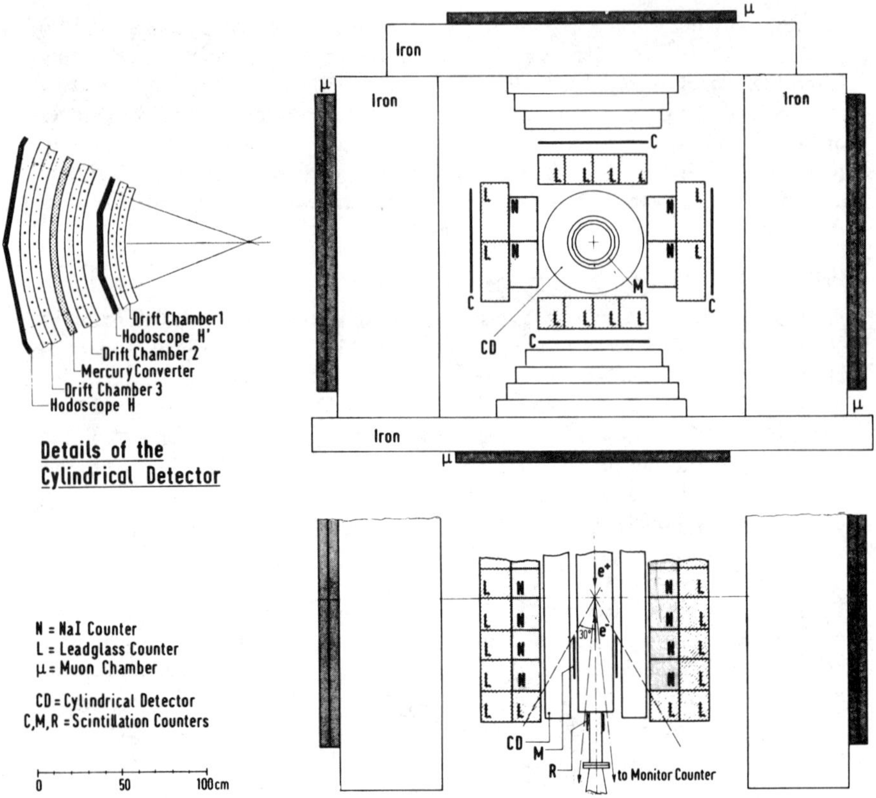

Figure 36 The DESY-Heidelberg detector apparatus at the e^+e^- intersecting storage rings DORIS (73).

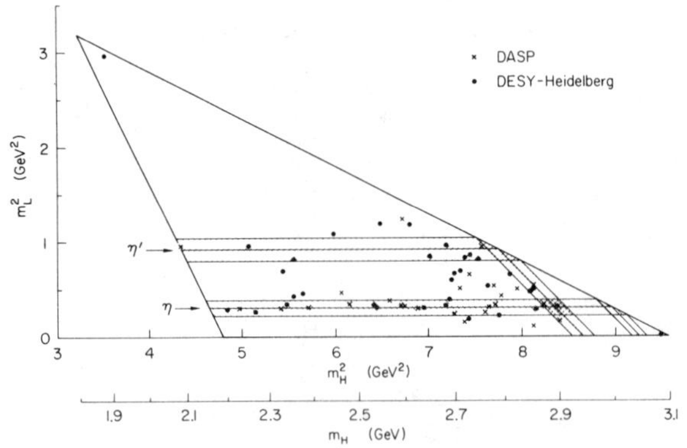

Figure 37 Dalitz plot of the smallest $\gamma\gamma$ invariant mass-squared vs the largest $\gamma\gamma$ mass-squared in $\psi(3095)$ decays to three photons. Data from DASP and DESY-Heidelberg (73).

Table 9 Properties of the intermediate states

State	Mass[a]	J^{PC} [b]	Ref.	Footnote(s)
$\chi(3415)$	3412 ± 3	0^{++}	50, 65, 67, 68	c, d
$\chi(3455)$	3454 ± 10	0^{-+}	65	e, f
$\chi(3500)$	3508 ± 4	1^{++}	50, 65, 67, 68	c
$\chi(3550)$	3553 ± 5	2^{++}	65, 67, 68	c, d

[a] Weighted averages of values quoted in the cited references.
[b] Even charge conjugation established by the radiative decay $\psi(3684) \to \psi(3095) + \gamma$.
[c] Preferred assignments of spin and parity consistent with observed angular correlations.
[d] Assignment of 0^- or 1^+ excluded by decays to $\pi^+\pi^-$ and K^+K^-.
[e] Four $\chi \to \gamma\psi$ decays observed; no evidence for hadron modes.
[f] Speculative assignment of spin and parity, neither supported nor contradicted by experimental evidence.

the DASP experiment are included. Again, the excess of events at mass near 2750 MeV/c^2 was attributed to $\gamma\gamma$ decay of a resonant state with production-decay branching-ratio product comparable to that due to η radiative production followed by $\gamma\gamma$ decay.

4.6 *Summary*

Five additional members of the ψ family have been established experimentally, albeit with reliability varying from firm [for $\chi(3415)$, $\chi(3500)$, and $\chi(3550)$] to mushy [for $X(2800)$ and $\chi(3455)$]. They are connected to the $\psi(3684)$, or $\psi(3095)$ via radiative transitions. Prodded by the suggestive level structure of a hypothesized bound quark-antiquark system, charmonium, these states have been assigned spectroscopic labels consistent with experimental results: $^3P_0$, $^3P_1$, and $^3P_2$ for the first three, and 1^1S_0 and 2^1S_0 for the others, respectively. Decays $\psi(3684) \to \chi(^3P) + \gamma$ occur with comparable branching fractions, $B(\psi' \to \gamma\chi) \simeq 0.1$, for each of the ^3P states. Properties of these states are listed in Table 9.

5 OTHER NARROW STATES, 1900 MeV/c^2 < M < 8000 MeV/c^2

A search for other narrow states coupled to lepton pairs was made by the LBL-SLAC group, which measured e^+e^--annihilation cross sections in small energy intervals from 3.2 to 7.6 GeV (75, 76). As demonstrated by the results plotted in Figure 38, none was discovered. Those results establish upper limits on the electron-pair decay widths of a resonance whose full width is much smaller than ~2 MeV (see Equation 3.1 with $\Gamma_h \simeq \Gamma$). Those limits are summarized in Table 10. Similar systematic searches at the storage ring ADONE, operated at energies between 1.9 and 3.1 GeV, also failed to unearth any new resonances. Limits on Γ_{ee} were established at ~5% of the ψe^+e^- width (77–81). Those limits are included in Table 10. Measurements at SPEAR of e^+e^--annihilation rates in coarser energy steps revealed the striking behavior near 4 GeV shown in Figure 39. At energies above the second ψ, the character of the energy dependence changes markedly.

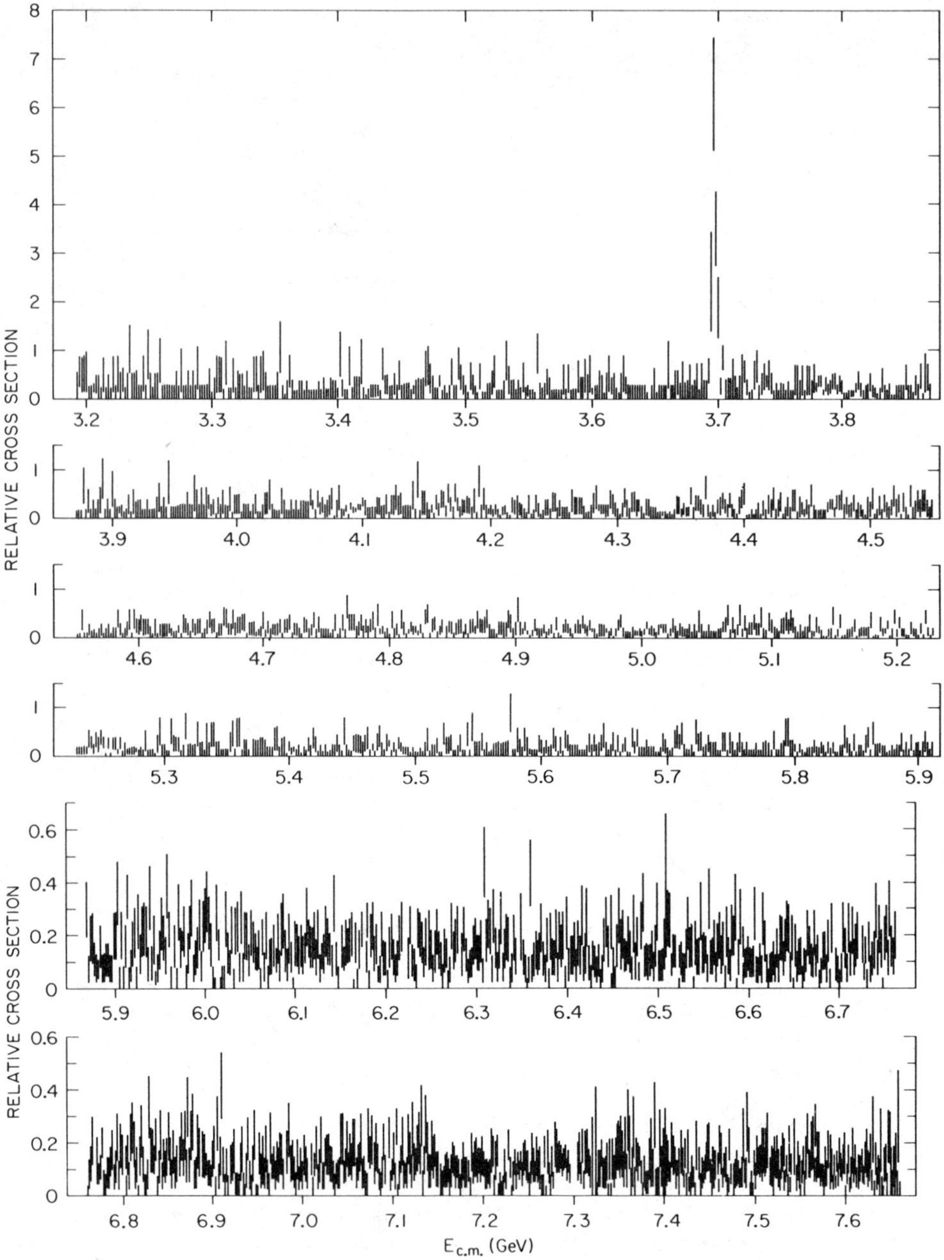

Figure 38 Energy dependence of the cross section for e^+e^- annihilation to hadrons measured at small intervals of total c.m. energy between 3.2 and 7.65 GeV; from LBL-SLAC (75, 76).

Table 10 Upper limits on parameters of narrow resonances other than $\psi(3095)$ and $\psi(3684)$

Mass range (MeV/c^2).	$\int \sigma_{\text{Res}}(E)\, dE$[a] (nb MeV)	Γ_{ee} (keV)[a]	Ref.
1910–2200	1150	0.21	77
2200–2545	800	0.20	77
2520–2595	690	0.20	81
2600–2694	1400	0.42	81
2700–2799	800	0.26	81
2800–2899	800	0.28	81
2900–3000	1600	0.62	81
2970–3090	800	0.32	78
3200–3500	970	0.47	75
3500–3680	780	0.44	75
3720–4000	850	0.55	75
4000–4400	620	0.47	75
4400–4900	580	0.54	75
4900–5400	780	0.90	76
5400–5900	800	1.11	76
5900–7600	450	0.87	76

[a] The results from ADONE (77, 78, 81) have been presented as ratios of upper limits on the integrated resonance cross section to the $\psi(3095)$ integrated cross section measured with the same apparatus. To eliminate effects of different detection efficiencies, we have used the SLAC-LBL value of the $\psi(3095)$ lepton-pair partial width (Table 3) in calculating the limits on the integrals and Γ_{ee} listed here (see Equation 3.4).

Below 4 GeV, the ratio of nonresonant hadronic to muon-pair cross sections, R, is essentially constant at $\simeq 2.5$. Above 4 GeV, R is also constant, but larger: $R \simeq 5.5$. In the transition interval, there appears to be a series of resonant structures that, however many there are, have widths more comparable to those of the lighter vector mesons than the ψ particles at 3095 and 3684 MeV/c^2.

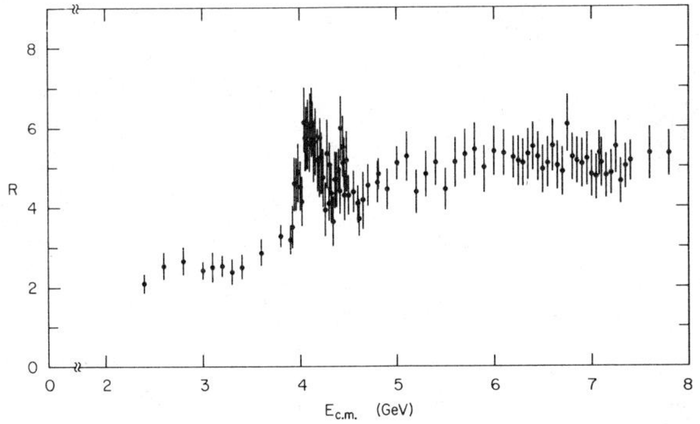

Figure 39 Energy dependence of the ratio R of the cross section for e^+e^- annihilation to hadrons to that for annihilation to muon pairs; from SLAC-LBL (76).

6 THEORETICAL ORIENTATION

Confronted with the necessity of understanding the mechanisms inhibiting the hadronic decay of the strongly interacting psions, appeal was made to new, hypothetical hadronic degrees of freedom. It had already been diversely suggested that the strong-interaction symmetry group is larger than SU(3). Certain suggestions, variously motivated, had been made to introduce a new additive quantum number (2, 82–85), charm (2), to be carried by a fourth quark, thus extending SU(3) to SU(4). Greenberg (86) and Han & Nambu (87) had argued for $SU(3)' \times SU(3)''$ as the proper symmetry group. In the first version cited (86), quarks are fractionally charged, while the Han-Nambu quarks have integral charges. Those schemes have nine quarks, each carrying one of three SU(3)″ "colors" as well as one of the three usual SU(3)′ "flavors." Ordinary hadrons must be SU(3)″ singlets. With $SU(4) \times SU(3)''$ symmetry, three colors, charm, and three other flavors equip the basic constituents of hadrons.

The fundamental hadronic entities now are the four flavored *u*, *d*, *s*, and *c* (charmed) quarks, whose properties are listed in Table 2. For these quarks, the Gell-Mann–Nishijuma formula now reads $Q = I_3 + (B+S+C)/2$. Each flavored quark has, in addition, the threefold color degree of freedom. The simple quark-parton model for e^+e^- annihilation to hadrons via the diagram of Figure 40, which predicts constant $R = \Sigma_j q_j^2$, the sum of squares of quark-parton charges, yields $R = 2$ below the threshold for annihilation to charmed quarks and $R = 3\frac{1}{3}$ above. The rough agreement with observation supports both the charm and color hypotheses of fractionally charged quarks.

It was suggested early that the narrow widths of ψ and ψ' could be accounted for with a color quantum number that is conserved in strong interactions (88, 89). In general, SU(3) color models of the Han-Nambu type suffer however, from the basic difficulty that radiative decays to ordinary hadrons conserve color, as do, in some versions, the cascade-decay processes $\psi' \to \psi\pi\pi$ and $\psi' \to \psi\eta$. Ad hoc suppression mechanisms must be imposed to obtain agreement with observation. Furthermore, since higher-dimensionality representations of color SU(3) are populated, there must exist many other colored states, for example colored ρ mesons.

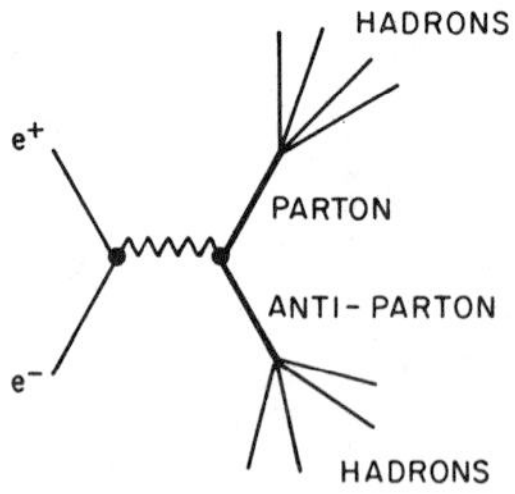

Figure 40 Feynman diagram for e^+e^- annihilation to hadrons via intermediate-quark–parton-antiparton pairs.

Table 11 Quark composition and properties of singly charmed hadrons

Particle[a]	Quark components	I, I_z	C	S
D^+, D^0	$c\bar{d}$, $c\bar{u}$	$(\frac{1}{2},\frac{1}{2}), (\frac{1}{2},-\frac{1}{2})$	+1	0
$D^-, \bar{D}^0$	$\bar{c}d$, $\bar{c}u$	$(\frac{1}{2},-\frac{1}{2}), (\frac{1}{2},\frac{1}{2})$	−1	0
F^+	$c\bar{s}$	0	+1	+1
F^-	$\bar{c}s$	0	−1	−1
Λ_c^+	cud	0	+1	0
Σ_c^0	cdd	1, −1	+1	0
Σ_c^+	cud	1, 0	+1	0
Σ_c^{2+}	cuu	1, 1	+1	0
Ξ_c^0	cds	$\frac{1}{2}, -\frac{1}{2}$	+1	−1
Ξ_c^+	cus	$\frac{1}{2}, \frac{1}{2}$	+1	−1
Ω_c^0	css	0	+1	−2

[a] This nomenclature is for states with lowest angular momentum, $J = 0$ for mesons, $J = \frac{1}{2}$ for baryons.

The decay $\psi(3684) \rightarrow \rho^c + \pi$ should, in some models (88), occur with a branching fraction comparable to that for decay to $\psi\pi\pi$. No evidence for such a particle has been revealed in the inclusive momentum distribution of pions from $\psi(3684)$ decay (50). Color models, at least in their simpler forms, are in serious contradiction with observed ψ and ψ' decay properties, and in any case have no place for the recently discovered mesons of mass ~ 1860 MeV/c^2 (see Section 7).

A generally successful description of the psionic states is obtained with a theoretical structure in which hadrons are colorless composites of the four fractionally charged, flavored quarks of Table 2, including charmed quarks. Strong interactions are mediated by exchanges of color gluons that serve to confine the quarks. An extended spectrum of states now occurs. Mesons still have $q\bar{q}$ content. Meson nonets grow to sixteen-dimensional multiplets, with the addition of new mesons containing one or more charmed quarks. Of these, one, $c\bar{c}$, has zero, or hidden, charm; the other six have exposed charm, $C = +1$ or -1. Table 11 lists these additional meson states, with designations D, F, etc introduced by Gaillard, Lee & Rosner (90).

Physically realized baryons, still with qqq content, are now grouped into two nonequivalent representations of dimensionality 20, one containing spin-$\frac{1}{2}$ and another spin-$\frac{3}{2}$ baryons. The additional states are charmed, with values $C = +1$, $+2$, or $+3$. Table 11 lists the charmed, spin-$\frac{1}{2}$ baryons that occur as composites of three quarks.

6.1 *Charmonium*

States found in e^+e^- annihilation, directly coupled to photons, are zero-charm vector mesons, and so in this picture, have $c\bar{c}$ quark content. Presuming that the χ states are reached in ordinary electromagnetic transitions from the heavier ψ states, the intermediate states and $X(2800)$ are also to be specified as $c\bar{c}$ composites.

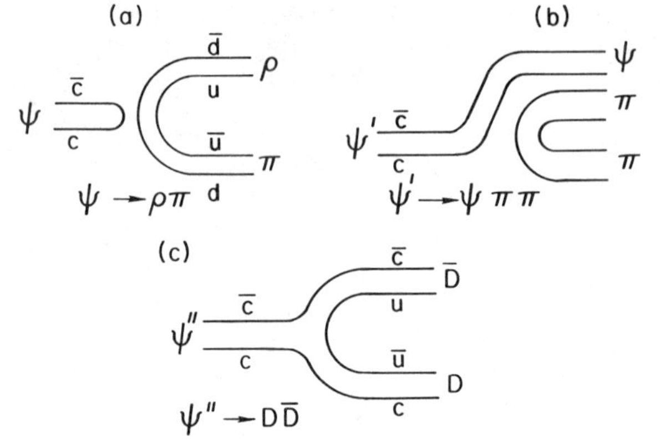

Figure 41 Duality diagrams of OZI-forbidden psion decays to (*a*) ordinary hadrons and (*b*) $\psi\pi\pi$; (*c*) OZI-allowed decays to charmed-meson pairs.

The empirical OZI rule must be invoked to provide a means of inhibiting the decays of the $c\bar{c}$ states to ordinary mesons composed only of *u*, *d*, or *s* quark-antiquark pairs. Strong, charm-conserving decays can proceed only to states containing *D* mesons, as depicted in the duality diagrams of Figure 41. Kinematics is invoked to prohibit decay to charmed mesons, whose masses must then be greater than 1842 MeV/c^2. Calculations have been made by De Rujula et al (91) of hadron masses in a general theory of quark confinement. Their estimates of the

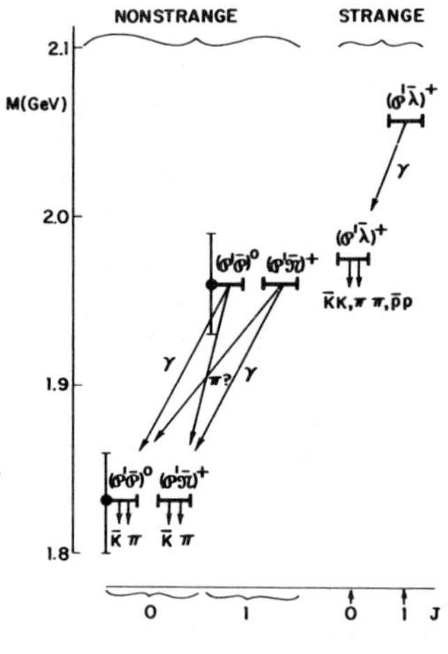

Figure 42 Spectrum of masses of lowest-lying charmed mesons as predicted by De Rujula, Georgi & Glashow (91). Their notation for the quark constituents may be readily translated to *u*, *d*, *s*, and *c*.

masses of charmed S-state quark-antiquark bound systems are shown as the energy level diagram of Figure 42. The results, admittedly not very precise, support the experimental requirement that $\psi(3684)$ be still below threshold for decay to charmed mesons. It should be noted, in this connection, that the energy dependence of the e^+e^--annihilation cross section (see Figure 39) behaves very much as if there were a threshold for pair production of particles of mass between 1.85 and 2.0 GeV/c^2. Also, the photoproduction results of Figure 13 invite a speculative interpretation in terms of a threshold at ~ 5 GeV energy for the inelastic, charm-conserving, OZI-allowed reaction $\psi N \rightarrow D\bar{D}N$, producing a pair of charmed mesons.

Identification of the psions with $c\bar{c}$ bound states requires comparison with the complete spectrum of such states. Even before the first ψ discovery, Appelquist & Politzer (5), reasoning from ideas of asymptotic freedom, expounded the view that this system of bound massive quarks could be described as a nonrelativistic atomic system, charmonium, that is analogous to positronium. Properties of charmonium states are determined by the combined effects of short-range $c\bar{c}$ interactions via exchange of massless vector gluons and some long-range confining potential. The pattern of charmonium energy levels that emerges, labeled with spectroscopic notation, is shown schematically in Figure 43. The pattern of observed energy levels is plotted suggestively in Figure 44, with $\psi(3095)$ the lowest-lying $^3S_1$ state and $\psi(3684)$ the first radially excited $^3S_1$ state. Remarkable qualitative agreement with the model predictions is apparent. That statement must be tempered with the reminder that the identities of the intermediate states have not been firmly established. Spins and parities of the intermediate χ states have been tentatively assigned values consistent with experimental data. Some theoretical arguments of Chanowitz & Gilman (92), based on quantitative predictions of radiative decay widths, have been brought to bear in assigning $\chi(3510)$, rather than $\chi(3445)$, the $^3P_1$ classification.

Numerous more or less quantitative predictions of level spacings, hadronic decay rates, leptonic decay rates, and radiative transition rates are available for comparison with the data. We briefly consider a selection from among the results of those calculations.

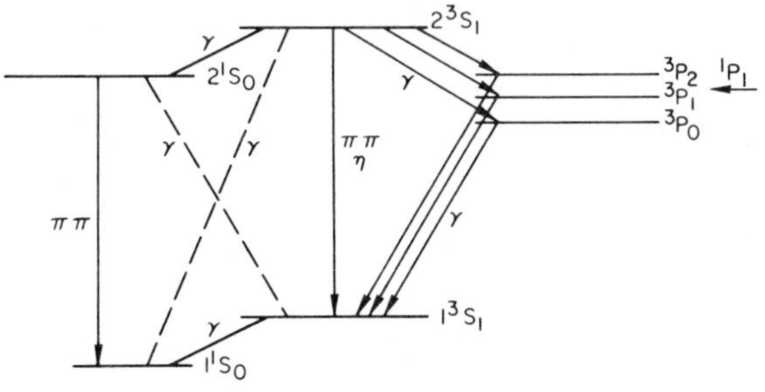

Figure 43 Schematic diagram of the energy levels of charmonium, the bound system of a charmed quark and its antiparticle. Some transition modes are indicated. See Appelquist & Politzer (5), for example.

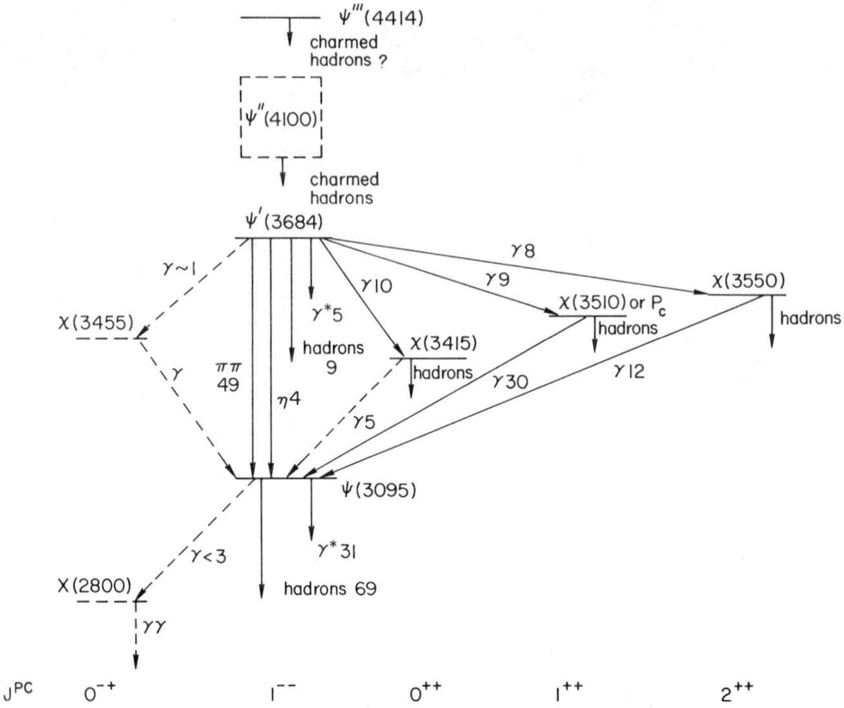

Figure 44 Energy-level diagram of observed ψ and χ states shown with spectroscopic notation corresponding to the levels of charmonium. Dashed lines indicate uncertain states. Numbers are now-outdated branching fractions. See text for current values.

6.1.1 LEVEL SPACINGS De Rujula, Georgi & Glashow (91) represent the quark-confinement interaction by a long-range, spin-independent potential common, in fact, to all quarks, ordinary as well as charmed ones. They argue that hadron mass splittings are determined by additional short-range perturbing forces resulting from exchange of a single, massless vector gluon. The corresponding interaction Hamiltonian has the form of the Fermi-Breit potential for the positronium atom (93), with the electrodynamic $\alpha = e^2/4\pi$ replaced by an effective hadronic coupling constant. Treating the Fermi-Breit interaction as a first-order perturbation, they obtain expressions for charmonium S- and P-state energy eigenvalues in terms of known hadron masses and the mass of the charmed quark. Applying their mass formula to $\psi(3095)$, the charmed-quark mass is estimated to be ~ 1650 MeV/c^2. With that, all other S- and P-state masses are determined. They obtain S-state hyperfine splittings

$$M(^3S) - M(^1S) \simeq 27 \text{ MeV}/c^2$$

for both the ground and first radially excited state, $\psi(3684)$; singlet P-state mass

$$M(^1P) \simeq 3650 \text{ MeV}/c^2\,;$$

triplet P-state mass

$$M(^3P_0) \simeq 3640 \text{ MeV}/c^2\,;$$

and hyperfine splittings

$$[M(^3P_1) - M(^3P_0)] \simeq 7.8 \text{ MeV}/c^2$$

and

$$[M(^3P_2) - M(\ P_1)] \simeq 6.2 \text{ MeV}/c^2.$$

These last values differ by roughly an order of magnitude from measured mass differences of the observed states given the charmonium classification shown in Figure 44. An alternative description, introduced by Eichten et al (94), assigns the $c\bar{c}$ system a potential that is partially Coulombic, $1/r$, and partially linearly increasing in its dependence on quark-antiquark separation. The linearly increasing part is meant to provide the long-range quark-confining force. They find a splitting of 230 MeV between the center of gravity of the 1^3P and the 2^3S levels, in reasonable agreement with observations. With that potential taken as the Fourier transform of $V(k^2)$ in a modified gluon propagator $g_{\mu\nu}V(k^2)$, Pumplin et al (95) and Schnitzer (96) arrive at hyperfine splitting

$$M(^3S) - M(^1S) \simeq 100 \text{ MeV}/c^2,$$

still about three times smaller than is observed with the current hesitant assignment of states. Better agreement is obtained with this prescription, but still, the agreement between model predictions of masses and observed values can hardly be characterized as better than at a semiquantitative level.

Further radial excitations, n^3S_1 bound states of charmonium, occur at energies determined by the characteristics of the $c\bar{c}$ binding potential and the quark mass. Using the 1^3S_1 and 2^3S_1 masses as inputs to fix the strength of a linear potential and the quark mass, Harrington et al (97) calculate eigenvalues

$$E(3^3S_1) = 4.18 \text{ GeV},$$

$$E(4^3S_1) = 4.61 \text{ GeV},$$

$$E(5^3S_1) = 5.00 \text{ GeV},$$

for the next three charmonium states accessible to e^+e^- annihilation in lowest order.

6.1.2 LEVEL WIDTHS Since annihilations to hadrons or photons occur at zero quark-antiquark separation, it may be expected that conditions are more favorable for validity of a perturbative, weak-coupling treatment of $c\bar{c}$ annihilation (22). Annihilations to lepton-pairs proceed via $c\bar{c}$ single-photon transitions, as indicated in Figure 45a; hadronic decays of $^3S_1$ proceed through the three-gluon intermediary of Figure 45b; $^1S_0$, $^3P_0$, and $^3P_2$ states decay to hadrons via two gluons,

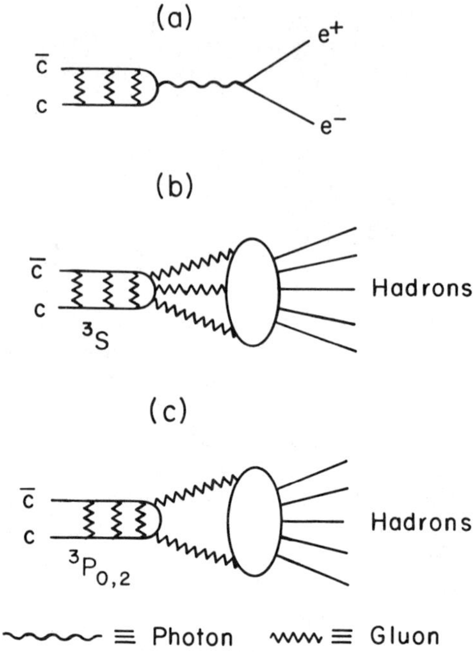

Figure 45 Diagrams for annihilation of charmonium states to: (*a*) electron pairs via intermediate photons; (*b*) and (*c*) hadrons via intermediate gluons.

shown in Figure 45*c*. Radiative decays are represented in Figure 45*b* and *c*, with gluons replaced by photons, as in positronium annihilation. The rate for each of these decay processes is proportional to $|\psi(0)|^2$, the square of the $c\bar{c}$ bound-state wave function at the origin, while the hadronic rates also depend upon α_s, the "running" gluon coupling constant. Radiative decay rates transpose directly to hadronic rates with the replacement of α by α_s. Formulas for these decay widths are (22)

$$\Gamma(\psi \to e^+e^-) = 4(Q_c\alpha)^2 \frac{1}{m_c^2}|\psi(0)|^2$$

$$\Gamma(\psi \to \text{hadrons}) = \frac{16}{9\pi}(\pi^2-9)\frac{5}{18}\alpha_s^3\frac{1}{m_c^2}|\psi(0)|^2,$$

where Q_c is the charge of the *c*-quark and m_c its mass. The ratio of these two partial widths determines the coupling constant $\alpha_s \simeq 0.2$ for $q^2 = (3095)^2$. Only the small variation in α_s between 3095 and 3684 MeV distinguishes the ratio $\Gamma[\psi(3684) \to e^+e^-]/\Gamma[\psi(3684) \to \text{hadrons}]$ from $\Gamma[\psi(3095) \to e^+e^-]/\Gamma[\psi(3095) \to \text{hadrons}]$. Neglecting the energy dependence of α_s, we obtain

$$\Gamma[\psi(3684) \to \text{hadrons}] = 2.1/4.8 \times 47 = 21 \text{ keV}.$$

The measured branching fractions are $\sim 53\%$ for $\psi\pi\pi+\psi\eta$, $\sim 3\%$ second-order electromagnetic, $\sim 30\%$ for $\gamma\psi$, and $\sim 2\%$ for lepton pairs, totaling 88%, so that $\Gamma[\psi(3684)\rightarrow \text{hadrons}] \simeq 27$ keV, perhaps fortuitously in rather good agreement with the prediction.

6.1.3 TRANSITIONS BETWEEN LEVELS Radiative transition rates provide another test of the bound-state model. Transitions from $\psi(3684)$, the 2^3S_1 level, to the 2^3P_2 states should be dominantly electric dipole in nature. With that presumption, the ratios of partial widths should be simply related to the transition energies

$$\Gamma[\psi(3684)\rightarrow {}^3P_2]/\Gamma[\psi(3684)\rightarrow {}^3P_1]/\Gamma[\psi(3684)\rightarrow {}^3P_0] \simeq 5k_2^3/3k_1^3/k_0^3,$$

where the k's are the photon momenta and the numerical factors are the statistical weights of the P states. The measured photon momenta give ratios 1.0/1.3/1.5, to be compared with the ratios of measured rates 1/1.1/1.25, the latter ratios with substantial errors. Early calculations with potential models, e.g. those of Eichten et al (94), produced a total width $\Gamma[\psi(3684)\rightarrow \gamma\chi] = 215$ keV for transitions to the three ^3P states, accounting for a major fraction of $\psi(3684)$ and in gross disagreement with the facts. Incorrect calculated values were used for the transition energies, however. Using the observed rather than the calculated values of photon momenta, we obtain the corrected values $\Gamma[\psi(3684)\rightarrow \gamma\chi_2] \simeq 23$ keV, $\Gamma[\psi(3684)\rightarrow \gamma\chi_1] \simeq 28$ keV, $\Gamma[\psi(3684)\rightarrow \gamma\chi_0] \simeq 36$ keV, which are in rather more acceptable agreement with observation. Later estimates, obtained with a modified calculation by Eichten et al (98), are also in satisfactory agreement with the data. It is not clear to what extent the improvement results from the more sophisticated, extended model, since experimental values of transition energies were used as input to the calculation. It appears that the $\psi(3684)\rightarrow\chi$ radiative transition rates can be accommodated within the context of the bound-state charmonium model and indeed, provide some support for the chosen spectroscopic classification.

Further predictions await definitive experimental confrontation. Among them we cite the rates for radiative transition from the ^3P states to $^3S_1$ ground state (98), $\Gamma(^3P\rightarrow {}^3S) \simeq 100$–$300$ keV; magnetic-dipole transition partial width (94) $\Gamma(2^3S_1\rightarrow 1^1S_0) \simeq 1$ keV; ^3P-state hadronic decay widths (99) $\Gamma(^3P_2\rightarrow \text{hadrons})/\Gamma(^3P_0\rightarrow \text{hadrons}) = 4/15$, obtained with a two-gluon annihilation process; and $\Gamma(^3P_1\rightarrow \text{hadrons})/\Gamma(^3P_2\rightarrow \text{hadrons}) \simeq 1/4$, gotten from the dominant gluon plus quark-pair annihilation (100).

6.2 *Charmed-Particle Decays*

It is necessary to depart from a strict delineation of the subject matter to consider properties of those ingredients crucial to the expanded hadronic substructure, states with exposed charm. Glashow, Iliopoulis & Maiani (3) recognized that the fourth quark, introduced to bring a symmetry to the fundamental entities of the weak interactions, provided the extra degree of freedom necessary for the natural proscription of strangeness-changing neutral currents. In fact, the prescribed neutral hadronic weak current

$$J^0_{h_\nu} = \bar{u}\gamma_\nu(1+\gamma_5)u+\bar{c}\gamma_\nu(1+\gamma_5)c-\bar{d}\gamma_\nu(1+\gamma_5)d-\bar{s}\gamma_\nu(1+\gamma_5)s$$

Table 12 Selection rules for weak decays of charmed particles

Decay-amplitude θ_c dependence	Selection rules	Examples
$\cos\theta_c$	$\Delta S = \Delta C = \Delta Q = \pm 1, \lvert\Delta I\rvert = 0$	$D^0 \to K^-\mu^+\nu$ $F^+ \to \mu^+\nu$ $\Lambda_c^+ \to \Lambda^0 e^+\nu$
$\sin\theta_c$	$\Delta S = 0, \Delta C = \Delta Q = \pm 1, \lvert\Delta I\rvert = \frac{1}{2}$	$D^+ \to \mu^+\nu$ $F^+ \to K^0\mu^+\nu$ $\Sigma_c^{2+} \to pe^+\nu$
$\cos^2\theta_c$	$\Delta S = \Delta C = \pm 1, \lvert\Delta I\rvert = 1$	$D^0 \to K^-\pi^+$ $D^\pm \to K^\mp\pi^\pm\pi^\pm$ $F^+ \to K^+K^-\pi^+$ $\Lambda_c^+ \to \Lambda^0\pi^+\pi^+\pi^-$
$\cos\theta_c \sin\theta_c$	$\Delta S = 0, \Delta C = \pm 1, \lvert\Delta I\rvert = \frac{1}{2}, \frac{3}{2}$	$D^0 \to \pi^+\pi^-$ $F^+ \to K^0\pi^+$ $\Lambda_c^+ \to n\pi^+$
$\sin^2\theta_c$	$\Delta S = -\Delta C = \pm 1, \lvert\Delta I\rvert = 0, 1$	$D^0 \to K^+\pi^-$ $D^\pm \to K^\pm\pi^-\pi^+$ $F^+ \to K^+K^+\pi^-$ $\Xi_c^0 \to p\pi^-$

is charm- as well as strangeness-conserving. Decay selection rules, or rather, ordering of amplitude strengths, are implicit in the form of the charged hadronic current

$$J_{h_\nu} = \bar{u}\gamma_\nu(1+\gamma_5)[d\cos\theta_c + s\sin\theta_c] + \bar{c}\gamma_\nu(1+\gamma_5)[-d\sin\theta_c + s\cos\theta_c],$$

where θ_c is the Cabibbo rotation angle ($\simeq 0.23$ radians). Decays involving quark-constituent transformations $c \to s$ are favored because of the presence of $\cos\theta_c$ in their transition amplitudes. We list in Table 12 selection rules corresponding to the θ_c dependence of the decay amplitudes and some selected examples of charmed-particle decay modes that obey those rules. Meson decay rates have been estimated to be $\sim 10^9$ sec^{-1} for leptonic, $\sim 10^{12}$ for semileptonic, and $\sim 10^{13}$ sec^{-1} for hadronic modes (90). There do not exist reliable estimates for partial widths for decays to the various allowed final states of hadrons, although naive phase-space considerations lead to the expectation of relatively low average multiplicity. Such simple-minded arguments may require significant modification on consideration of the role of resonant states as decay products.

7 STATES HEAVIER THAN 4000 MeV/c^2

The LBL-SLAC group has reported results of detailed measurements of the e^+e^--annihilation cross section in the energy interval 3.9–4.6 GeV (101). Detection efficiency determined from a Monte Carlo simulation of the apparatus' response varied from 0.53 at the lowest energy to 0.57 at the highest. Possible systematic errors in absolute scale are estimated to be $\sim \pm 15\%$, the result of uncertainties in the efficiency calculation. Those errors should introduce smaller relative varia-

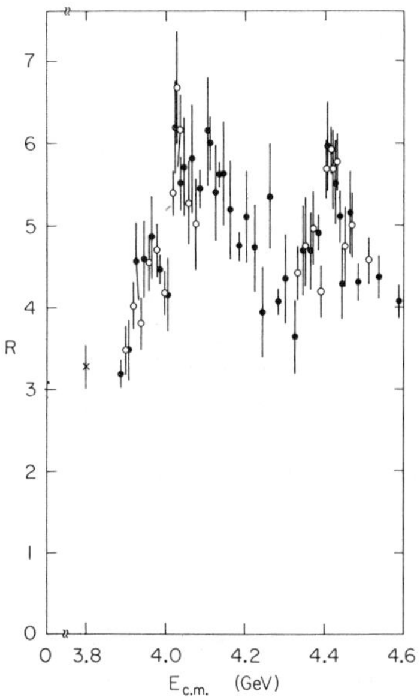

Figure 46 Energy dependence of the ratio $R = \sigma(e^+e^- \rightarrow \text{hadrons})/\sigma(e^+e^- \rightarrow \mu^+\mu^-)$ at c.m. energies between 3.8 and 4.6 GeV; from LBL-SLAC (101).

tions, less than 5%, among the values at the various energies. Figure 46 shows the data, plotted again as the ratio of hadronic-annihilation cross section to the muon-pair cross section, with statistical errors indicated. At the very least, two peaks are evident. It is, however, not possible to determine details of the structure in the variation of R with energy between 3.9 and 4.3 GeV. There appear to be a number of resonances; there may be thresholds for production of new particles. Interferences among resonant amplitudes may also account for some of the violent appearance of this substructure.

The data in the interval ~4.3–~4.6 GeV appear more consistent with the presence of a single, isolated resonant state, as seen in Figure 47. A fit was made

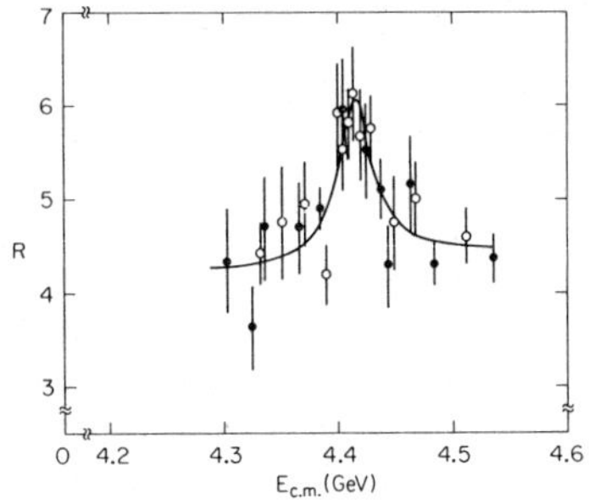

Figure 47 Resonance fit to the energy dependence of R near 4.4 GeV. See text for parameters of the state; from SLAC-LBL (101).

to a Breit-Wigner line shape, including radiative corrections, with a noninterfering background. Resonance parameters determined from the fit are shown in Table 3, together with those for $\psi(3095)$ and $\psi(3684)$, for comparison. The striking feature of this resonance is the magnitude of its full width, $\Gamma(4414) = 33 \pm 10$ MeV, more characteristic of strong decay than of the inhibited decay of $\psi(3095)$ and $\psi(3684)$. The electron width $\Gamma_{ee} = 0.44 \pm 0.14$ keV is one fifth that of $\psi(3684)$. Branching ratios to particular distinct hadronic final states have not as yet been determined.

A resonance at $\sim$4.4 GeV is nicely qualitatively compatible with an interpretation as an anticipated higher excitation of the charmonium structure. Indeed, as discussed above, two $^3S_1$ states may be expected in the energy interval 3.9–4.5 GeV. The smaller lepton-pair partial width of $\psi(4414)$ needs to be accounted for with a decreased modulus-squared of the wave function at zero separation, $|\psi(0)|^2$, of the higher-energy bound eigenstate. Unfortunately, all $^3S_1\, c\bar{c}$ eigenfunctions have the same value of $|\psi(0)|^2$, independent of principal quantum number, in the simplest model of a long-range binding potential, linearly rising with distance. Some "tuning" of the shape of the potential is required if the charmonium model is to give more than a qualitative description of the psionic states.

Two orders of magnitude increase in total width, compared to $\psi(3684)$, indicate that the OZI suppression mechanism has ceased to operate on $\psi(4414)$ decays. If the hidden charm interpretation of psions is to be maintained, it must be that this resonance rest energy is above the threshold for decay to pairs of particles bearing exposed charm.

8 CHARMED PARTICLES

Following some years of unsuccessful searches for charmed particles produced in hadron-hadron collisions (102) and in e^+e^- annihilations (103), all but conclusive evidence for their existence has been obtained by the LBL-SLAC group (104, 105), who observed decays of charmed mesons produced in e^+e^- annihilations, and by Knapp et al (106), who detected decays of photoproduced charmed baryons.

8.1 *Mesons*

In the LBL-SLAC experiment, evidence for $K\pi$ and $K\pi\pi\pi$ decay modes of charmed mesons was found in a study of 29,000 multiprong events produced in e^+e^- annihilations at c.m. energy between 3.90 and 4.60 GeV. The data are summarized succinctly in the effective-mass plots of Figure 48. A significant signal is evident near 1900 MeV/c^2 in the effective-mass distributions of all neutral combinations of two charged particles to which kaon mass and pion mass were arbitrarily assigned. Independent measurements of momentum and time of flight are made with the detection apparatus, but the timing resolution, 0.4 nsec, is insufficient to allow a distinction to be made between pions and kaons whose flight times in these events typically differ by 0.5 nsec. To make use of the time-of-flight information, the alternate mass assignments to each track were assigned relative weights appropriate to a Gaussian probability distribution with $\sigma = 0.4$ nsec:

$$W_i = \exp\left[-(t_i - t_T)^2/2\sigma^2\right],$$

where t_i is the flight time for mass m_i, determined from measured track parameters, and t_T is the directly measured time. The sum of track weights is made equal to unity and each mass combination, $\pi\pi$, πK, and KK, assigned a weight equal to the product of the track weights. With this method, the sum of weights assigned to all mass combinations just equals the total number of two-particle combinations in each event, so that there is no double counting. The second row of Figure 48 shows the invariant-mass spectra. A clear peak is evident in the $K^{\pm}\pi^{\mp}$ mass distribution, and the small residual signals in $\pi\pi$ and KK are understood to result from misidentification of true $K^{\pm}\pi^{\mp}$ events expected from inaccurate time-of-flight measurements. In the third row of Figure 48 are similarly weighted distributions, showing a peak in $K\pi\pi\pi$ at essentially the same mass as the $K\pi$ combinations. No significant signals were found in corresponding doubly charged two- or four-particle mass spectra.

Fits to the data with assumed Gaussian-shaped peaks above smoothly varying backgrounds yielded central values of mass 1870 and 1860 MeV/c^2 for $K\pi$ and $K\pi\pi\pi$, respectively. With consideration of the effects of systematic and random errors, it was concluded that both mass distributions are consistent with having a common source, decays of a particle of mass $M(D^0) = 1865 \pm 15$ MeV/c^2. The

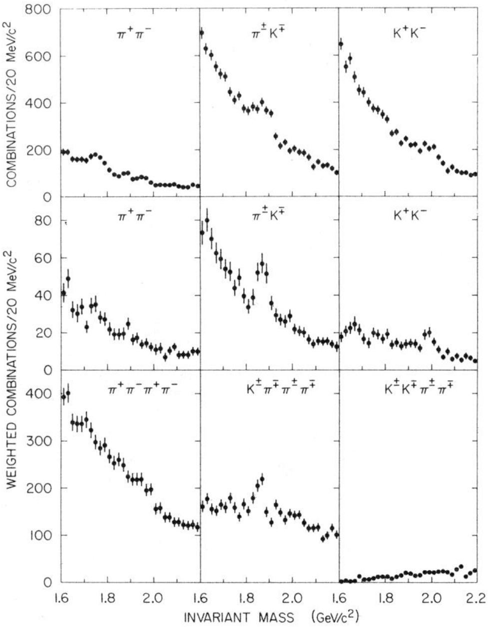

Figure 48 Inclusive neutral two- and four-particle invariant-mass spectra in multi-hadronic e^+e^- annihilation at c.m. energies 3.9–4.6 GeV. Top row, indiscriminate-mass assignments; middle row, mass assignments with time-of-flight weighting; bottom row, four-body mass assignments with time-of-flight weighting; from LBL-SLAC (104).

observed widths of the mass peaks are those expected from experimental resolution alone, 25 MeV/c^2 (rms) for the $K\pi$ system and 13 MeV/c^2 for the $K3\pi$ system, but a finite width of the state was not definitively excluded. A 90% confidence limit of 40 MeV/c^2 was deduced for the resonance decay width, based on quality of fits of a Breit-Wigner resonance shape convoluted with a Gaussian resolution function.

Peruzzi et al (105) report similar observations of peaks in effective-mass distribution of $K^-\pi^+\pi^+$ and $K^+\pi^-\pi^-$ combinations selected from multihadron e^+e^- annihilations at a c.m. energy of 4.03 GeV/c. Those data were obtained in a run at SPEAR with the LBL-SLAC magnetic detector, stimulated by the observation of the D^0 decays. Invariant mass spectra are shown in Figure 49 for three-body mass combinations, with $K\pi\pi$ mass assigned and each event weighted according to the prescription discussed above. Events with $\pi^+\pi^-$ pairs in the final state show no evidence of resonance peaks in their mass distributions, while a clear peak appears, at $M(D^\pm) = 1876 \pm 15$ MeV/c^2, in the exotic $K^-\pi^+\pi^+$ and $K^+\pi^-\pi^-$ states. That peak again has the appearance of resulting from decays of a resonance state whose width is a creature of merely experimental resolution. As with the neutral state, the 90% confidence-level upper limit to the width of the resonance is 40 MeV/c^2. Charged and neutral masses are equal within errors, but the measurements allow of a mass splitting of some 10 MeV/c^2, not atypical of splittings within isospin multiplets. It is natural and sensible then to identify the charged $K^\pm\pi^\mp\pi^\mp$ events as members of an isospin multiplet whose neutral partners are the D^0 mesons whose decays were revealed in $K^\pm\pi^\mp$ and $K^\pm\pi^\mp\pi^+\pi^-$ modes.

Charmed mesons must be produced in particle-antiparticle pairs in the annihilation of uncharmed positrons on electrons. Two pieces of experimental evidence

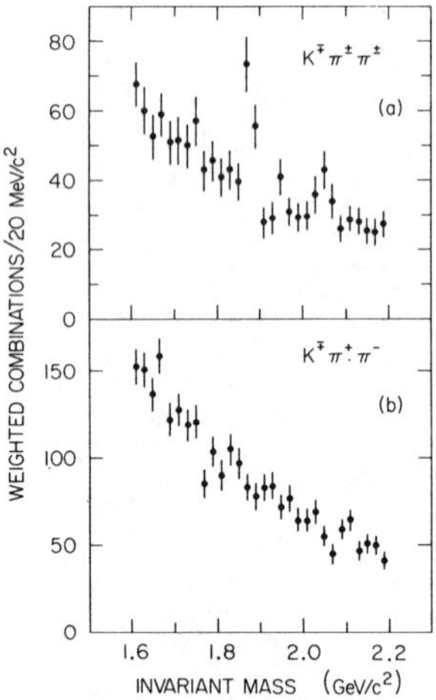

Figure 49 Inclusive mass spectra of charged three-particle combinations in multihadron products of e^+e^- annihilations at 4.03-GeV c.m. energy. Mass assigned according to time-of-flight weights in (*a*) exotic and (*b*) nonexotic systems; from SLAC-LBL (105).

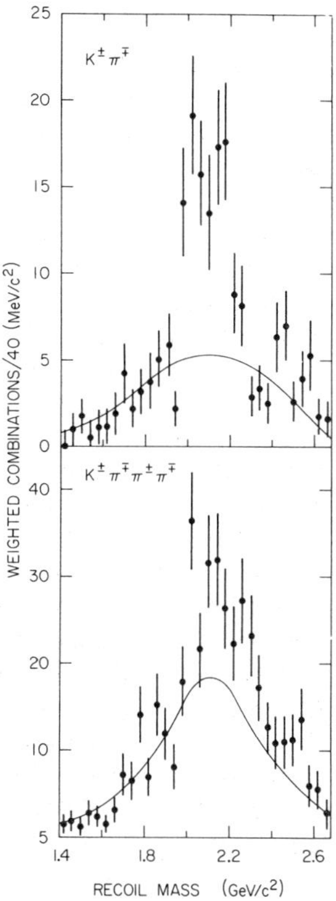

Figure 50 Distribution of masses accompanying $K^{\pm}\pi^{\mp}$ and $K^{\pm}\pi^{\mp}\pi^{+}\pi^{-}$ in events with $K\pi$, $K3\pi$-invariant masses in the peaks near 1865 MeV/c^2 seen in Figure 48. The smooth curves show background shapes obtained from events with $K\pi$ and $K3\pi$ masses adjacent to the peaks; from LBL-SLAC (104).

have been brought to bear on the question of the process by which the D^0 and $D^{\pm}$ mesons are produced. First, the observed spectra of masses recoiling against the $(K\pi)^0$, $(K3\pi)^0$, and $(K\pi\pi)^{\pm}$ systems with masses in the resonance region give evidence that the D mesons are produced in association with systems whose masses are greater than $\sim$1870 MeV/c^2. Figures 50 and 51 show those recoil-mass spectra for the D^0 and $D^{\pm}$ peaks, respectively, both with estimated background indicated. All three spectra show evidence of a recoil system with well-defined mass near 2000 MeV/c^2. A second peak near 2200 MeV/c^2 appears in the spectra of masses accompanying the neutral $K\pi$ and $K3\pi$ combinations. Since there is little signal above background near 1870 MeV/c^2 in any of the recoil-mass distributions, $\bar{D}D$-associated production must be a minor contributing process. Although the data are hardly conclusive, the results suggest that some D^0's and $D^{\pm}$ whose decays are observed are produced in the cascade decay of a heavier, narrow-width object.

It had previously been suggested (90, 91) that the $^3S_1$ vector charmed meson D^* is only some 100 MeV/c^2 heavier than the $^1S_0$ pseudoscalar, D, charmed meson. It is natural then to interpret the observations as indicating that major sources of the observed D mesons are the cascade decay of D^* mesons produced in the associated production processes $e^+e^- \to D\bar{D}^*, \bar{D}^*D$ and $e^+e^- \to D^*\bar{D}^*$. Comparable rates for these two production reactions is implied by the data. In addition to these results, which may be said to be direct observation of associated production, the LBL-SLAC group has searched for such $(K\pi)^0$, $(K3\pi)^0$ and $(K\pi\pi)^\pm$ mass peaks among the debris of e^+e^- annihilations at the ψ resonant energies, 3095 and 3684 MeV. Some 150,000 ψ and 300,000 ψ' decays were examined. Of those, some 72,000 are in fact decays via second-order electromagnetic decays of $\psi(3095)$. Since no statistically significant peaks were found near 1870 MeV/c^2 in the mass spectra of these states, it may be taken as established that the threshold energy is more than 3095 MeV for production of D mesons in e^+e^- annihilation via the one-photon intermediate state. Cross-section branching-ratio products were estimated to be 0.20 ± 0.05 nb for $K\pi$, 0.7 ± 0.1 nb for $K3\pi$, and 0.3 ± 0.15 nb for charged-$K\pi\pi$ modes. Allowing for numerous alternative decay modes, comparison with the total annihilation cross section at 4.03 GeV, $\sigma \simeq 30$ nb, makes it clear that D-meson production accounts for a substantial fraction of all annihilations at this energy.

These LBL-SLAC results indicate the existence of an isospin multiplet of heavy mesons of mass ~ 1870 MeV/c^2, which are produced, not in pairs, but very likely as decay products of somewhat more massive particles produced in pairs, or in association with the heavier particles. The mesons have, most likely, small widths as measured against typical or expected hadronic widths of the order 100 MeV/c^2. The characteristics of the charged decays show extreme departures from ordinary hadronic-decay norms. Only the exotic $K^\pm\pi^\mp\pi^\mp$-particle combinations occur; the isospin of the source, if conserved, is at least $\frac{3}{2}$, but the isospin siblings with charge 2 or more do not appear. While this behavior would be exceptionally aberrant in ordinary hadrons, it precisely conforms to the rules for decays of charmed mesons

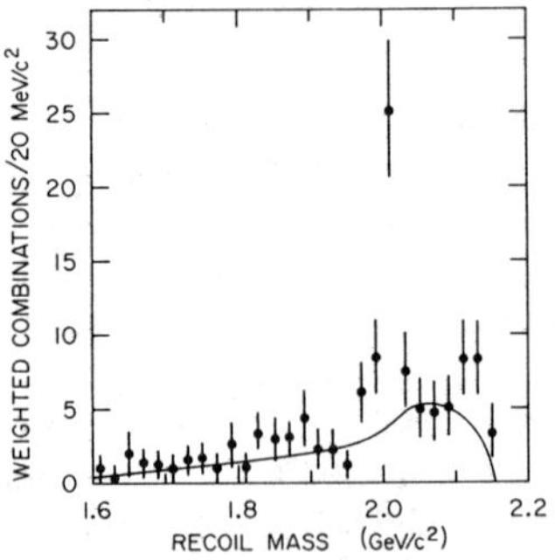

Figure 51 Spectrum of masses accompanying $K^\pm\pi^\mp\pi^\mp$ exotic combinations with mass in the peak near 1875 MeV/c^2 seen in Figure 49. The smooth curve is a background estimate based on the shape of the recoil-mass spectrum with nonexotic $K^\pm\pi^+\pi^-$ combinations; from SLAC-LBL (105).

in the Glashow-Iliopoulos-Maiani form of weak interaction theory with a charmed quark included among the four fundamental hadronic components of nature. Specifically,

$$D^0(c\bar{u}) \rightarrow K^-(s\bar{u}) + \pi^+(u\bar{d})$$

$$D^0(c\bar{u}) \rightarrow K^-(s\bar{u}) + \pi^+(u\bar{d}) + \pi^-(\bar{u}d) + \pi^+(u\bar{d})$$

$$D^+(c\bar{d}) \rightarrow K^-(s\bar{u}) + \pi^+(u\bar{d}) + \pi^+(u\bar{d})$$

are favored decays, with rates proportional to $\cos^4 \theta_c$, as are the corresponding antiparticle decays.

8.1.1 PARITY VIOLATION These already strong indications of the existence of the charmed quark are greatly reinforced by the observation that parity is not conserved in the decay of the charged D mesons. Unless very fundamental theoretical bases are to be overturned, parity violation is to be accepted as proof that decays of D mesons to hadrons proceed via weak interaction, violating the charm-conservation rule of the strong interaction. Wiss et al (107) have presented a discussion reminiscent of the τ-θ puzzle of 20 years ago, whose resolution came

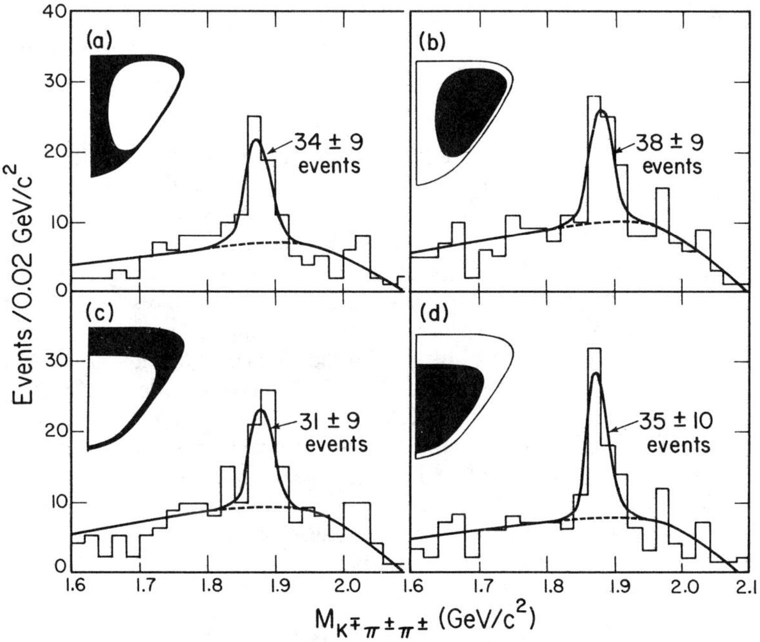

Figure 52 Invariant masses of $K^{\pm}\pi^{\mp}\pi^{\mp}$ combinations in multihadron states from e^+e^- annihilations at 3.9–4.25 GeV. Events restricted to those with recoil mass in the range 1.96–2.04 GeV/c^2. Distributions are for events that populate the dark regions of the Dalitz plots. Regions in (*a*) and (*b*) [(*c*) and (*d*)] are separated by contours of constant amplitude for decay of a 1^- [2^+] particle; from LBL-SLAC (107).

with the recognition of parity violation in the weak interaction. A sample of $K\pi\pi$ decays of $D^{\pm}$ was chosen from events produced in $e^{\pm}$ annihilations at c.m. energy between 3.9 and 4.25 GeV, restricted to those $K\pi\pi$ combinations accompanied by missing mass between 1.96 and 2.04 GeV, a procedure that produced 70 events in a narrow peak above a background of some 50 events. A Dalitz plot for those $K\pi\pi$ combinations with mass between 1860 and 1920 MeV/c^2 has density indistinguishable from uniform. Figure 52 shows the data divided into Dalitz-plot regions chosen to be equally populated by decays of 0^- mesons, but populated in the ratio $\sim 1/8$ for spin-parity 1^- and $\sim 1/6$ for spin-parity 2^+ meson decays. In both cases, the observed population ratio is unity to within errors; $J^P = 1^-$ is excluded to a confidence level 2×10^{-5}, 2^+ is ruled out with a confidence level of 0.002. Since three pseudoscalars cannot be in a zero-spin, even-parity state, only 0^- is a possible $D^{\pm}$ spin-parity assignment, unless the spin is more than two. However, the $K\pi$ state has natural spin parity. Presuming $D^{\pm}$ and D^0 to be members of an isospin multiplet and so to have the same parity, the observations show that the decay proceeds with comparable rates to states of opposite parity, and it follows that the decay interaction is parity-violating, another property of charmed mesons.

Presented with this assemblage of properties congruent to those of the predicted particles, it makes sense to end the reticence maintained by the experimentalists and recognize the discovery of charmed mesons.

8.2 *Baryons*

One other member of the charmed-particle family has been established with reasonable certainty, a charmed baryon. Knapp et al (106) have reported observation of a sharp peak in the mass spectrum of $\bar{\Lambda}\pi^+\pi^-\pi^-$ produced by photons interacting with beryllium. Some details of the experimental arrangement have been presented above in the discussion of ψ production by this group at Fermilab. Identification of Λ^0 and $\bar{\Lambda}^0$ decays is beyond question, as shown in the mass plots of Figure 53. Evidence of a negatively-charged narrow state decaying to $\bar{\Lambda}\pi^+\pi^-\pi^-$ is shown in the invariant-mass plots of Figure 54. Also shown there is the nonexistence of a positively charged partner, $\bar{\Lambda}\pi^+\pi^+\pi^-$. The state's mass was determined to be 2260 ± 10 MeV/c^2 and the measured full width of the peak is (40 ± 20) MeV/c^2. That width is consistent with the estimated experimental mass resolution of a zero-width state. Some evidence was found for a cascade decay of a heavier particle of mass ~ 2500 MeV/c^2 to the $\Lambda(3\pi)^-$ and a positive pion, as shown in Figure 55. There is a clear peak in the distribution of the difference in mass between the $\bar{\Lambda}(4\pi)^0$ combinations and those $\Lambda(3\pi)^-$ combinations with mass near the 2260 peak value. Those data indicate a cascade process

$$\bar{\Lambda}\pi^-\pi^-\pi^+\pi^+(\sim 2.5\ \text{GeV}/c^2) \rightarrow \bar{\Lambda}\pi^-\pi^-\pi^+(2.26\ \text{GeV}/c^2) + \pi^+.$$

A charmed antibaryon with charge -1 has quark components $\bar{c}\bar{u}\bar{d}$, i.e. one charmed antiquark and two uncharmed antiquarks. It is impossible to construct a charmed, positively charged antibaryon with a charmed antiquark and two ordinary antiquarks. It is natural to identify and classify the particle decaying to the

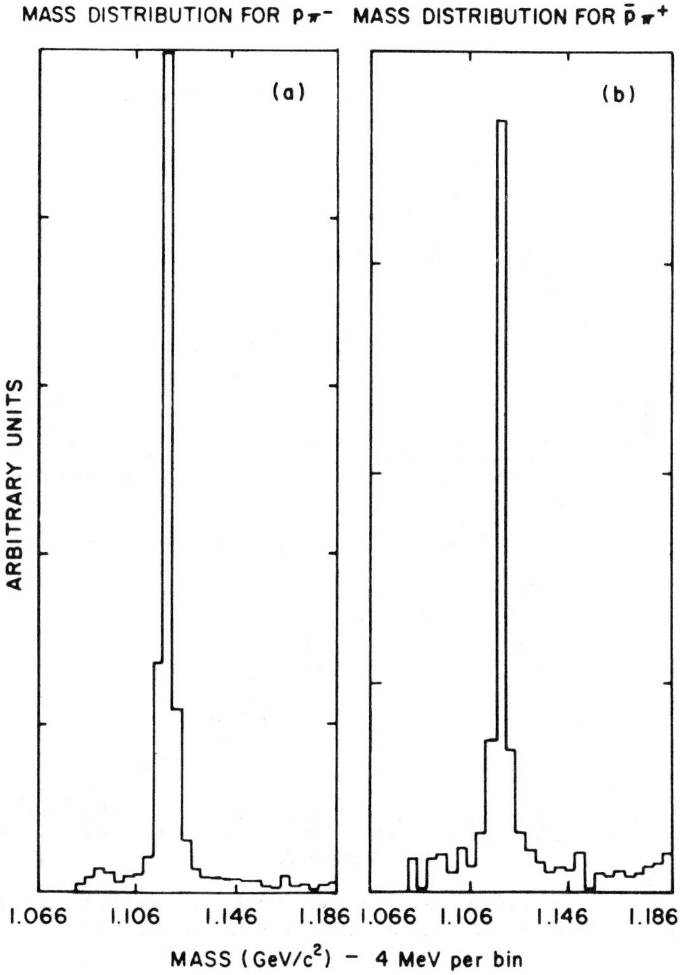

Figure 53 Invariant mass distributions of (*a*) $p\pi^-$ and (*b*) $\bar{p}\pi^+$ V^0's in multihadron states photoproduced at $\sim$250 GeV; from Knapp et al (106) at Fermilab.

$\bar{\Lambda}\pi^+\pi^-\pi^-$ system at 2260 MeV/c^2 as the charmed antibaryon $\bar{\Lambda}_c(\bar{c}\bar{u}\bar{d})$. Its higher-mass parent is then recognized as either $\bar{\Sigma}_c^0(\bar{c}\bar{d}\bar{d})$, a spin-$\frac{1}{2}$, $I = 1$ charmed baryon or $\bar{\Sigma}_c^{*0}(\bar{c}\bar{d}\bar{d})$ of spin $\frac{3}{2}$, isospin 1.

A single example of just such a production-decay sequence had previously been found among events initiated by neutrino interactions in a hydrogen bubble chamber (108). The event, shown in Figure 56, is interpreted as an example of the reaction $\nu p \rightarrow \mu^-\Lambda^0\pi^+\pi^+\pi^+\pi^-$. With those assignments of particle identities, the mass of the $(\Lambda 4\pi)^{2+}$ system is calculated to be 2426 ± 12 MeV/c^2 and the three possible

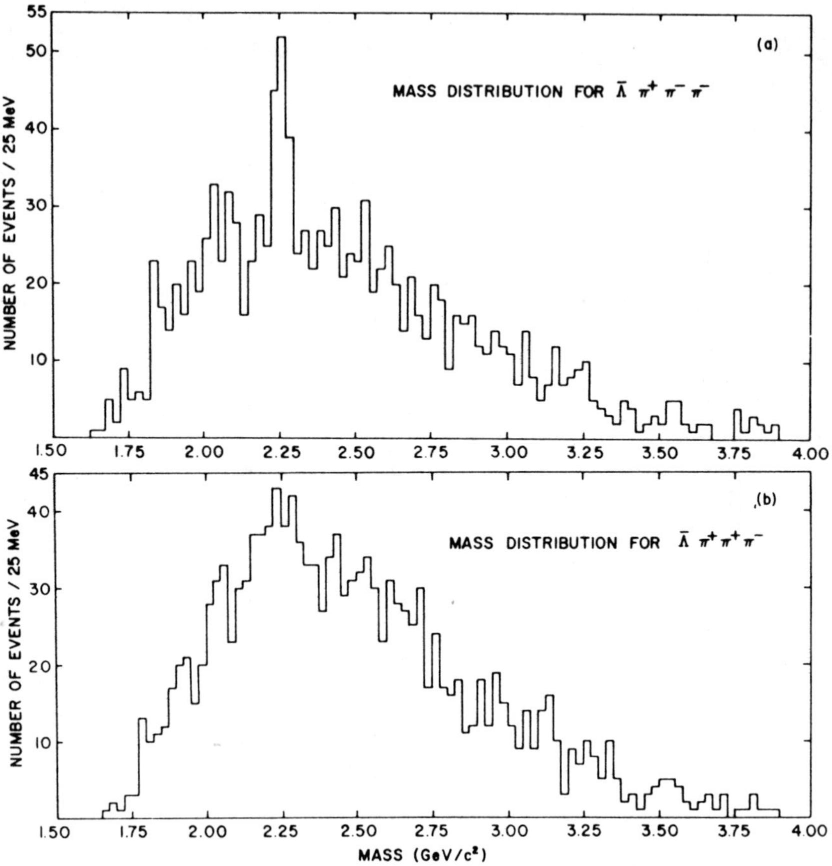

Figure 54 Distribution of invariant masses for (*a*) $\bar{\Lambda}(3\pi)^-$ and (*b*) $\bar{\Lambda}(3\pi)^+$ combinations in multihadron states photoproduced at ~250 GeV; from Knapp et al (106).

$(\Lambda\pi^+\pi^+\pi^-)$ combinations are found to be smaller in mass by 166 ± 15 MeV/c^2, 338 ± 12 MeV/c^2, and 327 ± 12 MeV/c^2. Noting that this reaction violates the selection rule $\Delta S = \Delta Q$ for ordinary hadrons, it was suggested that this event is an example of the sequential reactions, beginning with weak production of a charmed baryon

$$\nu p \rightarrow \Sigma_c^{2+}(cuu)\mu^- [\text{or } \Sigma_c^{*2+}(cuu)\mu^-],$$

Figure 55 Evidence for the cascade process $\bar{\Lambda}(4\pi)^0 \rightarrow \bar{\Lambda}(3\pi)^-\pi^+$ in multihadron states photoproduced at 250 GeV. (*a*) $\bar{\Lambda}(4\pi)^0$ invariant-mass spectrum; (*b*) $\bar{\Lambda}(3\pi)^-$ invariant-mass spectrum; (*c*) difference in mass between $\bar{\Lambda}(4\pi)^0$ and $\bar{\Lambda}(3\pi)^-$ combinations shown as the solid histogram for $\bar{\Lambda}(3\pi)^-$ masses in the peak at ~2.25 GeV/c^2 and as the dashed histogram for $\bar{\Lambda}(3\pi)^-$ just outside the peak; from Knapp et al (106). →

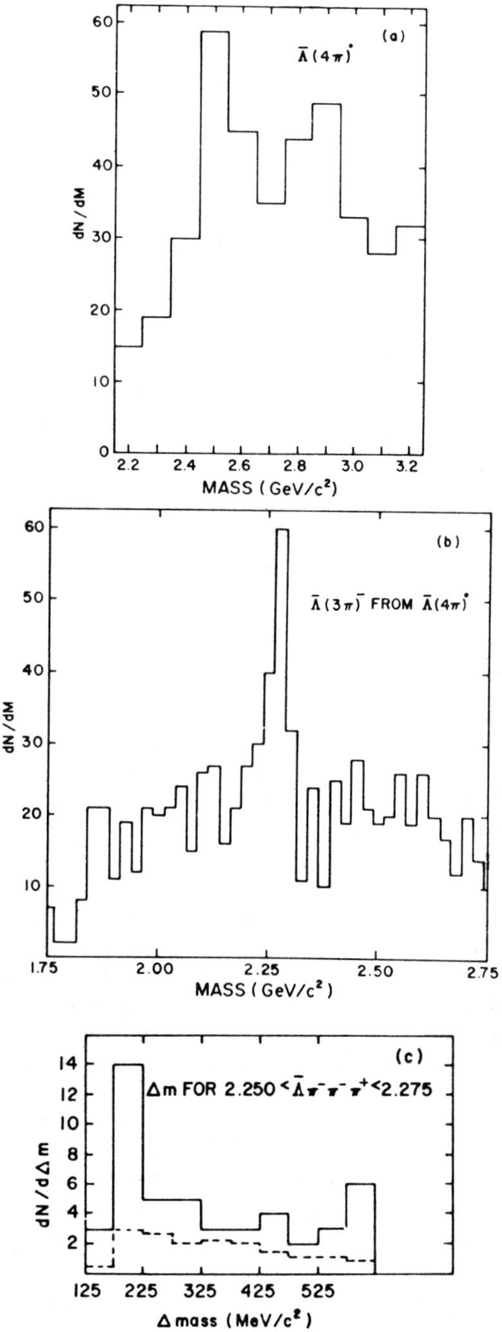
(a)
$\bar{\Lambda}(4\pi)^0$
dN / dM
MASS (GeV/c²)
(b)
$\bar{\Lambda}(3\pi)^-$ FROM $\bar{\Lambda}(4\pi)^0$
dN / dM
MASS (GeV/c²)
(c)
Δm FOR 2.250 < $\bar{\Lambda}\pi^-\pi^-\pi^+$ < 2.275
dN / dΔm
Δ mass (MeV/c²)

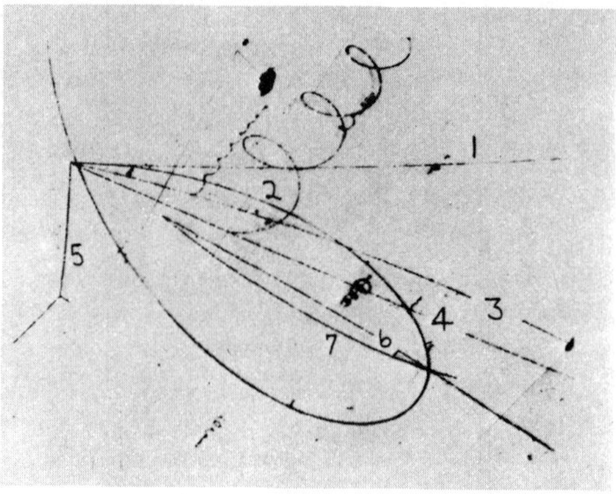

Figure 56 View of an event interpreted as an example of $\nu p \to \mu^- \Lambda^0 \pi^+ \pi^+ \pi^+ \pi^-$ produced in the BNL liquid-hydrogen bubble chamber. The Λ^0 is particularly conspicuous as tracks 6 and 7; observation of Cazzoli et al (108) at Brookhaven National Laboratory.

followed by strong decay

$$\Sigma_c^{2+} \text{ (or } \Sigma_c^{*2+}) \to \Lambda_c^+ (cud) \pi^+,$$

and the succeeding, favored weak decay

$$\Lambda_c^+ \to \Lambda^0 \pi^+ \pi^+ \pi^-.$$

The notation here is the same as above, indicating quark contents and spins. Particularly noteworthy are the remarkable coincidences between the masses of these objects and those predicted by De Rujula et al (91), i.e. 2250 MeV/c^2 for Λ_c, 2410 MeV/c^2 for Σ_c, and 2480 MeV/c^2 for Σ_c^*.

8.3 *Miscellany*

Other observations have provided more indirect evidence of the presence of charmed-quark components of hadrons. Events with two muons produced in high-energy neutrino-nucleon interactions have been interpreted in terms of production of new particles and their subsequent weak leptonic or semileptonic decays (109, 110). From the measured characteristics of the events, Benvenuti et al (109) deduce that the new particle, if a hadron, has mass between ~ 2 and ~ 4 GeV/c^2 and has a lifetime less than 10^{-10} sec.

Events with a negative muon, positron, and neutral strange particle in the final state have been observed as products of neutrino-initiated reactions in bubble chambers. Von Krogh et al (111) found four examples of the reaction $\nu_\mu N \to \mu^- e^+ K_s^0 X$ in an exposure of the Fermi National Accelerator Laboratory FNAL 15-ft bubble chamber filled with a neon-hydrogen mixture. Blietschau et al (112) have reported observation of three examples of $\nu_\mu N \to \mu^- e^+ V^0 (K^0 \text{ or } \Lambda^0) X$ in the freon-filled bubble chamber Gargamelle exposed to a neutrino beam at CERN.

One may surmise (both groups have) that these events show cascade production and semileptonic decay of charmed hadrons of mass about 2000 MeV/c^2.

Semileptonic decays of charmed hadrons have been invoked to explain the single-electron-plus-hadrons final states made in e^+e^- annihilations at c.m. energies between 4.0 and 4.2 GeV and detected in the DASP apparatus (113). For improved electron detection in this experiment, the apparatus of Figure 15 was supplemented with threshold Cerenkov counters on each side of the intersection region at DORIS. Of 87 electron+hadron events, 28 survived cuts on shower-counter pulse height, electron momentum, and vertex position. Of these, 22 had hadron multiplicity of four or greater. Calculated rates for conventional sources of these events are too low to account for any but a small fraction of those detected. In data taken at c.m. energies less than 3.7 GeV, events of this type did not occur with greater than calculated background rates. The electron-momentum distribution reproduced in Figure 57 disagrees with that expected of electron-decay products of a heavy lepton of mass 1.9 GeV/c^2, a conceivable source of the events. The posited new hadrons whose semileptonic decays are presumed to have been observed are to have mass between 1.8 and 2.1 GeV/c^2 and to be supplied with a new quantum number (charm, perhaps) conserved in strong and electromagnetic processes.

Rates for e^+e^- annihilations to states with a neutral K meson and an electron have been measured as a function of c.m. energy by Burmester et al (114) with the magnetic detector PLUTO at DORIS. That apparatus consists of fourteen cylin-

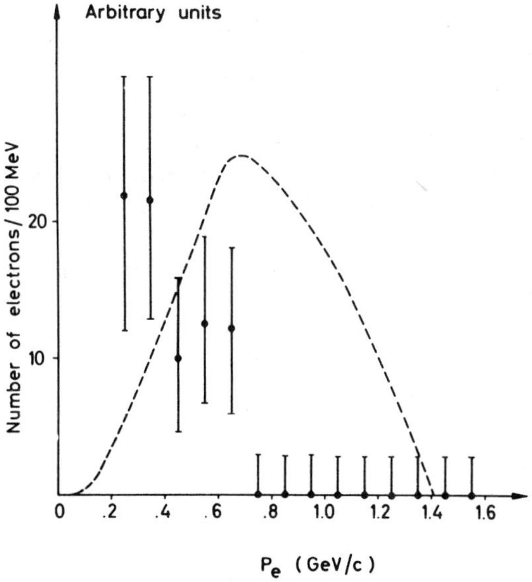

Figure 57 Inclusive momentum distribution of electrons in multiparticle e^+e^- annihilations at energies of 4.0–4.2 GeV. Events with multiplicity of four or greater are included. The dashed smooth curve shows a calculated distribution of electrons from decay of a 1.9-GeV/c^2 heavy lepton; from DASP (113).

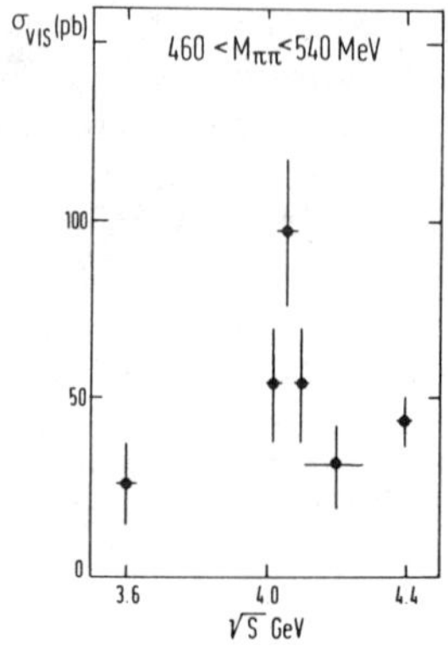

Figure 58 Energy dependence of the yield of detected correlated $K_S^0 e^{\pm}$ combinations in multiparticle states made in e^+e^- annihilations; from PLUTO (114) at DORIS.

drical proportional wire chambers arrayed coaxially about the direction of the intersecting beams. Two cylindrical lead converters, 0.44 and 1.7 radiation lengths thick, with following wire chambers provide for detection of electron-initiated electromagnetic showers. A super conducting coil surrounding the track-sampling chambers produces a 20,000-G magnetic field directed parallel to the axis of the detector array. Clear evidence for K^0 production was found in distributions of invariant mass of oppositely charged tracks in multiparticle events. Figure 58 shows the observed energy dependence of the detected yield of $e^{\pm} K_s^0(\rightarrow \pi^+\pi^-)$ events, the K_s^0's having been chosen as $\pi^+\pi^-$ pairs with mass in the observed peaks between 460 and 540 MeV/c^2. After background subtraction and correction for deficiencies, Burmester et al arrive at an estimated inclusive cross section at the maximum, $\sigma(e^+e^- \rightarrow e^{\pm} K^0 + \text{anything}) \simeq 3$ nb, accurate to about a factor of 2.

Table 13 Properties of observed narrow states with classifications as charmed hadrons

State	Mass (MeV/c^2)	Decay modes	Ref.	Footnote(s)
$D^0, \bar{D}^0$	1865 ± 15	$K^{\pm}\pi^{\mp}, K^{\pm}\pi^{\mp}\pi^+\pi^-$	104	a
$D^{\pm}$	1876 ± 15	$K^{\pm}\pi^{\mp}\pi^{\mp}$	105	a
$\bar{\Lambda}_c^-$	2260 ± 10	$\bar{\Lambda}\pi^-\pi^-\pi^+$	106	—
Λ_c^+	2260 ± 15	$\Lambda\pi^+\pi^+\pi^-$	108	b, c
$\bar{\Sigma}_c^0$	~ 2500	$\bar{\Lambda}_c^-\pi^+$	106	d
Σ_c^{2+}	2426 ± 12	$\Lambda_c^+\pi^+$	108	b, d

[a] Presuming these to be isospin partners, parity violation in their decays has been demonstrated (107).
[b] One event with a cascade decay.
[c] Mass not uniquely determined. One of three possible $\Lambda\pi^+\pi^+\pi^-$ from decay of the Σ_c^{2+} (see text for others).
[d] Not uniquely classified. May be excited state Σ_c^*.

These data are proffered as evidence for the presence of charmed mesons whose semileptonic decays, $D^{\pm} \to K^0 e^{\pm} \nu$ have been detected. Boldly accepting that interpretation, and presuming ~ 10 nb (of the 30 nb total) to be the cross section for production of charged D-meson pairs in e^+e^- annihilations at 4.05 GeV, we may make the rough estimate $\sim(3/2)/10 = 15\%$, for the $Ke\nu$ decay branching ratio. That value is not incompatible with crude theoretical estimates (90) of charmed-meson decay rates.

8.4 *Summary*

New mesons and baryons of long lifetime have been observed. Their currently established masses, modes of production, and decay characteristics, as listed in Table 13, correspond without exception to those of charmed hadrons manifesting a basic structure that includes the fourth, charmed quark.

9 SUMMARY AND CONCLUSIONS

In somewhat less than two years of intense experimental activity, at least 13 new unstable hadrons have been discovered. They fall into two separate but related classes. One, the psion family, includes, first, the new vector mesons $\psi(3095)$, $\psi(3684)$, and $\psi(4414)$; second, the intermediate states $\chi(3415)$, $\chi(3500)$, and $\chi(3550)$; and third, the less well-established $\chi(3455)$ and $X(2800)$. The more recent assembly of new states includes the neutral and charged D mesons, D^0 (and presumably $\bar{D}^0$) at 1865 MeV/c^2 and D^+, D^- at 1876 MeV/c^2, together with the first new baryons, Λ_c at 2260 MeV/c^2 and Σ_c at 2450 MeV/c^2. All the characteristics of these new particles, their quantum numbers, the pattern of masses, and the variety of decay modes and widths, are in close correspondence with those of states that occur in a four-quark model of hadronic substructure.

The body of experimental evidence presented in this review strongly supports the notion that this small number of fundamental objects is responsible for the structure and interactions of the hadrons; however, there are gaps and quantitative discrepancies. Charmonium spectroscopy is still in a relatively primitive state; the pseudoscalar $^1S_0$ states have only a tenuous connection with reality; properties of the intermediate states are not sufficiently well-determined to permit definitive identification as ^3P levels; radiative transition rates are poorly known and not in good agreement with theoretical values; the structure of the higher radially excited ^3S states is very confused and their decay rates, particularly to states with charmed mesons, are essentially unknown. Only a small fraction of the charmed hadrons required by the model has yet been discovered. One may, in fact, also still feel uneasy about the value of R for e^+e^- annihilations at energies above the charmed-quark threshold.

The new discoveries have provided some basis for optimism that a correct basic theoretical framework has been invented. It remains to be seen, and no doubt will be seen in experiments in the near future, whether that structure is complete, or, at least, needs expansion to include still more, heavier quarks.

ACKNOWLEDGMENT

Much of what I know of the subject of this review has been learned from my colleagues, past and present, in tne SLAC-LBL collaboration. They are: G. S. Abrams, M. S. Alam, J.-E. Augustin, A. M. Boyarski, M. Breidenbach, D. D. Briggs, F. Bulos, W. C. Carithers, S. Cooper, J. T. Dakin, R. G. DeVoe, J. M. Dorfan, G. E. Fischer, C. E. Friedberg, D. Fryberger, G. Goldhaber, D. Hitlin, G. Hanson, R. J. Hollebeek, B. Jean-Marie, J. Jaros, A. D. Johnson, J. A. Kadyk, R. R. Larsen, A. M. Litke, D. Lüke, B. A. Lulu, V. Lüth, H. L. Lynch, D. Lyon, R. J. Madaras, C. C. Morehouse, H. K. Nguyen, J. M. Paterson, M. L. Perl, F. M. Pierre, I. Peruzzi, M. Piccolo, T. Pun, P. Rapidis, B. Richter, B. Sadoulet, R. Schindler, R. F. Schwitters, J. Siegrist, W. M. Tanenbaum, G. H. Trilling, F. Vannucci, J. S. Whitaker, F. C. Winkelmann, J. Weiss, J. E. Wiss, and J. E. Zipse. I am greatly indebted to them for their efforts.

Particular thanks go to S. Cooper, M. Mandelkern, S. Shannon, M. Suzuki, and J. E. Wiss for corrections and clarifications that resulted from their careful reading of the text.

I am grateful for the hospitality and support of the Conselho Nacional de Pesquisas and the Universidade do Brasilia, Brazil where part of this review was prepared while I was a visiting professor in the Department of Physics.

Literature Cited

1. Particle Data Group, Trippe, T. G. et al 1976. Review of particle properties. *Rev. Mod. Phys.* 48:S1
2. Bjorken, J. D., Glashow, S. L. 1965. *Phys. Lett.* 11:225
3. Glashow, S. L., Iliopoulos, J., Maiani, L. 1970. *Phys. Rev. D* 2:1285
4. Carlson, C. E., Freund, P. G. O. 1972. *Phys. Lett. B* 39:349
5. Appelquist, T., Politzer, H. D. 1975. *Phys. Rev. Lett.* 34:43
6. Schwitters, R. F., Strauch, K. 1976. The physics of e^+e^- collisions. *Ann. Rev. Nucl. Sci.* 26:89
7. Cosme, G. et al 1974. *Phys. Lett. B* 48:159
8. Sakurai, J. J. 1969. *Currents and Mesons.* Chicago: Univ. Chicago Press
9. Nambu, Y., Sakurai, J. J. 1962. *Phys. Rev. Lett.* 8:79
10. High-Energy Reactions Analysis Group, Bracci, E. et al 1972. Compilation of cross sections $I-\pi^-$ and π^+ induced reactions. *CERN/HERA 72-1*, CERN, Geneva, Switzerland
11. Whitmore, J. 1976. *Phys. Rep. C* 27:188
12. Blobel, V. et al 1975. *Phys. Lett. B* 59:88
13. Anderson, K. J. et al 1976. *Phys. Rev. Lett.* 37:799
14. Silverman, A. 1975. *Proc. 1975 Int. Symp. Lepton Photon Interactions High Energies*, Stanford, Calif., p. 355
15. Ross, M., Stodolsky, L. 1966. *Phys. Rev.* 149:1172
16. Anderson, R. et al 1970. *Phys. Rev. D* 1:27
17. Behrend, H. J. et al 1975. *Phys. Lett. B* 56:408
18. Okubo, S. 1963. *Phys. Lett.* 5:165
19. Zweig, G. 1964. *CERN TH401, CERN TH412*, CERN, Geneva, Switzerland
20. Iizuka, J. 1966. *Suppl. Prog. Theor. Phys.* 37–38:21
21. Rosenzweig, C. 1976. *Phys. Rev. D* 13:3080
22. Appelquist, T., Politzer, H. D. 1975. *Phys. Rev. D* 12:1404
23. Aubert, J. J. et al 1974. *Phys. Rev. Lett.* 33:1404
24. Augustin, J.-E. et al 1974. *Phys. Rev. Lett.* 33:1406
25. Abrams, G. S. et al 1974. *Phys. Rev. Lett.* 33:1453
26. Bacci, C. et al 1974. *Phys. Rev. Lett.* 33:1408
27. Criegee, L. et al 1975. *Phys. Lett. B* 53:489
28. Becker, U. 1975. *New Directions in Hadron Spectroscopy, Proc. Summer*

Symp. Argonne Natl. Lab., July 7–10, 1975. Argonne, Ill., p. 209
29. Ting, S. C. C. 1975. See Ref. 14, p. 155
30. Aubert, J. J. et al 1975. *Nucl. Phys. B* 89:1
31. Knapp, B. et al 1975. *Phys. Rev. Lett.* 34:1044
32. Binkley, M. et al 1976. *Phys. Rev. Lett.* 37:571
33. Binkley, M. et al 1976. *Phys. Rev. Lett.* 37:574
34. Anderson, K. J. et al 1976. *Phys. Rev. Lett.* 36:237
35. Anderson, K. J. et al 1976. *Phys. Rev. Lett.* 37:799
36. Blanar, G. J. et al 1975. *Phys. Rev. Lett.* 35:346
37. Antipov, Y. M. et al 1976. *Phys. Lett. B* 60:309
38. Hom, D. C. et al 1976. *Phys. Rev. Lett.* 36:1236
39. Snyder, H. D. et al 1976. *Phys. Rev. Lett.* 36:1415
40. Hom, D. C. et al 1976. *Phys. Rev. Lett.* 37:1374
41. Büsser, F. W. et al 1975. *Phys. Lett. B* 56:482
42. Nagy, E. et al 1975. *Phys. Lett. B* 60:96
43. Boyarski, A. M. et al. 1975. *Phys. Rev. Lett.* 34:1357
44. Knapp, B. et al 1975. *Phys. Rev. Lett.* 34:1040
45. Camerini, U. et al 1975. *Phys. Rev. Lett.* 35:483
46. Gittelman, B. et al 1975. *Phys. Rev. Lett.* 35:1616
47. Nash, T. et al 1976. *Phys. Rev. Lett.* 36:1233
48. Ritson, D. M. 1976. *Particle Searches and Discoveries—1976, Vanderbilt. AIP Conf. Proc. No. 30, Part. Fields Subser. No. 11*, p. 75
49. Augustin, J.-E. et al 1975. *Phys. Rev. Lett.* 34:233
50. Wiik, B. H. 1975. See Ref. 14, p. 69
51. Lüth, V. et al 1975. *Phys. Rev. Lett.* 35:1124
52. Augustin, J.-E. et al 1975. *Phys. Rev. Lett.* 34:764
53. Jackson, J. D., Scharre, D. L. 1975. *Nucl. Instrum. Methods* 128:13
54. Braunschweig, W. et al 1976. *Phys. Lett B* 63:487
55. Jean-Marie, B. et al 1976. *Phys. Rev. Lett.* 36:291
56. Abrams, G. S. 1975. See Ref. 14, p. 25
57. Vannucci, F. et al 1977. *Phys. Rev. D* 15:1814
58. Gilman, F. J. 1975. *High Energy Physics and Nuclear Structure, Los Alamos, 9–13 June 1975. AIP Conf. Proc. No. 26*, p. 331
59. Gupta, V., Kögerler, R. 1975. *Phys. Lett. B* 56:473
60. Abrams, G. S. et al 1975. *Phys. Rev. Lett.* 34:1181
61. Tanenbaum, W. et al 1976. *Phys. Rev. Lett.* 36:402
62. Hilger, E. et al 1975. *Phys. Rev. Lett.* 35:625
63. Tanenbaum, W. et al 1975. *Phys. Rev. Lett.* 35:1323
64. Braunschweig. W. et al 1975. *Phys. Lett. B* 57:407
65. Whitaker, J. S. et al 1976. *Phys. Rev. Lett.* 37:1596
66. Feldman, G. et al 1975. *Phys. Rev. Lett.* 35:821
67. Trilling, G. H. 1976. Lawrence Berkeley Lab. Rep. LBL-5535
68. Biddick, C. J. et al 1977. *Phys. Rev. Lett.* 38:1324
69. Hughes, E. B. et al 1976. *Phys. Rev. Lett.* 36:76
70. Brown, L. S., Cahn, R. N. 1976. *Phys. Rev. D* 13:1195
71. Karl, G. et al 1976. *Phys. Rev. D* 13:1203
72. Kabir, P. K., Hey, A. J. G. 1976. *Phys. Rev. D* 13:3161
73. Heintze, J. 1975. See Ref. 14, p. 97
74. Bartel, W. et al 1976. *DESY 76/65*, DESY, Hamburg, Germany
75. Boyarski, A. M. et al 1975. *Phys. Rev. Lett.* 34:762
76. Schwitters, R. F. 1975. See Ref. 14, p. 5
77. Esposito, B. et al 1975. *Phys. Lett. B* 58:478
78. Bacci, C. et al 1975. *Phys. Lett. B* 58:481
79. Bacci, C. et al 1976. *Phys. Lett. B* 64:356
80. Barbiellini, G. et al 1976. *Phys. Lett. B* 64:359
81. Esposito, B. et al 1976. *Phys. Lett. B* 64:362
82. Tarjanne, P., Teplitz, V. L. 1963. *Phys. Rev. Lett.* 11:447
83. Amati, D. et al 1964. *Phys. Lett.* 11:190
84. Hara, Y. 1964. *Phys. Rev. B* 134:701
85. Maki, Z. et al 1964. *Prog. Theor. Phys. Kyoto* 32:144
86. Greenberg, O. W. 1964. *Phys. Rev. Lett.* 13:598
87. Han, M. Y., Nambu, Y. 1965. *Phys. Rev. B* 139:1006
88. Bars, I., Peccei, R. D. 1975. *Phys. Rev. D* 12:823
89. Feldman, G., Matthews, P. T. 1976. *Nuovo Cimento A* 31:447
90. Gaillard, M. K., Lee, B. W., Rosner, J. L. 1975. *Rev. Mod. Phys.* 47:277

91. De Rujula, A., Georgi, H., Glashow, S. L. 1975. *Phys. Rev. D* 12:147
92. Chanowitz, M., Gilman, F. J. 1976. *Phys. Lett. B* 63:178
93. Bethe, H. A., Salpeter, E. 1957. *Quantum Mechanics of One and Two Electron Atoms.* Berlin: Springer
94. Eichten, E. et al 1975. *Phys. Rev. Lett.* 34:369
95. Pumplin, J. et al 1975. *Phys. Rev. Lett.* 35:1538
96. Schnitzer, H. J. 1975. *Phys. Rev. Lett.* 35:1540
97. Harrington, B. J. et al 1975. *Phys. Rev. Lett.* 34:168
98. Eichten, E. et al 1976. *Phys. Rev. Lett.* 36:500
99. Barbieri, R. et al 1976. *Phys. Lett. B* 60:183
100. Barbieri, R. et al 1976. *Phys. Lett. B* 61:465
101. Siegrist, J. et al 1976. *Phys. Rev. Lett.* 36:700
102. Cester, R. et al 1976. *Phys. Rev. Lett.* 37:1178
103. Boyarski, A. M. et al 1975. *Phys. Rev. Lett.* 35:196
104. Goldhaber, G. et al 1976. *Phys. Rev. Lett.* 37:255
105. Peruzzi, I. et al 1976. *Phys. Rev. Lett.* 37:569
106. Knapp, B. et al 1976. *Phys. Rev. Lett.* 37:882
107. Wiss, J. et al 1976. *Phys. Rev. Lett.* 37:1531
108. Cazzoli, E. G. et al 1975. *Phys. Rev. Lett.* 34:1125
109. Benvenuti, A. et al 1975. *Phys. Rev. Lett.* 34:419
110. Barish, B. C. et al 1976. *Phys. Rev. Lett.* 36:939
111. Von Krogh, J. et al 1976. *Phys. Rev. Lett.* 36:710
112. Blietschau, J. et al 1976. *Phys. Lett. B* 60:207
113. Braunschweig, W. et al 1976. *Phys. Lett. B* 63:471
114. Burmester, J. et al 1976. *Phys. Lett. B* 64:369

Ann. Rev. Nucl. Part. Sci. 1978. 28: 387–499

3

CHARM AND BEYOND*

Thomas Appelquist[1]
J. Willard Gibbs Laboratory, Yale University, New Haven, Connecticut 06520

R. Michael Barnett[2]
Stanford Linear Accelerator Center, Stanford University, Stanford, California 94305

Kenneth Lane[3]
Lyman Laboratory of Physics, Harvard University, Cambridge, Massachusetts 02138

CONTENTS

* The US Government has the right to retain a nonexclusive, royalty-free license in and to any copyright covering this paper.

[1] Alfred P. Sloan Foundation Fellow. Research supported in part by the Department of Energy under contract EY-76-C-02-3075.

[2] Research supported in part by the Department of Energy.

[3] Research supported in part by the National Science Foundation under contract PHY77-22864.

1 INTRODUCTION

The physics of elementary particles has changed dramatically during the 1970s, especially during the last three or four years. A new quark carrying a new quantum number called charm has been discovered and there is mounting evidence for the existence of yet a heavier quark. A great deal has been learned about the structure of the weak interactions, and there is considerable optimism that we are beginning to understand strong interaction dynamics at a fundamental level. Indeed, some visionaries are already attempting grand syntheses of the strong, weak, and electromagnetic interactions. In this paper, we describe these theoretical and experimental developments, emphasizing the role of the new heavy quarks. It is largely a review for nonspecialists but specialists will find some new results or at least some new perspectives.

Hadrons, that is mesons and baryons, are made of quarks (1); after the events of the last three years there are no longer any skeptics. Since many of the details of the quark model of hadrons are discussed by O. W. Greenberg in a paper appearing in this issue of the Annual Review (2), we recall only some of the basic properties of the old quarks to prepare for our discussion of the new ones. Until 1974, all the known mesons and baryons could be understood as quark-antiquark and three-quark bound states respectively, where the quarks come in three varieties or "flavors," commonly called u, d, and s. The u and d (up and down) quarks form an isotopic spin SU(2) doublet and are the constituents of the nucleons and mesons of nuclear physics. The heavier s (strange) quark can bind with the others or with itself to produce the so-called strange particles that fill out the SU(3) multiplets of Gell-Mann (1). The relative heaviness of the strange quark means, of course, that this SU(3) symmetry is rather badly broken. All of these quarks have spin $\frac{1}{2}$ and carry fractional electric charge. The u quark has charge $\frac{2}{3}$ and the d and s quarks have charge $-\frac{1}{3}$.

1.1 *The Need For Charm*

With the success of SU(3) in the early 1960s, many people soon considered the possibility that yet heavier quarks might exist (3). They would presumably be constituents of hadrons too heavy and too short lived to have been seen at the time. It was partly a matter of "why not?" and partly motivated by primitive notions of quark-lepton symmetry. There were four leptons—the electron, the muon, and their respective neutrinos—so why shouldn't there be four quarks? Quark-lepton symmetry continues to be an important guiding principle in attempting to unify the fundamental forces of nature, but it seems likely that it will ultimately take a more subtle form than equal numbers of each.

It was a problem with weak interaction phenomenology that led Glashow, Iliopoulos, & Maiani (GIM), in their classic paper of 1970, to provide a genuine *raison d'être* for a fourth quark (4). It had been known for decades that the weak interactions, such as neutron β decay and $\mu^- \to e^- + \bar{\nu}_e + \nu_\mu$, could all be described by the interaction of two charge-changing currents with an interaction strength $G_F \approx 10^{-5}/M_P^2$. It is now known that neutral currents also play a role in the weak interactions. Processes like $\nu_\mu + p \to \nu_\mu + p$ have been observed and require both vector and axial-vector neutral current interactions with strength of order G_F. In 1970, the neutral weak currents had not been seen experimentally but, for the most part, neither had they been ruled out at the level G_F. It was expected by some people that they would appear at this level and this expectation was given a sound basis with the proof, a year later, of the renormalizability of gauge theories of the weak and electromagnetic interactions (see below).

There was one embarrassing problem with this state of affairs. One class of neutral current interactions, those involving strangeness-changing hadronic weak currents, was known to be tremendously suppressed. The branching ratios (5)

$$\frac{\Gamma(K_L^0 \to \mu^+\mu^-)}{\Gamma(K_L^0 \to \text{all})} \sim 10^{-8} \quad \text{and} \quad \frac{\Gamma(K^\pm \to \pi^\pm e^+ e^-)}{\Gamma(K^\pm \to \text{all})} = (2.6 \pm 0.5) \times 10^{-7} \qquad 1.1$$

exhibit the problem. That a strangeness-changing neutral current, coupling with strength G_F, might have been expected can be seen by examining the structure of the charge-changing hadronic weak current

$$\begin{aligned} J^\mu_\pm &= \bar{q}\gamma^\mu \tfrac{1}{2}(1-\gamma_5) T_\pm q \\ &\equiv \bar{q}_L \gamma^\mu T_\pm q_L. \end{aligned} \qquad 1.2$$

The quark spinor q contains $4 \times N$ components where N is the number of flavors and $T_\pm$ is an $N \times N$ matrix that changes the electric charge by ± 1 unit. Here $q_L \equiv \frac{1}{2}(1-\gamma_5)q$ is the left-handed part of the quark field[4]. It was expected by some and unified theories demanded, that there should exist a neutral partner of these two currents

$$J_0^\mu = \bar{q}\gamma^\mu \tfrac{1}{2}(1-\gamma_5)T_0 q \qquad 1.3$$

coupling with the same strength G_F, where

$$T_0 = [T_+, T_-]. \qquad 1.4$$

The charged current was known to have the Cabibbo form (6)

$$J_+^\mu = \bar{u}\gamma^\mu \tfrac{1}{2}(1-\gamma_5)(d \cos\theta_C + s \sin\theta_C) \qquad 1.5$$

where $\sin\theta_C = 0.23$. Therefore, the neutral current must contain a piece of the form

$$J_0^\mu = \bar{s}\gamma^\mu \tfrac{1}{2}(1-\gamma_5)\, d \sin\theta_C \cos\theta_C + \ldots, \qquad 1.6$$

which gives rise to $\Delta S = 1$ neutral interactions of order G_F.

The GIM solution (4) was to introduce a new, heavy, charge $\frac{2}{3}$ quark c with a left-handed weak coupling to the orthogonal Cabibbo combination $s \cos\theta_C - d \sin\theta_C$. The c quark was postulated to carry a new quantum number charm, conserved by the strong interactions. It can then easily be checked that with

$$J_+^\mu = \bar{u}\gamma^\mu \tfrac{1}{2}(1-\gamma_5)(d \cos\theta_C + s \sin\theta_C) + \bar{c}\gamma^\mu \tfrac{1}{2}(1-\gamma_5)(s \cos\theta_C - d \sin\theta_C), \qquad 1.7$$

the $\Delta S = 1$ piece of J_0^μ in Equation 1.6 is cancelled and the neutral current is, in fact, diagonal in all flavors. This current couples to itself and to leptonic neutral currents with strength G_F and, to lowest order in this interaction, the reactions in Equation 1.1 are forbidden. Higher order corrections might, of course, induce such reactions and because the branching ratios (Expression 1.1) are so small, this must be looked at carefully. It is only possible to do this in a renormalizable theory, so we turn next to a discussion of gauge theories of weak and electromagnetic interactions.

1.2 *The Weinberg-Salam Model*

Unified gauge theories provide the general framework for most modern work on weak and electromagnetic interactions. The prototype for all

[4] The spinor field q_L is not, strictly speaking, left-handed. The operator $\frac{1}{2}(1-\gamma_5)$ projects out the left-handed (negative helicity) part of the Dirac spinor $u(p)$ only in the zero mass limit. Thus for the large momentum ($p \gg m_q$) components of $q_L(x)$, $\frac{1}{2}(1-\gamma_5)$ can be thought of as a covariant version of a left-handed projection operator.

such models is based on the group SU(2) × U(1) and was first written down in detail by Weinberg in 1967 (7). It remains viable today in the face of a large amount of experimental data and will survive as at least a subgroup of the ultimate weak and electromagnetic theory.

Unified gauge theories are constructed by generalizing what is known about electromagnetic interactions to include the weak interactions. The inclusion of strong interactions in this framework is described in the next section. The forces are all mediated by the exchange of spin one bosons corresponding to vector gauge fields that are present to maintain the local gauge invariance of the Lagrangian. In electrodynamics, this means invariance under a space-time-dependent phase transformation $\psi(x) \to e^{iQ\theta(x)}\psi(x)$ on each matter field of charge Q, along with the transformation $A_\mu(x) \to A_\mu(x) + \partial_\mu\theta(x)$ on the electromagnetic field. Invariance is insured if A_μ enters the Lagrangian only in the gauge covariant derivative $D_\mu\psi = (\partial_\mu - iQA_\mu)\psi$ and if derivatives of A_μ enter only through the electromagnetic field tensor $F_{\mu\nu} = \partial_\mu A_\nu - \partial_\nu A_\mu$. By using these ingredients in a minimal way (excluding nonrenormalizable couplings), one obtains the Maxwell-Dirac theory for photons interacting with charged, spin $\frac{1}{2}$ particles of mass m

$$\mathscr{L} = \bar{\psi}[i\gamma^\mu D_\mu - m]\psi - \tfrac{1}{4}F_{\mu\nu}F^{\mu\nu}. \qquad 1.8$$

The phase transformations of electrodynamics form the group U(1), the group of unitary 1 × 1 matrices. Since both neutral and charge-changing currents play a role in the weak interactions, the corresponding group must be larger, but the gauge principle can be carried over by associating a separate gauge field $A_\mu^a(x)$ with each infinitesimal parameter of the group (8). The minimal possibility is SU(2) and, since this is a three-parameter group, there will be three gauge fields forming an isotopic triplet. The Lagrangian will take the form of Equation 1.8 except that now $F_{\mu\nu}$ will contain a term quadratic in A_μ^a and a coupling matrix will multiply the charge Q in $D_\mu\psi$. Such a theory can be proven to be renormalizable[5] (9).

The problem is that the weak bosons must be very massive—they have not yet been seen and the lower limit on the mass is about 25 GeV (10). The addition of a mass term $-\frac{1}{2}M^2 A_\mu^a A^{a\mu}$ to the Lagrangian, however, destroys the local gauge invariance and the renormalizability. The solution is to preserve the local gauge invariance of the Lagrangian, but allow it to be spontaneously broken, that is, not respected by the physical states. There is a general theorem that whenever this happens, massless

[5] This means that the infinities that appear in higher order computations can all be absorbed into the physical masses and coupling constants of the theory. The S matrix can then be computed to any order in terms of these few parameters.

scalar particles, known as Goldstone bosons, must be present (11). However, in gauge theories these quanta get mixed in with the longitudinal parts of the gauge fields. They allow the gauge bosons to become massive by providing the zero helicity degree of freedom forbidden to a massless vector boson. This is called the Higgs mechanism (12). A particularly simple way of realizing this is to introduce into the Lagrangian a multiplet of elementary spinless fields. After spontaneous symmetry breaking, some of these become the Goldstone bosons absorbed by the gauge fields, while the remaining members of the multiplet survive as physical, massive scalar particles known as Higgs bosons.

A unified model of this sort must incorporate both a weak gauge group and the electromagnetic gauge group. The SU(2) × U(1) model involves four gauge bosons. The spontaneous breakdown is arranged to preserve a local U(1) subgroup so that one boson stays massless and is identified with the photon, while the other three ($W^{\pm}$ and Z^0) become the intermediate bosons of the weak interactions. In the original form of the model (7), the left-handed leptons and quarks are put into SU(2) doublets and the right-handed components are in SU(2) singlets. The fermions are massless in the Lagrangian; their mass, along with the weak boson masses, arises from the spontaneous symmetry breakdown. There are two independent coupling constants g and g' for the SU(2) and U(1) groups respectively. The electric charge is given by $e = gg'/(g^2+g'^2)^{\frac{1}{2}}$ and since $W^{\pm}$ exchange describes the familiar process of β decay, the Fermi coupling constant G_F is given by

$$\frac{G_F}{2^{\frac{1}{2}}} = \frac{g^2}{8M_W^2}, \qquad 1.9$$

It follows that

$$M_W = \left(\frac{e^2}{32^{\frac{1}{2}}G_F}\right)^{\frac{1}{2}} \frac{1}{\sin\theta_W} = 37.3\ \text{GeV}/\sin\theta_W,$$

where θ_W is the Weinberg angle defined by $\tan\theta_W = g'/g$. In the simplest version of the theory, there is one complex SU(2) doublet of Higgs fields, and the Z^0 mass is given by $M_Z = M_W/\cos\theta_W$.

In this model, there remains one massive physical Higgs scalar. Its mass is a free parameter theoretically, but its presence in the theory is absolutely crucial for renormalizability. It has by now been proven that such theories are, in fact, renormalizable to all orders (9). Spontaneously broken gauge theories of weak and electromagnetic interactions are theories in the same sense that quantum electrodynamics itself is a theory.

Let us now return to the problem of $\Delta S = 1$ neutral currents in the four-quark GIM model. A process that is $O(G_F)$ in Born approximation will be

$O(G_F e^2)$ at the one-loop level in a renormalizable gauge theory. If a process is forbidden to lowest order, however, it is not hard to see that it will also vanish to $O(G_F e^2)$. Consider the decay $K_L^0 \to \mu^+\mu^-$ for example. It can proceed through an intermediate state consisting of two charged bosons, but the graph with an exchanged c quark is exactly cancelled by the graph with an exchanged u quark in the limit $m_u = m_c$. The amplitude is of order $G_F^2(m_c^2 - m_u^2)$ as long as $m_c, m_u \ll M_W$. This can be made consistent with rates and limits such as those in Expression 1.1 with $m_c \approx 1$–2 GeV. The mass difference between the K_L^0 and K_S^0 mesons arises from a similar interaction and led Gaillard & Lee (13) in 1973 to a similar estimate for m_c. These estimates are only approximate because of uncertainties associated with strong interaction corrections, but they later were found to be qualitatively correct.

1.3 *Experimental Indications*

The first direct experimental evidence for the existence of the charmed quark came from e^+e^- annihilation into hadrons. It was widely expected that $R(W) \equiv \sigma_{tot}(e^+e^- \to \text{hadrons})/\sigma(e^+e^- \to \mu^+\mu^-)$ would become constant above center-of-mass energy $W = 1$–2 GeV. The theoretical basis for this expectation and the value of the constant is given in Section 2. When this ratio was first measured above $W = 3$ GeV at the Cambridge Electron Accelerator (CEA) in 1972 (14), it was found to be well above the expected value—and apparently rising with W! These measurements, eventually confirmed at the Stanford Linear Accelerator Center (SLAC) (15) and the Deutsches Elektronen Synchrotron (DESY) (16), gave hope to the small band of charm enthusiasts.

The next piece of evidence for charm came from the neutrino scattering experiments at Fermilab. In the collision of muon-type neutrinos with matter, events were discovered that contained a $\mu^+\mu^-$ pair in the final state (17). These events, both in character and rate, could be naturally explained by charm excitation. The hadronic current (Equation 1.7) can couple to the muonic current $\bar{\mu}\gamma^\mu\frac{1}{2}(1-\gamma_5)\nu_\mu$ through W^+ exchange leading to a muon and c quark in the final state. The c quark can decay by this same interaction into a lepton-neutrino pair and preferentially a strange quark. This source of dimuon pairs requires the presence of K mesons in the final state in most events, and that was indeed verified somewhat later (17).

The case was strengthened considerably by the spectacular events of November 1974. The simultaneous discovery of the ψ/J resonance[6] by

[6] Although the $^3S_1$ $c\bar{c}$ ground state was simultaneously dubbed ψ (at SLAC) and J (at Brookhaven), the excited states have all been discovered at SLAC and DESY in e^+e^- annihilation. We refer throughout this paper to the $c\bar{c}$ states that can be directly produced in e^+e^- annihilation as ψ, ψ', ψ'', etc.

groups at SLAC (18) and Brookhaven (19) and the subsequent measurement of excited states (20) and radiative transitions (21) could only be understood in terms of the existence of a new quark of mass around 1.5 GeV. In the ψ it was bound to its own antiquark but, before long, particles made of one heavy and one or more old quarks were discovered (22), and their weak decay properties made it clear that the new quark was precisely the charmed quark predicted by GIM.

The discovery of the upsilon Υ in July 1977 (23) will perhaps be the beginning of a similar story. At least two and perhaps three states in the mass range 9.5–10.5 GeV have been discovered in the reaction $p+Be \rightarrow \Upsilon + \cdots \rightarrow \mu^+\mu^- + \cdots$, and it is irresistable to guess that they are the ground state and two radial excitations of yet another quark and its antiquark. In the case of the ψ/J, it is the e^+e^- colliding beam machines that have produced the most important discoveries. Unfortunately, the highest energy presently available with these machines is about 8 GeV, which puts the upsilon just out of range. However, the next generation of higher energy colliding beam machines at SLAC, DESY, and Cornell will be operating in the near future. They will, no doubt, produce a great deal of information about the upsilon and even heavier quark-antiquark systems if they exist.

It is important to mention one more experimental development here. In 1975, the SLAC–Lawrence Berkeley Laboratory (LBL) group announced the discovery (24) of a new particle τ with $m_\tau \approx 1.8$ GeV. The initial evidence came in the form of the "anomalous" events $e^+e^- \rightarrow e^\pm + \mu^\mp +$ missing energy, which indicates a primary process $e^+e^- \rightarrow \tau^+\tau^-$ followed by weak decay of the τ^+ and τ^-. With further study, it now seems clear that the τ is a lepton (25). These experiments and the τ properties are not discussed here, but it is important to keep the τ's existence in mind throughout. On the theoretical side, it must loom large in any attempts at grand synthesis or considerations of quark-lepton symmetry. Experimentally, there is the curious fact that its mass is so close to that of the charmed quark. In e^+e^- annihilation, charm threshold and $\tau^+\tau^-$ threshold come nearly on top of each other, so that the experimental study of each is complicated by the other.

2 QUANTUM CHROMODYNAMICS

2.1 *Background*

Even though charm was invented in response to problems with the weak and electromagnetic interactions, its discovery has had a considerable impact on strong interaction physics. The strong interactions have been a notoriously difficult problem for several decades and one might ask what impact the discovery of new strongly interacting quarks could have on our understanding of these forces. The answer is that early in 1973, a little more than a year before the experimental discovery of the ψ/J, a remarkable theoretical discovery was made (26), which has led to the growing consensus among theorists that we may finally have in hand a fundamental theory of the strong interactions. This has not yet been proven, but the measured properties of heavy quarks have reinforced this view and if yet heavier quarks are discovered, the reinforcement should become even stronger. The interplay of charm and this candidate theory of strong interactions called, quantum chromodynamics (QCD), is discussed throughout the paper.

2.2 *The Hidden Color Hypothesis*

The first ingredient in QCD is a new, completely hidden, quantum number known as color (27). This notion is explained in some detail in the accompanying article by Greenberg (2) and so we only summarize the essentials. Several problems with quark phenomenology have suggested that the number of quarks should be tripled so that each type of quark u, d, s, c, . . . comes in three so-called colors. In the most popular version of the color model, all the fundamental forces respect the symmetry under color interchange (28). This symmetry can be viewed as a new SU(3) symmetry and the observed hadrons are all singlets with respect to this new SU(3).

There are several reasons for believing in the color hypothesis. With color, the ground states of baryons can be simply understood in the quark model without abandoning Fermi-Dirac statistics. Consider the Ω^- for example; it consists of three s quarks in an orbital S state with spins aligned, leading to an overall symmetric wave function. With three colors, however, overall antisymmetry can be restored if the Ω^- is a color singlet. Color also plays a sort of counting role, raising the value of certain amplitudes to agree with experiment. One example is the decay $\pi^0 \rightarrow \gamma\gamma$ (29), and another is the value of $\sigma_{\text{tot}}(e^+e^- \rightarrow \text{hadrons})$ within the free quark (parton) approximation (30). There, neglecting fermion masses,

$$R(W) \equiv \frac{\sigma_{\text{tot}}(e^+e^- \rightarrow \text{hadrons})}{\sigma(e^+e^- \rightarrow \mu^+\mu^-)} = \sum_i Q_i^2 \qquad 2.1$$

where Q_i is the electric charge of a quark of type i in units of e. The inclusion of color in this sum raises the prediction from $\frac{2}{3}$ to 2 below charm threshold. This is a good first approximation to the experimental value of R (31–34) and within QCD it also turns out to be a good first approximation theoretically.

2.3 *The Color SU(3) Gauge Theory*

QCD is constructed by extending the global $SU(3)_c$ color symmetry to a local gauge symmetry. This requires the introduction of massless vector gauge fields $A^a_\mu(x)$, $a = 1, 2, \ldots, 8$, transforming according to the adjoint representation of $SU(3)_c$. The eight gauge fields are called colored gluons. The Lagrangian density is

$$\mathscr{L}(x) = \bar{\mathrm{q}}(x)\left[i\gamma^\mu \mathrm{D}_\mu - M_0\right]\mathrm{q}(x) - \tfrac{1}{4}F^a_{\mu\nu}(x)\,F^{a\mu\nu}(x), \qquad 2.2$$

where the quark field $\mathrm{q}(x)$ contains $4 \times 3 \times N$ components, corresponding to the three colors and the unknown number N of quark flavors; M_0 is the bare mass matrix, a product of Dirac and color unit matrices and a diagonal flavor matrix M_0^{AB}. The gauge field covariant derivatives are

$$\mathrm{D}_\mu = \partial_\mu - igT^a A^a_\mu$$

and 2.3

$$F^a_{\mu\nu} = \partial_\mu A^a_\nu - \partial_\nu A^a_\mu + gf^{abc}A^b_\mu A^c_\nu,$$

and we normalize the T matrices in the conventional way:

$$T_a = \tfrac{1}{2}\lambda_a, \qquad [\lambda_a, \lambda_b] = 2if_{abc}\lambda_c, \qquad \mathrm{Tr}\,\lambda_a\lambda_b = 2\delta_{ab}. \qquad 2.4$$

The theory is invariant under the gauge transformation

$$A_\mu(x) \equiv A^a_\mu(x)T^a \to U(x)A_\mu(x)U^{-1}(x) + \frac{i}{g}U(x)\partial_\mu U^{-1}(x)$$
$$\mathrm{q}(x) \to U(x)\mathrm{q}(x), \qquad 2.5$$

where $U(x) = \exp\{ig\theta^a(x)T^a\}$.

While QCD bears a superficial resemblance to quantum electrodynamics (QED), it is, in fact, considerably more complicated. It is a renormalizable theory, like QED, so that the ultraviolet divergences that appear in perturbation theory can be handled in the traditional way. Field theory Green's functions can be computed to any order in a power series expansion in a renormalized coupling constant. One expects this to be useful only if the coupling constant is small, however, and that depends on how the theory is renormalized. It is here that QCD and QED part company.

In quantum electrodynamics, the renormalized coupling constant α can be defined at zero momentum transfer (infinite distances), since the nonzero electron mass prevents infrared divergences in the photon propagator in this limit. The potential energy between two charged particles separated by a distance $r \gg 1/m_e$ is αr^{-1} with α determined experimentally to be approximately $\frac{1}{137}$. As a consequence of the r^{-1} behavior, the asymptotic states of the theory are the charged particles and photons corresponding to the fields in the Lagrangian. At distances $r \ll 1/m_e$, the charged shielding effect of vacuum polarization begins to disappear and the effective coupling strength begins to increase. The familiar one-loop result is

$$V(r) = \frac{\alpha}{r}\left[1 + \frac{\alpha}{3\pi}\log(rm_e)^{-1}\right] \tag{2.6}$$

and at distances $r \simeq m^{-1}e^{-3\pi/\alpha}$, the increase becomes substantial. Since the perturbation expansion breaks down at this point, the asymptotic behavior of QED as $r \to 0$ is unknown. Fortunately, physics at laboratory distance scales is insensitive to this asymptotic ignorance.

QCD differs from QED in many ways, in particular in the way the effective coupling strength as computed in perturbation theory varies with distance. This different behavior is the important feature of QCD called asymptotic freedom (26). We shall describe this property with an eye toward our later discussion of its role in heavy quark physics. At the end of this section some of the other known and conjectured properties of QCD will be summarized.

The Feynman rules for QCD can be generated using either functional methods or the canonical Hamiltonian formalism (35). In either case, a gauge must be chosen as in QED. In a general class of covariant gauges, the gluon propagator is

$$D_{\mu\nu}^{ab}(k) = \delta^{ab}\left(g_{\mu\nu} - \xi\frac{k_\mu k_\nu}{k^2}\right)\frac{1}{k^2 + i\varepsilon}, \tag{2.7}$$

where ξ is an arbitrary parameter specifying the gauge. The remaining Feynman rules, consisting of the quark propagator and the various vertices, can be read off directly from the Lagrangian (35). The only exception to this is that a careful treatment of unitarity demands the inclusion of fictitious scalar particles called Fadde'ev-Popov ghosts, which propagate only around closed loops (36).

For some purposes, it is convenient to adopt the physical Coulomb gauge (35). The propagator then consists of an instantaneous potential part and a transverse gluon part

$$\begin{aligned} D_{\mu\nu}^{ab}(k) &= \delta^{ab} i/\mathbf{k}^2 \qquad \mu = \nu = 0 \\ &= \delta^{ab}\frac{i}{k^2 + i\varepsilon}\left(\delta_{ij} - \frac{k_i k_j}{\mathbf{k}^2}\right) \qquad \mu = i,\ \nu = j. \end{aligned} \tag{2.8}$$

A Fadde'ev-Popov ghost must again be included, but it is instantaneous and it couples only to transverse gluons. In either covariant or Coulomb gauge, Ward identities can be established and renormalizability proven.

Now consider again the potential energy between two charge sources separated by a distance r (37, 38). The charge is now color and the sources can be taken to be a very heavy quark and antiquark. If they are in a quantum mechanical color singlet state, then the Born approximation to the potential is $-C_F\alpha_s/r$, where $T^a_{ik}T^a_{kj} = C_F\delta_{ij}$ and $\alpha_s = g^2/4\pi$ is the strong coupling constant. For SU(N), $C_F = (N^2-1)/2N$. The one-loop corrections to the potential are most conveniently computed in Coulomb gauge and, if light quarks are neglected for the moment, the two relevant diagrams are shown in Figure 1 (37). The necessary renormalization subtractions cannot be performed at infinite separation because the loops are infrared divergent at this point. Some other arbitrary point, say $r = \mu^{-1}$, must be chosen, and this necessarily brings in a new dimensional parameter μ. In QED, such a parameter *could* be introduced; in QCD, because of the self-coupling of the gauge field, it *must* be introduced. The static potential through one loop is

$$V(r) = -C_F\frac{\alpha_\mu}{r}\left[1+\frac{5}{6}\frac{\alpha_\mu}{\pi}C_A\ln(r\mu)^{-1} - \frac{16}{6}\frac{\alpha_\mu}{\pi}C_A\ln(r\mu)^{-1}\right], \qquad 2.9$$

where $\alpha_\mu = g_\mu^2/4\pi$ is the coupling strength corresponding to the scale $r = \mu^{-1}$ and $f_{acd}f_{bcd} = C_A\delta_{ab} = N\delta_{ab}$ for SU(N). The two logarithmic modification terms correspond to Figures 1*b* and 1*a* respectively. The first term is vacuum polarization of transverse gluon pairs, a charge-shielding effect similar to electron-positron vacuum polarization in QED. As expected, this contribution tends to make the effective coupling strength increase as $r \to 0$. The other contribution, unique to Yang-Mills theories, is a self-energy of the Coulomb field. It comes with the opposite sign and is larger in magnitude than vacuum polarization. The net result is that the effective coupling strength decreases as r decreases. This is the property called asymptotic freedom.

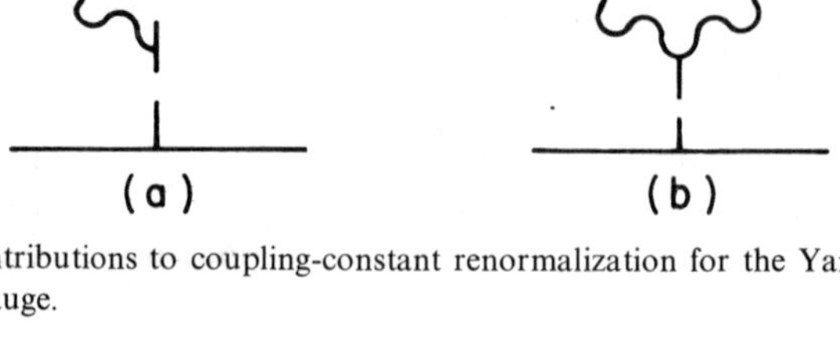

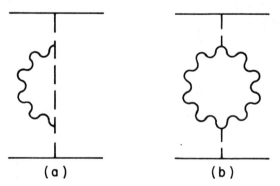

Figure 1 Contributions to coupling-constant renormalization for the Yang-Mills theory in Coulomb gauge.

It is not difficult to understand the physical mechanism behind asymptotic freedom (37). We do not describe it in detail except to say that the Coulomb field self-energy produces a collimation of the color electric field lines connecting the quark to the antiquark. The collimation becomes more pronounced as $r \to \infty$, increasing the energy stored in the field relative to the pure Coulomb potential. The decrease of the collimation as r decreases is the explanation of asymptotic freedom.

The behavior of the effective coupling strength as $r \to 0$ can be described to all orders in perturbation theory by a consideration of its scaling properties[7] (39). The potential can be written quite generally in the form

$$V(r) = -C_F \frac{\alpha(r\mu, \alpha_\mu)}{r}, \qquad 2.10$$

where $\alpha(r\mu, \alpha_\mu) \equiv \alpha_s(r)$ is the effective coupling strength, given through one loop by Equation 2.9. The potential must be independent of the choice of μ and the condition $(d/d\mu)\alpha_s(r) = 0$ can be used to show that

$$r \frac{\partial}{\partial r} \alpha_s(r) = \beta[\alpha_s(r)], \qquad 2.11$$

where

$$\beta(x) \equiv r \frac{\partial}{\partial r} \alpha(r\mu, x)\Big|_{r=\mu^{-1}}. \qquad 2.12$$

From Equation 2.9 we find

$$\beta(x) = \frac{11}{2}\frac{x}{\pi} + O(x^2) \qquad 2.13$$

and, therefore, to lowest order, Equation 2.11 can easily be integrated. The result is

$$\alpha_s(r) = \frac{\alpha_\mu}{1 + \frac{11}{2}\frac{\alpha_\mu}{\pi} \ln (r\mu)^{-1}}, \qquad 2.14$$

which agrees with Equation 2.9 to lowest order and which goes to zero as r decreases. As r increases Equation 2.14 shows that the theory becomes strongly coupled and the perturbation expansion breaks down. In this sense, QCD and QED are oppositely behaved.

[7] In a renormalizable field theory, the study of scaling properties is not just a simple matter of dimensional analysis. A new dimensional parameter is introduced through renormalization, and that complicates the scaling behavior. The formalism for studying this behavior is called the renormalization group.

2.4 *Properties of Quantum Chromodynamics*

2.4.1 COVARIANT FORMULATION For most purposes, it is best to work in a covariant gauge. The asymptotic form Equation 2.14 of the running coupling constant is gauge invariant, but its Feynman diagrammatic breakdown is completely different in a covariant gauge (26). As a result, the physical mechanism behind asymptotic freedom is not as transparent as in Coulomb gauge. The momentum space effective coupling constant $\alpha_s(-q^2)$ is defined in terms of both propagator and vertex functions at a Euclidean momentum $q^2 = q_0^2 - \mathbf{q}^2 < 0$. If there are f quark flavors for which $m \ll q$, then

$$\alpha_s(-q^2) = \frac{\alpha_\mu}{1+\left(\frac{11}{4}-\frac{1}{6}f\right)\frac{\alpha_\mu}{\pi}\ln\left(-\frac{q^2}{\mu^2}\right)}. \qquad 2.15$$

It is convenient to re-express $\alpha_s(-q^2)$ in the form

$$\alpha_s(-q^2) = \frac{\pi}{\left(\frac{11}{4}-\frac{1}{6}f\right)\ln\left(-\frac{q^2}{\Lambda^2}\right)}, \qquad 2.16$$

where $\Lambda = \mu \exp\left[-\pi/(11/2-f/3)\alpha_\mu\right]$. This explicitly exhibits the fact that $\alpha_s(-q^2)$ depends only on one parameter (α_μ and μ are not independent): Λ must be determined experimentally, and then $\alpha_s(-q^2)$ is completely specified.

2.4.2 APPLICATIONS OF ASYMPTOTIC FREEDOM It is possible to directly confront asymptotic freedom with experiment only in those few situations where the measured quantity is sensitive to the short-distance behavior of QCD alone. The most important example is deep inelastic electroproduction, where approximate Bjorken scaling (40a; in the context of QCD, see 40b) at momentum transfers larger than 1–2 GeV indicates that the theory is nearly free. In order to make this precise, that is, to truly isolate short-distance behavior from the data, one must make use of the Wilson operator product expansion (41). An explanation of this formalism falls beyond the purview of this paper and we simply summarize the situation: Short-distance physics can indeed be isolated and the experimental data suggests that $\alpha_s(-q^2) < 1$ for $(-q^2)^{\frac{1}{2}} > 1$–2 GeV. Analyses of electroproduction typically suggest a value of Λ around 500 MeV. This gives $\alpha_s(-q^2) \approx 0.5$ at $(-q^2)^{\frac{1}{2}} \approx 2$ GeV.

Another important application of asymptotic freedom is the total cross section for $e^+e^- \to$ hadrons (42). Moments of the cross section are related by dispersion relations to hadronic vacuum polarization at space-like q^2. This is sensitive only to short-distance behavior and is therefore computable in perturbation theory for $-q^2$ large enough. It has been conjectured that asymptotic freedom can be applied directly to $\sigma_{\text{tot}}(e^+e^- \to \text{hadrons})$ for $q^2 = W^2 > 0$ as though the final particles are quarks and gluons (43). This is true, at best, away from important thresholds such as charm-anticharm, and there the prediction for $R(W)$, Equation 2.1, is

$$R(W) = \sum_i Q_i^2 \left[1 + \frac{\alpha_s(W)}{\pi} \right]. \qquad 2.17$$

This result can be stated in a somewhat more solid form by averaging R over an interval in W (44). But where R shows no rapid variation with W, this should not be necessary and Equation 2.17 can be used reliably. The total-cross-section data is still rather poor for these purposes (31–34). Above charm threshold, the errors are large and the presence of the heavy τ lepton (24) complicates matters. Below charm threshold ($1 < W < 3$ GeV), the experimental value of R is somewhat above the parton model value of $\Sigma_i Q_i^2 = 2$. About the best that can be said now is that the data are consistent with $\Lambda \sim 500$ MeV.

There are many other situations where asymptotic freedom plays at least some role. Some examples are high momentum transfer hadron-hadron scattering (45) and the $\Delta I = \frac{1}{2}$ in nonleptonic weak decays (46). However, the short- and long-distance behavior of QCD cannot be clearly separated in these problems and they are perhaps less important for testing asymptotic freedom. There is one other possible application of asymptotic freedom that is very important for heavy quark physics. The ψ particle is very long lived and it has been suggested that this can be understood as a consequence of asymptotic freedom (47). This possibility will be discussed in detail in Section 3, but we emphasize here that if the narrow width of the ψ can be explained in this way, a value of $\alpha_s(M_\psi^2) \approx 0.2$ is required. Whether the analysis of electroproduction allows such a small effective coupling is not yet clear.

2.4.3 BEYOND PERTURBATION THEORY A widespread hope is that the spectrum of QCD consists only of the color singlet hadrons observed in the laboratory. The underlying quarks and gluons would then not be among the asymptotic states, and would exist only as the constituents of hadrons. To demonstrate this "confinement" and to compute the masses and other properties of hadrons starting from the QCD Lagrangian is an extraordinarily difficult and completely unsolved problem. A discussion of the many approaches to this problem is beyond the scope of this review and we only make a few comments to provide some perspective for the next sections.

None of these conjectured features of QCD can be seen in perturbation theory. As the distance scale r increases, the effective coupling (Equation 2.14 or Equation 2.16) increases, and perturbation theory is no longer directly useful. One might hope that general properties could still be extracted from the perturbation expansions or that the series could be summed to deal with large-distance effects, but even this seems unlikely. On the one hand, it can be shown that to any finite order of perturbation theory, there is no indication of confinement (48). The self-coupling of the gauge field suggests an infrared divergence structure much worse than QED, and one speculation was that this structure might have something to do with confinement. However, it has been shown (48) that properly defined transition probabilities are infrared finite order by order in a renormalized coupling constant α_μ. The situation is not unlike QED.

As far as summing the expansion is concerned, there is every indication that the series is not convergent. In fact, it would appear that the series is not even Borel summable (49) so that the perturbation expansion cannot be used to define the theory except for weak coupling. QCD is known to contain essential singularities at $\alpha_\mu = 0$ even at the classical level. The most important examples of this at the present are the instantons and other Euclidean field configurations with nontrivial topological structure (50). The ultimate role of these features of QCD in confinement is not yet known but they surely point up the inadequacy of perturbation theory. Perhaps the most ambitious attack on confinement and hadron structure has been the strong coupling expansion for QCD pioneered by Wilson (51). A short-distance cut-off in the form of a spatial or space-time lattice eliminates the weakly coupled sector of the theory. A linear confining potential $V(r) \sim \alpha r$ between quarks appears naturally as a consequence of the fact that the color electric flux is quantized on the lattice (52). The problem of taking the lattice spacing to zero is, however, unsolved and this is an important ingredient in computing the properties of hadrons on the lattice. The linear potential suggested by the lattice strong coupling expansion has been applied rather successfully to charmonium spectrum computations; an example of the interplay between charm and (in this case) suggested properties of QCD.

3 CHARMONIUM AND BEYOND

The existence of the narrow resonances discovered in November 1974 had been anticipated theoretically (47) prior to the experiments. If charm was real, then narrow $c\bar{c}$ bound states should exist below the threshold for charm production and the name charmonium was suggested, in analogy to positronium.

At the time it was, in fact, thought that the resemblance to positronium might be more than just an analogy (47, 53). Asymptotic freedom says that at short distances (less than about $\frac{1}{5}$ fermi) the strong interactions should behave nearly like α_s/r with $\alpha_s = g^2/4\pi \ll 1$ (but of course $\gg 1/137$). If the c and $\bar{c}$ were to spend most of their time within $\frac{1}{5}$ fermi of each other, then the similarity with positronium would become almost complete. It became clear, very quickly after the initial discoveries, that things were not going to be so simple. The radius of charmonium was closer to 1 fermi than $\frac{1}{5}$ fermi and the binding potential had a structure completely unlike the Coulomb potential of electrodynamics. Nevertheless, the qualitative resemblance to positronium is undeniable and the name charmonium has caught on.

This section reviews in some detail the major areas of theoretical research into this charmonium system. After a brief survey of the experimental situation in Section 3.1, we lay the groundwork for what we shall call the charmonium model (Section 3.2). This is the atomic model of heavy (c) quarks bound in a static, confining potential that, together with asymptotic freedom applied to short-distance processes, describes the spectrum and decay widths of states in the $c\bar{c}$ system.

The simplest version of the model is discussed in Section 3.3, with special emphasis on the choice of a phenomenological potential and the spectrum of states and transition probabilities resulting therefrom. At this naive level, the model already provides a fairly good description of the charmonium system with only two glaring exceptions—the even-charge-conjugation (even-C) states at 2.83 and 3.45 GeV. The main attempts to go beyond the basic model, by incorporating effects of quark spin dependence and of coupling $c\bar{c}$ states to charmed hadron decay channels, are reviewed critically in Sections 3.4 and 3.5. In both cases, we have tried to motivate the directions research has taken, evaluate the outcomes, and, by stressing the shortcomings of existing work, hopefully point the way for future improvement. Finally, Section 3.6 summarizes the straightforward applications of the potential model to bound systems of still heavier quarks. Here, the data is still too limited for critical evaluation of the theory, but we can look forward to rigorous experimental tests in just a year or two.

Lack of space prevents our discussing other charmonium research topics. Most notable are the attempts to compute charmonium properties without recourse to a potential model. These include Regge-trajectory analysis of the spectrum (e.g. 54); a topological S-matrix approach to understanding the small hadronic widths of charmonium (55); and the use of dispersion relations to (*a*) derive sum rules between leptonic widths of charmonium levels and integrals over the charm contribution to *R*, and (*b*) estimate two-photon widths of appropriate states by using sum rules derived from $\gamma\gamma$ scattering (56).

Table 1 The properties of charmonium. All data is taken from Reference 59, which cites original work; question marks indicate unknown values

Particle	$I^G(J^{PC})$	Mass (MeV)	Full width (MeV)	Decay mode	Fraction (%)
X(2830)	??(??$^+$)	2830 ± 30	?	$\gamma\gamma$	>0.8
ψ/J(3095)	$0^-(1^{--})$	3098 ± 3	0.069 ± 0.015	e^+e^-	7.3 ± 0.5
				$\mu^+\mu^-$	7.3 ± 0.5
				Direct hadrons	86 ± 2
				γX(2830)	<1.7
				γX(2830) $\rightarrow 3\gamma$	0.013 ± 0.004
				γX(2830) $\rightarrow \gamma p\bar{p}$	<0.004
χ(3415)	$0^+(0^{++})$	3415 ± 5	?	$\gamma\psi$(3095)	3 ± 3
				Hadrons	
P_c/χ(3510)	$0^+(1^{++})$	3508 ± 4	?	$\gamma\psi$(3095)	35 ± 16
				Hadrons	
χ(3455)	??(??$^+$)	3454 ± 7	?	$\gamma\psi$(3095)	$>24 \pm 16$
χ(3550)	$0^+(2^{++})$	3552 ± 6	?	$\gamma\psi$(3095)	14 ± 9
				Hadrons	
ψ'(3684)	$0^-(1^{--})$	3684 ± 4	0.228 ± 0.056	e^+e^-	0.88 ± 0.13
				$\mu^+\mu^-$	0.88 ± 0.13
				$\psi\pi^+\pi^-$	33.1 ± 2.6
				$\psi\pi^0\pi^0$	15.9 ± 2.8
				$\psi\eta$	4.1 ± 0.7
				$\gamma\chi$(3415)	7.3 ± 1.7
				$\gamma\chi$(3510)	7.1 ± 1.9
				$\gamma\chi$(3550)	7.0 ± 2.0
				$\gamma\chi$(3455)	<2.5
				$\gamma\chi$(3455) $\rightarrow \gamma\gamma\psi$	0.6 ± 0.4
				γX(2830)	<1.0
				Direct hadrons	~ 9
ψ''(3772)	$0^-(1^{--})$	3772 ± 6	26 ± 5	e^+e^-	0.0010 ± 0.0005
				$D^0\bar{D}^0$	56 ± 3
				D^+D^-	44 ± 3
ψ(4028)	?(1^{--})	4028 ± 20	~ 50	e^+e^-	~ 0.002
				Charmed mesons	~ 100
ψ(4414)	?(1^{--})	4414 ± 7	33 ± 10	e^+e^-	0.0013 ± 0.0003
				Charmed mesons	~ 100

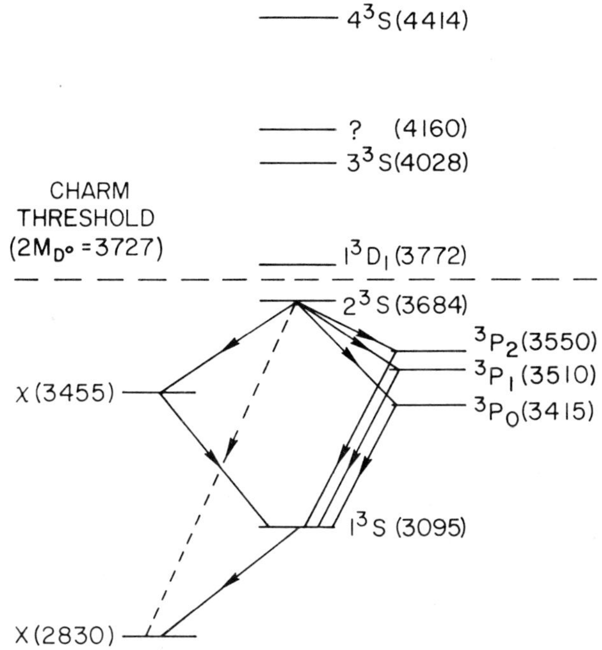

Figure 2 The observed charmonium levels with the notation $n^{2S+1}L_J$ and masses in MeV. The correct identification for X(2830) and χ(3455) is not known. Observed radiative transitions are shown as solid directed lines. The dotted line represents an M1 transition not yet observed. Splittings are approximately to scale.

3.1 *Experimental Overview*

There are several excellent, up-to-date reviews (57–59) on the production and decay characteristics of the charmonium states (58 and 59 reference experimental work not cited explicitly here). We content ourselves here with a brief survey of what has been seen, with special emphasis on production via e^+e^- annihilation; for other production mechanisms of ψ/J, see Section 6. Table 1 summarizes the known properties of charmonium. A level diagram[8] is shown in Figure 2.

Except for the simultaneous discovery of ψ/J in e^+e^- annihilation at SLAC (18) and in proton-beryllium collisions at Brookhaven (19), all the charmonium levels—including narrow states below charm threshold ($W_c = 2M_{D^0} = 3.727$ GeV) and broad ones above—have been discovered at the SLAC and DESY e^+e^- storage rings (SPEAR and DORIS). The

[8] We use the spectroscopic notation $n^{2S+1}L_J$, where n is the number of radial nodes plus one, S is the total $q+\bar{q}$ spin (0 or 1), L is the orbital angular momentum of the $q\bar{q}$, and J is the total angular momentum of the state.

experimental situation is incredibly beautiful and simple: States with $J^{PC} = 1^{--}$, the quantum numbers of the photon, are produced directly and copiously as resonances in e^+e^- collisions; narrow states lying below $\psi'(3684)$ and having even charge conjugation are observed in radiative transitions, $\psi' \to \gamma\chi$ and $\psi \to \gamma X$. It should be added that there remain additional charmonium levels to be discovered. An important one is the 1^1P_1 state with $J^{PC} = 1^{+-}$, expected near 3.45 GeV. In e^+e^- annihilation, this could be observed (with difficulty) only in the reaction chain

$$\begin{aligned} e^+e^- &\to \psi'(3684;\, 2^3S_1) \\ &\quad\hookrightarrow \eta_c'(?;\, 2^1S_0) + \gamma \\ &\qquad\hookrightarrow \chi(?;\, 1^1P_1) + \gamma \\ &\qquad\quad\hookrightarrow \text{hadrons.} \end{aligned} \tag{3.1}$$

The most striking feature of states lying below W_c is their very small total widths. In particular, first and second order electromagnetic decays such as $\psi' \to \gamma\chi$ and $\psi \to e^+e^-$ are competitive with strong interaction decays such as $\psi' \to \psi\pi\pi$ and $\psi \to$ hadrons. If we assume that these high-mass mesons are $c\bar{c}$ bound states, it follows that decay to ordinary hadrons, not containing charmed quarks, must proceed by mutual annihilation of the $c\bar{c}$ pair. The reluctance of any $q\bar{q}$ pair within a single hadron to annihilate, known as the Okubo-Zweig-Iizuka (OZI) rule (60), has been observed in the light hadrons for a long time. The most notable example is the suppressed decay $\phi\ (s\bar{s}) \to \pi^+\pi^-\pi^0$ (all containing u, d quarks). This empirical rule could be a simple consequence of asymptotic freedom in the case of heavy quarks.

Two features of e^+e^- annihilation make it ideal for discovery and study of the myriad of charmonium levels: First, the energy in a single beam, $E_b = \frac{1}{2}W$ (W = total center-of-mass energy), is known very precisely, to within 1–2 MeV. Second, the $J^{PC} = 1^{--}$ states are produced at rest in the center of mass. Together, these allow very precise measurement of the mass of the directly produced states. The even-C states below ψ' are produced by "sitting on" ψ' and ψ, i.e. setting $E_b = \frac{1}{2}M_{\psi'}$ or $\frac{1}{2}M_{\psi}$, and observing their radiative decays. For this, three methods have been used:

1. Measurement of the invariant mass of the charged hadrons in decays such as

$$\psi' \to \pi^+\pi^-\pi^+\pi^- + \text{missing neutrals,} \tag{3.2}$$

corresponding, perhaps, to $\psi' \to \chi + \gamma$, $\chi \to 4\pi$. If a peak is found in the charged hadron mass spectrum and if the mass of the missing neutral is consistent with its being a photon (rather than a π^0), a fairly precise determination of the χ mass results from constraining the mass of the parent to $M_{\psi'}$ and that of the neutral to zero. The X(2830) was found by a similar method (61), but with all neutral particles; in particular, this state has been seen only in the decay chain

$$\psi \rightarrow \gamma + \mathrm{X}(2830); \qquad \mathrm{X}(2830) \rightarrow \gamma\gamma. \tag{3.3}$$

Therefore, all we know about this meson is that it has even C and cannot have $J^P = 1^{\pm}$.

2. Measurement of the inclusive photon energy distribution in $\psi' \rightarrow \gamma +$ anything and $\psi \rightarrow \gamma +$anything. Here, the photon is definitely identified and its energy measured, usually with an energy resolution of (5–10%)/$E_\gamma^{\frac{1}{2}}$ (in GeV). Peaks in this distribution correspond to ψ or $\psi' \rightarrow \gamma +$a narrow $C = +1$ state. To date, only the states χ(3415), χ(3510), and χ(3550) have been detected by this method (as well as by the other two).

3. Observation of the double-cascade process,

$$\psi' \rightarrow \gamma\chi; \qquad \chi \rightarrow \gamma\psi. \tag{3.4}$$

Here, the ψ is identified by its decay to $\mu^+\mu^-$, and one photon is detected by its conversion to e^+e^-. The missing neutral energy is determined to be consistent with zero mass. There is a potential ambiguity in determining the χ mass because one does not know which photon came first in a given event of this type. The ambiguity is resolved neatly by plotting the two possible values of the $\psi\gamma$ invariant mass. The wrong solution shows a characteristic Doppler broadening induced by the motion of the χ. In addition to the well-established states at 3414, 3508, and 3552 MeV, this method has revealed the existence of a fourth intermediate state, χ(3455) (21). Seen in no other way, the only known decay mode of this even-C state is $\chi(3455) \rightarrow \gamma\psi$.

The comparative ease of detecting and identifying ψ and χ decay products in e^+e^- annihilation makes it also the best method for determining their spin-parities and branching fractions to individual final states. An outstanding example of this is the determination $J^{PC} = 1^{--}$ for ψ and ψ', assignments that are not obvious a priori. This was done by observing the characteristic destructive interference, just below $W = M_{\psi,\psi'}$, between

$$e^+e^- \rightarrow \gamma_v \rightarrow \mu^+\mu^- \qquad (\gamma_v = \text{virtual photon}) \tag{3.5a}$$

and

$$e^+e^- \rightarrow \gamma_v \rightarrow \psi,\psi' \rightarrow \gamma_v \rightarrow \mu^+\mu^-. \tag{3.5b}$$

The assignments of $J^P = 0^+$ to $\chi_0 = \chi(3415)$, $J^P = 1^+$ to $\chi_1 = \chi(3510)$, and $J^P = 2^+$ to $\chi_2 = \chi(3550)$ are based on the following considerations

(62): (*a*) χ_0 and χ_2 decay to $\pi^+\pi^-$ and K^+K^- and, therefore, both states have natural spin-parity; these modes have not been observed for χ_1, consistent with the 1^+ assignment; and (*b*) the angular distribution of the photon in $\psi' \rightarrow \gamma\chi_0$ is well fitted by $1+\cos^2\theta$, which is expected for $J = 0$. The angular distributions $1-\frac{1}{3}\cos^2\theta$ for $J = 1$ and $1+\frac{1}{13}\cos^2\theta$ for $J = 2$ are consistent with the rather meager measurements for χ_1 and χ_2, respectively.

Finally, the normal hadronic widths, on the order of 10–100 MeV, of the directly produced resonances above $\psi'(3684)$ are further dramatic confirmation of the OZI rule. Here, the charmed quarks need no longer annihilate since it is energetically possible for them to emerge (together with light quarks) as the charmed mesons D, D*, F, F*, and so on. All this is discussed in Section 4. Suffice it to say that these broad charmonium resonances were solely responsible for the unambiguous isolation of charmed mesons.

3.2 *Foundations of the Charmonium Model*

Perhaps the most important feature of the charmonium spectrum in Figure 2 is the fact that the level spacings are very small compared to the overall mass scale of the system. This suggests, at least for the states below charm threshold, that the system is nonrelativistic, with excitation energies small compared to the masses of the constituents. This is something completely new in strong interaction physics, and a great deal of theoretical work has gone into analyzing this system using a nonrelativistic Schrödinger equation formalism (63, 64).

In retrospect, this approach is somewhat too naive, especially with regard to the assumption of spin independence of the dominant $c\bar{c}$ interaction. The hyperfine splitting is not much smaller than the radial and orbital excitation energies. Nevertheless, the model has, at the very least, been a powerful predictive guide to the qualitative features of charmed quark physics.

The other aspect of the charmonium model is the attempt to understand the narrowness of the states below charm threshold as a consequence of asymptotic freedom (see Section 2.4). We now discuss the theoretical basis for this possibility in some detail, but it is important to keep in mind that it is largely a separate issue from the energy level structure. There, it is clear from Figure 2 that without a solution to QCD in the strong coupling regime, some phenomenological input is necessary. In the case of the decay widths, there is the possibility that perturbation theory may be directly relevant.

The idea is basically that with $c\bar{c}$ annihilation into light hadrons proceeding through gluons, the decay will be inhibited, since the effective

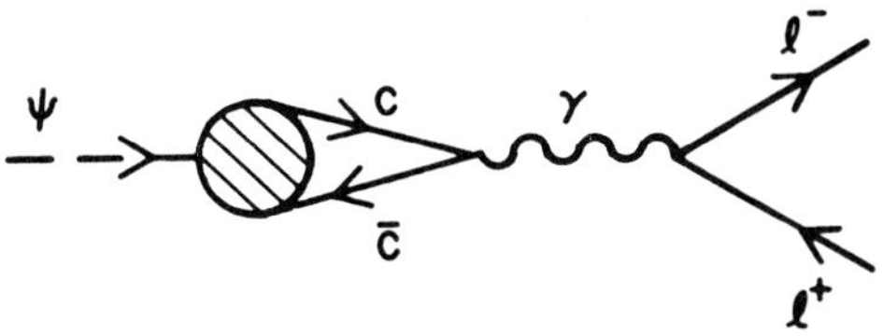

Figure 3 The electromagnetic decay of the $\psi(1^3S_1)$.

gluon coupling constant should be small at high energies (47, 65). The dominant contribution will come from the minimum number of intermediate gluons, which depends upon the quantum numbers of the charmonium state. Some rather striking experimental predictions can be made on the basis of this "gluon counting"; these are discussed in Section 3.3.

Consider the decay of the $\psi(^3S_1)$ state. Its dominant electromagnetic decay is shown in Figure 3. The $c\bar{c}$ pair must come together to annihilate into the virtual photon, and if the bound state is nonrelativistic, then, to first approximation, the decay width will be given by

$$\Gamma(\psi \rightarrow l^+ l^-) = \frac{4\alpha^2(2/3)^2}{M^2} |\Psi(0)|^2. \qquad 3.6$$

The charge of the charmed quark is $\frac{2}{3}|e|$, and M is the charmonium mass. The nonrelativistic radial wave function at the origin is $\Psi(0)$, and one cannot expect to be able to compute it in perturbation theory. The reason for this is that the mean radius $\langle r \rangle$ of charmonium is on the order of one fermi, a distance scale at which the effective coupling strength for the binding has become large. Thus $\Psi(0)$ will be determined in part by the nonperturbative, long-range part of the potential. The hadronic decay of the ψ must proceed through a minimum of three gluons. If this is indeed the dominant contribution, that is to say, if perturbation theory is truly relevant to this problem, the decay will proceed as shown in Figure 4. The $c\bar{c}$ annihilation will be essentially local—on the order of $1/m_c$ ($\ll \langle r \rangle$). The computation of the decay matrix element is then done

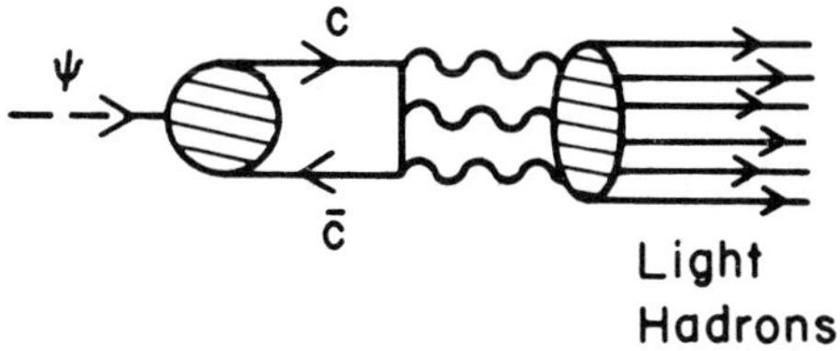

Figure 4 The hadronic decay of the $\psi(1^3S_1)$, assuming the minimal gluon mechanism.

in analogy to the parton model computation of $\sigma_{tot}(e^+e^- \to \text{hadrons})$ as if the final state consisted of three on-mass-shell gluons. This amounts to the statement that the transition from the three-gluon state into physical hadrons takes place with unit probability. A more satisfying theoretical justification of the three-gluon mechanism can be given and we turn to it shortly. If the mechanism just described is correct, then the total hadronic width of the ψ is given by (65)

$$\Gamma(\psi \to \text{hadrons}) = \frac{40}{81\pi}(\pi^2 - 9)\frac{\alpha_s^3(M^2)}{M^2}|\Psi(0)|^2. \tag{3.7}$$

The strong coupling constant is defined at the ψ mass and, as before, $\Psi(0)$ is the nonrelativistic wave function at the origin.

Before proceeding to the comparison of these expressions with experiment, we sketch the analysis that underlies Equation 3.7. A necessary condition for the applicability of lowest order perturbation theory is that no large dynamical factors enter in higher orders to make the expansion break down. One must analyze the quantity $\Sigma_n|M(c\bar{c} \to n)|^2$ where n is some quark-gluon final state; M is the decay matrix element and is defined to be two-particle ($c\bar{c}$) irreducible in the decay channel. Two-particle reducible contributions are absorbed into the definition of the wave function. If it can be shown that $\Sigma_n|M(c\bar{c} \to n)|^2$ is free of infrared singularities for the $c\bar{c}$ pair at rest, then there can be no large dynamical factors. This is because the result will involve no small energy or momentum factors, only the (large) charmed quark mass and the (large) renormalization scale. The infrared finiteness through order α_s^4 (the lowest order is α_s^3) for ψ decay has been checked and it is conjectured to be true to all orders. It is technically simpler to use the Coulomb gauge rather than covariant gauges in this analysis.

We make one last point before proceeding. The infrared analysis is necessary but not sufficient. It could well be that threshold singularities in high orders, or even completely nonperturbative effects prevent the use of perturbation theory in this simple way. The use of asymptotic freedom to explain the OZI rule is speculative. It is on much less solid footing than the conventional deep Euclidean application or even the direct computation of $\sigma_{tot}(e^+e^- \to \text{hadrons})$, since there the production is truly local, coming from the off-shell photon.

3.3 *The Basic Charmonium Model*

To go beyond qualitative predictions, a model of the $c\bar{c}$ interaction is necessary, and it is natural to realize this in terms of an instantaneous, central potential, $V_0(r)$. Such a model presupposes that, to a first approximation, one may neglect the influences of spin-dependent forces and of nearby open decay channels on spectroscopy and decay rates. The attempts to incorporate these effects in the basic model are described in Sections 3.4 and 3.5.

A simple possibility for V_0, motivated by asymptotic freedom at short $c\bar{c}$ separations and quark confinement at large ones, is (63, 64, 66)

$$V_0(r) = -\frac{\kappa}{r} + \frac{r}{a^2}. \qquad 3.8$$

Asymptotic freedom tells us to expect a rather small short-distance coupling, say $\kappa \sim 0.2$. The choice of a linear confining term is suggested by the lattice gauge theory (51, 52) and the dual string model (for a review, see 67). Then $1/a^2$ is related to the Regge slope and is $1/a^2 \approx 1$ GeV/fm ≈ 0.2 (GeV)^2. The third parameter of this model is the charmed quark mass, which is expected to be $m_c \simeq \frac{1}{2}M_\psi \simeq 1.5$ GeV.

We emphasize that these parameters are purely phenomenological. For example, m_c^{-1} is really the strength of the kinetic energy term in the $c\bar{c}$ Hamiltonian, and not necessarily equal to half the difference between the mass of a state and its energy eigenvalue.[9] Thus, within the general guidelines set by the above expectations, the three parameters will be determined by fitting to selected pieces of data. When this model is applied to bound systems of heavier quarks, the Υ for example, it is presumed that only the quark mass will change substantially; κ may decrease slightly, while the linear potential strength a^{-2} is thought to be independent of the quark mass (37).

A final word about this choice of V_0: From a purely phenomenological point of view, other forms may be equally reasonable and give as acceptable an account of the data. The advantages of V_0 are that it is well motivated and that it contains the minimum number of parameters needed to reach agreement with the observed spectra and decay rates in charmonium. There is evidence that V_0 does not adequately describe the Υ spectrum (23, 57), and some possible modifications are discussed in Section 3.6.

To obtain a "best" overall description of charmonium data, Eichten et al (63, 68) chose parameters κ, a, and m_c by (*a*) fitting to the $\psi'-\psi$ mass difference; (*b*) taking the electronic width of ψ to be 5.3 keV, one standard deviation above the measured value of 4.8 ± 0.5 keV; and (*c*) constraining $1.5 \lesssim m_c \lesssim 2.0$ GeV and $0.1 \lesssim \kappa \lesssim 0.4$. These constraints are consistent with the notion of heavy constituents moving nonrelativistically and

[9] The zero of energy of a system bound in an infinitely rising potential cannot be defined unambiguously.

with weak short-range interactions. Acceptable results are obtained with a range of parameters, and their preferred choice is

$$m_c = 1.65 \text{ GeV}, \qquad a = 2.07\,(\text{GeV})^{-1}, \qquad \kappa = 0.132. \qquad 3.9$$

A check on the self-consistency of the nonrelativistic approximation is provided by the mean-squared quark velocity in the 1S and 2S states; these are $\langle v^2 \rangle = 0.17$ and 0.28, respectively. It is worth emphasizing that the Coulomb part of the potential, although the most certain feature according to QCD, may not be very meaningful for charmonium. For this value of κ, it only becomes important below distance scales $\sim m_c^{-1}$, where a nonrelativistic potential picture ceases to be sensible.

3.3.1 CHARMONIUM SPECTRUM The most important consequence of the existence of a confining potential (of almost any shape) between c and $\bar{c}$ is that there will be ^{3}P states lying between the ^{3}S states, ψ and ψ' (63, 64, 69, 70). The reason for this is simple. Recalling that, in a purely Coulomb potential, the 2S and 1P states are degenerate, it is clear that the presence of a confining term will impart a greater kinetic energy to the 2S state, with its one radial node, than to the 1P, which has only an orbital node, thus lifting the degeneracy. The same remark applies to the relative ordering of 3S, 2P, and 1D states, and so only the amount of this splitting depends on details of the potential and quark mass. In charmonium, we are in an extremely fortunate position to observe this consequence of quark confinement because, lying below the OZI-allowed decay threshold, $2M_D$, the 1^3P states, like ψ and ψ', will be unusually narrow. Furthermore, they have even-charge conjugation ($C = +1$), and so may be detected by radiative transitions from ψ' and, if narrow enough,

Table 2 Predicted (68) and observed spin-triplet charmonium levels; assignment of $\psi(4028)$ to 3^3S_1, and existence of $\psi(4160)$ with its assignment to 2^3D_1 are open to question

State $n^3L_J(J^{PC})$	Predicted mass (MeV)	Candidate (measured mass)
$1^3S_1(1^{--})$	3095 (Input)	$\psi(3095)$
$1^3P_J(0^{++},1^{++},2^{++})$	3457	$\chi_{0,1,2}(3522 \pm 5)$
$2^3S_1(1^{--})$	3684 (Input)	$\psi'(3684)$
$1^3D_J(1^{--},2^{--},3^{--})$	3755	$\psi''(3772) = 1^3D_1$
$2^3P_J(0^{++},1^{++},2^{++})$	3957	Above charm threshold; difficult to produce
$3^3S_1(1^{--})$	4157	$\psi(4028)$
$2^3D_J(1^{--},2^{--},3^{--})$	4204	$\psi(4160) = 2^3D_1$
$4^3S_1(1^{--})$	4567	$\psi(4414)$

down to ψ. Again, branching ratios will depend on details of the model, but the existence of these states and their mode of observation is inescapable.

Using the potential V_0 with parameters in Equation 3.9, the Schrödinger equation may be numerically integrated to obtain the spectrum of low-lying spin-triplet states shown in Table 2 (68), together with the most likely candidates for these states. A number of explanatory remarks are in order: (*a*) For brevity only those states even remotely accessible to existing experimental techniques have been included. (*b*) With the neglect of spin-dependent forces, spin-singlet states such as $\eta_c = 1^1S_0$ and 1^1P_1 would be degenerate with the predicted center of gravity (c.o.g.) of the corresponding triplet states. And, by adjusting the constant zero of energy to give the correct ψ mass, that constant subsumes some of the hyperfine interaction ($\propto \mathbf{S}_1 \cdot \mathbf{S}_2$) for the low-lying ^3S states. (*c*) As noted previously, there is fairly strong evidence for the assignment of $\chi_0 = \chi(3415)$, $\chi_1 = \chi(3510)$, and $\chi_2 = \chi(3550)$ to $1^3P_{0,1,2}$ respectively; the c.o.g. of these levels is 3522 ± 5 MeV, somewhat higher than predicted. (*d*) Most model calculations of the splitting among the 1^3D_J levels expect it to be smaller than that observed for the 1^3P_J (see the next section), so that all three of these $L = 2$ states may be fairly close to the mass of $\psi(3772) = 1^3D_1$. (*e*) The region between c.m. energy $W = 4.0$ and 4.3 GeV in e^+e^- annihilation is quite complicated (see Section 3.5) and difficult to interpret. It is obvious that the peak $\psi(4028)$ is a resonance and, within the charmonium model, is assigned to the 3^3S_1 level. If it should become equally clear that the enhancement at $W \approx 4.16$ GeV is a resonance, the candidate assignment is 2^3D_1. That both these states and the $\psi(4414)$, assigned to 4^3S_1, appear $\sim$150 MeV lower than predicted shows that the approximations underlying the model are breaking down. Another sign of this is the sizeable electronic width, $\sim$0.2–0.4 keV, of $\psi(3772)$. In the nonrelativistic potential model, $\Gamma_e \propto |\Psi(0)|^2$ vanishes for all but $^3S_1$ states. The observed electronic width is fairly well understood as a coupled-channel effect, and is discussed in Section 3.5.

3.3.2 CHARMONIUM TRANSITION RATES A great deal can be said about the decay rates and branching ratios of the charmonium levels (below $D\bar{D}$ threshold), most of which requires some knowledge of the bound-state wave functions. The rates that can be computed in the nonrelativistic model fall into two classes: those that depend on the wave function at very short $c\bar{c}$ separations (electronic, two-photon, and the hadronic widths obtained from gluon counting), and those that involve overlaps of radial wave functions (E1 and M1 radiative widths). The reader should be aware of what is being neglected in these calculations. Those involving the wave function $\Psi_{nL}(r)$ near $r = 0$ certainly ignore possibly important relativistic and quantum effects such as spin dependence and pair creation, as well as the mixing among states (e.g. 2^3S_1 and 1^3D_1) induced by coupling to decay channels. The second type of calculation does not include almost certain reductions in the overlap integral due to differing gluon distributions in the initial and final states, nor does it take account of the spin-dependent and mixing effects noted above. Finally, the charmonium model has very little to say on the important question of $\psi' \to \psi\pi\pi$ (and $\Upsilon' \to \Upsilon\pi\pi, \Upsilon\eta$) because the rather low momentum imparted to the $\pi\pi$ or η makes gluon counting especially dubious. See, however, Gottfried (71) for scaling rules for these decays.

The leptonic decay width is readily computed from the graph in Figure 3. If Q is the charge of the quark q, M_{n0} the mass of the $J^{PC} = 1^{--}$ states $n\,^3S_1$, $\Psi_{n0}(0)$ its radial wave function at the origin, and m_l the mass of the charged lepton, the result is (72)

$$\Gamma(n^3S_1 \to l^+l^-) = 4Q^2\alpha^2 \frac{|\Psi_{n0}(0)|^2}{M_{n0}^2}\left(1+\frac{2m_l^2}{M_{n0}^2}\right)\left(1-\frac{4m_l^2}{M_{n0}^2}\right)^{\frac{1}{2}}. \qquad 3.10$$

The terms involving m_l are relevant for decays such as $\psi' \to \tau^+\tau^-$ and $\Upsilon \to \tau^+\tau^-$.

The rates for the OZI-forbidden direct hadronic decays of heavy quark systems via annihilation to the minimum possible number of gluons can be computed from graphs as in Figure 4. The results are (to lowest order in quark velocity) (65, 73)

$$\Gamma(n^3S_1 \to \text{hadrons}) = \frac{40}{81\pi}(\pi^2-9)\alpha_s^3 \frac{|\Psi_{n0}(0)|^2}{M_{n0}^2}, \qquad 3.11$$

$$\Gamma(n^1S_0 \to \text{hadrons}) = \frac{8}{3}\alpha_s^2 \frac{|\Psi_{n0}(0)|^2}{M_{n0}^2}, \qquad 3.12$$

$$\Gamma(n^3P_0 \to \text{hadrons}) = 96\alpha_s^2 \frac{|\Psi'_{n1}(0)|^2}{M_{n1}^4}, \qquad 3.13$$

$$\Gamma(n^3P_1 \to \text{hadrons}) = \frac{128}{3\pi}\alpha_s^3 \frac{|\Psi'_{n1}(0)|^2}{M_{n1}^4} \ln\left(\frac{4m_q^2}{4m_q^2 - M_{n1}^2}\right), \qquad 3.14$$

$$\Gamma(n^3P_2 \to \text{hadrons}) = \frac{128}{5}\alpha_s^2 \frac{|\Psi'_{n1}(0)|^2}{M_{n1}^4}, \qquad 3.15$$

$$\Gamma(n^1P_1 \to \text{hadrons}) = \frac{320}{9\pi}\alpha_s^3 \frac{|\Psi'_{n1}(0)|^2}{M_{n1}^4} \ln\left(\frac{4m_q^2}{4m_q^2 - M_{n1}^2}\right). \qquad 3.16$$

Here α_s is evaluated at the bound-state mass M_{nL}, and $\Psi'_{nL}(0)$ is the slope of the radial wave function at $r = 0$. The rate for ψ_{nL} to decay directly to a photon plus hadrons may also be obtained from gluon counting; the formula is (69, 74)

$$\Gamma(n^3\mathrm{S}_1 \to \gamma + \text{hadrons}) = \frac{32}{9\pi}(\pi^2 - 9)\,\alpha_s^2\,\alpha Q^2 \frac{|\Psi_{n0}(0)|^2}{M_{n0}^2}. \qquad 3.17$$

Two more direct decay widths, not involving gluon counting but of great importance nonetheless, are

$$\Gamma(n^3\mathrm{S}_1 \to \gamma \to \text{hadrons}) = R_{\text{bkgd}}\,\Gamma(n^3\mathrm{S}_1 \to \mathrm{e}^+\mathrm{e}^-) \quad \text{and} \qquad 3.18$$

$$\Gamma(n^1\mathrm{S}_0 \to \gamma\gamma) = 12Q^4\alpha^2 \frac{|\Psi_{n0}(0)|^2}{M_{n0}^2} \qquad 3.19\text{a}$$

$$= \frac{9}{2} Q^4 \left(\frac{\alpha}{\alpha_s}\right)^2 \Gamma(n^1\mathrm{S}_0 \to \text{hadrons}), \qquad 3.19\text{b}$$

where R_{bkgd} is the value of R just off the resonance peak.

Note that Equation 3.19a involves only the short-distance (positronium-like) assumption, while Equation 3.19b involves the much stronger gluon-counting assumption. The two-photon width of $^3\mathrm{P}_0$ and $^3\mathrm{P}_2$ states is related to their hadronic (two-gluon) widths by the same factor, $9Q^4\alpha^2/2\alpha_s^2$, as for $^1\mathrm{S}_0$ states. To the extent that wave functions are independent of total quark spin and angular momentum $\mathbf{J} = \mathbf{L} + \mathbf{S}$ (for fixed n, L), we have

$$\Gamma(n^3\mathrm{S}_1 \to \text{hadrons})/\Gamma(n^1\mathrm{S}_0 \to \text{hadrons}) = \frac{5(\pi^2 - 9)\alpha_s}{27\pi} \qquad 3.20$$

and

$$\Gamma(n^3\mathrm{P}_0 \to \text{hadrons}) : \Gamma(n^3\mathrm{P}_1 \to \text{hadrons}) : \Gamma(n^3\mathrm{P}_2 \to \text{hadrons})$$

$$= 1 : \frac{4\alpha_s}{9\pi} \ln\left(\frac{4m_q^2}{4m_q^2 - M_{n1}^2}\right) : \frac{4}{15}. \qquad 3.21$$

Equations 3.14 and 3.16 for $^3\mathrm{P}_1$, $^1\mathrm{P}_1 \to$ hadrons deserve some comment, In the spirit of gluon counting, a spin one state cannot decay to two massless, on-shell gluons, and so we expect these rates to be $O(\alpha_s^3)$, not $O(\alpha_s^2)$ (73, 75). In particular, χ_1 should be the narrowest of the $^3\mathrm{P}_J$ levels. In an actual calculation of these rates (73), the dominant contribution involves the exhibited logarithm, due to a threshold singularity at $2m_c = M_{\chi_1}$. Such a large logarithm is always worrisome in QCD calculations, since it may signal the breakdown of perturbation theory. So it is

Table 3 Direct decays in charmonium, calculated from Equations 3.10 to 3.21: $\Gamma(\chi_J \rightarrow \text{all}) = \Gamma(\chi_J \rightarrow \text{hadrons}) + \Gamma_\gamma(\chi_J \rightarrow \psi), \Gamma(\eta_c' \rightarrow \text{all}) = \Gamma(\eta_c' \rightarrow \text{hadrons}) + \Gamma(\eta_c' \rightarrow \eta_c \pi\pi)$, and $\Gamma(\eta_c' \rightarrow \eta_c \pi\pi) = \Gamma(\psi' \rightarrow \psi\pi\pi)$ are assumed

Mode	Rate (keV)	Branching ratio (%)
$\eta_c \rightarrow$ hadrons	5.1×10^3	100
$\eta_c \rightarrow \gamma\gamma$	7.1	0.14
$\psi \rightarrow \gamma +$ hadrons	6.1	8.0
$\chi_0 \rightarrow$ hadrons	1.8×10^3	90
$\chi_0 \rightarrow \gamma\gamma$	2.5	0.13
$\chi_1 \rightarrow$ hadrons	105	21
$\chi_2 \rightarrow$ hadrons	480	48
$\chi_2 \rightarrow \gamma\gamma$	0.66	6.6×10^{-2}
$\eta_c' \rightarrow$ hadrons	3.3×10^3	97
$\eta_c' \rightarrow \gamma\gamma$	4.5	0.13
$\psi' \rightarrow$ hadrons	31	14
$\psi' \rightarrow e^+e^-$	3.4	1.5
$\psi' \rightarrow \gamma +$ hadrons	3.9	1.7

difficult to take these results too seriously beyond the reasonable (and conservative) guess that $\Gamma(^3P_1 \rightarrow \text{hadrons})$ is $\sim \alpha_s \Gamma(^3P_{0,2} \rightarrow \text{hadrons})$.

The predictions of these formulae for the charmonium system are listed in Table 3. Experimental comparisons are best delayed until the discussions of radiative transition rates. The value of α_s used here is determined by fitting to the total width of ψ,

$$\begin{aligned}\Gamma(\psi \rightarrow \text{all}) &\cong \Gamma(\psi \rightarrow e^+e^-)[2 + R_{\text{bkgd}}] + \Gamma(\psi \rightarrow \gamma + \text{hadrons}) \\ &\qquad + \Gamma(\psi \rightarrow \text{hadrons}) \\ &= 69 \pm 15 \text{ keV},\end{aligned} \tag{3.22}$$

where the small width for $\psi \rightarrow \gamma X(2830)$ has been ignored. Using $R_{\text{bkgd}} \approx 2.2$ (58, 59) and $\Gamma(\psi \rightarrow e^+e^-) = 5.3$ keV gives[10]

$$\alpha_s(M_\psi) \approx 0.19. \tag{3.23}$$

The wave functions for states other than ψ are determined using the parameters in Equation 3.9.

The important qualitative features of these calculations are: (*a*) The ground-state pseudoscalar η_c is expected to have a total width in the MeV range, about 100 times greater than the width of ψ, and its branching ratio to two photons should be $\sim 10^{-3}$. Similar remarks apply to its first radial excitation η_c', which may also decay to $\eta_c + \pi\pi$ [presumably, $\Gamma(\eta_c' \rightarrow \eta_c\pi\pi)$

[10] This value of $\alpha_s(M_\psi)$ is about a factor of two smaller than that deduced from analyses of electroproduction data.

$\approx \Gamma(\psi' \to \psi\pi\pi)$]. (*b*) As discussed above, gluon counting implies that $^3P_1$ and $^1P_1$ have considerably less hadronic width than $^3P_0$ and $^3P_2$. An immediate consequence is that the branching ratio $B(\psi' \to \chi_J\gamma)\, B(\chi_J \to \psi\gamma)$ should be largest for the 1^{++} state χ_1. (*c*) The potential model predicts

$$\Gamma(\psi' \to e^+e^-) = 3.4\ \text{keV} \quad \text{and} \quad \Gamma(\psi' \to 3\ \text{gluons} \to \text{hadrons}) = 31\ \text{keV},$$

both in fair agreement with the measured values (59), 2.0 keV and ~20 keV respectively. These results lend some support to both the presence of a linear confining term in V_0 (since for $\kappa = 0$, $|\Psi_{10}(0)| = |\Psi_{20}(0)|$) and to the gluon-counting calculation of the direct hadronic width.

We turn now to the radiative decays. The E1 transition rate between S- and P-wave states having the same total quark spin is

$$\Gamma_\gamma(\mathrm{S} \leftrightarrow \mathrm{P}) = \frac{4}{9}\left(\frac{2J_\mathrm{f}+1}{2J_\mathrm{i}+1}\right) Q^2\alpha\,|E_\mathrm{if}|^2\,\omega^3, \qquad 3.24$$

where ω is the photon energy, J_i (J_f) the total angular momentum of the initial (final) state, $Q = \frac{2}{3}$ for charmed quarks, and E_if is the transition dipole matrix element,

$$E_\mathrm{if} = \int_0^\infty r^2\,\mathrm{d}r\,\Psi_\mathrm{i}(r)\,\Psi_\mathrm{f}(r)\,r. \qquad 3.25$$

Here, $\Psi_{\mathrm{i,f}}$ are initial (final) state radial wave functions. The E1 rates between $^3D_1$ and ^3P_J states are given by

$$\Gamma_\gamma(^3\mathrm{D}_1 \leftrightarrow {}^3\mathrm{P}_J) = \frac{4}{9}\left(\frac{2J_\mathrm{f}+1}{2J_\mathrm{i}+1}\right) D_\mathrm{if}Q^2\alpha\,|E_\mathrm{if}|^2\,\omega^3, \qquad 3.26$$

where $D_\mathrm{if} = 1, \frac{1}{4}, \frac{1}{100}$ for $J = 0, 1, 2$ respectively. For the charmonium system, the 1^3D_1 state lies above charm threshold and so it has a very small branching ratio to $\chi_J+\gamma$. Therefore, Equation 3.26 is useful only indirectly, through the mixing between 2^3S and 1^3D_1 (see Section 3.5). In the Υ and heavier quark bound systems, however, there is an excellent chance for direct observation of the $^3D_1 \leftrightarrow {}^3P_J$ E1 transitions.

The M1 transition rate between $^3S_1$ and $^1S_0$ states is taken to be

$$\Gamma_\gamma(^3\mathrm{S}_1 \leftrightarrow {}^1\mathrm{S}_0) = \frac{16}{3}\left(2J_\mathrm{f}+1\right)\left(\frac{Q}{2m_\mathrm{q}}\right)^2 \alpha\,|M_\mathrm{if}|^2\,\omega^3, \qquad 3.27$$

where a Dirac moment is assumed for the quark, and

$$M_\mathrm{if} = \int r^2\,\mathrm{d}r\,j_0(\tfrac{1}{2}\omega r)\,\Psi_\mathrm{i}(r)\,\Psi_\mathrm{f}(r); \qquad 3.28$$

where j_0 is the spherical Bessel function of order zero.

For the "allowed" M1 transitions between hyperfine partners, M_{if} is very nearly unity because $\Psi_i = \Psi_f$ and $\frac{1}{2}\omega\langle r\rangle_{if} \ll 1$, so that $j_0 \approx 1$. For the same reason, M1 transitions between S states corresponding to different radial quantum numbers $n_i \neq n_f$ are strongly suppressed. It is still true that $\frac{1}{2}\omega\langle r\rangle_{if} \ll 1$, and

$$M_{if} \approx -\frac{\omega^2}{24}\int r^2\, dr\, \Psi_i(r)\, \Psi_f(r)\, r^2 \ll 1 \qquad (i \neq f). \qquad 3.29$$

The immediate consequence is that the "hindered" M1 transitions $\psi'(2^3S_1) \to \eta_c(1^1S_0)+\gamma$ and $\eta_c'(2^1S_0) \to \psi(1^3S_1)+\gamma$ are expected to be very rare compared to allowed M1 and E1 transitions.

The predictions of the potential model for E1 rates and branching fractions are compared with experimental observations in Table 4. Following the custom of traditional spectroscopy, experimental values of the χ masses are used, so what is being tested here is the theoretical strength Γ_γ/ω^3.

Given the naiveté of the simple potential model, the agreement is rather good, with theory lying within 1–2 standard deviations of experiment. Especially noteworthy are: (*a*) The normalized experimental rates are (with large errors) (59)

$$\frac{\Gamma(\psi' \to \gamma\chi_2)}{5\omega_2^3} : \frac{\Gamma(\psi' \to \gamma\chi_1)}{3\omega_1^3} : \frac{\Gamma(\psi' \to \gamma\chi_0)}{\omega_0^3} = 1:0.7:0.6, \qquad 3.30$$

with unity expected for pure E1 transitions from $^3S_1$ to ^3P_J states. (*b*) The measured branching ratios for $\chi_J \to \gamma\psi$ are quite consistent with predictions based on both the potential model and gluon counting for

Table 4 E1 decays in charmonium for theory (68) and experiment (59); predicted total widths for $\chi_{0,1,2}$ of 2.0, 0.5, and 1.0 MeV, respectively have been assumed

Datum	Theory (keV for Γ_γ, % for B_γ)	Experiment (keV for Γ_γ, % for B_γ)
$\Gamma_\gamma(\psi' \to \chi_2)$	27	16 ± 9
$\Gamma_\gamma(\psi' \to \chi_1)$	38	16 ± 8
$\Gamma_\gamma(\psi' \to \chi_0)$	44	16 ± 9
$\Gamma_\gamma(\chi_2 \to \psi)$	525	—
$B_\gamma(\chi_2 \to \psi)$	52	14 ± 9
$\Gamma_\gamma(\chi_1 \to \psi)$	395	—
$B_\gamma(\chi_1 \to \psi)$	79	35 ± 16
$\Gamma_\gamma(\chi_0 \to \psi)$	190	—
$B_\gamma(\chi_0 \to \psi)$	9.5	3 ± 3

the χ hadronic widths, with χ_1 considerably more narrow than χ_0 and χ_2. These facts strengthen the J^P assignments discussed in Section 3.1.

The M1 transition rates are compared with experiment in Table 5, where we have made the tentative assignments of $X(2830) = \eta_c(1^1S_0)$ and $\chi(3455) = \eta'_c(2^1S_0)$. If these identifications are found to be correct, they will represent a serious failure of the charmonium model:

1. From Table 1,

$$B[\psi \to X(2830)\gamma]\, B(X \to \gamma\gamma) = 1.3 \pm 0.4 \times 10^{-4} \qquad 3.31a$$

$$B[\psi \to X(2830)\gamma] < 0.017 \quad (90\% \text{ confidence level}). \qquad 3.31b$$

From these, one infers

$$B[X(2830) \to \gamma\gamma] \gtrsim 8 \times 10^{-3}, \qquad 3.31c$$

which is at least a factor of five larger than the predicted value (Table 3). This is to be contrasted with the apparent success of calculated direct decay rates for the ^3P states and for ψ'.

2. The model fails by at least an order of magnitude in predicting $\psi \to \gamma\eta_c$. This is especially puzzling when one recalls that M1 transitions among the light mesons are fairly well described by the nonrelativistic quark model (72, 76). Particularly relevant to $\psi \to \gamma\eta_c$ are the predictions $\Gamma(\phi \to \gamma\eta) = 70$ keV and $\Gamma(\phi \to \gamma\pi^0) = 6.9$ keV, to be compared to the measured widths of 82 ± 16 keV and 5.7 ± 2.0 keV, respectively. To add to the puzzle, there is the apparently successful prediction of a strongly suppressed $\psi' \to \gamma\eta_c$ transition.

Table 5 M1 decays in charmonium; observed total widths of ψ and ψ' (Table 1) and predicted total widths of η_c and η'_c (Table 2) are used in determining branching ratios

Datum	Theory (keV for Γ_γ, % for B_γ)	Experiment (keV for Γ_γ, % for B_γ)
$\Gamma_\gamma(\psi \to \eta_c)$	26	<1.2
$B_\gamma(\psi \to \eta_c)$	37	<1.7
$B_\gamma(\psi \to \eta_c)B(\eta_c \to \gamma\gamma)$	0.052	0.013 ± 0.004
$\Gamma_\gamma(\psi' \to \eta_c)$	1.9	<2.3
$B_\gamma(\psi' \to \eta_c)$	0.83	<1.0
$B_\gamma(\psi' \to \eta_c)B(\eta_c \to \gamma\gamma)$	1.2×10^{-3}	$<3.4 \times 10^{-2}$
$\Gamma_\gamma(\psi' \to \eta'_c)$	17	<5.7
$B_\gamma(\psi' \to \eta'_c)$	7.5	<2.5
$B_\gamma(\psi' \to \eta'_c)B(\eta'_c \to \gamma\gamma)$	1.0×10^{-2}	$<3.1 \times 10^{-2}$
$\Gamma_\gamma(\eta'_c \to \psi)$	0.53	—
$B_\gamma(\eta'_c \to \psi)$	1.6×10^{-2}	$>24 \pm 16$
$B_\gamma(\psi' \to \eta'_c)B_\gamma(\eta'_c \to \psi)$	1.2×10^{-3}	0.6 ± 0.4

3. While the prediction $B(\psi' \to \gamma\eta_c') = 0.075$ is only three times larger than the observed branching ratio limit, the inferred lower limit, $B(\eta_c' \to \gamma\psi) \gtrsim 0.15$, is 2–3 orders of magnitude greater than what one expects theoretically for this hindered M1 transition.

If X(2830) really is the η_c, the resolution to these difficulties must lie partly in correcting the assumption of identical radial wave functions for ψ and η_c, i.e. that gluons play an important role in suppressing both the M1 overlap integral and $\Gamma(\eta_c \to 2 \text{ gluons})/\Gamma(\eta_c \to \gamma\gamma)$. On the other hand, given the successes of gluon counting for direct decays of spin-triplet states, the verified suppression of the hindered M1 transition $\psi' \to \gamma\eta_c$, and the experimental fact that $\Gamma(\psi \to \gamma\eta_c \gtrsim 1$ keV, there is no way to understand the identification $\chi(3455) = \eta_c'$ with such a large branching ratio to $\gamma\psi$.

It is always possible, of course, that η_c and η_c' have not been discovered yet and that they lie $\lesssim 100$ MeV below their hyperfine partners, as originally expected (47, 53). In that case, theoretical estimates of the M1 rates are greatly reduced and, in fact, lie within experimental limits for states at such masses.

The natural question then is: What are these two states? Various conjectures abound including:

1. They are four-quark ($c\bar{c}\, q\bar{q}$) or molecular states (η_c and π^0 bound in an S wave, say) (77) or, perhaps $c\bar{c}$ states with a gluon excitation (78). There are no convincing models for such relatively low-mass systems that are not pure $c\bar{c}$, nor is there even the ability to make convincing estimates of transition rates. Such techniques are sorely needed.

2. Another interesting speculation is that they are Higgs mesons H (79). If this is the case, H certainly does not have the "conventional" coupling to quarks $\sim G_F^{\frac{1}{2}} m_q$, for then one would estimate (with $m_c \approx \frac{1}{2} M_{\psi,\psi'}$)

$$\frac{\Gamma(\psi,\psi' \to \gamma\mathrm{H})}{\Gamma(\psi,\psi' \to e^+e^-)} \approx \frac{G_F M_{\psi,\psi'}^2}{2^{\frac{1}{2}}\pi\alpha} \frac{M_{\psi,\psi'}^2(M_{\psi,\psi'} - M_{\mathrm{H}}^2}{(M_{\psi,\psi'}^2 + M_{\mathrm{H}}^2)^2} \approx 10^{-4} \qquad 3.32$$

Not only does one need the coupling of H to charmed quarks to be anomalously large, but the coupling to light quarks must be anomalously small if $\mathrm{H} \to \gamma\gamma$ is to be a sizable decay mode (80). [Actually, $\chi(3455) = \mathrm{H} \to \psi\gamma$ might be a dominant decay mode of a more-or-less conventional Higgs meson.]

3. One final possibility, suggested by Harari (81), is that $\chi(3455)$ is the ${}^1D_2$ level of charmonium. Because of the strongly hindered nature of ${}^3S_1 \to {}^1D_2$ radiative transitions, such a possibility is viable only if ψ' and ψ contain a sizable admixture ($\sim$5–10%) of ${}^3D_1$ and if $\Gamma({}^1D_2 \to \text{hadrons}) \approx \Gamma(\psi \to \text{hadrons})$ (82). Perhaps the most serious objection to this identification, based as it is on the presumed large splitting between

all triplet and singlet states of given L, is that $\Gamma(\psi' \to \gamma\eta'_c)$ is expected to be $\sim 10\,\Gamma(\psi' \to \gamma\,^1D_2)$ and yet, on this hypothesis, η'_c has not been seen yet.

To summarize, while the basic model gives a very good qualitative and creditable quantitative description of the spectrum and transition rates of the spin-triplet charmonium levels, it fails to account for practically all observed features of the proposed singlet levels. Either something very important (and largely unknown) is missing from the model or, more happily, the model is telling us that new degrees of freedom—which it was never intended to handle—have been discovered.

3.4 *Including Spin Dependence*

Even before the states between ψ and ψ' were discovered and their splittings measured, many people began trying to incorporate quark spin dependence into the charmonium model. The very earliest work utilizing a Coulomb model (47, 53, 69) was much too naive and was soon abandoned. However, the nonrelativistic character of the low-lying $c\bar{c}$ states suggested that the Bethe-Salpeter equation would be a useful formalism and that some analog of the Breit-Fermi Hamiltonian for positronium would continue to be relevant. Almost nothing was known about the structure of the Bethe-Salpeter kernel and some educated guesses were needed to make the computation of splittings possible. Some perturbation-theoretic analyses of QCD already suggested that the spin-dependent part of the Hamiltonian would be strongly modified away from a Coulomb form just as the spin-independent confining part was (83). These same investigations further indicated that the modifications of the two pieces might not be simply related.

One early guess, however, was that the spin-dependent interaction would have only a short-range Coulomb-type structure (84). Very small fine structure splittings were predicted, such as $M_\psi - M_{\eta_c} \approx 30$ MeV and $M(^3P_2) - M(^3P_1) \approx 5$–$10$ MeV. This is in sharp disagreement with experiment, and such an approach now seems inadequate. In particular, Johnson (85) has recently emphasized that at least one part of the spin effect is necessarily long range, namely that part of the spin-orbit interaction arising from Thomas precession (see Equation 3.37 below). Indeed, almost all treatments of the spin forces in charmonium have focused on the long-range part of the $c\bar{c}$ interaction. One exception is the model of Celmaster et al (86), mentioned below in Section 3.6. They assume an r-dependent short-distance coupling $\kappa(r)$, whose form is suggested by asymptotic freedom, and they use $V_{AF}(r) = -\kappa(r)/r$ to generate the Breit-Fermi interaction. Although they obtained much larger splittings than those in Reference 84 (see Table 6), they have neglected to take account of the long-range contribution from Thomas precession.

The most popular approach, pioneered by Schnitzer (87) and by Pumplin, Repko & Sato (88), and since generalized by many authors (89), has been to assume that heavy quark binding is effectively due to "single gluon exchange with renormalization group improvement," summarized by an instantaneous Bethe-Salpeter kernel consisting of vector and scalar interaction terms:

$$V_{\text{coul}}(\mathbf{k}^2)\gamma_1^{\mu}\gamma_{2\mu}+V_{\text{v}}(\mathbf{k}^2)\Gamma_1^{\mu}(k)\Gamma_{2\mu}(k)+V_{\text{s}}(\mathbf{k}^2)\mathbf{1}_1\mathbf{1}_2. \qquad 3.33$$

The subscripts 1 and 2 refer to the c and $\bar{\text{c}}$ quarks, **1** is a unit matrix in Dirac space, k is the four-momentum carried by the exchanged gluon, and (with $\sigma_{\mu\nu}=(2i)^{-1}[\gamma_\mu,\gamma_\nu]$)

$$\Gamma_\mu(k)=\gamma_\mu-\frac{i\lambda}{2m_{\text{c}}}\sigma_{\mu\nu}k^\nu, \qquad 3.34$$

where λ is the color magnetic moment of the quark—an adjustable parameter. In the spin-independent, nonrelativistic limit of this interaction, the potential is

$$V_0=V_{\text{coul}}+V_{\text{v}}+V_{\text{s}}. \qquad 3.35$$

In most discussions, it has been assumed that

$$V_{\text{v}}=\eta V_{\text{lin}} \qquad V_{\text{s}}=(1-\eta)V_{\text{lin}}, \qquad 3.36$$

where V_{lin} is the linear confining potential r/a^2 in coordinate space, and η is another adjustable parameter.

Using Equations 3.33 to 3.36, the authors obtained the spin-dependent potential by following the same steps used to convert the kernel for positronium into the Breit interaction. The result is:

$$\begin{aligned} V_{\text{spin}}(r)=&\frac{1}{2m_{\text{c}}^2}\left[\frac{4\kappa}{r^3}+4(1+\lambda)\frac{1}{r}\frac{\text{d}V_{\text{v}}}{\text{d}r}-\frac{1}{r}\frac{\text{d}V_0}{\text{d}r}\right]\mathbf{L}\cdot\mathbf{S}\\ &+\frac{2}{3m_{\text{c}}^2}\left[4\pi\kappa\delta(\mathbf{r})+(1+\lambda)^2\,\nabla^2V_{\text{v}}(r)\right]\mathbf{S}_1\cdot\mathbf{S}_2\\ &+\frac{1}{3m_{\text{c}}^2}\left[\frac{3\kappa}{r^3}+\frac{1}{r}\frac{\text{d}V_{\text{v}}}{\text{d}r}-\frac{\text{d}^2V_{\text{v}}}{\text{d}r^2}\right]S_{12}, \end{aligned} \qquad 3.37\text{a}$$

where

$$S_{12}=3\mathbf{S}_1\cdot\hat{\mathbf{r}}\,\mathbf{S}_2\cdot\hat{\mathbf{r}}-\mathbf{S}_1\cdot\mathbf{S}_2. \qquad 3.37\text{b}$$

The last term in the spin-orbit part of V_{spin} is the Thomas precession contribution, and contains the only influence of the scalar interaction on spin dependence. For the potentials in Equation 3.36, V_{spin} is given by

$$V_{\text{spin}}(r) = \frac{1}{2m_c^2}\left\{\frac{3\kappa}{r^3} + \frac{1}{ra^2}\left[\eta(3+4\lambda) - (1-\eta)\right]\right\}\mathbf{L}\cdot\mathbf{S}$$

$$+\frac{2}{3m_c^2}\left[4\pi\kappa\delta(\mathbf{r}) + \frac{2\eta}{ra^2}(1+\lambda)^2\right]\mathbf{S}_1\cdot\mathbf{S}_2$$

$$+\frac{1}{3m_c^2}\left[\frac{3\kappa}{r^3} + \frac{\eta}{ra^2}(1+\lambda)^2\right]S_{12}. \qquad 3.38$$

When used perturbatively, this interaction generates the following mass formulae (below, M_L = bare mass of the level ψ_{nL} determined by V_0 and $\langle r^{-c}\rangle_L = \langle\psi_{nL}|r^{-c}|\psi_{nL}\rangle$):

$$M(^3S_1) = M_0 + \frac{1}{6m_c^2}\left[4\pi\kappa|\Psi_{n0}(0)|^2 + 2(1+\lambda)^2\,\eta a^{-2}\langle r^{-1}\rangle_0\right]$$

$$M(^1S_0) = M_0 - \frac{1}{2m_c^2}\left[4\pi\kappa|\Psi_{n0}(0)|^2 + 2(1+\lambda)^2\,\eta a^{-2}\langle r^{-1}\rangle_0\right]$$

$$M(^3P_2) = M_1 + \frac{1}{5m_c^2}\left\{7\kappa\langle r^{-3}\rangle_1 + \left[\left(9+13\lambda+\frac{3}{2}\lambda^2\right)\eta - \frac{5}{2}\left(1-\eta\right)\right] a^{-2}\langle r^{-1}\rangle_1\right\}$$

$$M(^3P_1) = M_1 + \frac{1}{m_c^2}\left\{-\kappa\langle r^{-3}\rangle_1 + \left[\left(-1-\lambda+\frac{1}{2}\lambda^2\right)\eta + \frac{1}{2}\left(1-\eta\right)\right] a^{-2}\langle r^{-1}\rangle_1\right\}$$

$$M(^3P_0) = M_1 - \frac{1}{m_c^2}\left\{4\kappa\langle r^{-3}\rangle_1 + \left[(3+4\lambda)\eta - (1-\eta)\right] a^{-2}\langle r^{-1}\rangle_1\right\}$$

$$M(^1P_1) = M_1 - \frac{(1+\lambda)^2\eta}{m_c a^2}\langle r^{-1}\rangle_1. \qquad 3.39$$

For completeness, we include formulae for the splitting of D levels and mixing of $^3D_1$ with $^3S_1$:

$$M(^3D_3) = M_2 + \frac{1}{7m_c^2}\left\{20\kappa\langle r^{-3}\rangle_2 + \left[(23+32\lambda+2\lambda^2)\eta - 7(1-\eta)\right] a^{-2}\langle r^{-1}\rangle_2\right\}$$

$$M(^3\mathrm{D}_2) = M_2 + \frac{1}{m_c^2}\left\{-\kappa\langle r^{-3}\rangle_2 + \left[\left(-1-\lambda+\frac{1}{2}\lambda^2\right)\eta + \frac{1}{2}\left(1-\eta\right)\right] a^{-2}\langle r^{-1}\rangle_2\right\}$$

$$M(^3\mathrm{D}_1) = M_2 + \frac{1}{6m_c^2}\left\{-30\kappa\langle r^{-3}\rangle_2 + \left[(-26-34\lambda+\lambda^2)\eta + 9(1-\eta)\right] a^{-2}\langle r^{-1}\rangle_2\right\}$$

$$M(^1\mathrm{D}_2) = M_2 - \frac{(1+\lambda)^2\eta}{m_c^2 a^2}\langle r^{-1}\rangle_2$$

$$\langle n^3\mathrm{S}_1 \,|\, V_{\mathrm{spin}} \,|\, m^3\mathrm{D}_1\rangle = \frac{8^{\frac{1}{2}}}{12m_c^2}\left[3\kappa\langle nS \,|\, r^{-3} \,|\, m\mathrm{D}\rangle + (1+\lambda)^2 \eta a^{-2}\langle nS \,|\, r^{-3} \,|\, m\mathrm{D}\rangle\right]. \qquad 3.40$$

The splittings among the P and S states determined by various authors from Equation 3.39 are listed in Table 6, together with the measured mass differences; for comparison purposes, we assumed X(2830) = $1^1\mathrm{S}_0$ and

Table 6 Spin-dependent splittings in charmonium measured in MeV. The numbers in brackets are values of κ, $1/a^2$ (GeV^2), m_c (GeV), η, and λ used by various authors. In Reference 86 the Coulomb parameter is not constant (see Equation 3.70) and only this short-range part of the potential is kept in computations of spin dependence

Author parameters	$M_{\chi_2} - M_{\chi_1}$	$M_{\chi_1} - M_{\chi_0}$	$M_\psi - M_{\eta_c}$	$M_{\psi'} - M_{\eta_c'}$
Experiment	44 ± 6	95 ± 5	265 ± 14	230 ± 7
Schnitzer (87) [0.2, 0.19, 1.6, 1.0, 0.0]	87	63	70	58
Pumplin et al (88) [0.0, 0.30, 1.5, 1.0, 0.0]	152	117	119	92
Henriques et al (89) [0.8, 0.18, 1.6, 0.0, 0.0]	40 (input)	80	95	—
Schnitzer (87) [0.2, 0.19, 1.6, 1.0, 1.1]	182	170	268 (input)	225
Chan (89) [0.2, 0.19, 1.6, 0.12, 5.0]	40 (input)	98	262 (input)	225
Carlson & Gross (89) [0.27, 0.20, 1.37, 0.08, 4.4]	41 (input)	98	265 (input)	181
Celmaster et al (86) [—, —; 1.98, 1.0, 0.0]	92	100	150	80

$\chi(3455) = 2^1S_0$. For the D states we content ourselves with two remarks. First the 1^3D_1 level comes out about 40 MeV below its unperturbed mass[11] (which is 3.755 GeV in Table 2). Thus, the agreement between the coupled-channel calculation (in Section 3.5) and the observed mass 3.772 GeV of ψ'' is perhaps fortuitous. Second, the predicted mixing angle between 1^3D_1 and 2^3S_1 is

$$\varepsilon = \frac{1}{2}\tan^{-1}\frac{2\langle 2^3S_1 | V_{\text{spin}} | 1^3D_1\rangle}{M(2^3D_1) - M(2^3S_1)} \approx 6^\circ. \qquad 3.41$$

This is much less than one infers from measurements of the ψ'' electronic width. Using

$$\varepsilon = \tan^{-1}\left\{\frac{M_{\psi''}}{M_{\psi'}}\left[\frac{\Gamma(\psi'' \to e^+e^-)}{\Gamma(\psi \to e^+e^-)}\right]^{\frac{1}{2}}\right\}, \qquad 3.42$$

these are $\varepsilon = (26 \pm 3)^\circ$ for $\Gamma_e(\psi'') = 0.37 \pm 0.09$ keV (90), and $\varepsilon = (19 \pm 3)^\circ$ for $\Gamma_e(\psi'') = 0.18 \pm 0.05$ keV (91).

The lessons of this attack on the problem of spin dependence may be summarized thus: Insofar as one is willing to extend the hypothesis of an instantaneous interaction between heavy quarks beyond the realm of the simple spin-independent potential V_0, the ansatz Equation 3.33 is the basis of a quite reasonable first effort. Clearly, the assumption of a purely short-range origin for spin forces is inadequate. Economy of parameters then demands that one attribute spin forces to the long-range part, here assumed to be r/a^2. As we have seen, this still leaves some freedom in the Dirac structure of the kernel, and the work of Henriques et al (89) first suggested that the P-state splittings are best fit if the long-range interaction is scalar. The reason for this is that the Thomas precession term, which is the most important part of the spin-orbit interaction in this case (η small), orders the ^3P_J levels oppositely from the other terms in V_{spin}. This feature is needed to explain the unexpectedly small ratio $[M(^3P_2) - M(^3P_1)]/[M(^3P_1) - M(^3P_0)] \approx 0.42$.

If one now assigns X(2830) to $\eta_c(1^1S_0)$, the splitting from ψ may be fit, as Schnitzer found (87), by assuming a rather large quark color anomalous moment, $\lambda(\eta)^{\frac{1}{2}} \approx 1$. All this was put together by Chan and by Carlson & Gross (89), who combined the successful features of the last two named pieces of work to obtain excellent agreement with the data.

But, in this flush of success, one must not lose sight of two important facts. First, all that has been accomplished so far is to fit four mass splittings with two parameters, λ and η. In fact, as Carlson & Gross point out, $M(^3P_1) - M(^3P_0)$ is far less sensitive to parameters than is

[11] For this, we have used the parameters of Carlson & Gross (89).

$M(^3P_2) - M(^3P_1)$. Therefore, in a fit to the latter plus the $\psi - X(2830)$ splitting, the former is almost automatically correct. And, it is not surprising that a large $1^3S_1 - 1^1S_0$ implies a comparable $2^3S_1 - 2^1S_0$ splitting. The real test of this phenomenology will come when the multitude of intramultiplet splittings in the Υ system are measured—a potentially difficult task if they are ~5–10 times smaller, as expected. Second, given the difficulty of accommodating the X(2830) and $\chi(3455)$ in the simple charmonium scheme, it may well be that it has been a great mistake all along to use these states in determining the parameters in V_{spin}. Again, only time will tell.

From a theoretical standpoint, the procedure outlined above suffers from the lack of a "first principles" justification for its starting point, the Bethe-Salpeter kernel defined in Equations 3.33 and 3.36. More ambitious approaches are in progress in a number of dynamical models that incorporate gluon degrees of freedom more or less explicitly. These include lattice gauge theories (51, 52) the MIT bag model (85, 92), and the quark-string models (78, 93). In each of these models, the long-range spin-independent potential between heavy quarks is shown to be linear. Further there is a reasonably well-defined procedure for extracting the spin-dependent interaction. While it is too early to assess these approaches, it is hoped that they will provide insight to this most puzzling aspect of charmonium dynamics.

3.5 *Coupling Charmonium to Its Decay Channels*

It was recognized very early (63) that the quantum mechanical coupling between charmonium states and their OZI-allowed decay channels could modify the predictions of the naive potential model. The development of a model for this coupling and the resulting predictions have been presented in a series of papers by Eichten et al (68, 75, 94; see also 95).[12] The main issues one wants to address in such a model are these:

1. Renormalization (shifts) of the bare spectrum generated by $V_0(r)$ and the widths of $c\bar{c}$ states above charm threshold, $W_c = 2M_{D^0} = 3.727$ GeV.
2. Renormalizations of the wave functions deduced from V_0. This includes leakage from the $c\bar{c}$ sector to the charmed hadron sector, as well as the mixing among charmonium levels having the same J^{PC}. Both of these will affect rates for all the transitions discussed in Section 3.3. Of special interest is the fact that decays forbidden or strongly suppressed in the potential model, such as $^3D_1 \rightarrow e^+e^-$ and $2^3S_1 \rightarrow 1^1S_0 + \gamma$,

[12] Kogut & Susskind (95) implement the coupling to decay channels in a quite different way from Eichten et al. In particular, charmed mesons never appear explicitly.

can be enhanced through mixing between 1^3D_1 and 2^3S_1, and between 2^3S_1 and 1^3S_1 (or 2^1S_0 and 1^1S_0), respectively.

3. A description of e^+e^- annihilation in the charm threshold region, $W_c \leqq W \lesssim 4.4$ GeV. In particular, one wants to interpret the structure of $R = \sigma(e^+e^- \to \text{hadrons})/\sigma(e^+e^- \to \mu^+\mu^-)$ in this region, and to discuss the general (and sometimes peculiar) features of the exclusive channel cross sections $\sigma(e^+e^- \to D^0\bar{D}^0)$, $\sigma(e^+e^- \to D^+D^-)$, $\sigma(e^+e^- \to D^0\bar{D}^{*0})$, etc.

3.5.1 FORMALISM AND DESCRIPTION OF THE MODEL The description of what happens to a discrete set of states in one part of Hilbert space when it is immersed in and coupled to a continuum of states belonging to another subspace is a classic problem in quantum mechanics and was, formally, solved long ago (96a; more recent discussions are given in 96b).

Let the total Hamiltonian of this system be

$$\mathscr{H} = \mathscr{H}_0 + \mathscr{H}_I, \tag{3.43}$$

where $\mathscr{H}_0$ is responsible for the binding of the discrete states $|n\rangle$ and the continuum states $|v\rangle$:

$$\begin{aligned} &\mathscr{H}_0|n\rangle = \varepsilon_n|n\rangle, \qquad \langle n|m\rangle = \delta_{nm}, \\ &\mathscr{H}_0|v\rangle = \omega_v|v\rangle, \qquad \langle v|\mu\rangle = \delta(v-\mu), \\ &\langle v|n\rangle = 0, \end{aligned} \tag{3.44}$$

while $\mathscr{H}_I$ is responsible for their coupling:

$$\langle v|\mathscr{H}|n\rangle = \langle v|\mathscr{H}_I|n\rangle. \tag{3.45}$$

In the present shorthand notation, $|n\rangle$ stands for any of the pure $c\bar{c}$ levels $|\psi_{nLJ}\rangle$, and $|v\rangle$ for any state of charmed hadrons having zero net charm, zero total momentum, and total energy ω_v. (States with more than one c and one $\bar{c}$ are ignored.)

The problem we face is to describe the eigenstates $|N\rangle$ of $\mathscr{H}$ with eigenvalue E_N. These states include the observed bound states (ψ, X_J, ψ', etc) below charmed threshold as well as the continuum states and resonances in e^+e^- annihilation above W_c. (Because $\mathscr{H}$ allows for decay, E_N need not be real—and certainly won't be for the resonances.) We begin by expanding $|N\rangle$ in the complete basis formed by the eigenstates $|n\rangle$ and $|v\rangle$ of $\mathscr{H}_0$:

$$\begin{aligned} &|N\rangle = \sum_n a_{nN}|n\rangle + \int dv\, a_{vN}|v\rangle, \\ &a_{nN} = \langle n|N\rangle, \qquad a_{vN} = \langle v|N\rangle. \end{aligned} \tag{3.46}$$

From Equations 3.44 to 3.46, the expansion coefficients and the energy E_N are solutions of the eigenvalue problem

$$\sum_m \left(E_N - \varepsilon_n\right) \delta_{nm} - \Omega_{nm}(E_N)\right] a_{mN} = 0 \quad \text{and} \tag{3.47}$$

$$\det\left[W - \varepsilon - \Omega(W)\right] = 0 \quad (\text{roots } E_N), \tag{3.48}$$

where

$$\Omega_{nm}(W) = \int \frac{d\nu}{W - \omega_\nu + i0} \langle n | \mathscr{H}_I | \nu \rangle \langle \nu | \mathscr{H}_I | m \rangle$$

$$\equiv \Delta_{nm}(W) - \frac{i}{2} \Gamma_{nm}(W). \tag{3.49}$$

For $E_N < W_c$, the solutions of Equations 3.47 and 3.48 correspond to the bound charmonium levels, the matrix $\Delta_{nm}(E_N)$ describes the shift in the mass of these levels from the "bare" masses ε_n, and the width matrix $\Gamma_{nm}(E_N < W_c)$ vanishes. For $|E_N| > W_c$, the state $|N\rangle$ is a resonance that decays almost exclusively to charmed hadrons, having mass M_N, width Γ_N, and $E_N = M_N - \frac{1}{2} i \Gamma_N$. Given the coefficients a_{nN}, one may determine $a_{\nu N}$ from

$$a_{\nu N} = \frac{1}{E_N - \omega_\nu + i0} \sum_m a_{mN} \langle \nu | \mathscr{H}_I | m \rangle \tag{3.50}$$

Finally, the continuum eigenstates $|N\rangle$ may be determined from integral equations similar in structure to Equations 3.47 and 3.48. For reasons that will become clear shortly, the model makes little use of these. Rather, the recalculation of the transition rates of the bound states and of

$$\Delta R \equiv \frac{\sigma(e^+e^- \to \text{charmed hadrons})}{\sigma(e^+e^- \to \mu^+\mu^-)} \tag{3.51}$$

requires only a knowledge of Ω, which in turn requires a model for $\mathscr{H}_I$.

The assumptions and approximations defining the model for $\mathscr{H}_I$ and Ω are (68, 75):

1. The Hamiltonian is taken to be

$$\mathscr{H} = -\tfrac{3}{8} \sum_{a=1}^{8} \int d^3x\, d^3y \rho_a(\mathbf{x},t)\, V(\mathbf{x}-\mathbf{y})\, \rho_a(\mathbf{y},t) \tag{3.52}$$

$$+ \text{quark kinetic energy terms,}$$

where

$$\rho_a(x) = \sum_{\text{flavors}(i)} q_i^\dagger(x) \frac{\lambda_a}{2} q_i(x)$$

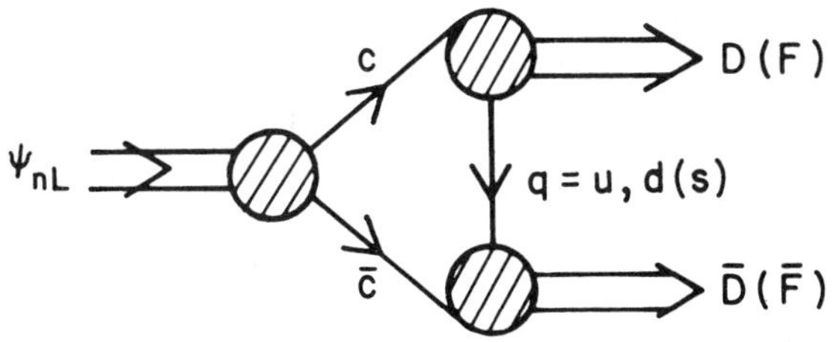

Figure 5 The transition amplitude (Equation 3.53) for $\psi_{nL} \rightarrow D\bar{D}$ or $F\bar{F}$.

is the octet of color charge densities and $V(\mathbf{x}-\mathbf{y}) = |\mathbf{x}-\mathbf{y}|/a^2$ is the instantaneous confining potential in Equation 3.8. The Coulomb piece has been dropped for simplicity.[13]

2. As the form of V implies, calculations with this Hamiltonian necessarily are carried out in the nonrelativistic approximation. Therefore, $\mathscr{H}$ is explicitly spin independent. The only dependence on quark total spin that enters the computation of Ω is through the use of "spin-split thresholds" for the continuum states $|\nu\rangle$, i.e. the mass difference between D^* and D, F^* and F, is put in by hand.

3. When decomposed into creation and annihilation operators, $\mathscr{H} = \mathscr{H}_0$ (binding of $q_i\bar{q}_j$)$+\mathscr{H}_I$ (pair emission: $q_i \rightarrow q_i + \bar{q}_j + q_j$) + other terms (e.g. emission of two pairs from the vacuum) that are discarded. While the nonrelativistic binding mechanism is presumably consistent only for $c\bar{c}$ states (where it reproduces the results of the spin-independent potential model described in Section 4.3), it is also used to generate bound-state wave functions for charmed meson states $c\bar{q}$ and $\bar{c}q$ (q = u,d,s from now on).

4. The model assumes that the transition $\psi_{nLJ} \rightarrow$ charmed mesons is a sequential quasi-two-body process,[14] e.g. $\psi_{nLJ} \rightarrow \bar{D}D^*$; $D^* \rightarrow D\pi$. Accordingly, the only terms kept in $\mathscr{H}_I$ are those describing light-pair emission, $c \rightarrow c+q+\bar{q}$ and $\bar{c} \rightarrow \bar{c}+q+\bar{q}$, which governs $\psi_{nLJ}(c\bar{c}) \rightarrow D(c\bar{q})+\bar{D}(\bar{c}q)$, as depicted in Figure 5. There, the shaded circles denote bound-state vertex functions (simply related to the wave functions in the nonrelativistic limit) and they emphasize that the model incorporates the extended nature of the parent and its decay products. While certain features (to be mentioned below) of the transition amplitudes computed with $\mathscr{H}_I$ may be model independent, the nonrelativistic approximation used in the computation is very questionable.

[13] Note that the Hamiltonian in Equation 3.52 corresponds to the nonrelativistic limit of a pure vector kernel in Equation 3.33.

[14] The experimental justification for the quasi-two-body hypothesis and for the neglect of charmed baryons is presented in Section 4.

5. The final approximation made in References 68 and 75 is a drastic truncation of the continuum states $|v\rangle$ to include only the ground-state charmed mesons D, D*, F, and F*. Consequently, the model is reliable (even semiquantitatively) only where the effects of higher thresholds (e.g. charmed P states) may be ignored. For the calculation of ΔR, the breakdown due to neglect of higher thresholds is already apparent at $W \approx 4.1$ GeV.

The general form of the transition amplitude for $\psi_{nLJ} \rightarrow$ pair of ground state charmed mesons is

$$\langle c\bar{q}(J_1; E_1, \mathbf{p}), \bar{c}q(J_2; E_2, -\mathbf{p})|\mathscr{H}_I|c\bar{c}(nLJ; W, \mathbf{0})\rangle$$

$$= \left(m_q \frac{m_c m_q}{m_c + m_q}\right)^{-1} \times \text{spin factor } (J_i) \times \text{form factor } (n, L; |\mathbf{p}|; m_c, m_q, a), \qquad 3.53$$

where $(E_i, \mathbf{p}_i)$ are the four-momenta of the outgoing pair, $W = E_1 + E_2$, and J_i are total angular momenta. The parameters entering the calculation of this amplitude are the quark masses m_c, m_q and the linear potential strength a^{-2}. Because of the spin independence of the $q\bar{q}$ production mechanism, all dependence on quark spin appears in a Clebsch-Gordon coefficient, the second factor in Equation 3.53. The first factor, which implies a suppressed production of $F = c\bar{s}$ relative to $D = c\bar{u}$, arises from the S-wave nature of the production mechanism and from the charmed meson wave function. The P-wave form factors for the first

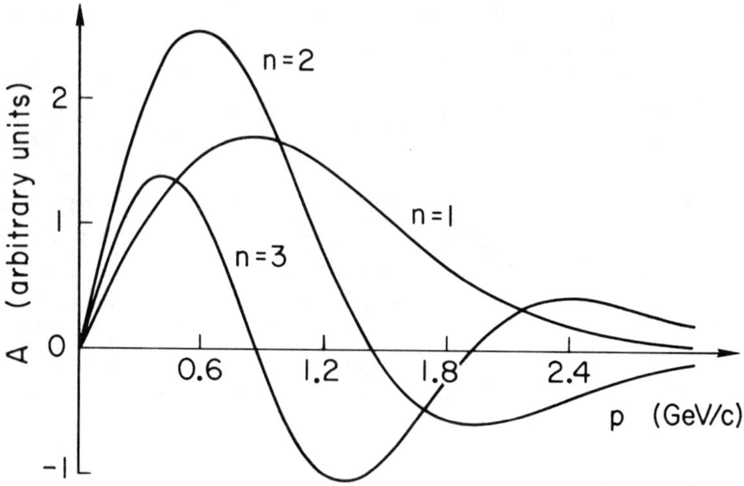

Figure 6 The P-wave form factor in Equation 3.53 for $\psi_{n0} \rightarrow D\bar{D}$ ($n = 1, 2, 3$) as a function of the momentum of the outgoing charmed meson, $p = \frac{1}{2}(W^2 - 4M_D^2)^{\frac{1}{2}}$ (from 68). See Table 7 for parameters used in this calculation.

three $L = 0$ $c\bar{c}$ levels to go to a $D\bar{D}$ pair are plotted in Figure 6 as a function of the D momentum, $|\mathbf{p}|$. The high-momentum fall-off and the oscillations reflect the finite extent of the bound states and the nodes in the wave functions of radial excitations. These zeroes in the form factor play an important role in the behavior of the exclusive cross sections $\sigma(e^+e^- \to D\bar{D}, D\bar{D}^*, D^*\bar{D}^*)$ as a function of center-of-mass energy W.

The calculation of level positions, radiative transition rates, and ΔR in this model proceeds as below.

3.5.2 RENORMALIZATION OF SPECTRUM The parameters of the model are m_c and a, the light quark masses $m_u = m_d$ and m_s, and the charmed meson masses M_D, M_{D^*}, M_F, and M_{F^*}. The last named are now chosen to have their measured values. Both m_c and a generate a bare $c\bar{c}$ spectrum, which is renormalized by the coupling to the continuum. For fixed $m_u \approx \frac{1}{3}$ GeV and $m_s \approx \frac{1}{2}$ GeV, m_c and a are adjusted until the Green's function

$$\mathcal{G}_{mn} = [W - \varepsilon - \Omega(W)]^{-1}_{mn} \qquad 3.54$$

has $J^{PC} = 1^{--}$ poles at 3.095 and 3.684 GeV (corresponding to ψ and ψ'), and a residue at 3.095 GeV such that the ψ electronic width is 5.3 keV. The computed ψ' electronic width is 3.4 keV, which is the same as found in the basic model (Table 3) and about 1.7 times the observed value. Table 7 contains a list of the bare and renormalized masses of the spin-triplet states below charm threshold. Especially noteworthy are the large downward shifts in masses and the very small splitting of the ^3P_J states. The first effect shows that $\mathcal{H}_I$ is by no means a weak perturbation. The second is a consequence of the fact that the only spin dependence comes from the split thresholds and will be sizable only for states very near (within ~50–100 MeV) these thresholds.

Table 7 Mass shifts in the coupled-channel model (68); parameters used are $m_c = 1.69$ GeV, $a = 1.80$ GeV^{-1}, $m_u = m_d = 0.33$ GeV, and $m_s = 0.50$ GeV. D and D* masses are taken from experiment, while $M_F = 2.00$ GeV and $M_{F^*} = 2.14$ GeV are assumed

State	Bare mass (MeV)	Mass shift (MeV)	Renormalized mass (MeV)
ψ	3191	−96	3095
ψ'	3893	−209	3684
ψ''	3976	−208	3768
χ_2	3622	−170	3451
χ_1	3622	−180	3442
χ_0	3622	−191	3431

3.5.3 RADIATIVE TRANSITIONS With the parameters determined by renormalization, the radiative transition rate for $\psi' \to \chi_J + \gamma$, say, is computed as follows: Using the (oversimplified) notation of Equation 3.46, let $|N\rangle = |\psi'\rangle$, $|M\rangle = |\chi_J\rangle$, and $j_\lambda = \frac{2}{3}(\bar{c}\gamma_\lambda c + \bar{u}\gamma_\lambda u) - \frac{1}{3}(\bar{d}\gamma_\lambda d + \bar{s}\gamma_\lambda s)$ be the electromagnetic current. The E1 transition amplitude is (68)

$$\langle M|j_\lambda|N\rangle = \sum_{m,n} a^*_{mM}\, a_{nN}\, \langle m|j_\lambda|n\rangle + \int d\mu\, d\nu\, a^*_{\mu M}\, a_{\nu N}\, \langle \mu|j_\lambda|\nu\rangle$$

$$+ \sum_n \int d\nu \left(a^*_{\nu M}\, a_{nN}\, \langle \nu|j_\lambda|n\rangle + a^*_{nM}\, a_{\nu N}\, \langle n|j_\lambda|\nu\rangle \right). \qquad 3.55$$

The first term on the right in Equation 3.55 includes only the parts of ψ' and χ_J in the discrete ($c\bar{c}$) sector, with $\langle m|j_\lambda|n\rangle$ computed just as in the potential model without coupling to the continuum. The second term, involving the continuum components of ψ' and χ_J, contains electromagnetic transition matrix elements of charmed mesons; these are taken from standard quark model calculations. The third (cross) term involves a transition between the discrete and continuum sectors under the action of j_λ.

In lieu of some long and not-very-illuminating formulae for the terms in Equation 3.55, a few remarks on their relative importance are offered. The most important contribution to the discrete-sector terms is obviously the diagonal one: $|n\rangle = |2^3S_1\rangle$ and $|m\rangle = |1^3P_J\rangle$. The next single most important contribution to this set comes from $|n\rangle = |1^3D_1\rangle$, i.e. the mixing of $^3S_1$ and $^3D_1$ states due to nearby spin-split thresholds, and this is rather sensitive to the precise position of the $D\bar{D}$ and $D\bar{D}^*$ thresholds. The S-D mixing is most important for $\psi' \to \chi_0 + \gamma$ because of a large Clebsch-Gordan coefficient for $1^3D_1 \to 1^3P_0 + \gamma$ (see Equation 3.26). Because of energy denominators, the continuum-sector terms are dominated by the nearest threshold accessible to both ψ' and χ_J. Thus, the continuum is considerably more important (roughly a factor of two in amplitude) for χ_0 than for χ_1 and χ_2 because $\chi_0 \to D\bar{D}$ in an S wave, while $\chi_1 \to D\bar{D}$ is forbidden and $\chi_2 \to D\bar{D}$ is suppressed by a D-wave factor. In amplitude, the continuum contribution to χ_0 is about half the diagonal contribution and of the same sign. Finally, the mixed terms are practically negligible.

For the M1 transitions, only the discrete-sector terms have been computed so far. They are obtained from the standard formula (68, 72)

$$\Gamma(\psi_N \to \eta_{c,M} + \gamma) = \frac{16}{27}\,\alpha\,\frac{\omega^3}{m_c^2}\left|\sum_n a_{nN}\, a^*_{nM}\right|^2, \qquad 3.56$$

i.e. only nonhindered terms (same principal quantum number) are kept.

Table 8 Radiative transition rates in the coupled-channel model (68); parameters used are given in Table 7

Mode	Width (keV)
$\psi' \to \gamma\chi_2$	19
$\psi' \to \gamma\chi_1$	28
$\psi' \to \gamma\chi_0$	37
$\psi \to \gamma\eta_c$	21
$\psi' \to \gamma\eta_c$	12
$\psi' \to \gamma\eta_c'$	8
$\eta_c' \to \gamma\psi$	0.1

The overlap factor $\sum_n a_{nN}\, a_{nM}^*$ is 0.7 for $\psi' \to \eta_c'$, -0.13 for $\psi' \to \eta_c$, -0.05 for $\eta_c' \to \psi$, and 0.9 for $\psi \to \eta_c$, where η_c and η_c' were taken to lie at 2.8 and 3.45 GeV for the purpose of this calculation. For the hindered M1 transition amplitudes, one may reasonably expect the neglected terms to be comparable to those so far computed.

The final results for E1 and M1 transition rates are displayed in Table 8. Compared with the results of the potential model (Tables 4, 5), the E1 rates show a modest improvement, though they are still one to two standard deviations from experiment. Once again the M1 rates bear no resemblance to those observed for $\psi' \to \gamma\chi(3455)$, $\chi(3455) \to \gamma\psi$, and $\psi \to \gamma X(2830)$. Taking this together with the unexpectedly large hyperfine splittings, there can no longer be any doubt that something very important is missing from the charmonium model *or* that the identification of these states as hyperfine partners of ψ' and ψ is wrong.

3.5.4 CHARMED MESON PRODUCTION IN e^+e^- ANNIHILATION The essence of the model for ΔR is that charmed meson production is a quasi-two-body process mediated by those $c\bar{c}$ states that couple to the photon. It is thus in the spirit of vector-meson dominance (97), generalized to include coupled-channel mixing. The quasi-two-body hypothesis, which has proven to be correct for c.m. energy $W \lesssim 4.4$ GeV, implies that charmed meson spectroscopy can be readily and accurately carried out by studying the invariant mass recoiling against D (or F) in the reaction $e^+e^- \to D(F) + \text{anything}$.

For a charmed quark charge of $\frac{2}{3}$, ΔR is given by (68)

$$\Delta R(W) = \frac{32\pi}{W^2} \sum_{m,n} \Psi_m(0) \left[\frac{\mathscr{G}^\dagger(\Omega^\dagger - \Omega)\mathscr{G}}{2i} \right]_{mn} \Psi_n(0), \qquad 3.57$$

where the quantity in brackets is the absorptive part of the Green's function in Equation 3.54. The $\Psi_n(0)$ are the wave functions at zero $c\bar{c}$ separation of the discrete-sector states. Since Ω is really a sum over the allowed continuum channel types, $\nu = D^0\bar{D}^0$, D^+D^-, etc, ΔR may be written as a sum over exclusive-channel ratios, R_ν. Since $\Psi(0) \neq 0$ only for S-wave states in this model,

$$R_\nu(W) = \frac{32\pi}{W^2} \sum_{m,n} \sum_{l,l'=0,2} \Psi_{m0}(0)\, \mathscr{G}^*_{m0,m'l} \Gamma^\nu_{m'l,n'l'}\, \mathscr{G}_{n'l',n0} \Psi_{n0}(0), \qquad 3.58$$

where the orbital quantum number $l,l' = 0,2$ has been made explicit. Equation 3.49 defines $\Gamma^\nu_{m'l,n'l'}$, and it is the only factor in Equation 3.58 that varies from one channel to the next—through its dependence on the momentum p_ν, the intrinsic angular momenta, and the constituent quark masses of the outgoing charmed mesons (see Equation 3.53).

The reason for including $^3D_1$ levels, l or $l' = 2$, in the orbital sum in R_ν is this: As mentioned, the only dependence on total quark spin in the calculation of Ω and $\mathscr{G}$ enters through the use of spin-split thresholds. This induces a mixing between nearby S and D states that can emerge as a D-state resonance pole in off-diagonal elements such as $\mathscr{G}_{20,12}$. This mixing is strongest when a $^3D_1$ pole sits in the middle of a set of spin-

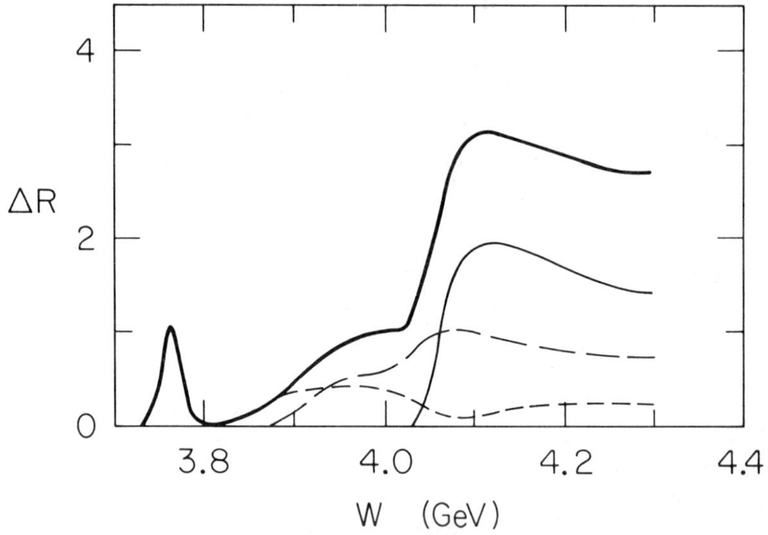

Figure 7 The charm contribution to R as computed in the coupled-channel model (94). The heavy solid curve is the sum of the contributions from $D\bar{D}$ (*short-dashed curve*), $D\bar{D}^* + D^*\bar{D}$ (*long-dashed*), and $D^*\bar{D}^*$ (*light solid*); F-meson production makes a negligible contribution. The thresholds used are given in the text; other parameters are very similar to those in Table 7.

split thresholds (e.g. at $W \approx 3.8$ GeV), and is considerably weaker when it is far from such thresholds (so that, for example, D and D* look degenerate).

Figure 7 shows a graph of ΔR taken from Reference 94. Completed some time before charmed meson masses were accurately measured, it assumed the thresholds $W_{DD} = 3.730$, $W_{DD^*} = 3.885$, $W_{D^*D^*} = 4.040$, $W_{FF} = 4.00$, $W_{FF^*} = 4.15$, and $W_{F^*F^*} = 4.30$ GeV. For comparison, the most recent data from the various collaborations at SPEAR (31, 32) and at DORIS (33, 34) is shown in Figures 8–11.

The prediction of the parameters of $\psi''(3772)$ more than a year before its discovery must be regarded as the greatest success of the coupled-channel model, especially in view of the fact that all attempts so far to understand the spin-dependent forces in charmonium have failed to give the requisite $2^3S_1 - 1^3D_1$ mixing by more than an order of magnitude.

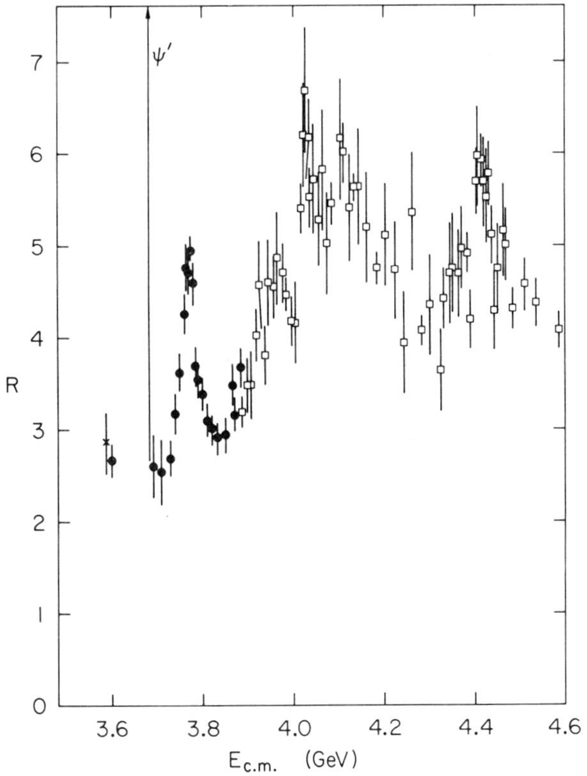

Figure 8 R in the charm threshold region as measured by the SLAC-LBL collaboration (31) at SLAC. Radiative corrections have been applied.

Typical predictions of Γ_e for this state based on the tensor force in a Breit Hamiltonian are $\sim$20 eV. The predicted mass and hadronic width of ψ'' agree, within errors, with the measured values (see Table 1). The predicted electronic width of about 150 eV is 2.5 times smaller than that reported by the SLAC-LBL collaboration (370$\pm$90 eV) (90) while nearly the same as that measured by the DELCO group (180$\pm$45 eV) (91). As we discuss in Section 4, the most important feature of ψ'' is that it decays exclusively to $D\bar{D}$, providing a unique, high-precision setting in which to study these mesons.

Comparison of the theoretical curve with *R* data for energies *W* between 3.8 and 4.2 GeV shows only qualitative agreement between the two.

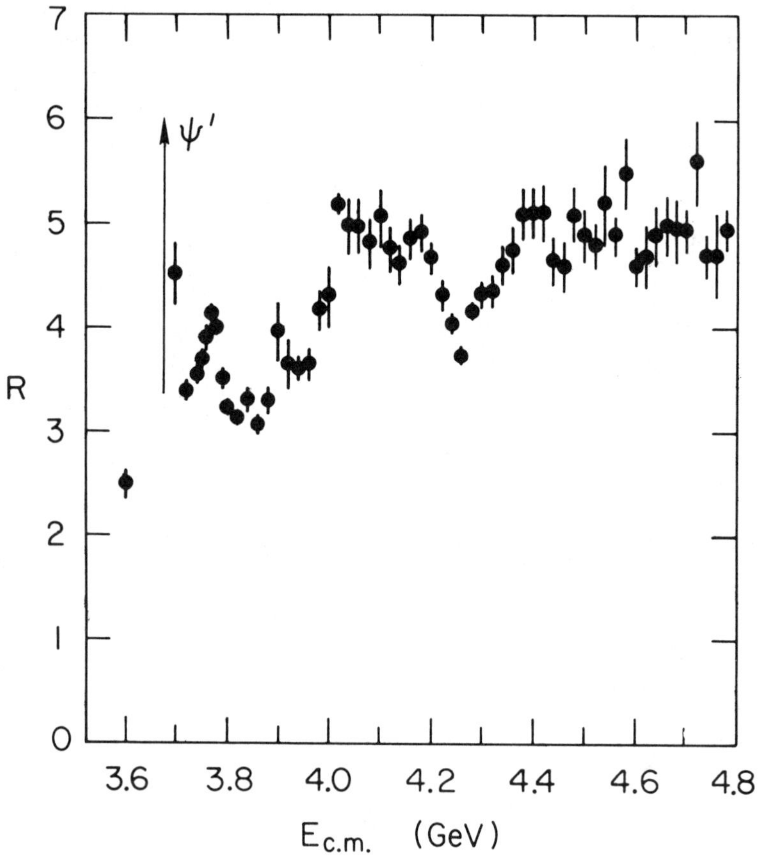

Figure 9 *R* as measured by the DELCO collaboration (32) at SLAC. Errors shown are statistical and the vertical scale may have an overall systematic uncertainty of $\pm$20%.

Points of agreement include: (*a*) The dip in ΔR to zero near 3.8 GeV due, in the model, to the vector-meson dominated production; (*b*) The rise in ΔR near 3.95 GeV. This is the $D\bar{D}^*$ threshold in the model calculation,

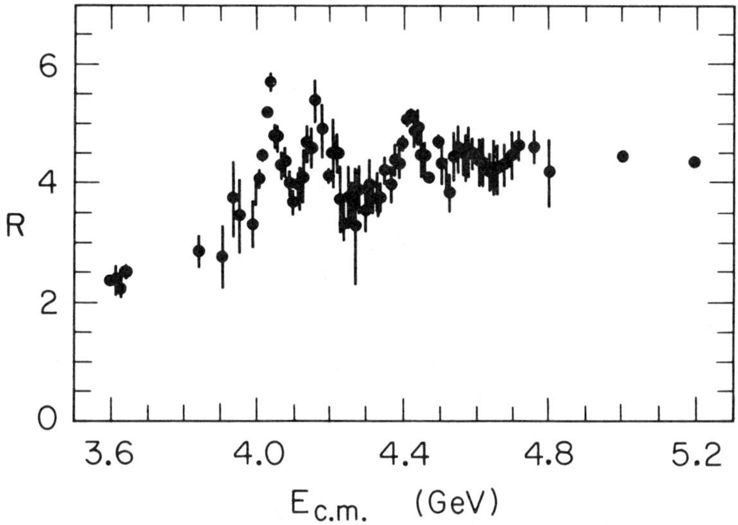

Figure 10 R as measured by the DASP collaboration (33) at DESY. The absolute normalization is estimated to have a $\pm 10\%$ systematic error.

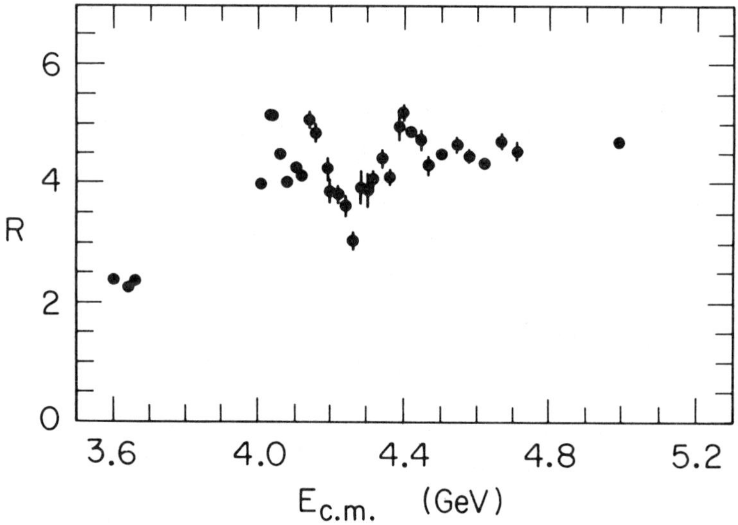

Figure 11 R as measured by the PLUTO collaboration (34) at DESY. Errors shown are statistical only.

but there ΔR (≈ 1) is only about one half the measured value;[15] (*c*) The sharp rise shown in Figure 7 is due to the concurrence of the important $D^*\bar{D}^*$ threshold with the 3^3S_1 charmonium level—the corresponding rise in the data, culminating in $\psi(4028)$, is considerably sharper ($\Delta R \approx 3$ in about 30 MeV); and (*d*) The dip in exclusive $D\bar{D}$ production at this resonance is due to a zero in the $3^3S_1 \rightarrow D\bar{D}$ form factor near $p_D \approx 750$ MeV. This striking prediction of the model (68, 94, 98) is confirmed experimentally. Study of the mass distribution recoiling against observed D's at 4.028 GeV gives the relative exclusive-channel ratios as (99):

$$R(D^{*0}\bar{D}^{*0}) : R(D^{*0}\bar{D}^0 + D^0\bar{D}^{*0}) : R(D^0\bar{D}^0) \\ = 1.00 \pm 0.10 : 0.85 \pm 0.09 : 0.10 \pm 0.06 \qquad 3.59$$

This preference of $\psi(4028)$ for $D^*\bar{D}^*$, despite the limited phase space, has been interpreted by some authors (77, 100) as an indication that $\psi(4028)$ is an almost bound state of these two mesons—a $D^*\bar{D}^*$ molecule. It is difficult to test this rather ad hoc hypothesis because no model of such objects exists that can be relied upon for further predictions. In the meantime, the existence of a near zero in $D\bar{D}$ production near 4.028 GeV can be tested by careful study of this region, and will further establish the notion of quarks through the observation of a node in their bound-state wave functions.

Above $W \approx 4.1$ GeV, the model calculation breaks down badly, and bears little resemblance to the data. In particular, the enhancement near 4.15 GeV, the dip at 4.3 GeV, and the obvious resonance $\psi(4414)$ are all beyond the reach of the model as presently constituted. If $\psi(4028)$ is indeed the (highly distorted) 3^3S_1 charmonium level, then the spectroscopy of the naive potential model would lead one to interpret the enhancement at 4150 MeV as the 2^3D_1 state and the resonance at 4414 MeV as the 4^3S_1 state. But this is perhaps pushing the naive model too far. Most of its assumptions are questionable for such high excitations, and even the nonrelativistic spectroscopic notation may be meaningless.

Finally, it should be mentioned that several other models predict states in this region beyond those expected in the linear potential model. To mention two examples: (*a*) The model of Giles & Tye (78), in which the $c\bar{c}$ pair is bound by a string with dynamical degrees of freedom, expects a number of levels corresponding to vibrational excitations of the string. No prediction is made for the leptonic width of these new states, so that their observability is an open question. (*b*) The (essentially) logarithmic potential proposed by a number of authors (86, 101, 102), has a greater

[15] To extract ΔR from experiment, one should subtract from R a constant background of about 2.2 due to production of noncharmed hadrons plus the contribution of τ production, which is approximately $(1+2m_\tau^2/W^2)(1-4m_\tau^2/W^2)^{\frac{1}{2}}$ at c.m. energy W.

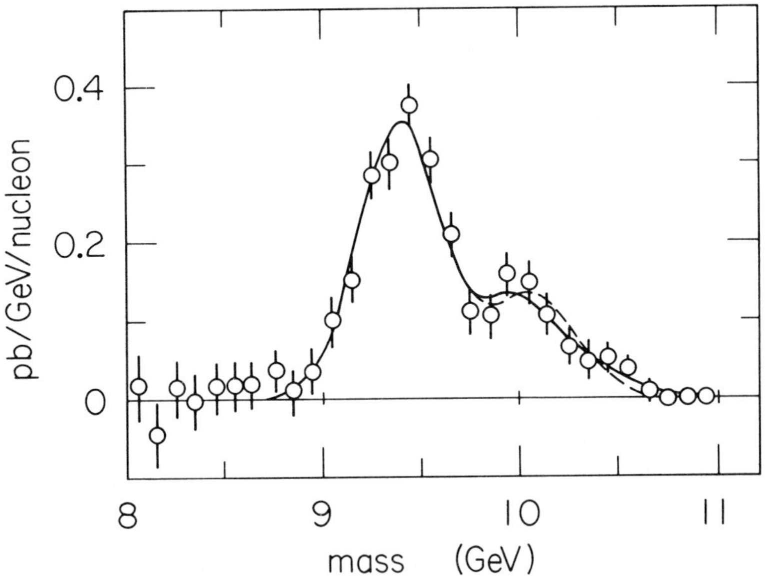

Figure 12 The $\mu^+\mu^-$ invariant mass spectrum (background subtracted) in the Υ region (23). The solid curve is a fit to two zero-width resonances (smeared by resolution); the dashed curve is a fit to three resonances.

level density than does the linear model, and predicts the 4^3S_1 and 5^3S_1 levels at 4.25 and 4.41 GeV, respectively.

3.6 *Beyond Charmonium*

The recent discovery at Fermilab of enhancements in the μ-pair invariant mass near $M_{\mu^+\mu^-} \approx 10$ GeV (23) is widely interpreted as solid evidence for the existence of a new quark, Q, with mass $m_Q \simeq 5$ GeV. The data (Figure 12) shows clearly two, and possibly three, resonances, called Υ, Υ', and Υ''. Assuming these to have zero width (consistent with the experimental resolution of ~200 MeV), a fit to the data gives their masses as

$$M_{\Upsilon} = 9.40 \pm 0.013 \text{ GeV}$$
$$M_{\Upsilon'} = 10.01 \pm 0.04 \text{ GeV}$$
$$M_{\Upsilon''} = 10.40 \pm 0.12 \text{ GeV}. \quad 3.60$$

While the statistical significance of Υ'' is still not large, the obvious interpretation is that these are the ground and first two radially excited $^3S_1$ states of a $Q\bar{Q}$ system.[16]

[16] We do not discuss here the possibility that Υ and Υ' are the ground states of two distinct, but nearly degenerate, $Q\bar{Q}$. See, for example, Cahn & Ellis (103).

Since they are narrow enough to have appreciable $\mu^+\mu^-$ branching ratios, they must all lie below the threshold for decay into a pair of mesons $Q\bar{q}+\bar{Q}q$ containing one new and one old quark (q = u,d) each. Two other striking features of these states readily inferred from the data are these: (*a*) The 2^3S-1^3S mass difference

$$M_{\Upsilon'}-M_{\Upsilon} = 610 \pm 50 \text{ MeV} \qquad 3.61$$

is, within errors, the same as in charmonium, $M_{\psi'}-M_{\psi} = 589$ MeV. As we shall see, this is about 150 MeV larger than expected if one adheres to the "standard" potential V_0 in Equation 3.8. (*b*) The ratio of the observed $\mu^+\mu^-$ signals at Υ' and Υ is

$$R_\mu = \left.\frac{B(\Upsilon' \to \mu^+\mu^-)\, d\sigma/dy}{B(\Upsilon \to \mu^+\mu^-)\, d\sigma/dy}\right|_{y=0} = 0.37 \pm 0.04. \qquad 3.62$$

The corresponding value of R_μ for ψ' and ψ production is about 0.02 (59). This strongly suggests (103) that $B(\Upsilon' \to \mu^+\mu^-) \gg B(\psi' \to \mu^+\mu^-)$ and, therefore, that $\Gamma(\Upsilon' \to \Upsilon + \text{anything}) \ll \Gamma(\psi' \to \psi + \text{anything}) \approx 130$ keV. These features of the Υ system receive considerable attention in the following discussion.

It hardly need be emphasized that bound systems of quarks heavier than charm will provide critical tests of the foundations of the charmonium model—gluon counting and the use of a nonrelativistic potential. Furthermore, the observed spectrum and branching ratios for radiative and direct decays will sharpen our knowledge of the form of this potential, since it is expected to be largely independent of m_Q. And, of course, the relative strength of radiative and leptonic decays to purely hadronic ones will help determine the new quark's charge.

These issues and more have already sparked considerable theoretical interest in the Υ system where, as we just mentioned, the prediction (104) of the standard potential for $M_{\Upsilon'}-M_{\Upsilon}$ appears to have failed. But a complete test of the form of the potential requires a comparison with experiment of its expectations for the myriad of branching ratios and absolute widths, as well as the details of the spectrum accessible only to e^+e^- storage ring experiments. And preliminary to making meaningful predictions, one must decide the relative positions of the ground state $Q\bar{Q}$ and the threshold for OZI-allowed decays. Only then can one know in a given model how many states are bound (i.e. narrow) and what transitions among these should be observed.

Eichten & Gottfried (104) have addressed the question of the m_Q dependence of the threshold and ground-state energies. While their arguments are, strictly speaking, valid only in the $m_Q \to \infty$ limit, errors should be small so long as $m_Q \gg m_q = m_{u,d,s}$, which is already true for

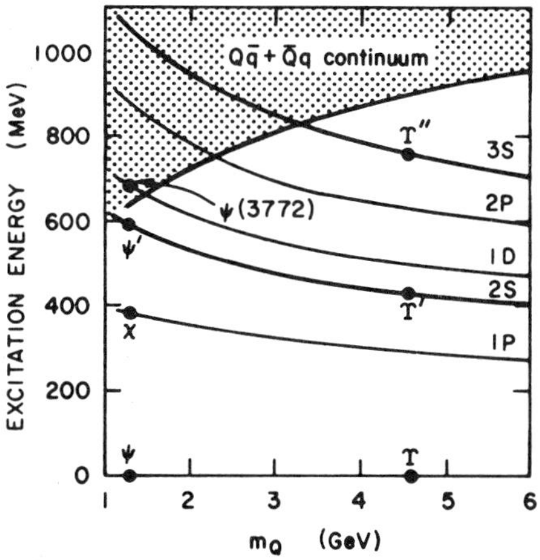

Figure 13 Predicted excitation spectrum of $Q\bar{Q}$ levels and the threshold for OZI-allowed decays, as a function of the heavy quark mass, m_Q, in the linear + Coulomb potential model (104) with $m_c = 1.37$ GeV.

the charmed quark. In the $Q\bar{q}$ system, for example, the reduced mass $\mu = m_Q m_q/(m_Q + m_q) \approx m_q$, and the dynamics is essentially independent of m_Q. The mass of the $Q\bar{q}$ ground-state pseudoscalar is then the sum of $m_Q + m_q$, the binding energy (a function of m_q only), and a correction due to the $^3S - {}^1S$ hyperfine splitting. This last term decreases like m_Q^{-1} in a heavy-light system, so that the threshold energy W_Q may be written as

$$W_Q = 2\left[M_D + m_Q - m_c + \frac{3}{4}\left(1 - \frac{m_c}{m_Q}\right)\left(M_{D*} - M_D\right)\right]. \qquad 3.63$$

In the $Q\bar{Q}$ system, the mass M_1 of the ^3S ground state is

$$M_1 = E_1(Q\bar{Q}) + E_0(m_Q), \qquad 3.64$$

where E_1 is the ground state eigenvalue and E_0 is the zero of energy, whose definition is not completely obvious in an infinitely rising potential such as V_0. Writing

$$E_0 = 2m_Q + \Delta(m_Q), \qquad 3.65$$

all that is known about Δ is (*a*) $\Delta(m_c) \approx -205$ MeV and (*b*) $m_Q^{-1}\Delta(m_Q) \rightarrow 0$ as $m_Q \rightarrow \infty$. Eichten & Gottfried interpolate[17] with $\Delta(m_Q) = \Delta(m_c)m_c/m_Q$. Using the potential V_0 in Equation 3.8, with essentially the same parameters as in Equation 3.9, Eichten & Gottfried computed the excitation spectrum shown in Figure 13. Using Equations 3.63 to 3.65, the quark

[17] Quigg and Rosner (105) have recently argued that the number of bound $^3S_1$ levels grows approximately as $2(m_Q/m_c)^{\frac{1}{4}}$.

mass appropriate to the Υ system is $m_Q = 4.6$ GeV, and the threshold for OZI-allowed decays is 900 MeV above the Υ. Thus, they predicted three bound ^{3}S states, as seems to be the case, with masses

$$M_{\Upsilon'} - M_{\Upsilon} \approx 450 \text{ MeV}, \qquad M_{\Upsilon''} - M_{\Upsilon} \approx 750 \text{ MeV},$$
$$M(\text{Q}\bar{\text{u}}; 1^1\text{S}_0) \approx 5.16 \text{ GeV} \qquad (\text{for } m_c = 1.37 \text{ GeV}),$$
$$M(\text{Q}\bar{\text{u}}; 1^3\text{S}_1) - M(\text{Q}\bar{\text{u}}; 1^1\text{S}_0) \approx 100 \text{ MeV}. \qquad 3.66$$

One possible explanation for the apparent failure of the highly motivated linear + Coulomb model to predict correctly the $\Upsilon' - \Upsilon$ and $\Upsilon'' - \Upsilon$ separations is this: While good arguments exist for both the small and large r behavior of the potential, it may well be that the systems under investigation, ψ and Υ, see mainly an intermediate-range portion of the $\text{Q}\bar{\text{Q}}$ potential. This may be neither linear nor Coulomb and, in any case, no arguments exist that give a clue to its shape.

Whatever the reason for failure, the large $\Upsilon' - \Upsilon$ mass difference has caused renewed interest in alternative forms of the potential. One choice, the logarithmic potential, studied some time ago by Machacek & Tomozawa (101) and more recently by Quigg & Rosner (102), is currently in vogue because of its peculiar property that the $\text{Q}\bar{\text{Q}}$ excitation spectrum is independent of the Q mass. Thus, by fitting to the ψ system, Quigg & Rosner find

$$V_1(r) = 0.733 \text{ GeV} \ln (r/r_0) \qquad 3.67$$

with r_0 an arbitrary constant, and $m_c = 1.1$ GeV. The spectrum of the first few excited $\text{Q}\bar{\text{Q}}$ levels may be found from Figure 13 by drawing horizontal lines through the dots corresponding to members of the charmonium family. The predictions of the two models for charmonium start to deviate around the 3^3S_1 state; in the logarithmic potential,

$$M(\text{c}\bar{\text{c}}; 3^3\text{S}_1) = 4.03 \text{ GeV}, \qquad M(\text{c}\bar{\text{c}}; 4^3\text{S}_1) = 4.25 \text{ GeV} \qquad 3.68$$

compared to 4.17 and 4.6 GeV, respectively, in the V_0 model. The separation between the ground state and OZI threshold will not be the same as in the Eichten-Gottfried calculation because quark masses appropriate to V_1 will differ somewhat from those for V_0. In particular, Quigg & Rosner find $m_Q \cong 4.5$ GeV so that, using Equation 3.63, they predict

$$M(\text{Q}\bar{\text{u}}; 1^1\text{S}_0) = 5.33 \text{ GeV} \qquad (\text{for } m_c = 1.1 \text{ GeV}) \qquad 3.69$$

and that three to four ^{3}S states will be bound.

Celmaster, Georgi & Machacek (86) have proposed still another potential inspired by the linear + Coulomb model:

$$V_2 = \frac{-8\pi}{27r}\left(\ln\frac{1}{r\Lambda e^{\gamma}}\right)^{-1} + \frac{(1-e^{-Ar})r}{2\pi} + E_0; \qquad 3.70$$

$$\Lambda = 0.50 \text{ GeV}, \quad \gamma = 0.577, \quad A = 0.16 \text{ GeV}, \quad E_0 = 0.39 \text{ GeV}.$$

The logarithm and coefficient in the short-range part of V_2 is motivated by appeal to asymptotic freedom (however, the argument of the logarithm has been modified). The linear potential strength, $(2\pi)^{-1}$ $(\text{GeV})^2$, is taken from the slope of the Regge trajectory. The new, intermediate-range part, $(r/2\pi)\exp(-Ar)$, is chosen arbitrarily. The parameters in Equation 3.68 are determined by fitting to the spectra of light as well as heavy ($c\bar{c}$) mesons—a very questionable procedure for any nonrelativistic potential. The charmed quark mass that results is $m_c = 1.98$ GeV. With all parameters determined (including the zero of energy, E_0, which is assumed m_Q independent), the Υ mass fixes $m_Q = 5.4$ GeV. This, in turn, leads to

$$M(Q\bar{u}; 1^1S_0) = 5.35 \text{ GeV} \qquad (\text{for } m_c = 1.98 \text{ GeV}) \qquad 3.71$$

and the prediction that three to four ^3S levels will be bound below OZI threshold.

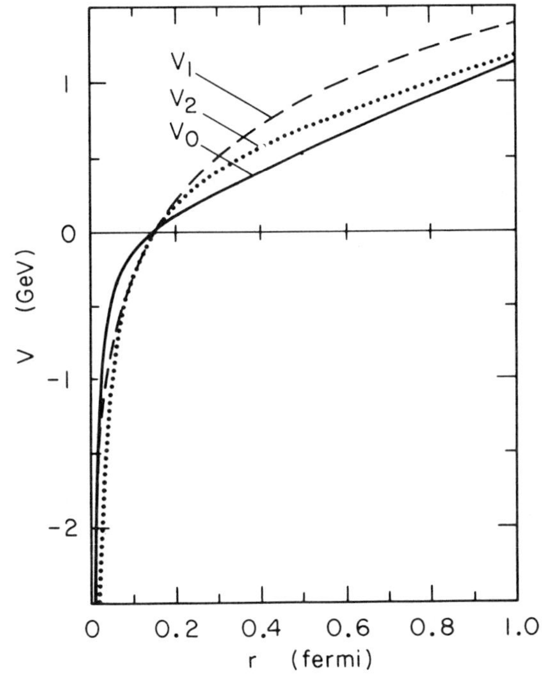

Figure 14 The $Q\bar{Q}$ potentials V_0 (68), V_1 (102), and V_2 (86). The zeros of energy have been chosen to make them cross at the same value of r.

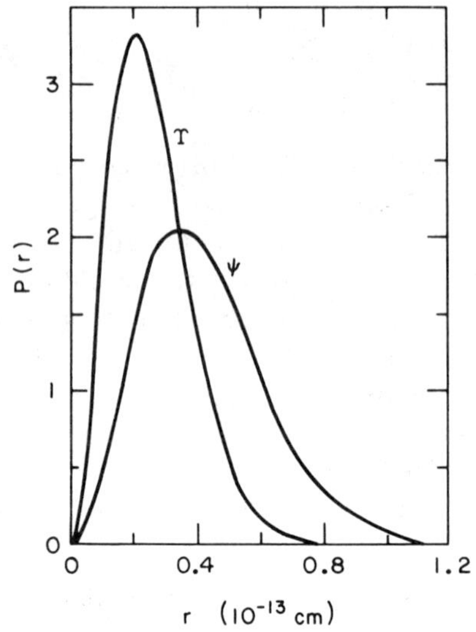

Figure 15 Radial probability for the ψ and Υ ground states, computed from V_0 (68), as a function of the $Q\bar{Q}$ separation r. We thank K. Gottfried for the use of his figures.

The potentials V_0, V_1, V_2 are plotted in Figure 14, and the radial probability $P(r)\,dr$ for the $\psi(1^3S_1)$ and $\Upsilon(1^3S_1)$ states as determined by V_0 is shown in Figure 15. Several remarks immediately follow from these figures: (*a*) V_1 and V_2 are practically congruent, and so the level spacing determined by V_2 is almost independent of m_Q and fits well what is known about the Υ spectrum. There will be small differences in the wave functions, hence the rates, predicted by the two models because of the different values of m_Q. (*b*) All three potentials are nearly congruent in the region in which the ψ wave function is large, $0.2 \lesssim r \lesssim 0.6$ fm. Thus, it is not surprising that all three give the same spectrum of low-lying $c\bar{c}$ levels and roughly comparable strengths for radiative decays. (*c*) Over most of the region in which the Υ wave function is concentrated, $0.1 \lesssim r \lesssim 0.3$ fm, there is considerable difference between the shape of V_0 and of $V_{1,2}$. It is not surprising, therefore, that the V_0 spectrum is quantitatively different from the V_1, V_2 spectra here.

The predictions in References 86, 102, and 104 for Υ' and Υ branching ratios and total widths are listed in Table 9. To estimate $\Upsilon' \to \Upsilon\pi\pi + \Upsilon\eta$, we have used Gottfried's scaling relation (71)

$$\Gamma(\Upsilon' \to \Upsilon\pi\pi + \Upsilon\eta) \approx \left(\frac{m_c}{m_Q}\right)^2 \Gamma(\psi' \to \psi\pi\pi + \psi\eta) \qquad 3.72$$

Table 9 Predicted Υ and Υ' transition rates (86, 102, 104)

		V_0		V_1		V_2	
Particle	Mode	$\lvert e_Q \rvert = \frac{1}{3}$	$\frac{2}{3}$	$\frac{1}{3}$	$\frac{2}{3}$	$\frac{1}{3}$	$\frac{2}{3}$
Υ'	$l^+l^-(l = e, \mu, \tau)$	0.45	1.8	0.5	1.9	1	4
	$\sum_J(\gamma + 1^3P_J)$	6.8	27	10	40	8.8	35
	$\Upsilon\pi\pi + \Upsilon\eta$	~12	~12	~12	~12	~12	~12
	Hadrons (direct)	8.6	8.6	9.1	9.1	19	19
	Hadrons (2nd order E.M.)	1.9	7.6	2.0	8.0	4.2	17
	All	~31	~60	~35	~75	~47	~95
Υ	$l^+l^-(l = e, \mu, \tau)$	0.7	2.7	1.1	4.3	2.5	10
	Hadrons (direct)	14.5	14.5	23	23	52	52
	Hadrons (2nd order E.M.)	2.8	11.4	4.5	18	10	40
	All	~19	~34	~31	~54	~70	~122

with $(m_c/m_Q)^2 \approx 1/10$. Since cascade radiative decays will constitute only a very small part of $\Upsilon' \rightarrow \Upsilon$ transitions, the suppression in Equation 3.72 goes a long way toward explaining the unexpectedly large value of R_μ in Equation 3.62.

While very little is known of the details of the Υ system at present, we can look forward in the next few years to a flood of data from the new e^+e^- storage rings at Cornell, DESY, and SLAC. It can only be hoped that out of all this will come a clearer picture of the "correct" phenomenological potential for heavy quark binding and, indeed, of the foundations of the charmonium model. Beyond this, we need a better understanding of the corrections to the naive model due to quark spin and the inevitable presence of light quark and gluonic degrees of freedom. The experimental study of $Q\bar{Q}$ systems is essential, but equally so is progress in understanding the structure of quantum chromodynamics itself—a theory still very much in its infancy.

4 CHARMED HADRONS

Direct evidence for the existence of charmed particles was announced in the summer of 1976, a year and a half after discovery of the ψ/J (22). For an early discussion of charm phenomenology that anticipated many of the recent discoveries, the reader is referred to the paper of Gaillard, Lee & Rosner (106). The properties of the charmed hadrons observed so far are shown in Table 10. Several other recent theoretical and experimental reviews are available (59, 107, 108).

4.1 *Theory*

4.1.1 SPECTROSCOPY Since all the charmed particles contain one or more light quarks, relativistic motion will very likely make them much more difficult to treat with a potential model than charmonium. A detailed discussion of attempts at light quark dynamics falls outside the scope of our paper. The point we want to make here is that nothing terribly surprising seems to be going on. The experiments have largely confirmed the qualitative expectations of the charm model.

Table 10 Charmed particle properties: All data is taken from Reference 108, which cites original work; quark content is shown below particle name; D^{*+} branching ratios are estimated (108) as described in the text for $m_u/m_c = m_\rho/m_\psi$ (parenthetical numbers are for $m_u/m_c = 0$)

Particle	$I(J^P)$	Mass (MeV)	Decay mode	Branching fraction (%)
D^0	$\frac{1}{2}(0^-)$	1863.3 ± 0.9	$K^-\pi^+$	2.2 ± 0.6
$(c\bar{u})$				
			$\bar{K}^0\pi^+\pi^-$	4.0 ± 1.3
			$K^-\pi^+\pi^0$	12 ± 6
			$K^-\pi^+\pi^-\pi^+$	3.2 ± 1.1
			$e^+\nu_e +$ hadrons	~ 10
D^+	$\frac{1}{2}(0^-)$	1868.3 ± 0.9	$\bar{K}^0\pi^+$	1.5 ± 0.6
$(c\bar{d})$				
			$K^-\pi^+\pi^+$	3.9 ± 1.0
			$e^+\nu_e +$ hadrons	~ 10
		$\delta = M_{D^+} - M_{D^0}$ $= 5.0 \pm 0.8$		
D^{*0}	$\frac{1}{2}(1^-)$	2006 ± 1.5	$D^0\pi^0$	55 ± 15
			$D^0\gamma$	45 ± 15
D^{*+}	$\frac{1}{2}(1^-)$	2008.6 ± 1.0	$D^0\pi^+$	68 ± 8 (63 ± 9)
			$D^+\pi^0$	30 ± 8 (27 ± 7)
		$\delta^* = M_{D^{*+}} - M_{D^{*0}}$ $= 2.6 \pm 1.8$	$D^+\gamma$	2 ± 1 (10 ± 5)
		$\delta - \delta^*$ $= 2.4 \pm 2.4$		
F^+	$0(0^-)$	2039.5 ± 1.0	$\eta\pi^+$	?
$(c\bar{s})$				
			$K^+K^-\pi^+$	?
			$K^+\bar{K}^0$	?
			$K^+K^-\pi^+\pi^-\pi^+$	?
F^{*+}	$0(1^-)$	2140 ± 60	$F^+\gamma$	100
Λ_c^+ $[c(ud)_A]$	$0(\frac{1}{2}^+)$	2260 ± 10	$\Lambda\pi^+\pi^+\pi^-$	?
Σ_c^{*++} $[c(ud)_S]$	$1(\frac{3}{2}^+)$	2426 ± 12	$\Lambda_c^+\pi^+$	?

The masses in Table 10 can, in fact, be qualitatively understood in the most naive form of the nonrelativistic quark model by adding the appropriate masses:

$$m_c = \tfrac{1}{2}M_\psi \approx 1.55 \text{ GeV} \qquad m_u = m_d \approx 0.33 \text{ GeV} \qquad m_s \approx 0.46 \text{ GeV} \tag{4.1}$$

This reproduces the pseudoscalar and $J^P = \frac{1}{2}^+$ baryon masses reasonably well and the vector mesons require roughly an extra 150 MeV hyperfine energy. Given the observed masses $M_D \approx 1.865$ GeV and $M_{D^*} \approx 2.005$ GeV, a better estimate is

$$\begin{aligned} M_{F^*} &= M_{D^*} + M_\phi - M_{K^*} = 2.13 \text{ GeV} \\ M_F &= M_{F^*} - (M_{D^*} - M_D) = 1.99 \text{ GeV}. \end{aligned} \tag{4.2}$$

The reported F and F* masses in Table 10 are reproduced nicely by this sum rule.

Mass splittings between D^0 and D^+ and between D^{*0} and D^{*+} have been measured (108) and are also interesting theoretically. In fact, because the $D^* - D$ mass difference leads to an extremely small Q value for $D^* \to D\pi$, the $D^0 - D^+$ and $D^{*0} - D^{*+}$ splittings have important experimental consequences in sorting out the spectroscopy of the D's. In the nonrelativistic quark model, the splitting is the sum of the down-up mass difference and a contribution from single-photon exchange:

$$\begin{aligned} M_{D^+} - M_{D^0} &= m_d - m_u + \frac{2}{3}\alpha\left[\left\langle\frac{1}{r}\right\rangle_D + \frac{2\pi}{m_c m_u}\left|\Psi_D(0)\right|^2\right] \\ M_{D^{*+}} - M_{D^{*0}} &= m_d - m_u + \frac{2}{3}\alpha\left[\left\langle\frac{1}{r}\right\rangle_D - \frac{2\pi}{3m_c m_u}\left|\Psi_D(0)\right|^2\right]. \end{aligned} \tag{4.3}$$

These expressions can be evaluated using a current algebraic extraction of $m_d - m_u$ (109) and an atomic quark model analogous to charmonium for the D mesons. The result is $M_{D^+} - M_{D^0} \approx 7.0$ MeV and $M_{D^{*+}} - M_{D^{*0}} \approx 6.5$ MeV to be compared with the experimental values of 5.0 ± 0.8 MeV and 2.6 ± 1.8 MeV, respectively. An alternative estimate of these splittings using the Massachusetts Institute of Technology (MIT) Bag model (92, 110) gives essentially the same result.

In addition to the S-wave charmed mesons, P states should exist as well. Their masses have been estimated by Eichten et al (68) to be

$$\begin{aligned} M_D(1^3P_0) &= 2.44 \text{ GeV} \qquad M_D(1^3P_1) = 2.58 \text{ GeV} \\ M_D(1^1P_1) &= 2.45 \text{ GeV} \qquad M_D(1^3P_2) = 2.58 \text{ GeV}. \end{aligned} \tag{4.4}$$

The evidence for these states (as well as the beautiful measurements of the D and F masses) are discussed shortly.

4.1.2 DECAYS In the standard WS-GIM model (4, 7), the hadronic current Equation 1.7 leads to the selection rule $\Delta C = \Delta S = \pm 1$ for charm decays. Ignoring QCD renormalization corrections, the complete effective Hamiltonian for charm decays in this model is

$$\begin{aligned}\mathscr{H}_{\Delta C} = 2^{-\frac{1}{2}} G_{\mathrm{F}} [\cos\theta_{\mathrm{C}}\, \bar{\mathrm{c}}\gamma_\lambda(1-\gamma_5)\mathrm{s} - \sin\theta_{\mathrm{C}}\, \bar{\mathrm{c}}\gamma_\lambda(1-\gamma_5)\mathrm{d}] \\ \times [\cos\theta_{\mathrm{C}}\, \bar{\mathrm{d}}\gamma^\lambda(1-\gamma_5)\mathrm{u} + \sin\theta_{\mathrm{C}}\, \bar{\mathrm{s}}\gamma^\lambda(1-\gamma_5)\mathrm{u} \\ + \sum_{l=\mathrm{e},\mu} \bar{l}\gamma^\lambda(1-\gamma_5)\nu_l] + \mathrm{h.c.}\end{aligned} \qquad 4.5$$

If one naively assumes that charmed hadron decays are processes in which only the constituent c quark participates, one estimates from Equation 4.5 the following relative rates (ignoring questions of phase space):

$$\begin{aligned}&\mathrm{c} \to \mathrm{s u \bar{d}} = 3\cos^4\theta_{\mathrm{C}} \\ &\mathrm{c} \to \mathrm{d u \bar{d}} = 3\cos^2\theta_{\mathrm{C}}\sin^2\theta_{\mathrm{C}} \\ &\mathrm{c} \to \mathrm{s u \bar{s}} = 3\cos^2\theta_{\mathrm{C}}\sin^2\theta_{\mathrm{C}} \\ &\mathrm{c} \to \mathrm{d u \bar{s}} = 3\sin^4\theta_{\mathrm{C}} \\ &\mathrm{c} \to \mathrm{s} l^+ \nu_l = \cos^2\theta_{\mathrm{C}} \qquad (l = \mathrm{e},\mu) \\ &\mathrm{c} \to \mathrm{d} l^+ \nu_l = \sin^2\theta_{\mathrm{C}} \qquad (l = \mathrm{e},\mu).\end{aligned} \qquad 4.6$$

The factor of 3 is due to a sum over the color of the quarks in the non-charmed piece of the hadronic current. Ignoring all but $\Delta C = \Delta S$ transitions, one expects D (and F) decays will be 60% nonleptonic and 40% semileptonic, divided equally between e and μ. All of these will involve a single kaon, which provides the outstanding signal for the presence of charm. Note, in particular, that $\Delta C = \Delta S$ implies that one should see a D^+ signal in the exotic channel $K^-\pi^+\pi^+$, but not in the nonexotic $K^+\pi^+\pi^-$. Finally, the D lifetime is estimated in this model to be

$$\tau(\mathrm{D} \to \mathrm{all}) = \tfrac{1}{5}(m_\mu/M_{\mathrm{D}})^5\, \tau(\mu \to \mathrm{e}\nu\bar{\nu}) \approx 10^{-13}\ \mathrm{sec}. \qquad 4.7$$

Thus visual observation of D decays can be made only with high resolution techniques using emulsions or streamer chambers (111).

More detailed studies of charmed meson decays have been made by Einhorn & Quigg (112) and by Ellis et al (113), as well as by several other groups (114). These are based on the analyses of the operator structure of the nonleptonic weak Hamiltonian carried out by Gaillard & Lee (46) and by Altarelli & Maiani (46). The nonleptonic Hamiltonian is found to consist of two pieces, one transforming as the [**20**] representation under SU(4) (flavor), the other as [**84**]. When decomposed with respect to SU(3) subgroups (the symmetry group of u, d, s quarks), one finds

$$[\mathbf{20}] = \mathbf{6} \oplus \mathbf{8} \oplus \mathbf{6^*}$$
$$[\mathbf{84}] = \mathbf{6} \oplus \{\mathbf{3} \oplus \mathbf{15_M}\} + \mathbf{6^*} \oplus \{\mathbf{3^*} \oplus \mathbf{15^*_M}\} + \{\mathbf{1} \oplus \mathbf{8} \oplus \mathbf{27}\} \qquad 4.8$$

with square brackets used to distinguish representations of SU(4) from those of SU(3), and the subscript M denoting a representation of mixed symmetry. The octet in the decomposition of [**20**] is the $\Delta C = 0$, $|\Delta S| = 1$ operator responsible for nonleptonic K decay; its matrix elements are enhanced relative to the octet and 27-plet in [**84**].

On this basis, Einhorn & Quigg argued that the $\Delta C = \pm 1$ pieces, **6** and **6*** in [**20**] should have enhanced matrix elements relative to the $\Delta C = \pm 1$ parts of [**84**], namely $\mathbf{3} \oplus \mathbf{15_M} + \mathbf{3^*} \oplus \mathbf{15^*_M}$. (Actually, only $\mathbf{15_M} \oplus \mathbf{15^*_M}$ appear in the Hamiltonian.) Now, part of this octet enhancement is due to the sign of anomalous dimensions appearing in the operator product expansion (41), while an appreciable further enhancement is due to incalculable matrix elements of the operator, i.e. it is of uncertain origin. Furthermore, some of the octet enhancement arises from the choice of renormalization point in the evaluation of the anomalous dimensions of the operators; this was taken to be 1 GeV for K decay (46), and assumed to be the same by Einhorn & Quigg (112). Ellis et al argue that this renormalization point should be taken higher when dealing with decays of charmed hadrons, say $m_c \sim 2$ GeV. This, they claim, diminishes sextet enhancement of the $|\Delta C| = 1$ Hamiltonian.

Now, all of this has measurable consequences. Under the reasonable assumption that decays of high-mass states such as D and F are quasi-two-body, Einhorn & Quigg point out that sextet enhancement implies that D^+ has no Cabibbo-enhanced decays ($\propto \cos^4 \theta_c$) to a pair of pseudoscalars (such as $\bar{K}^0\pi^+$) or a pair of vectors (like $\bar{K}^{*0}\rho^+$). The only Cabibbo-enhanced decays of D^+ then would be to a pseudoscalar plus a vector, say $D^+ \to \bar{K}^0\rho^+ \to \bar{K}^0\pi^+\pi^0$ and $D^+ \to \bar{K}^{*0}\pi^+ \to K^-\pi^+\pi^+$ or $\bar{K}^0\pi^0\pi^+$.

To the contrary, Ellis et al find $\Gamma(D^+ \to \bar{K}^0\pi^+) \approx \Gamma(D^0 \to K^-\pi^+)$, a "sextet enhanced" rate. Using a variety of techniques, they estimate the following ranges of branching ratios for charmed meson decay:

$$B(\mathrm{D,F} \to l + \nu_l + \text{hadrons}) = 0.1\text{–}0.25$$

$$B(\mathrm{D} \to l + \nu_l + \mathrm{K}) = 0.03\text{–}0.08$$

$$B(\mathrm{F} \to l + \nu_l + \eta) = 0.02\text{–}0.05$$

$$B(\mathrm{D}^0 \to \mathrm{K}^-\pi^+ + \bar{\mathrm{K}}^0\pi^0) = 0.03\text{–}0.18$$

$$B(\mathrm{D}^+ \to \bar{\mathrm{K}}^0\pi^+) = 0.02\text{–}0.10$$

$$B(\mathrm{F}^+ \to \eta\pi^+ + \mathrm{K}^+\bar{\mathrm{K}}^0) = 0.02\text{–}0.12. \qquad 4.9$$

Arguments over operator enhancement aside, all authors (112–114) agree that, because of the large number of modes available for decay, no single branching ratio is expected to be more than a few percent.

One other interesting aspect of D decays has to do with the possibility of $D^0\bar{D}^0$ mixing (115). If this is induced by charm-changing neutral currents such as $\bar{c}\gamma_\lambda\frac{1}{2}(1\pm\gamma_5)u$ coupled to the Z^0 weak boson, then

$$\Delta M(D^0,\bar{D}^0) \sim G_F M_D^3 \gg \Gamma(D^0 \to \text{all}) \sim G_F^2 M_D^5$$

and $D^0\bar{D}^0$ mixing will be complete. One then will see $D^0 \to K+\ldots$ as often as $D^0 \to \bar{K}+\ldots$. If $|\Delta C| = 2$ transitions are mediated by second order (G_F^2) processes or by neutral Higgs bosons, mixing may be less than complete but still appreciable. Thus, a measurement of $\Gamma(D^0 \to K+\ldots)/\Gamma(D^0 \to \bar{K}+\ldots)$ gives us important information about the structure of weak currents (both charged and neutral) as well as constraints on the couplings of Higgs mesons to quarks.

Charmed baryon decays are considerably more complicated and correspondingly uncertain. The reader is referred to the above papers (and enclosed references) for details beyond the gross estimates one can make from Equations 4.6 and 4.7.

As we hinted earlier, the masses of D's and D*'s are so delicately arranged that they cause an unprecedented complication in sorting out D spectroscopy. The problem is that the Q values for $D^* \to D\pi$ are so small that the electromagnetic (M1) decay $D^* \to D\gamma$ is competitive with the strong one. Therefore, when studying the invariant mass recoiling against D^0 produced in e^+e^- annihilation at 4.028 GeV, say, one sees a very rich structure corresponding to

$$\begin{array}{lll} e^+e^- \to D^0\bar{D}^0 & \text{(Recoil mass } M_x = M_{\bar{D}^0}) & \\ e^+e^- \to D^0\bar{D}^{*0} & (M_x = M_{\bar{D}^{*0}}) & \\ e^+e^- \to D^{*0}\bar{D}^0, & D^{*0} \to D^0\pi^0 \text{ or } D^0\gamma & (M_x = M_{\bar{D}^0\pi^0} \text{ or } M_{\bar{D}^0\gamma}) \\ e^+e^- \to D^{*0}\bar{D}^{*0}, & D^{*0} \to D^0\pi^0 \text{ or } D^0\gamma & (M_x = M_{\bar{D}^{*0}\pi^0} \text{ or } M_{\bar{D}^{*0}\gamma}) \\ e^+e^- \to D^{*+}D^-, & D^{*+} \to D^0\pi^+ & (M_x = M_{D^-\pi^+}) \\ e^+e^- \to D^{*+}D^{*-}, & D^{*+} \to D^0\pi^+ & (M_x = M_{D^{*-}\pi^+}). \end{array} \quad 4.10$$

The decay $D^{*0} \to D^+\pi^-$ is energetically forbidden. The relative strength of each component of the recoil distribution is determined by the product of the exclusive-channel cross section and the appropriate D* branching ratio. So these branching ratios are of great importance in charmed meson spectroscopy.

In addition to their model for calculating exclusive-channel cross sections, Eichten et al (68) have estimated the various D* branching ratios as follows: For the M1 decays they use the naive quark model formula

$$\Gamma[\mathrm{D}^*(\mathrm{c\bar{q}}) \to \mathrm{D}(\mathrm{c\bar{q}})+\gamma] = \frac{4}{3}\alpha\left(\frac{e_\mathrm{c}}{2m_\mathrm{c}}+\frac{e_\mathrm{q}}{2m_\mathrm{q}}\right)^2 p^3, \qquad 4.11$$

where $e_\mathrm{c} = e_\mathrm{u} = \frac{2}{3}$, $e_\mathrm{d} = -\frac{1}{3}$, $p = (M^2_{\mathrm{D}^*} - M^2_\mathrm{D})/2M_{\mathrm{D}^*}$, and they use quark masses of $m_\mathrm{u} = m_\mathrm{d} = 0.33$ GeV and $m_\mathrm{c} = 1.87$ GeV [determined from $\Gamma(\psi \to \mathrm{e^+e^-})$ and $M_{\psi'} - M_\psi$ in the pure linear potential model].

The $\mathrm{D}^* \to \mathrm{D}\pi$ width is obtained by assuming a form suggested by their model calculation of $\psi_{nL} \to \mathrm{D\bar{D}}$; it is

$$\Gamma(\mathrm{D}^* \to \mathrm{D}\pi) = \frac{p^3}{72\pi M^2_{\mathrm{D}^*}} C^2 \left|(M_{\mathrm{D}^*}E_\mathrm{D}E_\pi)^{\frac{1}{2}} A\right|^2, \qquad 4.12$$

where $E_{\mathrm{D},\pi} = (p^2 + M^2_{\mathrm{D},\pi})^{\frac{1}{2}}$, C is an isospin Clebsch-Gordan coefficient, and A is an amplitude depending only on m_u in the limit that heavy quark mass $m_\mathrm{c} \to \infty$. Assuming further that m_s is large enough so that A can be estimated from $\mathrm{K}^* \to \mathrm{K}\pi$ decays gives

$$A = 47.8\ \mathrm{GeV}^{-\frac{3}{2}}. \qquad 4.13$$

Using the measured (108) D and D* masses, they obtain the widths and branching ratios listed in Table 11. As we will see shortly, the results in Table 11 are in remarkable agreement with those determined from experiment under much less model-dependent assumptions.

To conclude this discussion, we mention first that the expected (and measured) $\mathrm{F}^* - \mathrm{F}$ mass difference ($\lesssim M_\pi$) implies the $\mathrm{F}^* \to \mathrm{F}\gamma$ is the only decay mode of this $C = S = 1$ vector meson. Using $m_\mathrm{s} = 0.50$ GeV in Equation 4.11 gives

$$\Gamma(\mathrm{F}^{*+} \to \mathrm{F}\gamma) = 0.2\ \mathrm{keV}. \qquad 4.14$$

Table 11 Predicted D* widths and branching ratios (68)

Mode	Width (keV)	Branching ratio (%)
$\mathrm{D}^{*0} \to \mathrm{D}^0\pi^0$	39.7	53.0
$\mathrm{D}^{*0} \to \mathrm{D}^0\gamma$	35.2	47.0
$\mathrm{D}^{*0} \to$ all	74.9	—
$\mathrm{D}^{*+} \to \mathrm{D}^+\pi^0$	22.2	28.4
$\mathrm{D}^{*+} \to \mathrm{D}^0\pi^+$	53.4	68.4
$\mathrm{D}^{*+} \to \mathrm{D}^+\gamma$	2.5	3.2
$\mathrm{D}^{*+} \to$ all	78.0	—

Second, the apparent success of the M1 formula, Equation 4.11, for D* decays stands in sharp contrast to its apparent failure in the charmonium system.

Finally, it is unfortunate that the formula, Equation 4.12, is unlikely to be testable in bound systems of a still heavier quark Q with u, d, s. With the hyperfine splitting decreasing like M_Q^{-1}, the only energetically allowed decays of possible new mesons M*(Q$\bar{u}$) to M(Q$\bar{u}$), will be the radiative ones. Looking on the bright side, this situation will make M, M* spectroscopy a little easier, and—if the M* width can be measured—provide further tests of the M1 formula.

4.2 *Experiment*

What follows is a brief discussion of the properties summarized in Table 10. For more detail and reference to experimental work not cited explicitly, the reader is referred to Feldman's recent review (108).

To date, charmed mesons have been identified directly only in e^+e^- annihilation experiments, where their production cross sections are $\sim 50\%$ of the total and the kinematics is especially simple. The D and D* mesons have been positively identified by the SLAC-LBL collaboration (22). We hasten to add that before and since their discovery, there has been plenty of indirect evidence for charmed mesons in neutrino experiments (see Section 6) in e^+e^- annihilation at CEA (14), at SLAC (31, 32), and DESY (33, 34), and in photoproduction at SLAC (see Section 6.3). The F^+ and F^{*+} (decaying to $F^+\gamma$) were discovered by the double-arm spectrometer (DASP) collaboration (116) at DESY at c.m. energy 4.414 GeV. And the F^+ is tentatively confirmed by the SLAC-LBL collaboration in data taken at 4.16 GeV (108). All of these discoveries and all of the precision data on D^0 and D^+ were obtained at the peaks in the annihilation cross section—3.772, 4.028, 4.16, and 4.414 GeV—which are charmonium resonances above threshold. [There is one exception: the beautiful measurement (108) of $M_{D^{*+}} - M_{D^0}$ required high energy, 6.8 GeV, to detect the π^+ in $D^{*+} \to D^0\pi^+$.]

Evidence for charmed baryons comes from two sources. The first is a single neutrino event in the Brookhaven 7-ft bubble chamber (117):

$$\nu_\mu p \to \mu^-\Lambda\pi^+\pi^+\pi^+\pi^-. \tag{4.15}$$

This is interpreted as $\nu_\mu p \to \mu^-\Sigma_c^{*++}$; $\Sigma_c^{*++} \to \Lambda_c^+\pi^+$; $\Lambda_c^+ \to \Lambda\pi^+\pi^+\pi^-$ because the event violates the $\Delta S = \Delta Q$ rule. The second comes from a peak in the inclusive $\bar{\Lambda}\pi^-\pi^-\pi^+$ spectrum at 2.26 GeV in a photoproduction experiment at Fermilab (118). This group also reports evidence for the sequence $\bar{\Sigma}_c^{*0} \to \bar{\Lambda}_c^-\pi^+$, with a Σ_c^* mass of 2.43 GeV. These masses are exactly those determined in the Brookhaven experiment and expected in the quark model. Very indirect evidence for charmed baryons in e^+e^- annihilation comes from the sharp rise in proton and Λ production in the 4.4–5.0 GeV region (108). But upper limits on cross section times branching ratio to observable modes (σB) are typically an order of magnitude lower than for D production at the same energies. Most interesting in these studies is that Λ production is consistently only 10–15% of proton production at all energies, which suggests that charmed baryons preferentially decay to K + nucleon + ... rather than to strange baryons.

The isospin assignments in the table are made purely on theoretical grounds; no experimental information exists other than the fact that $D^* \to D\pi$ precludes $I = 0$ for both these charmed mesons.[18] Similarly, no measurements of J^P exist for F, F*, and charmed baryons. Assuming that D^0 and D^+ have the same J^P, observation of $D^0 \to K^-\pi^+$ and study of the Dalitz plot for $D^+ \to K^-\pi^+\pi^+$ proves that parity is violated in their decays (119), which suggests that they decay via weak interactions. This fact of parity violation is now obvious (without the assumption of equal J^P) from the observed decay modes of D^0 and D^+. Assuming that the parity of D is -1, observing that $D^* \to D\pi$ and $e^+e^- \to D\bar{D}^*$, and measuring the angular distributions for

$$e^+e^- \to D\bar{D}^* \quad (D \to \bar{K}\pi), \qquad 4.16$$

the SLAC-LBL collaboration is able to rule out $J^P_D = J^P_{D^*}$ for $J^P_D = 0^\pm$ and $J^P_D = 1^-$, $J^P_{D^*} = 0^-$, whereas they find high confidence levels for the hypothesis $J^P_D = 0^-$, $J^P_{D^*} = 1^-$.

The remarkably precise measurements (108) of the D^0 and D^+ masses come from data taken at the peak of $\psi'' = \psi(3772)$, the 1^3D_1 charmonium level. This accuracy is possible because: (*a*) ψ'' decays exclusively to $D^0\bar{D}^0$ and D^+D^- ($D\bar{D}^*$ is energetically forbidden), and (*b*) it lies just about 40 MeV above threshold, so that the D's are moving very slowly. Thus, small errors in the measurement of p_D are unimportant in determining M_D from

$$M_D = [(M_{\psi''}/2)^2 - p_D^2]^{\frac{1}{2}} = (E_b^2 - p_D^2)^{\frac{1}{2}}, \qquad 4.17$$

where the beam energy E_b is very well measured. Of course, p_D is determined from the momenta of the D-decay products. Figure 16 shows the

[18] Of course, the fact that only D^0 and D^+ (with no D^- or D^{++}, say) are observed in the decays $D^{*+} \to D^+\pi^0$, $D^0\pi^+$, and the fact that the D^- is observed to decay to states of positive strangeness, are rather convincing evidence in favor of $I = \frac{1}{2}$ as well as $Q = +\frac{2}{3}$ for the charmed quark.

invariant mass spectra for D^0 and D^+. The clearly visible $\delta = M_{D^+} - M_{D^0} = 5.0 \pm 0.8$ MeV is only slightly less than predicted in References 109 and 110.

The D^{*0} mass is measured by a similar trick at $W = 4.028$ GeV. Here, $e^+e^- \rightarrow D^{*0}\bar{D}^{*0}$ is picked out and $p_{D^{*0}}$ is measured as follows: The momentum spectrum of the D^0's detected at $\psi(4.028)$ is measured. This is shown in Figure 17. The small Q value for $D^{*0} \rightarrow D^0\pi^0$ makes these

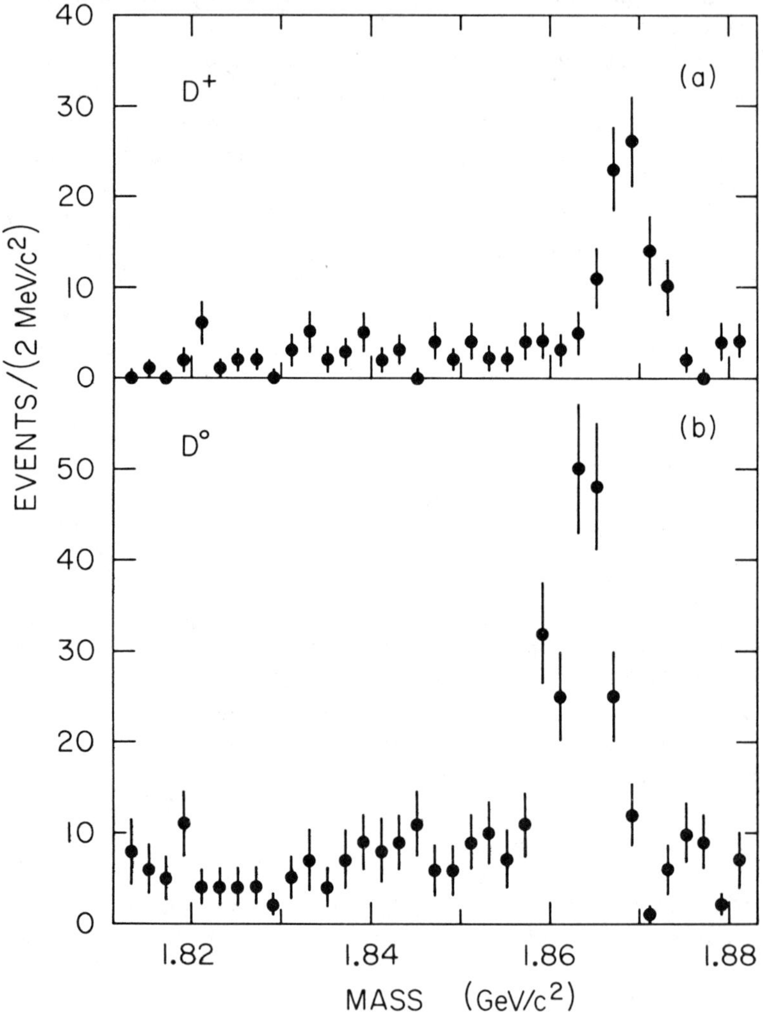

Figure 16 Invariant mass spectra for the sum of all observed (*a*) D^+ and (*b*) D^0 decay modes yielding all-charged-particle final states (108).

components of the distribution (curves B and E in the figure) rather sharply peaked. The lower peak (B) obviously corresponds to $D^{*0}\bar{D}^{*0}$, and a simple kinematical exercise gives the center of this peak as

$$p_D^{ctr}(D^{*0}\bar{D}^{*0};\, D^{*0} \to D^0\pi^0) = p_{D^{*0}}E_{D^0}/M_{D^{*0}}. \qquad 4.18$$

Here, $E_{D^0} = M_{D^0}$, and $M_{D^{*0}} = W/2$ to a very good approximation, so that

$$M_{D^{*0}} = [(W/2)^2 - (p_D^{ctr})^2]^{\frac{1}{2}} \qquad 4.19$$

determines the mass quite accurately. With the $D^{*+} - D^0$ mass difference accurately determined from high energy data as noted above, there results

$$\delta^* = M_{D^{*+}} - M_{D^{*0}} = 2.6 \pm 1.8 \text{ MeV}, \quad \text{and} \quad \delta - \delta^* = 2.4 \pm 2.4 \text{ MeV}. \qquad 4.20$$

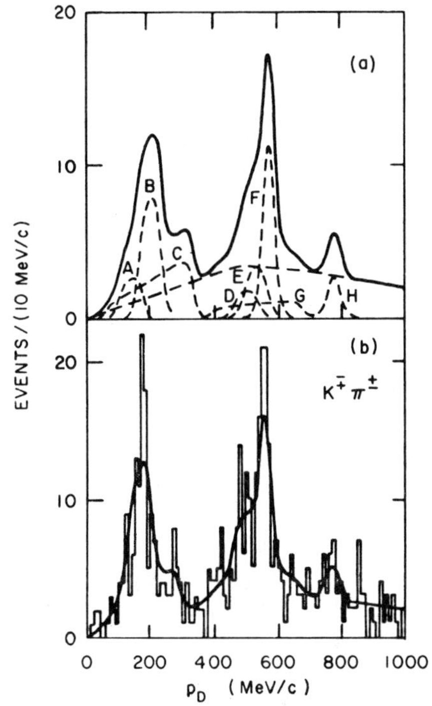

Figure 17 The D^0 momentum spectrum at 4.028 GeV, for $D^0 \to K^{\mp}\pi^{\pm}$ (from 99). The solid curves represent an isospin-constrained fit to the data. Part (*a*) shows the various contributions to the fit in (*b*). Curves A, B, C are from $e^+e^- \to D^*\bar{D}^*$ with (A) $D^{*+} \to D^0\pi^+$, (B) $D^{*0} \to D^0\pi^0$, and (C) $D^{*0} \to D^0\gamma$. Curves D, E, F, G are from $D^*\bar{D} + \bar{D}^*D$ production with (D) $D^{*+} \to D^0\pi^+$, (E) $D^{*0} \to D^0\pi^0$, (F) direct D^0, and (G) $D^{*0} \to D^0\gamma$. Curve H is the contribution from $D^0\bar{D}^0$ production.

This is purely an electromagnetic hyperfine splitting and is expected to be ~1 MeV in most theoretical estimates. Finally, the Q values used in constructing Table 11 are shown in Figure 18.

Masses of the other charmed mesons are determined by similar techniques, with the most precise measurements always coming from $e^+e^- \to M_c\bar{M}_c$ where M_c is a charmed meson, $\bar{M}_c$ its antiparticle. These masses are determined in standard ways dictated by the experimental arrangement. And, finally, at $W = 4.4$ GeV, there is some evidence for peaking in the recoil mass distribution against D^0 (99). The peak occurs near $M_x = 2.4$ GeV, possibly corresponding to the quasi-two-body production of charmed P states with D or D*. If we use the masses in Equation 4.4, the process is

$$e^+e^- \to D\,\bar{D}(1^1P_1) + D(1^1P_1)\,\bar{D}. \tag{4.21}$$

Before the discovery of ψ'', it was possible to measure with certainty only σB for the various D-meson decay modes. With this pure source of $D^0\bar{D}^0$ and D^+D^- comes the ability to measure absolute branching fractions. This, in turn, permits absolute determination of the charmed component of the total annihilation cross section. The main features of the nonleptonic fractions in Table 10 are: (*a*) they are all only a few

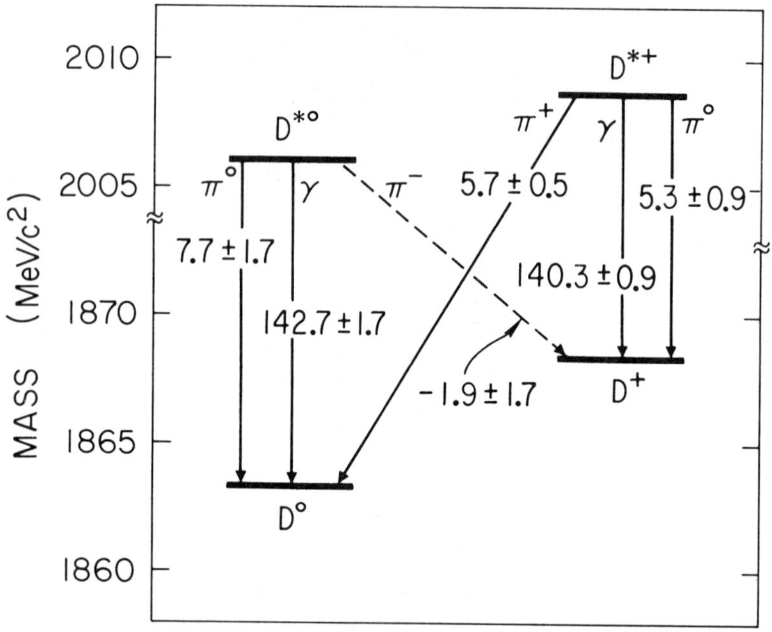

Figure 18 Q values for $D^* \to D$ transitions.

percent, as expected; (*b*) the decay $D^+ \to K^0\pi^+$ does not seem suppressed, so that the nonleptonic $\Delta C = \Delta S$ Hamiltonian may not be as simple (or as mysterious—depending on one's point of view) as the one governing K decay; and (*c*) so far there is little evidence (119) for quasi-two-body decay, e.g. $D^0 \to (K^{*-}\pi^+, \bar{K}^0\rho^0) \to \bar{K}^0\pi^+\pi^-$. There is no evidence for Cabibbo-suppressed decays; present limits are somewhat above theoretical expectations of order $\tan^2\theta_c$ (59).

The semielectronic decay fractions measured to date are really an average determined from measurement of

$$\sum_{M_c = D^0, D^+, F^+, \ldots} \sigma(e^+e^- \to M_c + \ldots) \cdot B(M_c \to e^+\nu_e + \ldots). \qquad 4.22$$

To reduce contamination from F's and charmed baryons (so that their semileptonic fractions can be unfolded in future experiments), one wants data taken at the lowest possible energies. This still gives an average over D^0 and D^+, a problem that can be resolved using "tagged" D's from ψ'' decays (108). At any rate, three experiments have now measured the average semielectronic branching ratio at low energies: the DASP collaboration at DESY found $\langle B_e \rangle = 0.10 \pm 0.03$ at $W = 3.99$–4.08 GeV (120); an LBL-SLAC-Stanford-Northwestern-Hawaii (LSSNH) collaboration found $\langle B_e \rangle = 0.072 \pm 0.028$ from running at ψ'' (121); and, at the same energy, the DELCO collaboration at SLAC has measured 0.11 ± 0.03 (122). It is worth remarking that DELCO has an order of magnitude more solid angle (60%) for electron detection than do either of the other two experiments. The average of these three is

$$\langle B_e \rangle = 0.093 \pm 0.017. \qquad 4.23$$

In Table 10 we arbitrarily took $B_e \approx 10\%$ for both D^0 and D^+. This result is a factor of two smaller than expected in a naive quark model calculation, but within the wide range estimated by Ellis et al (113).

The DASP, LSSNH, and DELCO collaborations also have measured the electron momentum spectrum in multiparticle events, presumably corresponding to $D \to e^\pm + X$ (rather than $\tau \to e + X$). Figure 19 shows the DELCO results; the results of the other two groups are quite similar but with poorer statistics. All of the spectra show the characteristic shape expected from $D \to Ke\nu$ and $K^*e\nu$, with good fits obtained by assuming sizable fractions of these two modes. None of the experiments were sensitive to the V, A structure of the amplitude for the decay $D \to K^*e\nu$. Certainly, this is one of the most important questions for future study.

The question of $D^0\bar{D}^0$ mixing has been studied by two methods. The first (99) is to observe $D^0 \to K^-\pi^+$ and look at the charge of the kaon resulting from decay of the $\bar{D}$ in recoil. The second (108) is to tag D^0 by observing the π^+ in $D^{*+} \to D^0\pi^+$ and then count the number of times D^0 decays to $K^+\pi^-$ instead of $K^-\pi^+$. In both cases, the apparent $\Delta C = -\Delta S$ decays are consistent with what is expected from $\pi - K$ misidentification; and the violation of the $\Delta C = \Delta S$ rule is $\lesssim 17\%$ at the 90% confidence level. This certainly rules out maximal $D^0\bar{D}^0$ mixing, i.e. $|\Delta C| = 2$ currents coupled to Z^0, but not necessarily a small mixing due to second order weak or Higgs boson effects.

D* branching ratios may be determined as follows (108): Fitting the relative contributions of curves B and C in Figure 17, there results (99)

$$B(D^{*0} \to D^0\gamma) = 0.45 \pm 0.15. \qquad 4.24$$

Hence, $B(D^{*0} \to D^0\pi^0) = 0.55 \pm 0.15$, and D^{*+} branching ratios may be obtained under more general assumptions than those made in Equation 4.12, namely (*a*) isospin conservation in $D^* \to D\pi$ decays; (*b*) $\Gamma(D^* \to D\pi)$

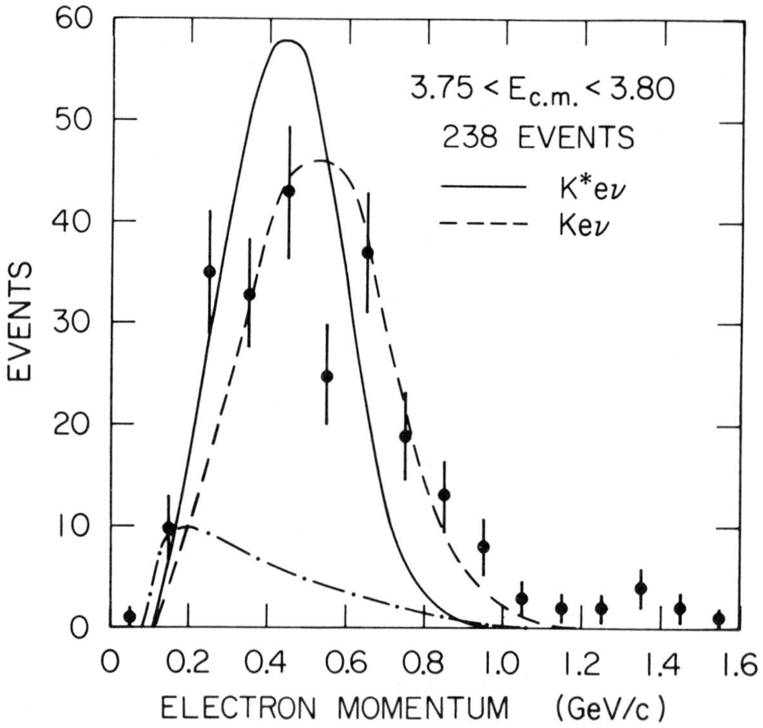

Figure 19 The momentum spectrum of inclusive electrons in multiparticle decays of the ψ'' as measured by the DELCO collaboration (32) at SLAC. Solid and dashed curves are theoretical spectra expected for $D \to K^*e\nu$ and $D \to Ke\nu$. The dot-dashed curve indicates the estimated background remaining in the data.

proportional to p_D^3; (*c*) Equation 4.11 for M1 rates. It is then clear that if $I = \frac{1}{2}$ for D and D* and $m_u/m_c \lesssim \frac{1}{4}$, the resulting widths and branching fractions are nearly indistinguishable from the predicted ones in Table 11.

All in all, the gross characteristics of charmed hadrons are just what were expected on theoretical grounds. But the theory is a long way from being well tested, and a great deal more experimental study of the details of charmed particle weak interactions is needed. It is to be hoped that this "bread and butter" physics (best carried out at the ψ'') will not be overlooked in the rush for new physics at the Υ and still higher energies.

5 BEYOND CHARM

5.1 *Motivations*

The existence and characteristics of the quarks u, d, s, and c are well established. However, there are reasons for considering the existence of further quarks, even heavier than the c. The quarks to be considered here are those of charge $\frac{2}{3}$ (called t quarks) and of charge $-\frac{1}{3}$ (called b quarks). While there is no known reason that quarks of charges $-\frac{4}{3}$, $+\frac{5}{3}$, etc. are forbidden, there is no theoretical or experimental motivation for them; they present no essentially different features and are not discussed here.

The theoretical motivation for quarks beyond charm comes from gauge theories for the weak and electromagnetic interactions. Within the group SU(2) × U(1) there is little purely theoretical motivation for more than four quarks. However, if for aesthetic or other reasons, one required the existence of both left- and right-handed charged currents (for quarks), then no SU(2) × U(1) model with only four quarks is even remotely consistent with the data. One six-quark model (123) with the coupling[19] $(ub)_R$ (and $m_b > 11$ or 12 GeV) is possible; all other SU(2) × U(1) models (e.g. see 124) with right-handed couplings for u and/or d quarks appear to be inconsistent with the data. Of course, there is absolutely no experimental evidence requiring any right-handed charged currents among quarks.

Many models based on higher gauge groups require six (or more) quarks. If an SU(3) of "flavor" is contained in the group, then triplets of quarks with charges $\frac{2}{3}$, $-\frac{1}{3}$, $-\frac{1}{3}$ can be found.[20] Some argue that the noninteger nature of quarks can be understood in a natural manner in

[19] Our notation $(q_1 q_2)_L$ [or $(q_1 q_2)_R$] refers to the charge-changing weak current, Equation 1.2, with $(1-\gamma_5)$ [or $(1+\gamma_5)$] implied. The $W^\pm$ boson couples the two fermions, q_1 and q_2.

[20] The SU(3) used in such models refers to the weak interactions: a triplet might contain u, d, and b quarks. It is quite distinct from the old SU(3) associated with the light quarks (u, d, and s).

such theories where the sum of quark charges is zero. In theories involving SU(3), the fifth and sixth quarks are both expected to have charges $-\frac{1}{3}$. It should be mentioned that some higher groups contain SU(2) × U(1) as a physically relevant subgroup and are able to reproduce all of the WS-GIM predictions for neutrino-hadron interactions. SU(3) generalizations of SU(2) × U(1) models are discussed in Section 5.2.

There have been attempts to find theories that unify the strong interactions with the electromagnetic and weak interactions. Among such theories are those based on the exceptional groups E_6 and E_7, studied by Gürsey, Ramond, and Sikivie (125). The group E_6 has SU(3) × SU(3) × $SU(3)_c$ as a maximal compact subgroup, while E_7 has SU(6) × $SU(3)_c$. The E_6 theory can be reduced to a model very similar to the WS-GIM model, and E_7 to the SU(2) × U(1) model with $(ub)_R$. Both models require six quarks (two with charges $\frac{2}{3}$ and four with $-\frac{1}{3}$). The quarks found in these theories based on exceptional groups automatically have fractional charges.

There are further motivations for new quarks. Following the discovery of the charmed quark, four quarks and four leptons were known to "exist." Later a fifth lepton τ (of mass 1.8 GeV and charge ± 1) was discovered (24). Horn & Ross (126) showed that in the WS-GIM model, existing data require the existence of a neutral lepton coupled to τ (as ν_e is to e). Some have speculated that these additional leptons may indicate the need for additional quarks. Within the WS-GIM model, it is necessary to have equal numbers of quark doublets and lepton doublets in order to cancel VVA triangle anomalies (127), which otherwise prevent renormalizability of the gauge theory. In any unified gauge theory, the proof of renormalizability makes strong use of current conservation through the associated Ward identities (9). The formal conservation of axial currents, however, is not necessarily true in the presence of fermions. Triangle diagrams with one axial and two vector vertices will destroy the axial conservation unless cancellations are arranged among the different fermions that can circulate in the loop. In the WS-GIM model, the anomaly cancels between the $(\nu_e\ e)_L$ and $(u\ d)_L$ doublets, and between the $(\nu_\mu \mu)_L$ and $(cs)_L$ doublets. Therefore the presence of a doublet associated with τ requires a quark doublet $(tb)_L$ in that model. It may be relevant to mention that the present, limited data show the branching ratio (128) for the decay $\tau \to \nu\pi$ to be substantially lower than expected; this apparent failure of a firm theoretical prediction clouds the interpretation of τ, but more data are needed before taking this result seriously. In other models the triangle anomalies are cancelled by different means so that conclusions concerning new quarks can be different.

An early experimental motivation for new quarks was the report of anomalous energy dependencies for antineutrino scattering cross sections and distributions (129). However, more recent experiments (130) with higher statistics find no large anomalies. Another relevant observation in neutrino scattering has been the discovery (131) of events with three outgoing muons ($\mu^-\mu^-\mu^+$). The number of events reported at this time is quite small, and it is impossible to determine their origin now. Three possible sources involving new heavy particles have been suggested; however, two "background" sources could also provide a significant rate of "trimuon" production. One source of trimuons in neutrino scattering could be the production of a charged heavy lepton (heavier than the τ lepton) that has a sequential decay (involving another new heavy lepton) into three muons and other particles (132). Another source (133) involves the simultaneous production of a neutral lepton (which decays into $\mu^-\mu^+\nu$) and a quark b (which decays into a negative muon and other particles). Finally, a heavy quark t could be produced and decay sequentially through either a quark b or a neutral heavy lepton (134). Alternatively, trimuon events could simply be the result of ρ, ω, ϕ, and ψ production and decay to $\mu^+\mu^-$, or muon-pair bremsstrahlung (135) off quarks or the muon. At this time one cannot, therefore, say whether or not trimuon events are an indication of the existence of new heavy quarks, but the amount of data should increase sharply in the near future.

There is, of course, one substantial motivation for quarks beyond charm. Upon its discovery (23) in pp scattering, the $\Upsilon(9.4)$ was immediately interpreted as a $\bar{q}q$ meson (103). Analyses indicated that the associated quark had charge $-\frac{1}{3}$; however, these analyses involve significant assumptions, and it should be emphasized that one cannot reach reliable conclusions concerning the charge in hadronic collisions.

In e^+e^- annihilation it should be easy to determine the nature of $\Upsilon(9.4)$ and the charge of its constituent quark. The charge will be evident by determination of the leptonic width, $\Gamma(\Upsilon \to \mu^+\mu^-)$, [about 0.7 keV for b quarks and 2.8 keV for t quarks of mass 5 GeV, according to Eichten & Gottfried (104)]. Use of the branching ratio of Υ to $\mu^+\mu^-$ is not completely reliable, since theoretical calculation of the total width is difficult. The cross sections expected for Υ are much smaller than those for ψ (see for example 136). The integrated area under a resonance in e^+e^- annihilation is given by

$$\sum = \frac{6\pi^2}{m_\Upsilon^2} B_{\text{had}}\, \Gamma(\psi \to e^+e^-), \qquad 5.1$$

where $\Gamma(\psi \to e^+e^-) = \Gamma(\psi \to \mu^+\mu^-)$. For a $-\frac{1}{3}$ charge quark one finds $\Sigma \approx 150$ nb-MeV compared with 10^4 nb-MeV for ψ, and the signal-to-background ratio may be only 2 to 1. The maximum ratio of the cross section in the resonance Υ (for charge $-\frac{1}{3}$) to background would be approximately 10 compared to 300 for ψ. While Υ will not be as dramatic as ψ, it will be quite noticeable in e^+e^- experiments, and its discovery there will be an important confirmation of a new quark.

5.2 *Extending the Standard Model*

The simplest extension of the WS-GIM model within SU(2) × U(1) is the addition of a new left-handed doublet with t and b quarks (137), which, together with a new doublet for τ leptons, would cancel triangle anomalies:

$$\begin{pmatrix} u \\ d \end{pmatrix}_L \quad \begin{pmatrix} c \\ s \end{pmatrix}_L \quad \begin{pmatrix} t \\ b \end{pmatrix}_L$$
$$\begin{pmatrix} \nu_e \\ e \end{pmatrix}_L \quad \begin{pmatrix} \nu_\mu \\ \mu \end{pmatrix}_L \quad \begin{pmatrix} \nu_\tau \\ \tau \end{pmatrix}_L \qquad 5.2$$

with all right-handed components in singlets. These new doublets have little impact on the phenomenology of the lighter particles. The d, s, and b quarks in these doublets are actually mixtures similar to the Cabibbo mixture for the four-quark model (see the discussion in Section 8). As discussed in Sections 6 and 7 and elsewhere, there are almost no data in conflict with this expanded WS-GIM model.

Within the gauge group SU(2) × U(1) it is also possible to construct models with right-handed charged currents. The relevant couplings are those to u and d quarks. Some models (123, 124) have $(u\ b)_R$ or $(t\ d)_R$ or both. If one is willing to consider quarks of charge $-\frac{4}{3}$ or $+\frac{5}{3}$, then models with $(d\ v)_R$ or $(r\ u)_R$ can be obtained. Of models with such right-handed doublets, only one (123) is consistent with present neutral current data (see Section 7):

$$\begin{pmatrix} u \\ d \end{pmatrix}_L \quad \begin{pmatrix} c \\ s \end{pmatrix}_L \qquad\qquad \begin{pmatrix} u \\ b \end{pmatrix}_R \quad \begin{pmatrix} c \\ g \end{pmatrix}_R$$
$$\begin{pmatrix} \nu_e \\ e \end{pmatrix}_L \quad \begin{pmatrix} \nu_\mu \\ \mu \end{pmatrix}_L \quad \begin{pmatrix} \nu_\tau \\ \tau \end{pmatrix}_L \quad \begin{pmatrix} N_e \\ e \end{pmatrix}_R \quad \begin{pmatrix} N_\mu \\ \mu \end{pmatrix}_R \quad \begin{pmatrix} N_\tau \\ \tau \end{pmatrix}_R, \qquad 5.3$$

where $m_b > 11$ or 12 GeV (see Section 6) but the g quark could be the constituent of Υ. The N_e, N_μ and N_τ are heavy neutral leptons. For this model, the cancellation of triangle anomalies occurs separately within the quark sector and within the lepton sector.

One can modify such models to include a coupling $(c\ s)_R$ but not $(c\ d)_R$; this has been discussed by Golowich & Holstein and others (138).

There is no reason, a priori, that quarks (or leptons) must be in doublets. However, SU(2) triplets (or higher representations) require quarks of charge $-\frac{4}{3}$ or $+\frac{5}{3}$, and will not be considered here.

We have mentioned a variety of theoretical reasons for considering other weak and electromagnetic gauge groups beyond SU(2) × U(1). Furthermore it is possible that future data will rule out SU(2) × U(1) as the full group. However the present phenomenological success of the WS-GIM model indicates that SU(2) × U(1) will be a good subgroup of any larger group.

Various authors (139) have noted that there are models within $SU(2)^L \times SU(2)^R \times U(1)$ [where L = left and R = right and $SU(2)^L$ is the same SU(2) as above] that reproduce virtually all the neutrino-hadron scattering results of the WS-GIM model. Georgi & Weinberg (140) have generalized these results and shown that at zero momentum transfer, the neutral current interactions of neutrinos in an SU(2) × G × U(1) gauge theory are the same as in the corresponding SU(2) × U(1) theory if neutrinos are neutral under G. They also noted that one of the neutral gauge bosons in the expanded group must have a mass below that of the Z^0 (80 GeV) of the SU(2) × U(1) model.

In $SU(2)^L \times SU(2)^R \times U(1)$ there are seven gauge bosons, $W_L^{\pm}$, $W_R^{\pm}$, Z_1^0, Z_2^0, γ, in contrast to the four in SU(2) × U(1) ($W^{\pm}$, Z^0, γ). It can be arranged so that Z_1^0 has purely axial-vector couplings to all particles (except neutrinos) and that Z_2^0 has purely vector couplings; this assures the absence of parity violation in neutral current interactions (see Section 7). One version of the model has the couplings:

$$\begin{pmatrix} u \\ d \end{pmatrix}_L \quad \begin{pmatrix} c \\ s \end{pmatrix}_L \quad \begin{pmatrix} t \\ b \end{pmatrix}_L \qquad (u\ b)_R \quad (c\ s)_R \quad (t\ d)_R$$

$$\begin{pmatrix} \nu_e \\ e \end{pmatrix}_L \quad \begin{pmatrix} \nu_\mu \\ \mu \end{pmatrix}_L \quad \begin{pmatrix} \nu_\tau \\ \tau \end{pmatrix}_L \qquad (N_e\ e)_R \quad (N_\mu\ \mu)_R \quad (N_\tau\ \tau)_R, \qquad 5.4$$

where column doublets are coupled by W_L (the usual W) and row doublets by W_R. Since $(u\ b)_R$ is coupled by W_R, which has no direct couplings to ν_μ, the usual lower limits on the mass of b do not apply (and $\Upsilon \equiv \bar{b}b$ is possible).

Another gauge group that has received considerable attention is SU(3) × U(1). For the models (141) considered, extreme values of the parameters can be chosen, which will reduce these models to conventional SU(2) × U(1) models. For intermediate values of the parameters, the phenomenological results are somewhat different. One version of the models resembles the WS-GIM model, while another resembles the SU(2) × U(1) model with $(u\ b)_R$.

One extension (125, 142) of the standard WS-GIM model, which has $SU(2) \times U(1)$ as a good subgroup in a fairly natural way, is based on the group $SU(3) \times SU(3)$. The neutrino-hadron scattering results are essentially the same as for the WS-GIM model although the value $\sin^2 \theta_W = \frac{3}{8}$ predicted in this model seems somewhat larger than present experimental indications. There are 16 gauge bosons, including the usual $W^{\pm}$, Z^0, and γ. Many of these bosons must be three (or more) times as heavy as the $W^{\pm}$ for phenomenological purposes. Among these are the right-handed equivalents of $W^{\pm}$ and most of the bosons carrying diagonal neutral currents. In this model the leptons are placed in two $(\bar{3},3)$ representations. The quarks are in triplets such as

$$\begin{pmatrix} u \\ d \\ b \end{pmatrix}_L \quad \begin{pmatrix} c \\ s \\ g \end{pmatrix}_L \qquad (u\,s\,b)_R \qquad (c\,d\,g)_R, \tag{5.5}$$

where the first two quarks in each column triplet are coupled by $W^{\pm}$ and all other quarks are coupled by different bosons. One of the most interesting features of this model is that the lightest new quark, b, always decays semileptonically, including modes such as $b \to u l^- \bar{\nu}$ and $b \to d\nu\bar{\nu}$. Thus, this model predicts a large amount of missing neutral energy in $e^+e^- \to (b\,\bar{q}) + (\bar{b}\,q)$.

One of the questions in constructing new models concerns the weak coupling of the b quark where $\Upsilon(9.4) \equiv b\bar{b}$. While the standard assumption places the b quark in a left-handed doublet with a t quark, there are several other couplings that are consistent with all data (see Sections 6 and 7). In $SU(2) \times U(1)$ models, the couplings $(t\,b)_R$ and $(c\,b)_R$ are allowed. In a model such as $SU(2)^L \times SU(2)^R \times U(1)$ the b quark can even have a right-handed coupling to u quarks since that interaction is mediated by W_R, which does not couple to ν_μ. Those models with quarks in SU(3) triplets can have couplings such as $(u\,d\,b)_L$, $(t\,b\,d)_R$, or $(c\,b\,d)_R$. There certainly is no evidence that b quarks have left-handed couplings to t quarks. In fact, some of the models mentioned here have no t quarks.

Not all theories involve quarks with fractional charge. Pati & Salam and others (143) have proposed models with quarks of integer charge [following Han & Nambu (144)], which nonetheless reproduce many of the results of conventional gauge theories. In this theory, however, quarks and gluons can exist as free particles (before decaying). In the basic model, there are 16 fermions:

$$\begin{pmatrix} u_R & u_Y & u_B & \nu_e \\ d_R & d_Y & d_B & e^- \\ s_R & s_Y & s_B & \mu^- \\ c_R & c_Y & c_B & \nu_\mu \end{pmatrix} \text{ with charges } \begin{pmatrix} 0 & 1 & 1 & 0 \\ -1 & 0 & 0 & -1 \\ -1 & 0 & 0 & -1 \\ 0 & 1 & 1 & 0 \end{pmatrix}, \tag{5.6}$$

where R, Y, B are the "colors" red, yellow, blue, and the leptons are considered to be the fourth color. The model can be expanded to include other quarks and leptons. One of the problems with this model is that it predicts free, massive gluons that have not been observed. This and other aspects of the Pati-Salam model are discussed critically in Reference 145.

6 PRODUCTION BY NEUTRINOS, HADRONS, AND PHOTONS

Although most experimental information about new quarks has come from e^+e^- annihilation, other methods of production have played an extremely important role. The ψ/J was produced hadronically at the same time that it appeared in e^+e^- annihilation, and charmed baryons have been observed only in neutrino and photoproduction. In this section, some of these other methods are discussed. They can yield important information about the weak and electromagnetic theory and about QCD as a possible strong interaction theory.

6.1 *Production by Neutrinos*

In neutrino scattering in the WS-GIM model, where u quarks have a left-handed coupling to d quarks $(\mathrm{u\,d})_\mathrm{L}$, one expects $\nu_\mu \mathrm{d} \rightarrow \mu^- \mathrm{u}$ or $\bar{\nu}_\mu \mathrm{u} \rightarrow \mu^+ \mathrm{d}$ to be the usual charged-current processes. Most results are consistent with this hypothesis, and one must look at rare processes in order to learn more.

In the scattering of neutrinos off nucleons, it is possible to produce single charmed mesons (or baryons). However, since there is no large coupling of valence (u or d) quarks to c quarks, this additional cross section is not large. The coupling ($\bar{\mathrm{c}}\mathrm{d} \sin \theta_\mathrm{c}$) with $\sin^2 \theta_\mathrm{C} \approx 0.05$ leads to a 5% rise in the expected cross section for neutrinos (above the threshold energy). There is no similar Cabibbo-suppressed ($\sin \theta_\mathrm{C}$) process possible for antineutrinos. The coupling ($\bar{\mathrm{c}}\mathrm{s} \cos \theta_\mathrm{C}$) leads to an increase in both neutrino and antineutrino cross sections; however, the amount of strange quarks in the sea (i.e., of $\mathrm{s}\bar{\mathrm{s}}$ pairs in the nucleon) is quite small, of order 5% (146), so that resulting effects are small. Since 5% effects are difficult to measure experimentally and since comparable or larger QCD effects may occur, little evidence for charm is found in total cross sections. Similarly, little effect is seen in y distributions $[y \equiv (E_\nu - E_\mu)/E_\nu]$.

It might be helpful to give a simplified description of several of the features of QCD that should result in similar effects in neutrino scattering (147). With increasing $Q^2 (\equiv -q^2 = 4E_\nu E_\mu \sin^2 \theta_\mathrm{lab}/2)$ one expects that: (*a*) the x distributions of quarks [where $x \equiv Q^2/2M_\mathrm{N}(E_\nu - E_\mu)$] will shrink (i.e. become more peaked toward zero); (*b*) the fraction of the struck nucleon's momentum carried by valence quarks will decrease slowly; (*c*) the fraction carried by sea quarks ($u\bar{u}$, $d\bar{d}$ and $s\bar{s}$ pairs in the nucleon) will increase. There are helicity arguments that show $\sigma(\nu q_1 \rightarrow \mu^- q_2) = 3\sigma(\nu\bar{q}_2 \rightarrow \mu^-\bar{q}_1) = 3\sigma(\bar{\nu}q_2 \rightarrow \mu^+ q_1) = \sigma(\bar{\nu}\bar{q}_1 \rightarrow \mu^+\bar{q}_2)$, where $\bar{q}$ indicates antiquark. In neutrino reactions then, scattering off valence quarks is enhanced relative to that off sea quarks, while in antineutrino reactions scattering off sea quarks is enhanced (although most momentum is always carried by valence quarks). As a result, one expects neutrino cross sections to decrease with increasing E_ν (which is proportional to $\langle Q^2 \rangle$) and antineutrino cross sections to increase slightly. A related effect is the increase of $\langle y \rangle$ for antineutrinos with increasing E_ν (for neutrinos there is little effect).

Although charm is difficult to detect in total cross sections and distributions, evidence for charm is quite clear in other aspects of neutrino experiments. Charmed particles decay into muons and into electrons 10–20% of the time, and these leptons can be detected. If charm production is 5–10% of the total, and the branching ratio to muons (electrons) is 10–20%, then 0.5–2% of all neutrino-induced events should contain an extra muon (electron). This rate of "dilepton" production (17) is in fact roughly what is observed (because of experimental cuts and efficiencies the exact rate is not easy to determine directly). Furthermore, in distributions of variables such as y, E_μ, and various angles, one finds (148) strong evidence for the additional lepton coming from the decay of a produced heavy quark (with mass of approximately 2 GeV).

Neither the rate nor the distributions show that this heavy quark is charm. However, since charm usually decays to s quarks, in bubble chamber experiments one can see if events with two leptons also have a K meson or a Λ^0 baryon. When neutrinos change d quarks into c quarks, one strange particle should result. However, when an s quark in the sea is changed into a c quark, there is always the remaining $\bar{s}$ quark from the pair in addition to the s quark from c quark decay, so that two strange particles result. Since antineutrino scattering lacks a Cabibbo-suppressed mode of charm production, the number of strange particles (two) is expected to be greater than for neutrinos (roughly 1.5). At present, results have been reported only for neutrinos. Two experiments (149) have reported about 3.5 K mesons per $\mu^- e^+$ event, while one other (150) with much higher statistics has reported about 1.0 K mesons per event. The latter corresponds closely to the predictions for charm.

In all of these features (cross sections, y distributions, dilepton rates, presence of strange particles) little room remains for significant production of any heavier quarks. Of course, for sufficiently massive quarks, all production would be deferred until higher energies. Present data

(129, 130) indicate that any b quark (charge $-\frac{1}{3}$) that has a right-handed coupling to u quarks (through W bosons) must have $m_b \geqq 11$ or 12 GeV, certainly excluding the quark in $\Upsilon(9.5)$. If given that $m_b = 5$ GeV, then the coupling squared for $(\mathrm{u\,b})_R$ must be 0.1 (or less) of that for $(\mathrm{u\,d})_L$. Any t quark (charge $\frac{2}{3}$) that has a right-handed coupling to d quarks (through W bosons) must have $m_t \geqq 5$ or 6 GeV. For the left-handed couplings $(\mathrm{u\,b})_L$ and $(\mathrm{t\,d})_L$, the limits from analysis of the data are $m > 8$ GeV in both cases if the couplings are full strength. For 5-GeV t or b quarks, the (left-handed) couplings squared must be 0.3 (or less) of that for $(\mathrm{u\,d})_L$.

In the WS-GIM model, the additional coupling $(\mathrm{t\,b})_L$ (see Section 5) would lead to little t or b quark production, because the mixing angles between heavy quarks and light quarks must be small (see Section 8.1). From the universality of quark and lepton couplings and from the $K_L^0 - K_S^0$ mass difference, one finds (151) that the $\bar{\mathrm{t}}\mathrm{d}$ coupling ($\bar{\mathrm{u}}\mathrm{d}$ coupling) is not likely to be more than 10% (5%) of the $\bar{\mathrm{u}}\mathrm{d}$ coupling [i.e. rates at the 1% (0.3%) level]. Clearly even at high energies there will be little impact on cross sections.

There are many other couplings possible for t and b quarks besides $(\mathrm{t\,b})_L$, and some of these also have few observable consequences in neutrino physics. Among such couplings are the right-handed couplings $(\mathrm{t\,b})_R$ and $(\mathrm{c\,b})_R$. Also the couplings $(\mathrm{u\,b})_R$ and $(\mathrm{t\,d})_R$ are possible (even for relatively light t and b) in models such as $SU(2)^L \times SU(2)^R \times U(1)$, where those couplings occur not through the usual W boson but through a new boson that does not couple ν_μ to μ and is heavier than W.

One can look for signals for new quarks in multilepton events, although in the WS-GIM model the rates will not be high. Given the coupling squared ($\lesssim 0.01$), the branching ratio to muons (~ 0.2), and the phase space suppression F (see Table 12), the rate for dilepton events from new heavy quarks is less than $0.2\text{-}F$ times the rate for dileptons from charm in this model. Clearly then, detection of t and b quarks will be difficult. One possible method of distinguishing dileptons for t or b

Table 12 The phase space suppression F (relative to zero mass quarks) for production in neutrino scattering of quarks of given mass

Quark mass (GeV)	F for Fermilab Quad triplet flux: All E	F for Fermilab Quad triplet flux: $E > 100$ GeV	F for CERN Wide-band flux: All E	F for CERN Wide-band flux: $E > 100$ GeV
5	10^{-1}	3×10^{-1}	10^{-1}	2×10^{-1}
10	6×10^{-3}	3×10^{-2}	10^{-3}	10^{-2}
15	10^{-4}	6×10^{-4}	10^{-5}	10^{-4}

decay from dileptons for charm decay involves the energy of the secondary (decay) lepton. Dileptons from heavy quark decay should be significantly more energetic (152).

Another means of finding evidence for the production of t or b quarks comes from examination of events in which three leptons are produced. In the WS-GIM model, b quarks are likely to decay into c quarks unless $m_t < m_b$ (see Section 8.1). In some cases, both the b and c quark decays would involve leptons. In antineutrino scattering, b quarks can then be produced (along with the usual μ^+) and decay sequentially into two muons (or electrons). The resultant "trilepton" events would occur at less than $10^{-4} F$ (see Table 12) of the total rate. If $m_b > m_t$, it is clear that trilepton events can also occur at the same rate, but with b decay to t instead.

If $m_t > m_b$, then (in this model) t quarks can decay into b quarks, which decay into c quarks. In neutrino scattering t quarks can be produced and decay into two, three, or more leptons. For such trilepton events the rate would be less than $8 \times 10^{-4}F$ of the total rate (counting both $\mu^-\mu^-\mu^+$ and $\mu^-\mu^+\mu^+$ events).

Trimuon events have been reported by three groups (131). At Fermilab, one group (131) reports a rate of 10^{-4} of the total rate, while at CERN a rate of 5×10^{-5} has been reported (131) (in both cases $E_\nu > 100$ GeV and $E_\mu > 4$ GeV is required). Presumably some (and possibly all) of these events come from background sources (such as ρ decay). The rates given above for the WS-GIM model are upper bounds (since upper bounds on $t\bar{d}$ and $u\bar{b}$ couplings are used) and even then appear to be lower than these reported rates, but the question of backgrounds should be resolved. Various studies (132–135) have been made concerning the expected characteristics of such trilepton events; more data are needed before conclusions can be reached.

The multilepton events found in b or t quark decays are expected (in the WS-GIM model) to be accompanied by the presence of strange particles. If $m_b < m_t$, then b quarks decay into c quarks, which decay into strange particles; and the t quarks decay into b quarks. If $m_t < m_b$, then t quarks decay dominantly into s quarks directly; and the b quarks decay into t quarks.

While the above remarks were taken in the context of the WS-GIM model, similar conclusions follow in many other models. There are a large variety of left- and right-handed couplings possible for b and t quarks in both SU(2) × U(1) and other models, and frequently these result in detectable signals in neutrino experiments as described above.

6.2 *Hadroproduction of Heavy Particles*

Heavy vector mesons such as ψ and Υ, which decay to $\mu^+\mu^-$ or e^+e^-, have been observed in pp, pN, $p\bar{p}$, and πp scattering experiments (153).

Several different approaches to calculating the cross sections have been advocated. Some of them have been modified as further data were reported, and here most attention will be given to the later versions.

One obvious way to produce ψ mesons is by the fusion of a c quark from one of the incoming hadrons with a $\bar{c}$ quark from the other hadron, where the c and $\bar{c}$ quarks come from the "sea" of their respective hadrons (154). The magnitude of such a process is difficult to estimate since assumptions are needed about the validity of SU(4) and/or about the manner of decay of the ψ (used to estimate the fusion coupling). However, this process has an unavoidable consequence that is easy to test: All ψ production should be accompanied by the simultaneous production of two charmed particles. The experiments done report (155) no evidence for this process, and it must be quite suppressed compared to other processes.

One modification of such an approach is to argue that the ψ is produced by the fusion of light quarks and antiquarks (154). However, since there are not many antiquarks in a proton compared to an antiproton, one expects ψ production in $p\bar{p}$ scattering to be at least twenty times that in pp scattering (depending on the shape of quark and antiquark distributions). Experiment indicates a factor of about seven. While this approach appears to be inadequate, such diagrams are included in two other (more successful) methods.

In one of these other approaches it is argued that the P-wave states (χ) of the ψ family are produced much more frequently than are ψ's, and that the observed ψ's are primarily decay products of those P-wave states (156a; for Υ production, see 130, 156b). The production of ψ's (relative to χ's) is said to be suppressed because ψ couplings require at least three gluons, so that the resulting effective coupling is much smaller than that for χ, which requires only two gluons. In such a picture, the χ is produced by two processes: (*a*) the fusion of two gluons, one from each of the colliding hadrons, and (*b*) the fusion of a quark and an antiquark from the colliding hadrons. The latter process can be assumed negligible in pp scattering (where there are few antiquarks), but is important in $p\bar{p}$ scattering. This approach can obtain the correct ratio (157) for ψ production in pp relative to $p\bar{p}$ scattering, which is 0.15 ± 0.08 at $s^{\frac{1}{2}} = 8.75$ GeV. The cross section (via gluons, which are labeled g below) can be written as:

$$\sigma(\mathrm{A}+\mathrm{B} \to \psi + \mathrm{X}) = \int \mathrm{d}x_1 \mathrm{d}x_2 f^{\mathrm{A}}(x_1) f^{\mathrm{B}}(x_2) \sigma(\mathrm{gg} \to \chi) B(\chi \to \psi + \gamma)$$

$$= \frac{8\pi}{m^3} \Gamma(\chi \to \mathrm{gg}) B\tau \int_\tau^1 \frac{\mathrm{d}x}{x} f^{\mathrm{A}}(x) f^{\mathrm{B}}\left(\frac{\tau}{x}\right), \qquad 6.1$$

where $f(x)$ are gluon distribution functions, $\tau \equiv m^2/s$ and m is the mass of χ. Using

$$\Gamma(\chi \to gg)B \approx \Gamma(\chi \to \psi + \gamma) = \tfrac{4}{3}\alpha e_Q^2 \omega^3 |\langle\psi|\mathbf{r}|\chi\rangle|^2 \propto e_Q^2 m^{-\frac{5}{3}}, \qquad 6.2$$

where e_Q is the charge of the quark ($\frac{2}{3}$ for charm), ω is the $\chi - \psi$ mass difference, and the last proportionality holds for a linear potential model, then one finds

$$\sigma(A + B \to \psi + \chi) \propto e_Q^2 m^{-\frac{14}{3}} F(\tau). \qquad 6.3$$

In this approach both the form and magnitude of $F(\tau)$ can be calculated, and the results shown in Figure 20 are in reasonable agreement with the data. Even if $F(\tau)$ was not calculable, one could take $F(\tau)$ from the data for ψ and apply Equation 6.3 to the production of Υ and heavier particles.

One motivation for this approach is that it provides an obvious explanation for the observed large suppression of ψ' relative to ψ production (153). Since there are no P-wave states that can decay into ψ', it can only be produced directly; but direct production was assumed to be suppressed. If ψ production is really an indirect process involving $\chi \to \psi + \gamma$, then the observation of γ's associated with ψ production is a

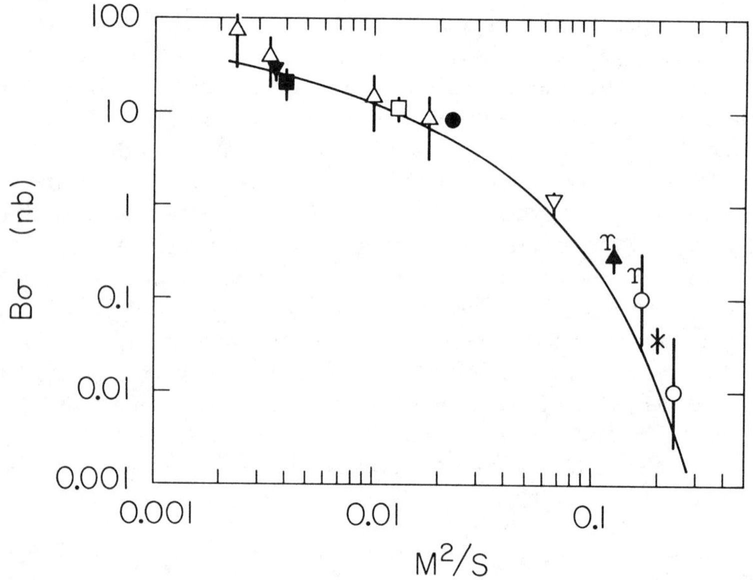

Figure 20 The cross section for ψ production in pp scattering times the branching ratio to muons as a function of τ (where $\tau \equiv m_\psi^2/s$). The data is from Reference 153. The curve is a theoretical prediction (156a) of Carlson & Suaya. The two symbols Υ are the data for Υ production adjusted according to Equation 6.3.

crucial test of this approach [a recent experiment may in fact see such γ's (158)].

In another approach (159a; for Υ production, see 159b), the production of a pair of quarks, c and $\bar{c}$, is calculated. When the invariant mass of the pair is less than two times the mass of D mesons, it is assumed that ψ's (or other ψ family members) can be produced. There are three types of diagrams that contribute to $c\bar{c}$ production in this approach: (*a*) a quark from one of the colliding hadrons can annihilate with an antiquark from the other hadron to give a single gluon, which produces a $c\bar{c}$ pair; (*b*) a gluon from each of the hadrons can couple to a c quark line, producing a $c\bar{c}$ pair; and (*c*) gluons from each of the hadrons can fuse to a single gluon, which produces a $c\bar{c}$ pair. In each case a color singlet is obtained via final-state interactions neglected in calculations. For diagram (*a*) the cross section is given by

$$\sigma = \int_{(2m_c)^2}^{(2m_D)^2} \frac{ds'}{s'} \sigma(q\bar{q} \to g \to c\bar{c})\tau \int_{\tau}^{1} \frac{dx}{x} f(x) f\left(\frac{\tau}{x}\right), \qquad 6.4$$

where s' is the subenergy, $\tau = s'/s$, and $f(x)$ are quark distributions. One way to estimate σ in the integrand is to assume an analogy to Drell-Yan calculations (160) of $\mu^+\mu^-$ production. Then σ is given by

$$\sigma \approx \int_{(2m_c)^2}^{(2m_D)^2} ds' \frac{d\sigma'_{DY}}{ds'} \frac{\alpha_s^2}{\alpha^2} \frac{2}{3}, \qquad 6.5$$

where α_s is the strong coupling constant and σ'_{DY} is the Drell-Yan cross section calculated without the quark charges. Diagrams (*b*) and (*c*) can be calculated similarly. When all three types of diagrams are included, one obtains the correct ratio for ψ production in pp relative to $p\bar{p}$ scattering.

In Figure 20 the data for the production of ψ in proton-nucleon scattering are shown. The curve is a theoretical calculation of Carlson & Suaya (156a) using the indirect production ($\chi \to \psi$) approach. Other authors using this and other approaches have obtained similar results for ψ production. To test the basic hypotheses of the approaches discussed, one can examine whether they can account for the observed cross section for $\Upsilon(9.4)$ production. For the indirect production approach, one sees from Equation 6.3 that if the data are adjusted for e_Q^2, $m^{-\frac{14}{3}}$ and the branching ratio $B(\Upsilon \to \mu^+\mu^-)$, then the Υ data should lie on the same curve as the ψ data. In Figure 20 these adjusted data are shown with the symbol Υ and do, in fact, lie on the same curve. It was assumed that the quark associated with Υ had charge $-\frac{1}{3}$; otherwise the adjusted data points would lie a factor of about 16 lower (both e_Q^2 and B change by a factor of about 4). For the direct production approach, adjusted data points lie below the ψ data by a factor (according to one calculation) of about 5 for charge $-\frac{1}{3}$ quarks and 20 for $\frac{2}{3}$ charge quarks (only B is different for different charges). Since it is probably unreasonable to expect these models for Υ production to be accurate to better than an order of magnitude, these results do not distinguish the two approaches nor are they completely reliable determinations of the charge of the quark in Υ.

It is interesting to ask what are the highest mass vector mesons ($Q\bar{Q}$) that can be produced in hadronic collisions at existing and future accelerators and storage rings. In Table 13 [from Carlson (161)] it is assumed that at least 10 events per year must be observed. Of course, these results are only crude estimates, since extrapolation from 3-GeV particles (ψ) to particles of enormous mass is difficult. There could be unforeseen complications; one such complication suggested by Bjorken and by Nieh (162) is that very massive quarks could have very significant weak decay modes so that the branching ratio to $\mu^+\mu^-$ (or e^+e^-) would decrease. The weak decay width becomes a sizable fraction of the total when the quark mass becomes comparable to the W boson mass.

The production of charmed particles ($D^+ \equiv c\bar{d}$, $D^0 \equiv c\bar{u}$, etc.) in hadronic collisions has not yet been observed. However, various theoretical estimates suggest that the actual cross sections are not far below the present experimental limits. There are some simple methods to estimate cross sections at Fermilab energies. For example, one could guess

$$\frac{\sigma_D}{\sigma_\psi} \approx \frac{\sigma_K}{\sigma_\phi} \approx 10. \qquad 6.6$$

Table 13 The highest mass of $Q\bar{Q}$ mesons that can be produced in pp collisions at existing and proposed facilities (161)[a]

Facility	$s^{\frac{1}{2}}$	Luminosity	Mass (GeV) $-\frac{1}{3}$ charge	Mass (GeV) $\frac{2}{3}$ charge
ISR (31+31)	62	10^{31}	21 (21)	26 (26)
TRISTAN (180+180)	360	10^{33}	76 (86)	92 (115)
ISABELLE (400+400)	800	10^{33}	87 (114)	108 (165)
FNAL (270+1000)	1040	10^{33}	90 (122)	112 (181)
POPAE (1000+1000)	2000	10^{33}	94 (140)	117 (218)
UNK (2000+2000)	4000	10^{33}	96 (151)	121 (243)
VBC (10^4+10^4)	2×10^4	10^{33}	98 (161)	124 (276)
VBA (fixed target)	140	10^{37}	89 (96)	96 (103)

[a] The mass calculations include corrections for the existence of weak decay modes (see text); the parenthetical numbers are without such corrections. The first seven facilities are storage rings with the energy of each colliding beam given.

This gives $\sigma_D \approx 1\ \mu b$ for pp scattering. Sivers (163) has suggested that one could assume that the appropriate transverse momentum scaling variable is $(p_\perp^2 + m^2)^{\frac{1}{2}}$ rather than $p_\perp$; then the cross section for D meson production might be related to that for pions with $p_\perp \approx 2$ GeV so that

$$\sigma_D \approx 0.1 \ln\left(\frac{s}{4m_D^2}\right) \exp(26 m_D s^{-\frac{1}{2}})\ \text{mb}. \qquad 6.7$$

This method requires an assumption about the charmed quark content of nucleons; for the above estimate $c/s = 0.2$ (where c and s are charmed and strange quark content, respectively) was assumed, which gives $\sigma_D \approx 10\ \mu b$. But there is reason to believe c/s is perhaps an order of magnitude smaller, so that σ_D should be much smaller. The present experimental limit is $\sigma_D < 1.5\ \mu b$ at $s^{\frac{1}{2}} = 27$ GeV (164).

More sophisticated calculations have been carried out (151, 154, 156a,b, 159a,b, 165). These are usually extensions of the "direct production" approach to ψ production where the limits of integration over s' (such as in Equation 6.4) are changed to $4m_D^2$ and s. As for ψ production, the diagrams (*a*), (*b*), and (*c*) can all contribute. Babcock, Sivers & Wolfram (165) (among others) discuss the results of such QCD calculations and conclude that diagrams (*b*) and (*c*) are more important than (*a*). They also discuss higher order effects and argue that it is reasonable to neglect them for most purposes. With standard assumptions Babcock et al estimate for pp scattering $\sigma_D \approx 1\ \mu b$ at $s^{\frac{1}{2}} = 27$ GeV (also $\sigma_D \approx 10\ \mu b$ at $s^{\frac{1}{2}} = 54$ GeV and 100 μb for $s^{\frac{1}{2}} \geqq 200$ GeV). They argue that the present experimental limit on D production in hadronic collisions favors either smaller values of α_s (than expected from leptoproduction experiments) or gluon distributions that are more peaked toward small x.

For heavier mesons such as $Q\bar{u}$ and $Q\bar{d}$ where $Q\bar{Q} \equiv \Upsilon(9.4)$, most estimates (165) are that for a very large range of energies the cross sections for production of $Q\bar{d}$ or $Q\bar{u}$ mesons will be two orders of magnitude lower than those for D mesons. This clearly makes observation of such mesons very difficult in hadronic collisions.

In addition to the above calculations, which are based on behavior expected for $y \approx 0$, there have been calculations of peripheral production (for $x > 0.5$) of charmed particles. For example, the cross section (166) for $\pi^- p \to D^- C_0^+$ (C is a charmed baryon) has been estimated as 0.5 nb while triple Regge calculations (166) of $\pi p \to DX$ (for $x > 0.5$) give $\sigma \approx 60$ nb.

6.3 *Photoproduction of ψ and Charm*

One of the earliest papers on ψ (prior to its discovery) was written by Carlson & Freund (167), who discussed the photoproduction of a then hypothetical $c\bar{c}$ vector meson. The photoproduction of ψ is usually assumed to be a dominantly diffractive process that can be understood with a modified vector-dominance model (97). The modification allows for the $\gamma\psi$ coupling to be different at $q^2 = 0$ and $q^2 = m_\psi^2$. With this assumption the cross section (see 168) is

$$\frac{d\sigma}{dt}(\gamma N \to \psi N) = \frac{3\lambda^2}{\alpha M_\psi} \Gamma(\psi \to e^+e^-) \frac{d\sigma}{dt}(\psi N \to \psi N), \qquad 6.8$$

where λ measures the variation of the $\gamma\psi$ coupling $g_{\gamma\psi}$ with q^2 and the off-mass-shell extrapolation of the invariant amplitude ($\lambda = 1$ for the "naive" vector-dominance model). Making use of the optical theorem, one finds:

$$\frac{d\sigma}{dt}\left(\gamma N \to \psi N\right)\bigg|_{t_{\min}} = \frac{1}{e^{-bt_{\min}}} \frac{3\Gamma(\psi \to e^+e^-)}{16\pi\alpha M_\psi} \lambda^2(1+\rho^2)\sigma_{\text{tot}}^2(\psi N), \qquad 6.9$$

where $d\sigma/dt$ was assumed to have t dependence e^{-bt} [which is consistent with data (169) for $b = 2.9$ GeV^{-2}] and $\rho \equiv (\text{Re}\mathscr{A}/\text{Im}\mathscr{A}) \to 0$ as $s \to \infty$, with $\mathscr{A}$ the amplitude for $\psi N \to \psi N$. An independent determination of $\sigma_{\text{tot}}(\psi N)$ can be extracted from the observed A dependence (170) (where A is the effective number of nucleons per nucleus) of ψ photoproduction; experiments on Be and Ta give $\sigma_{\text{tot}}(\psi N) = 3.5 \pm 0.8$ mb at $E_\gamma = 20$ GeV. To avoid consideration of threshold factors and of ρ, we will assume that this value stays approximately constant up to higher energies ($E_\gamma \approx 80$ GeV). Next, an assumption about the value of λ is needed. If the value of the naive vector-dominance model ($\lambda = 1$) is taken, then for $\rho \approx 0$, $d\sigma/dt(t = 0) \approx 400$ nb/GeV^2, which is far above the experimental values (169, 171) of about 60 nb/GeV^2 at $E_\gamma \approx 80$ GeV. Some theoretical models give $\lambda \approx 0.5$, which gives $d\sigma/dt$ $(t = 0) \approx 100$ nb/GeV^2; choosing $\lambda = 0.3$ or 0.4 gives 40 or 60 nb/GeV^2. Clearly, the naive vector-dominance model must be modified to account for ψ photoproduction.

From knowledge of ψ photoproduction, there are some immediate implications for the photoproduction of charmed particles. By use of unitarity, it can be shown (168) that for a given energy:

$$16\pi \frac{d\sigma}{dt}\left(\gamma N \to \psi N\right)\bigg|_{t_{\min}} \leq (1+\varepsilon)^2(1+\rho^2)\left(\frac{q^{\psi N}}{q^{\gamma N}}\right)\sigma(\gamma N \to \text{charm}) \; \sigma(\psi N \to \text{charm}), \qquad 6.10$$

where ε measures violations of the OZI rule and

$$\frac{q^{\psi N}}{q^{\gamma N}} = \frac{[s-(m_p+m_\psi)^2]^{\frac{1}{2}}[s-(m_p-m_\psi)^2]^{\frac{1}{2}}}{s-m_p^2} \qquad 6.11$$

(which is 0.33, 0.72, 0.93 for $E_\gamma = 10, 20, 80$ GeV). Application of the OZI rule again implies that $\sigma_{\text{tot}}(\psi\text{N}) \approx \sigma(\psi\text{N} \rightarrow \text{charm})$. Using $E_\gamma = 20$ GeV data (169, 171) for $(\text{d}\sigma/\text{d}t)$ and $\sigma_{\text{tot}}(\psi\text{N})$ in Equation 6.10, one finds

$$\sigma(\gamma\text{N} \rightarrow \text{charm}) \gtrsim 115 \text{ nb}/(1+\varepsilon)^2(1+\rho^2). \qquad 6.12$$

Since ε and ρ are presumably small, it is safe to say $\sigma(\gamma\text{N} \rightarrow \text{charm}) \gtrsim$ 100 nb.

It is possible to estimate the photoproduction of charm by QCD techniques. Gluons from the nucleon and the incoming photon can each couple to a c quark line producing a $c\bar{c}$ pair. This has been discussed by various authors (172, 173), who find $\sigma(\gamma\text{N} \rightarrow \text{charm})$ increasing from about 100 nb at $E_\gamma = 20$ GeV to about 400 nb at $E_\gamma = 80$ GeV. Some assumptions are required to apply perturbative QCD to this problem, but the approach is not implausible. Roughly speaking, the large mass of the charmed quark is expected to set the scale for the effective coupling constant. The use of QCD here is similar in some ways to its use in explaining the total width of the $c\bar{c}$ states, and it surely needs further theoretical analysis.

These results are in approximate agreement with a sum rule of Shifman, Vainshtein & Zakharov (173), and may be consistent with the reported observation (118) of a charmed antibaryon of mass 2.26 GeV (whose cross section has not been reported yet). This experiment of Knapp et al (118) saw evidence for what may have been $\Lambda_c^- \rightarrow \bar{\Lambda}\pi^-\pi^-\pi^+$.

7 NEUTRAL CURRENT INTERACTIONS

The weak interactions provide an important probe in the study of new quarks (and new leptons). The neutral current interactions are a crucial measure not only of the existence of new quarks but also of the structure of the gauge theories of weak and electromagnetic interactions. Were the neutral current predictions of the WS model to fail, one would be forced to consider other models. In fact, however, the WS model is in good agreement with most neutral current data, as will be discussed below. The importance of neutral current phenomenology can be seen in the successful prediction of the existence of the c quark from the absence of strangeness-changing neutral currents [via the GIM mechanism (4)]. For much of the study of neutral current interactions the neutrino is used as a probe; it is uniquely suited to this purpose, since it is the only particle that has only weak interactions.

It is possible, a priori, that c quarks (or b or t quarks) could be produced directly by neutral current processes in which u quarks were changed into c quarks. In the WS model (with all left-handed quarks in doublets, none in singlets) such charm (or other flavor) changing neutral currents are forbidden by the GIM mechanism. Two types of experiments indicate that the neutral weak boson Z^0 [of SU(2) × U(1) models] does not change charm: (*a*) e^+e^- annihilation experiments (174) find no $D^0\bar{D}^0$ mixing (such mixing should be found if charm-changing currents exist); (*b*) neutrino scattering experiments (17) also find no evidence of

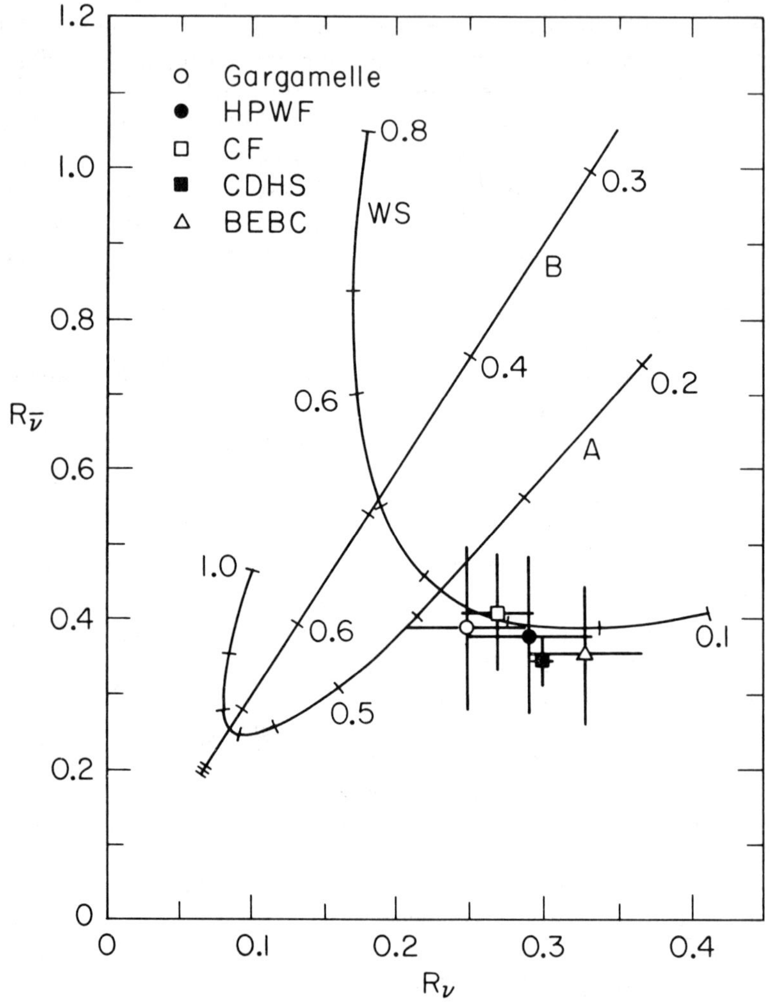

Figure 21 The ratio $\sigma(\nu N \to \nu + X)/\sigma(\nu N \to \mu + X)$ for antineutrinos vs that ratio for neutrinos. The tenth values of $\sin^2\theta_W$ are shown with tick marks on the theoretical curves. The curve labeled A refers to the model with $(u\ b)_R$: the curve labeled B refers to the model with both $(u\ b)_R$ and $(t\ d)_R$. The data are from Reference 175.

$D^0\bar{D}^0$ mixing (which would lead to $\mu^-\mu^-$ events). In a model for a group other than SU(2) × U(1) it may be possible to have flavor-changing neutral currents if they occur via another boson that is quite heavy or does not have flavor-conserving couplings to light fermions. For SU(2) × U(1) (in which many models are possible) there is little likelihood that there would be t- or b-changing neutral currents, given that there are no strangeness- or charm-changing currents. In SU(2) × U(1) models with quarks of charges $\frac{2}{3}$ and $-\frac{1}{3}$ only, there are no flavor-changing neutral currents if all quarks of a given handedness are in doublets (or are all in singlets) of SU(2). If most quarks were in doublets but one were in a singlet, mixing among quarks would lead to flavor-changing neutral currents for all quarks of that charge (unless there were some reason why mixing was prevented) (115).

There are four types of neutrino experiments commonly used to test the diagonal (flavor-conserving) neutral current structure of gauge theories. These are inclusive scattering off heavy nuclei (175), elastic scattering off protons (176), semi-inclusive (single pion) scattering off heavy nuclei (177), and elastic scattering off electrons (178). The first three can be used to calculate the neutral current couplings of u and d quarks. In SU(2) × U(1) models these quark couplings are given by:

$$q_L = \tau_3^L - Q\sin^2\theta_W$$
$$q_R = \tau_3^R - Q\sin^2\theta_W \qquad 7.1$$

where L (R) refers to left-handed (right-handed), τ_3 is the weak isospin ($\pm\frac{1}{2}$ for quarks in doublets, 0 in singlets), Q is the charge of the quark, and θ_W is the Weinberg angle, which is a free parameter of this theory. The inclusive and elastic scattering results are usually reported as ratios of neutral current to charged-current cross sections for both neutrinos and antineutrinos; the semi-inclusive scattering experiments give ratios of π^+ to π^- (in the current fragmentation region). With these six numbers, the possible couplings (u_L, u_R, d_L, d_R) are severely limited. Some of the data are shown in Figures 21 and 22. Analyses have been done by many authors (179). An analysis of Hung & Sakurai (179) [who make use of conclusions of Sehgal (179)] finds that there are only two sets of couplings for u and d quarks allowed by the data. These are shown in Table 14 (where the uncertainties are always ± 0.15). Note that if all four signs are changed in set A or in set B, the resulting sets of couplings are, of course, equally allowed. If $\sin^2\theta_W = 0.3$ is chosen, then the WS model predicts that u_L, d_L, u_R, d_R are 0.3, −0.4, −0.2, 0.1. This is very close to set A of allowed couplings. There may be other models such as the SU(2) × U(1) model with $(u\ b)_R$ (see Section 5) that have values similar to those of set B. Note, however, that the parameter θ_W is attributable only to a specific model, and other models may fit the data for different values of that parameter. If there is any shortcoming to the above analysis, it is that these results depend crucially on use of specific parton model assumptions in the analysis of the semi-inclusive data. Since that data was taken at very low energies where parton model assumptions could be questioned, it would be best to confirm the conclusions by independent means. A new analysis, which is near completion, by Abbott & Barnett (179) (based on very new data) will try to use independent methods to further isolate the allowed values of the neutral current couplings of u and d quarks.

There are three types of neutrino-electron elastic scattering experiments

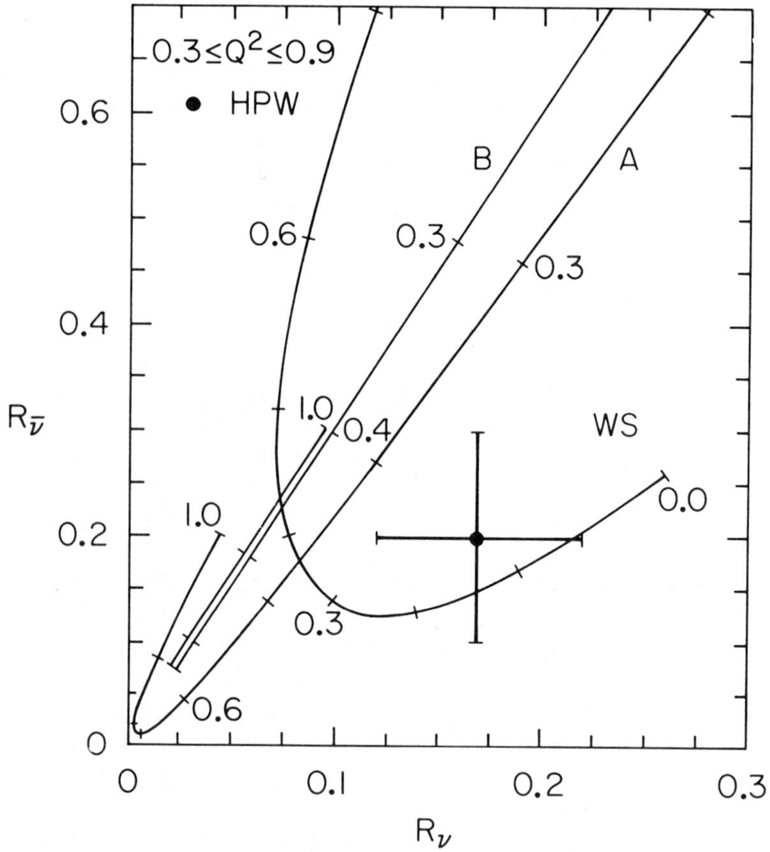

Figure 22 The ratio $\sigma(\bar{\nu}p \to \bar{\nu}p)/\sigma(\bar{\nu}p \to \mu^+n)$ vs the ratio $\sigma(\nu p \to \nu p)/\sigma(\nu n \to \mu^- p)$. The notation is the same as for Figure 21. The data shown are from D. Cline et al (176).

Table 14 Allowable couplings for u and d quarks (uncertainties are ± 0.15)

	u_L	d_L	u_R	d_R
A	+0.29	−0.40	−0.24	0
B	+0.29	−0.40	+0.24	0

(178) that have been reported: they are with ν_μ, $\bar{\nu}_\mu$, and $\bar{\nu}_e$ beams. With each cross section one can determine a locus of points in the $g_A - g_V$ plane consistent with the 90% confidence level upper and lower bounds for that cross section. Each of these is an annulus, and in Figure 23 the intersection of these three regions (which is shaded) is the allowed region. The WS model with $\sin^2 \theta_W = 0.25$–0.3 lies within the lower part of the allowed region. Some other models lie in the upper part of the allowed region.

There are experiments testing weak neutral currents that do not involve neutrinos. These concern effects that arise from parity violation, which is

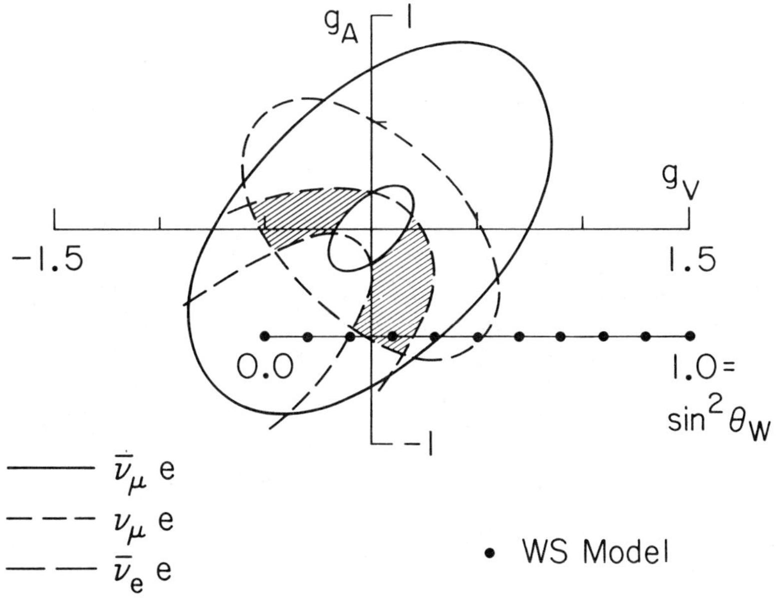

Figure 23 The limits placed on g_A and g_V by data for νe scattering. The outer (inner) lines indicate 90% confidence upper (lower) limits. The shaded regions are the overlap or allowed regions for g_A and g_V. The line with dots for tenth values of $\sin^2 \theta_W$ is the prediction of the WS model. The data are from Reference 178 (the data of Reithler, which were not used, would give somewhat larger values of g_A and g_V).

possible in weak neutral currents in contrast to electromagnetic currents (these are purely vector and conserve parity). Among such experiments are those parity-violating transitions in heavy atoms (bismuth, thalium, cesium), in light atoms (hydrogen and deuterium), and in nuclei. There are also experiments that measure polarization asymmetries in electron-nucleon deep inelastic scattering and in $e^+e^- \to \mu^+\mu^-$.

The experiments involving bismuth are already reporting results that are consistent with zero parity violation. The Oxford group finds the measured optical rotation to be $(2.7 \pm 4.7) \times 10^{-8}$ rad, while the Washington group finds $(-0.7 \pm 3.2) \times 10^{-8}$ rad (180). The optical rotation measured by this type of experiment on heavy nuclei should be dominated by the interference term $A_{\text{electron}} V_{\text{hadron}}$ rather than by $V_{\text{electron}} A_{\text{hadron}}$. There is some controversy concerning the atomic and nuclear theory calculations (181, 182); however, the best estimates are that for the WS model the Oxford experiment should find -15×10^{-8} rad and the Washington experiment -12×10^{-8} rad. Within $SU(2) \times U(1)$ one can obtain a zero result for the bismuth experiment if the electron is given vector couplings ($A_e = 0$); the electron is vector if it has a coupling $(N_e e^-)_R$ in addition to $(\nu_e e^-)_L$ (where N_e is a heavy neutral lepton).

There are some models, notably $SU(2)^L \times SU(2)^R \times U(1)$ models (see Section 5), that expect both $A_e V_h$ and $V_e A_h$ to be zero (to first order). These models have two weak bosons (Z_V^0 and Z_A^0), one with purely axial-vector couplings to all fermions (except neutrinos) and one with purely vector couplings. There are experiments that are sensitive to both of the VA terms and have the added feature that they lack the theoretical difficulties of experiments on heavy nuclei (182). These experiments are performed on hydrogen and deuterium, and involve either atomic transitions or electron-nucleon deep inelastic scattering. While the theory is clear, it will be difficult for these experiments to obtain sufficient sensitivity to distinguish among various models. Were the absence of parity violation to be confirmed by these experiments (which expect results in the next year or two), it would be a serious problem for the WS-GIM model.

The neutral current interactions present a serious challenge to any gauge theory of quarks and leptons. A theory that could account for the wide range of data discussed here would be most impressive.

8 CONSERVATION LAWS

In the construction of gauge theories that incorporate more than four quarks (and leptons), the question of mixing among fermions must be considered again. It was already clear from Cabibbo mixing that the weak interaction eigenstates were not identical to the mass eigenstates. While a deep understanding of the cause of this mixing is still lacking, the phenomenological consequences of it should not be overlooked. One important consequence can be the breakdown of certain conservation laws. Two relevant conservation laws are those for *CP* (the product of charge conjugation and parity) and for muon number. The violation of these quantities is quite small: *CP*-violating decays of K^0 mesons are about 10^{-3} of *CP*-conserving decays ("milliweak"), and muon-number-violating decays of muons have never been observed and are less than 10^{-8} of muon-number-conserving decays. The understanding of such conservation laws and their breakdown is a crucial step in building a theory of quarks and leptons.

8.1 *CP Violation*

The theory of *CP* violation has been studied (183) for many years. A variety of approaches has been considered involving left-handed and sometimes right-handed currents. Here attention will be limited to the case of the WS model, although some results are applicable to other models. Consideration of *CP* violation in weak interactions involves not only the question of how it occurs, but also of why it is milliweak. In the WS model, the possibility of *CP* violation depends in an important way on the number of quarks. If there were only four quarks (u, c, d, s) and only left-handed currents, then *CP* would be completely conserved in the quark sector. Weinberg (184) and Sikivie (185) have proposed that *CP* violation could occur only in Higgs exchange in such models, which can automatically give a milliweak violation.

If there are six (or more) quarks, then one expects to find *CP* violation, which a priori need not be small. In contrast to the four-quark case where the weak coupling matrix has one parameter (the Cabibbo angle θ_C), the WS model with six quarks [discussed first by Kobayashi & Maskawa (137)] has four parameters. They can be taken to be four angles, θ_C, θ_1, θ_2, and δ, in terms of which the weak coupling matrix is:

$$\begin{pmatrix} C_C & -S_C C_2 & -S_C S_2 \\ S_C C_1 & C_C C_1 C_2 - S_1 S_2 e^{i\delta} & C_C C_1 S_2 + S_1 C_2 e^{i\delta} \\ S_C S_1 & C_C S_1 C_2 + C_1 S_2 e^{i\delta} & C_C S_1 S_2 - C_1 C_2 e^{i\delta} \end{pmatrix}, \qquad 8.1$$

where the rows correspond to the quarks u, c, and t, the columns to d, s, and b, and $C_C \equiv \cos\theta_C$, $C_1 \equiv \cos\theta_1$, etc. *CP* violation cannot be calculated since three angles are not known; however, there are experimental results that limit the possible values of θ_1 and θ_2 and allow some comment on the expected magnitude of *CP* violation.

In this generalized case, θ_C must still have the usual value ($\theta_C \approx 13°$). From the universality of quark and lepton couplings, Ellis et al (151) find that the $u\bar{b}$ coupling $\sin^2\theta_C \sin^2\theta_2 < 0.003$, so that $\sin^2\theta_2 < 0.06$. Given this limit, the fact that charmed particles decay dominantly to strange particles leads to no useful limit on θ_1 (only $\sin^2\theta_1 \lesssim 0.8$). Following the method of Gaillard & Lee (13), the $K_L - K_S$ mass difference can set some bounds on $\sin^2\theta_1$ (151), depending on several factors: the c quark mass, the t quark mass, and the quantitative accuracy of the Gaillard-Lee estimate. If $\cos^2\theta_2 \approx 1$, then

$$\sin^2\theta_1 \approx \{a+[a^2+(f-1)\eta b]^{\frac{1}{2}}\}b^{-1}, \tag{8.2}$$

where $a \equiv \eta+\eta \ln \eta$, $b \equiv 1+\eta+2\eta \ln \eta$, $\eta \equiv m_c^2/m_t^2$, and f is a factor measuring the multiplicative deviation from the Gaillard-Lee estimate. Clearly if that estimate were exact ($f=1$), then $\sin^2\theta_1 = 0$ (note that a is negative). If $f=2$, one finds $\sin^2\theta_1 = 0.24$ for $\eta = 0.1$ ($m_t \approx 5$ GeV) and $\sin^2\theta_1 = 0.07$ for $\eta = 0.01$ ($m_t \approx 15$ GeV). If $f=5$, one finds $\sin^2\theta_1 = 0.61$ ($\eta = 0.1$) and 0.25 ($\eta = 0.01$).

With this information plus a guess for δ, one can estimate (see 151, 186) the ratio of the CP-violating to the CP-conserving parts of the K^0 mass matrix:

$$|\varepsilon| \approx 2^{-\frac{1}{2}}\left|\frac{Im\, M_{12}^{K}}{\Delta m^{K}}\right| \approx 2^{\frac{1}{2}} \sin\delta \sin\theta_1 \sin\theta_2 \left(\frac{b\sin^2\theta_1 - a}{b\sin^4\theta_1 - 2a\sin^2\theta_1 + \eta}\right), \tag{8.3}$$

where a, b and η are defined above and δ is the phase in the weak coupling matrix 8.1. If one chooses $\theta_C = \theta_1 = \theta_2 = \delta$ (which puts all angles below the experimental upper limits), then one finds the calculated $|\varepsilon|$ to be 10 times the observed value (187) (which is about 2×10^{-3}) for $\eta = 0.1$ and 40 times it for $\eta = 0.01$. Alternatively, given the observed CP violation and choosing intermediate values for θ_1 and θ_2, one can determine δ. For $\sin^2\theta_2 = 0.03$ and $f = 1.5$ (which gives $\sin^2\theta_1 = 0.15$ for $\eta = 0.1$ and 0.05 for $\eta = 0.01$), the values obtained are $\sin^2\delta = 2 \times 10^{-4}$ for $\eta = 0.1$ and $\sin^2\delta = 5 \times 10^{-5}$ for $\eta = 0.01$.

To summarize, the WS model with six quarks does give CP violation. By choosing the angles in the weak coupling matrix to be sufficiently small, one certainly can obtain the correct magnitude for CP violation. If, however, a random choice of angles is made (within the bounds described above), the predicted CP violation can be one or two orders of magnitude larger than the observed violation. While the magnitude of CP violation cannot be predicted accurately, the CP-violating terms are clearly much smaller than those for nonrare decays, so that the usual K^0 decay phenomenology is obtained qualitatively. It is possible, of course, that there are symmetry arguments or other reasons why θ_1, θ_2, and/or δ must be small.

This analysis also gives information concerning the coupling strengths for various charged-current terms useful in other sections of this review. Using the coupling matrix 8.1, the $u\bar{b}$ coupling squared is proportional to $\sin^2\theta_C \sin^2\theta_2$, which is less than 0.003 compared to the $u\bar{d}$ coupling. The $t\bar{d}$ coupling squared is proportional to $\sin^2\theta_C \sin^2\theta_1 \lesssim 0.03$. Furthermore, the ratios of couplings squared for $t\bar{s}/t\bar{d}$ and $c\bar{b}/u\bar{b}$ are both greater than 10 for most but not all angles θ_1 and θ_2. The small *CP* violation indicates that at least one of the angles θ_1, θ_2, and δ must be even smaller, but it does not indicate which one(s).

8.2 *Muon-Number Nonconservation*

Among the interesting tools for understanding the structure of the weak and electromagnetic interactions are experiments searching for processes such as $\mu \to e\gamma$, $\mu \to eee$, and $\mu^- N \to e^- N$. While the standard theories expect lepton number to be conserved, muon number may be violated in higher order diagrams. It is assumed that μ and ν_μ have muon-number one and all other particles have zero. Here, three means of finding muon-number violation in SU(2) × U(1) models are discussed.

In the context of the WS model (although it is applicable elsewhere) Bjorken & Weinberg (188) consider the interactions of leptons with Higgs scalars:

$$H = -g_1 \overline{\begin{pmatrix} \nu_\mu \\ \mu^- \end{pmatrix}}_L \begin{pmatrix} \phi_1^+ \\ \phi_1^0 \end{pmatrix} \mu_R^- - g_2 \overline{\begin{pmatrix} \nu_e \\ e^- \end{pmatrix}}_L \begin{pmatrix} \phi_2^+ \\ \phi_2^0 \end{pmatrix} \mu_R^-$$
$$- g_3 \overline{\begin{pmatrix} \nu_\mu \\ \mu^- \end{pmatrix}}_L \begin{pmatrix} \phi_3^+ \\ \phi_3^0 \end{pmatrix} e_R^- - g_4 \overline{\begin{pmatrix} \nu_e \\ e^- \end{pmatrix}}_L \begin{pmatrix} \phi_4^+ \\ \phi_4^0 \end{pmatrix} e_R^- + \text{h.c.} \qquad 8.4$$

where the ϕ_i are linear combinations (not necessarily independent) of several scalar fields of definite mass. Since the μ and e are defined as the physical states found in the diagonalization of the mass matrix, if there is only one Higgs doublet (as is sometimes assumed), then g_2 and g_3 must be zero. However, if there is more than one Higgs doublet, then in general it is possible that g_2 and/or g_3 are nonzero, and virtual Higgs scalars will give physical transitions between μ and e such as those shown in Figure 24. Because the Higgs coupling to the light leptons is so weak, the two-loop diagrams (Figure 24*b*), in general, dominate one-loop diagrams (Figure 24*a*):

$$\frac{1\text{ loop}}{2\text{ loops}} \approx \frac{2\pi}{\alpha}\left(\frac{m_\mu}{m_H}\right)^2. \qquad 8.5$$

Bjorken & Weinberg roughly estimate

$$\frac{\mu \to e\gamma}{\mu \to e\nu\bar{\nu}} \lesssim 10^{-8}, \tag{8.6}$$

depending on the amount of mixing among the Higgs scalars.

No muon-number violation has been observed yet. The present experimental limit (90% confidence level) for $B(\mu \to e\gamma)$ is 3.6×10^{-9} (189).

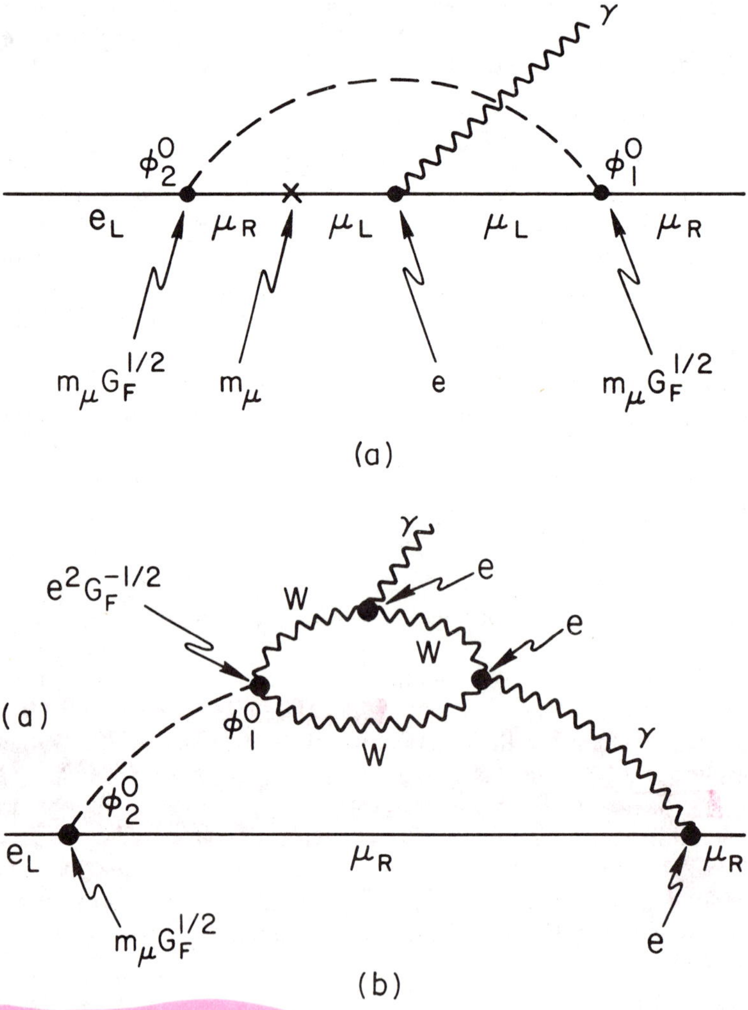

Figure 24 (*a*) One- and (*b*) two-loop diagrams in which virtual Higgs exchange leads to the decay $\mu \to e\gamma$. This figure was taken from Reference 188.

In the model, the decay $\mu \to 3e$ was expected to be very small. The decay $K_L \to \mu e$ is forbidden in lowest order (or one would get strangeness-changing neutral currents). They predict

$$\frac{\sigma(\mu^- N \to e^- N)}{\sigma(\mu^- N \to \nu N')} \sim 4 \times 10^{-9}, \qquad 8.7$$

where N is a nucleus, and the experimental limit is 1.6×10^{-8} (190).

In models that also have right-handed currents there is another source of muon-number violation. This source, discussed first by Cheng & Li and by Bilenkii et al (191), involves the mixing of massive neutral leptons that have right-handed couplings to the electron and muon, $(N'_e e^-)_R$ and $(N'_\mu \mu^-)_R$. In analogy with the Cabibbo mixing of the d and s quarks, they suggest:

$$\begin{aligned} N'_e &= N_e \cos\phi + N_\mu \sin\phi \\ N'_\mu &= -N_e \sin\phi + N_\mu \cos\phi. \end{aligned} \qquad 8.8$$

Then clearly if one considers the simple one-loop diagram of Figure 25, there will be a GIM-like cancellation. The cancellation is not complete, to the extent that N_e and N_μ have unequal masses; the amplitude for this $\mu \to e\gamma$ process is proportional to

$$\cos\phi \sin\phi \,(m^2_{N_\mu} - m^2_{N_e}). \qquad 8.9$$

Bjorken, Lane & Weinberg (192; this paper refers to much of the literature on muon-number violation) argue that the Higgs couplings that give masses and lead to the above mixing also cause small but finite mixing of the left-handed parts of N_e and N_μ with ν_e and ν_μ. This mixing is of order m_μ/m_N. There are, as a result, left-right diagrams in addition to the right-right diagram, Figure 25. These left-right terms have the same form as the right-right terms, but their amplitude is multiplied by -6. If the value of Expression 8.9 is 1 GeV^2, then (incorporating the Bjorken-Lane-Weinberg modification)

$$\frac{\mu \to e\gamma}{\mu \to e\nu\bar{\nu}} \approx 4 \times 10^{-10}. \qquad 8.10$$

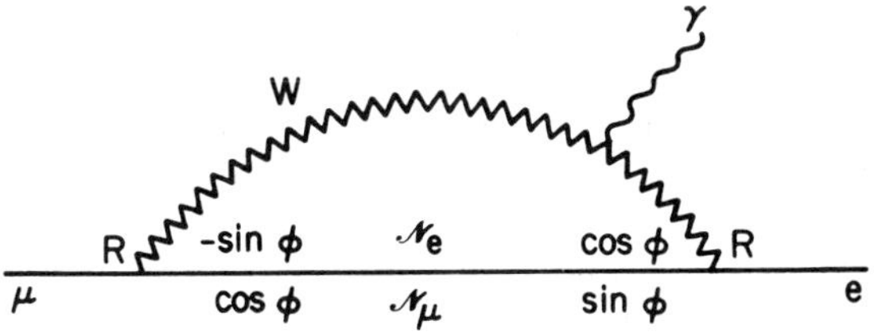

Figure 25 One of the diagrams in which N_e and N_μ exchange leads to the decay $\mu \to e\gamma$.

Cheng & Li estimate the branching ratio for $\mu \to \mathrm{ee\bar{e}}$ to be about 10^{-11} [where the experimental limit (193) is 6×10^{-9}], and for $\mu^- \mathrm{N} \to \mathrm{e}^- \mathrm{N}$ to be as large as 10^{-9}. For $m_{\mathrm{N}_\mu}/m_{\mathrm{N}_\mathrm{e}} \approx 4$, they find the branching ratio for $\mathrm{K_L} \to \mathrm{e}\bar{\mu}$ to be about 10^{-10} [the experimental limit is 2×10^{-9} (194)].

Glashow (195) and Fritzsch (196) have shown that muon number can be violated in models without right-handed currents and with only one Higgs doublet. If the charged heavy lepton τ has a left-handed coupling to a massive neutral lepton N_τ, then N_τ can mix with ν_e and ν_μ. Decays such as $\mu \to \mathrm{e}\gamma$ could occur in the same fashion as proposed by Cheng & Li and by Bilenkii et al, where Δm^2 is replaced with $m_{\mathrm{N}_\tau}^2$.

The mixed states can be written as:

$$\nu_\mathrm{e}' = \nu_\mathrm{e} \cos\theta + \mathrm{N}_\tau \sin\theta$$

$$\nu_\mu' = \nu_\mu \cos\phi + (-\nu_\mathrm{e} \sin\theta + \mathrm{N}_\tau \cos\theta)\sin\phi$$

$$\mathrm{N}_\tau' = (\mathrm{N}_\tau \cos\theta - \nu_\mathrm{e}\sin\theta)\cos\phi - \nu_\mu \sin\phi. \qquad 8.11$$

Both angles can be shown to be small by the requirement of universality (seen through μ and β decays) and by the lack of ν_e in ν_μ' (ν_μ do not produce electrons in scattering). If $\mu \to \mathrm{e}\gamma$ were observed at the 10^{-10} to 10^{-9} level, then the smallness of the angles θ and ϕ requires that m_{N_τ} be quite large (much larger than m_τ in fact). Since $m_\tau < m_{\mathrm{N}_\tau}$, the heavy lepton τ can only decay through the mixing of N_τ with ν_e and ν_μ.

It would be possible to rule out this mode of muon-number violation by measuring the lifetime of τ carefully, but it probably will be difficult to obtain a better experimental limit than the present value of about 10^{-11} sec. Since the angle ϕ is so small, it would be very rare for ν_μ scattering to produce τ leptons.

A precise measurement of the conservation or nonconservation of muon number would be a valuable tool for studying the existence, mixing, and currents of new leptons.

9 ATTEMPTS AT A GRAND SYNTHESIS

Although charm and many predictions of gauge theories for neutral and charged currents seem well verified, the experimental evidence in favor of strong and weak gauge theories remains somewhat indirect. From a theoretical point of view, however, the importance of gauge theories cannot be overemphasized. They present the first possibility for a theoretically consistent description of both the weak and electromagnetic interactions, and the strong interactions.

In fact, if the gauge theory framework applies generally, then there is every reason to believe that within it, a grand unification of all three interactions can be attained. It is an important enterprise to begin working on this, even though it is some ways premature. New questions can be raised and a framework provided for studying unsolved problems such as mass generation and the existence of mixing angles like θ_c.

We do not enter into a detailed discussion of the many models proposed for grand unification. However, there are several features and problems necessarily common to any specific model and it is possible to discuss the subject in general terms. This will give the reader some appreciation of these general features and will also serve as a conclusion for the entire review.

First of all, it is worth repeating and underscoring the two assumptions that form the foundation for the approach to grand unification to be discussed here.

1. The weak and electromagnetic interactions are described by a spontaneously broken gauge theory based on some Lie group G_w, perhaps one of the several models we have discussed. The success of the WS model in dealing with neutrino neutral current interactions indicates that it will contain $SU(2) \times U(1)$ as a subgroup. If G_w is larger than $SU(2) \times U(1)$, then presumably some of the gauge bosons of G_w will be considerably heavier than M_W and M_Z. The quarks and leptons are assumed to fill out low dimensional representations of G_w, with the total number "reasonably" small. If G_w is simple (or semisimple in the form $G'_w \times G'_w$, with a reflection symmetry relating the two factors), then a single coupling constant is involved. If $SU(2) \times U(1)$ is a subgroup, θ_W will then be determined before proceeding on to grand unification.

With the fermion content arranged to eliminate triangle anomalies, this theory will be renormalizable, and therefore exhibit only logarithmic growth with energy. The coupling strength at laboratory energies will be of order $\alpha = \frac{1}{137}$. At higher energies E, the effective coupling constant $\alpha(E)$ (see Section 2) can change logarithmically but the effect is not significant until $\alpha \log (E/E_{\text{lab}})$ becomes of order unity.

2. The underlying theory of strong interactions is QCD (Equation 2.2). QCD is renormalizable, and providing that the total number of quark flavors f is less than 16.5 (Equation 2.15), it is asymptotically free. The running coupling constant $\alpha_s(-q^2)$ for $q^2 > 1$–$2\ \text{GeV}^2$ takes the form

$$\alpha_s(-q^2) = \frac{\alpha_\mu}{1+\dfrac{11\alpha_\mu}{4\pi}\ln\left(\dfrac{-q^2}{\mu^2}\right)-\dfrac{\alpha_\mu}{\pi}\displaystyle\sum_{i=1}^{f}\int_0^1 dz\, z(1-z)\ln\left(\dfrac{m_i^2-q^2z(1-z)}{m_i^2+\mu^2z(1-z)}\right)}, \quad 9.1$$

which agrees with Equations 2.15 and 2.16 if $-q^2, \mu^2 \gg m_i^2$. If $-q^2, \mu^2 \ll m_i^2$ for some flavors, then those quarks can be seen to "decouple" (197) in $\alpha_s(-q^2)$. Even if $f > 16.5$, a temporary asymptotic freedom could sustain itself until a q large enough to vacuum polarize the seventeenth flavor is attained. Nevertheless, in the spirit of assumption 1, we shall take $f < 16.5$ so that the asymptotic freedom is truly asymptotic.

The essential notion in superunification is that both QCD and the weak and electromagnetic theory must be viewed as low energy theories. Since they are both renormalizable theories with only logarithmic variation in energy, the range of energies over which they can be viewed as independent theories is necessarily very large. Nevertheless, it is possible to imagine that at some extremely large energy, the strong, weak and electromagnetic interactions will be described by a single theory—a gauge theory, of course—based on some Lie group G, which contains $G_w \times SU(3)_c$.

If a spontaneous symmetry breakdown takes place at some extremely large mass scale M, then it is possible that some subset of the gauge bosons acquires a mass of order M leaving the subgroup $G_w \times SU(3)_c$ unbroken. The G_w and $SU(3)_c$ gauge bosons will remain massless at this level. At energy scales somewhat below M, the exchange of the superheavy bosons of mass M will be suppressed, and it will appear as if QCD and the weak and electromagnetic theory are two separate field theories. It is possible that the spontaneous symmetry breakdown is a multistep process. If G_w is larger than $SU(2) \times U(1)$, it could break down at some scale $M' \ll M$ (but $M' \gg M_W, M_Z$). The final step at mass scale M_W, M_Z would then leave only the U(1) subgroup intact, with a massless photon.

That is the scenario in rough outline. Its actual implementation depends on finding the right group G and probably on understanding spontaneous symmetry breaking and the Higgs mechanism much more deeply than we do now. Nevertheless, several features and consequences of the program seem to be understood. We offer a list of those we consider to be most important and then conclude with a partial and subjective list of the many unsolved problems.

1. If the grand unification group G is simple (or semisimple in the form $G' \times G'$, with a reflection symmetry relating the two factors), the unified theory will involve a single coupling constant. It is then possible to estimate the order of magnitude of M, the unification mass scale (198, 199). The QCD effective coupling constant $\alpha_s(-q^2)$ is already small at $-q^2 = 10\ \text{GeV}^2$. Estimates range from about 0.2 based on charmonium decay to about 0.5 based on analyses of electroproduction. This, however, is still much larger than the weak coupling constant (of order $\alpha = \frac{1}{137}$), and if these theories are to coalesce into a single-coupling-constant

theory, then something must bring the coupling strengths together. It is primarily the logarithmic decrease of $\alpha_s(-q^2)$ with $-q^2$ that does this. This decrease can bring $\alpha_s(-q^2)$ down to weak and electromagnetic strength at very high momentum scales, but where the effective weak and electromagnetic coupling strength still has not changed much from laboratory energies. The grand unification mass M can be estimated roughly by equating the effective coupling strengths. The result depends on several unknown factors, such as the starting value $\alpha_s(10\ \mathrm{GeV}^2)$, the number of quark flavors, the contribution of Higgs bosons, and the mass scales characterizing the various possible levels in the symmetry-breaking chain. Most estimates (198, 199), however, have ranged between $M \approx 10^{16}$ GeV and $M \approx 10^{19}$ GeV, far beyond laboratory energies. One intriguing feature of these estimates is that they imply the existence of elementary particles with masses on the order of the Planck mass $G^{-\frac{1}{2}} = 1.22 \times 10^{19}$ GeV. This suggests that grand unification may necessarily involve the gravitational interaction, a possibility to which we return shortly.

2. The prototype for grand unification into a simple group is the G = SU(5) model of Georgi & Glashow (200). It has a maximal subgroup structure of SU(2) × U(1) × $\mathrm{SU(3)_c}$, so that the usual WS model and QCD are naturally incorporated. If experiments force us to go beyond SU(2) × U(1) for $\mathrm{G_w}$, then some larger group will have to be used for G. Some possibilities are SO(10) (201) and the exceptional groups $\mathrm{E_6}$ and $\mathrm{E_7}$ proposed by Gürsey and collaborators (125). The group, $\mathrm{E_6}$ for example, contains SU(3) × SU(3) × $\mathrm{SU(3)}_c$ and can accommodate six quarks and four charged leptons in two 27-plets. The possibility of $\mathrm{G_w}$ = SU(3) × SU(3) was already discussed in Section 5 and it might be an attractive possibility. However, the study of symmetry breakdown in the $\mathrm{E_6}$ model indicates that some of the SU(3) × SU(3) bosons will become superheavy so that $\mathrm{G_w}$ is some proper subset of SU(3) × SU(3) (202).

3. Once SU(2) × U(1) is embedded in a larger simple group, the value of the Weinberg angle is determined. The value of $\sin\theta_W$ depends on the structure of G and on the energy scales characterizing the stages of symmetry breakdown to SU(2) × U(1). In the SU(5) model, for example, the value of $\sin^2\theta_W$ at the unification mass scale M is $\frac{3}{8}$. This will be reduced by renormalization effects at laboratory energies since the effective coupling constants of the SU(2) and U(1) subgroups scale differently with q. Numerical estimates lead to $\sin^2\theta_W \approx 0.2$ (198, 199) and the range of values allowed by experiment is now $0.2 \leqq \sin^2\theta_W \leqq 0.3$ (see Section 7). A similar prediction is obtained in the $\mathrm{E_6}$ model (202). It is premature to take these estimates of $\sin^2\theta_W$ too seriously, but they

might be useful in at least excluding some groups G. For example, the choice $G = E_7$ leads to $\sin^2 \theta_W = \frac{3}{4}$ at the grand unification mass M, and it would appear that this is too large to be brought into agreement with experiment by renormalization effects (203).

4. An important feature of grand unification is that quarks and leptons are placed together in single representations of G. Thus, there will be gauge bosons (sometimes called leptoquarks) that connect leptons to quarks and lead to the breakdown of separate lepton and quark conservation. This is a potential disaster since the experimental lower bound on the lifetime for such transitions is incredibly large. Quark nonconservation can lead to baryon nonconservation, and the lower limit on the proton lifetime is 2×10^{30} years (204)! However, if the leptoquark mass is on the order of the grand unification mass M predicted from renormalization group considerations, then lifetimes at least this long can be obtained. The decay of a proton into a lepton plus pions can proceed by single leptoquark exchange (205), so that the lifetime will be proportional to M^4/M_H^5 where M_H is some typical hadronic mass scale. Furthermore, the constant of proportionality might be expected to be of order $\alpha^{-2}(M^2) \geqq 10^3$ (199). With $M_H \leqq 1$ GeV and $M \geqq 10^{16}$ GeV, one obtains proton lifetimes well in excess of 10^{30} years. Although it appears to be sufficiently suppressed, baryon nonconservation is a natural feature of many grand unification models[21] (205).

The grand unification scenario is attractive, but there are too many unanswered questions and loose ends to be sure that it is the wave of the future. A short list of some of these problems should point up the limitations of the present theoretical framework, both for dealing with grand unification and perhaps even for understanding the weak and electromagnetic interactions alone.

1. The nature of spontaneous symmetry breakdown is very poorly understood. The only known way of implementing the breakdown and the Higgs mechanism is by explicitly introducing multiplets of elementary Higgs fields into the Lagrangian. For the SU(2) × U(1) model, this required only one complex doublet and led to only one physical Higgs boson. However, for the large groups G required for grand unification, or even for larger weak and electromagnetic groups G_w, several very big multiplets of Higgs fields are needed to get a reasonable pattern of symmetry breaking (125, 201). The notion of fundamental Higgs fields becomes uneconomical, if not downright unpalatable.

A possibility with some attraction is that the spontaneous symmetry

[21] It is possible to push the decay to higher orders in the coupling constant. For an extensive discussion of grand unification and proton stability, see 205.

breakdown is "dynamical," that is, not induced by fundamental Higgs fields in the Lagrangian. The Goldstone bosons incorporated into the gauge fields would presumably be bound states of quarks. Whether and how this kind of dynamical symmetry breakdown takes place is unknown, and, in fact, it is hard to see what kind of force could produce the necessary binding. Even if dynamical breakdown is a possibility, in moving up from SU(2) × U(1) to G, it is perhaps reasonable to introduce Higgs fields at each stage. They may appear fundamental at one level if not at all energy scales.

2. There seems to be almost no understanding of quark and lepton masses and weak mixing angles. Hopefully, these fundamental questions will find their solution within the framework of spontaneous symmetry breakdown, but just how is far from clear. (For a recent discussion along with some speculation and references to other work, see 206.) Many ingredients will presumably enter the solution: the unifying groups G_w and G; spontaneous symmetry breakdown; and the effects of strong, weak, and electromagnetic renormalization. Empirical relations among these parameters, such as $\theta_c^2 \approx m_d/m_s$ and $m_e/m_\mu \approx m_u/m_c$, are tantalizing but they can be insidiously misleading.

3. Perhaps grand unification into a Lie group G, without also an incorporation of gravity, is impossible. The natural appearance of mass scales on the order of the Planck mass $G^{-\frac{1}{2}}$ suggests this possibility. At such extremely-small-distance scales, the gravitational interaction is probably comparable to the other fundamental forces, and it could play an important role, for example, in driving spontaneous symmetry breakdown. Just how this might work is very poorly understood, but at least one development offers promise. The concept of supersymmetry (207), which relates fermions and bosons, has recently been combined with the notion of local gauge invariance to produce a class of theories known as supergravity (208). These theories contain a graviton, one or more spin-$\frac{3}{2}$ particles known as gravitinos, and a host of lower spin particles. They have the possibility of being renormalizable (209), but whether or not realistic theories of grand (supergrand?) unification can be constructed along these lines is not yet clear. If supergravity is the road to grand unification, then the group G will not be a simple Lie group. Instead, it will be a graded (or super) Lie group, with an algebra containing both commutation and anticommutation relations.

It is important to think about these deep theoretical questions, but if the recent past is something of a guide, progress will come in more modest steps with experiment playing an important, and possibly leading, role. It is hard to think of another period in particle physics when so many important experiments were either under way or in the planning stages. Within the next few years we may well be able to offer plausible, if not completely accepted, answers to some questions of limited scope but of great import:

1. Is everything "in order" with charmonium and the charmed particles? The pseudoscalar $c\bar{c}$ states are especially puzzling.
2. To what extent is the upsilon a new, improved charmonium system? Will it give us important information about the quark-antiquark potential in the transition region from linearity to short distances ($\lesssim 1\ \text{GeV}^{-1}$)?
3. Will QCD remain a viable theory of strong interactions?
4. Will spontaneously broken gauge theories remain a viable theoretical framework for weak and electromagnetic interactions? The most urgent need here is for some evidence in favor of the existence of intermediate vector bosons—perhaps in the mass range 60–100 GeV. Do Higgs bosons, either elementary or composite, exist in an accessible mass range?
5. What is the correct gauge group of weak and electromagnetic interactions G_w? How many quarks and leptons are there and how do they fill out representations of G_w? The weak interaction properties of the τ lepton and the b quark and the question of parity violation in atomic physics are central here.
6. What about masses and mixing angles? Are the neutrinos ν_e, ν_μ, and ν_τ massless? Are muon and tau lepton number separately conserved at more stringent levels? Will the pattern of quark and lepton masses offer some clue to the understanding of mass generation?

If the near future provides us with answers to some of these questions, it may be an even more exciting era in particle physics than the recent past.

ACKNOWLEDGMENTS

We wish to thank many of our colleagues for conversations that were useful in preparation of this review, especially J. Bjorken, C. Carlson, E. Eichten, J. Ellis, G. Feldman, F. Gilman, S. Glashow, K. Gottfried, F. Gürsey, F. Martin, C. Quigg, J. Rosner, R. Suaya, and S. Weinberg. Two of us (T.A. and K.L.) wish to acknowledge the hospitality of the SLAC theory group and two of us (M.B. and K.L.) wish to acknowledge the hospitality of the Yale theory group. It has been a privilege for all three of us to have known and to have learned from Ben Lee.

Literature Cited

1. Gell-Mann, M. 1964. *Phys. Lett.* 8:214; Zweig, G. 1964. *CERN Rep. TH.401, TH.412*
2. Greenberg, O. W. 1978. *Ann. Rev. Nucl. Part. Sci.*: 28:327–86
3. Bjorken, J. D., Glashow, S. L. 1964. *Phys. Lett.* 11:255; Amati, D., et al. 1964. *Phys. Lett.* 11:190; Tarjanne, P., Teplitz, V. L. 1963. *Phys. Rev. Lett.* 11:447; Hara, Y. 1964. *Phys. Rev. B* 134:701; Maki, Z., Ohnuki, Y. 1964. *Prog. Theor. Phys.* 32:144
4. Glashow, S. L., Iliopoulos, J., Maiani, L. 1970. *Phys. Rev. D* 2:1285
5. Carithers, W. C. 1973. *Phys. Rev. Lett.* 30:1336, 31:1025; Fukushima, Y. 1976. *Phys. Rev. Lett.* 36:348
6. Cabibbo, N. 1963. *Phys. Rev. Lett.* 10:531
7. Weinberg, S. 1967. *Phys. Rev. Lett.* 19:1264; Salam, A. 1968. *Elementary Particle Physics: Relativistic Groups and Analyticity* (Nobel Symp. 8), ed. N. Svartholm, p. 367. Stockholm: Almquist & Wiksell.
8. Yang, C. N., Mills, R. 1954. *Phys. Rev.* 96:191
9. 't Hooft, G. 1971. *Nucl. Phys. B* 33:173; Lee, B. W. 1972. *Phys. Rev. D* 6:1188; Lee, B. W., Zinn-Justin, J. 1972. *Phys. Rev. D* 5:3132, 3137, 3155
10. Barish, B. C. et al. 1977. *Phys. Rev. Lett* 39:1595
11. Goldstone, J. 1961. *Nuovo Cimento* 19: 154; Goldstone, J., Salam, A., Weinberg, S. 1962. *Phys. Rev.* 127:965
12. Higgs, P. W. 1964. *Phys. Rev. Lett.* 12:132, 13:508; 1966. *Phys. Rev.* 145: 1156; Guralnik, G. S., Hagen, C. R., Kibble, T. W. B. 1964. *Phys. Rev. Lett.* 13:585; Englert, F., Brout, R. 1964. *Phys. Rev. Lett.* 13:321
13. Gaillard, M. K., Lee, B. W. 1974. *Phys. Rev. D* 10:897
14. Litke, A. et al. 1973. *Phys. Rev. Lett.* 30:1189; Tarnopolsky, G. et al. 1974. *Phys. Rev. Lett.* 32:432
15. Richter, B. 1974. *Proc. Int. Conf. High Energy Phys., 17th, London, 1–10 July 1974*, p. IV-37. Rutherford Laboratory: Sci. Res. Counc.
16. Burmester, J. et al. 1977. *Phys. Lett. B* 66:395
17. Benvenuti, A. et al. 1975. *Phys. Rev. Lett.* 35:1199, 1203, 1249; 1977. *Phys. Rev. Lett.* 38:1183; Barish, B. C. et al. 1976. *Phys. Rev. Lett.* 36:939; 1977. *Phys. Rev. Lett.* 39:981; von Krogh, J. et al. 1976. *Phys. Rev. Lett.* 36:710; Holder, M. et al. 1977. *Phys. Lett. B* 69:377, 70:396; Deden, H. et al. 1977. *Phys. Lett. B* 67:474; Baltay, C. et al. 1977. *Phys. Rev. Lett.* 39:62; Bosetti, P. et al. 1977. *Phys. Rev. Lett.* 38:1248; Ballagh, H. C. et al. 1977. *Phys. Rev. Lett.* 39:1650
18. Augustin, J.-E. et al. 1974. *Phys. Rev. Lett.* 33:1406
19. Aubert, J. J. et al. 1974. *Phys. Rev. Lett.* 33:1404
20. Abrams, G. S. et al. 1974. *Phys. Rev. Lett.* 33:1453
21. Braunschweig, W. et al. 1975. *Phys. Lett. B* 57:407; Feldman, G. J. et al. 1975. *Phys. Rev. Lett.* 35:821; Tanenbaum, W. M. et al. 1975. *Phys. Rev. Lett.* 35:1323
22. Goldhaber, G. et al. 1976. *Phys. Rev. Lett.* 37:255; Peruzzi, I. et al. 1976. *Phys. Rev. Lett.* 37:569
23. Herb, S. W. et al. *Phys. Rev. Lett.* 39: 252; Innes, W. R. et al. 1977. *Phys. Rev. Lett.* 39:1240; Kephart, R. D. et al. 1977. *Phys. Rev. Lett.* 39:1440
24. Perl, M. L. et al. 1975. *Phys. Rev. Lett.* 35:1489
25. Perl, M. L. et al. 1977. *Phys. Lett. B* 70:487
26. Politzer, H. D. 1973. *Phys. Rev. Lett.* 26:1346; Gross, D., Wilczek, F. 1973. *Phys. Rev. Lett.* 26:1343
27. Greenberg, O. W. 1964. *Phys. Rev. Lett.* 13:598
28. Fritzsch, H., Gell-Mann, M. 1973. *Proc. Int. Conf. High Energy Phys., 16th, Chicago-Batavia, Ill., 1972*, ed. J. D. Jackson, A. Roberts, Vol. 2, p. 135. Natl. Accel. Lab.: Batavia, Ill.
29. Bellettini, G. et al. 1970. *Nuovo Cimento A* 66:243; Kryshkin, V. I. et al. 1970. *J. Exp. Theor. Phys.* 30:1037; Browman, A. et al. 1974. *Phys. Rev. Lett.* 33:1400
30. Cabibbo, N., Parisi, G., Testa, M. 1970. *Nuovo Cimento Lett.* 4:35
31. Rapidis, P. et al. 1977. *Phys. Rev. Lett.* 39:526
32. Kirkby, J. 1977. *1977 Proc. Int. Symp. Lepton Photon Interactions High Energies*, ed. F. Gutbrod, p. 3. Hamburg, Germany: DESY
33. Yamada, S. 1977. See Ref. 32, p. 69
34. Burmester, J. et al. 1977. *Phys. Lett. B* 66:395; Knies, G. 1977. See Ref. 32, p. 93
35. Abers, E. S., Lee, B. W. 1973. *Phys. Rep.* 9C:1; For a review of many features of QCD see Marciano, W. J., Pagels, H. R. 1978. *Phys. Rep. C* 36:137
36. Fadde'ev, L. D., Popov, V. N. 1967. *Phys. Lett. B* 25:29

37. Appelquist, T., Dine, M., Muzinich, I. J. 1977. *Phys. Lett. B* 69:231; 1978. *Phys. Rev. D* 17:2074
38. Feinberg, F. 1977. *Phys. Rev. Lett.* 39: 316; *MIT Rep. CTP-687*; Fischler, W. 1977. *Nucl. Phys. B* 129:157
39. Gell-Mann, M., Low, F. 1954. *Phys. Rev.* 95:1300; Callan, C. 1970. *Phys. Rev. D* 2:1541; Symanzik, K. 1970. *Commun. Math. Phys.* 18:227
40a. Bjorken, J. D. 1969. *Phys. Rev.* 179: 1547
40b. Nachtmann, O. 1977. See Ref. 32, p. 811
41. Wilson, K. 1964. *Cornell Rep. LNS-64-15*. Unpublished; 1969. *Phys. Rev.* 179: 1499
42. Adler, S. L. 1974. *Phys. Rev. D* 10:3714
43. Appelquist, T., Georgi, H. 1973. *Phys. Rev. D* 8:4000; Zee, A. 1973. *Phys. Rev. D* 8:4038
44. Poggio, E., Quinn, H., Weinberg, S. 1976. *Phys. Rev. D* 13:1958; Shankar, R. 1977. *Phys. Rev. D* 15:755
45. Field, R. D. 1978. *Phys. Rev. Lett.* 40: 997
46. Gaillard, M. K., Lee, B. W. 1974. *Phys. Rev. Lett.* 33:108; Altarelli, G., Maiani, L. 1974. *Phys. Lett. B* 52:351; Shifman, M. A., Vainshtein, A. I., Zakharov, V. I. 1977. *Nucl. Phys. B* 120:316
47. Appelquist, T., Politzer, H. D. 1975. *Phys. Rev. Lett.* 34:43
48. Yao, Y.-P. 1976. *Phys. Rev. Lett.* 36:653; Appelquist, T. et al. 1976. *Phys. Rev. Lett.* 36:768; Kinoshita, T., Ukawa, A. 1976. *Phys. Rev. D* 13:1573; 1977. *Phys. Rev. D* 15:1596; Poggio, E., Quinn, H. 1976. *Phys. Rev. D* 14:578
49. 't Hooft, G. 1977. *Deeper Pathways in High Energy Physics—Orbis Scientiae 1977*, ed. A. Perlmutter, L. F. Scott. New York: Plenum. 699 pp.
50. Polyakov, A. 1975. *Phys. Lett. B* 59:82; Belavin, A. et al. 1975. *Phys. Lett. B* 59:85; 't Hooft, G. 1976. *Phys. Rev. Lett.* 37:8
51. Wilson, K. 1975. *Phys. Rev. D* 10:2445
52. Kogut, J., Susskind, L. 1975. *Phys. Rev. D* 11:395
53. De Rújula, A., Glashow, S. L. 1975. *Phys. Rev. Lett.* 34:46
54. Chang, N.-P., Nelson, C. A. 1975. *Phys. Rev. Lett.* 35:1492
55. Chew, G. F., Rosenzweig, C. 1976. *Nucl. Phys. B* 104:290
56. Novikov, V. A. et al. 1977. *Phys. Rev. Lett.* 38:626; 1977. *Phys. Lett. B* 67:409
57. Gottfried, K. 1977. See Ref. 32, p. 667
58. Feldman, G. J., Perl, M. L. 1975. *Phys. Rep. C* 19:234; Schwitters, R. F., Strauch, K. 1976. *Ann. Rev. Nucl. Sci.* 26:89–149; Wiik, B. H., Wolf, G. 1977. *DESY Rep. 77/01*; Chinowsky, W. 1977. *Ann. Rev. Nucl. Sci.* 27:393–464
59. Feldman, G. J., Perl, M. L. 1977. *Phys. Rep. C* 33:285
60. Okubo, S. 1963. *Phys. Lett.* 5:165; Zweig, G. 1964. See Ref. 1, Zweig; Iizuka, J. 1966. *Prog. Theor. Phys.* 37–38: Suppl. 21
61. Braunschweig, W. et al. 1977. *Phys. Lett. B* 67:243
62. Chanowitz, M. S., Gilman, F. J. 1976. *Phys. Lett. B* 63:178; Lane, K. 1975. Presented at Meet. Div. Part. Fields, Am. Phys. Soc., Seattle, WA, Aug., 1975; Eichten, E. et al. 1976. *Phys. Rev. Lett.* 36:500
63. Eichten, E. et al. 1975. *Phys. Rev. Lett.* 34:369
64. Kang, J. S., Schnitzer, H. J. 1975. *Phys. Rev. D* 12:841, 2791; Harrington, B. J., Park, S. Y., Yildiz, A. 1975. *Phys. Rev. Lett.* 34:168, 706
65. Appelquist, T., Politzer, H. D. 1975. *Phys. Rev. D* 12:1404
66. Tryon, E. P. 1972. *Phys. Rev. Lett.* 28:1605; Kogut, J., Susskind, L. 1974. *Phys. Rev. D* 9:3501; Wilson, K. G. 1974. *Phys. Rev. D* 10:2445
67. Willemsen, J. F. 1974. *Proc. Summer Inst. Part. Phys., July 29–Aug. 10, 1974*, ed. M. C. Zipf, Vol. 1, p. 445. SLAC, Stanford Univ., CA
68. Eichten, E. et al. 1978. *Phys. Rev. D* 17:3090
69. Appelquist, T. et al. 1975. *Phys. Rev. Lett.* 34:365; Callan, C. G. et al. 1975. *Phys. Rev. Lett.* 34:52
70. Martin, A. 1977. *Phys. Lett. B* 67:330; Grosse, H. 1977. *Phys. Lett. B* 68:343
71. Gottfried, K. 1978. *Phys. Rev. Lett.* 40:598
72. Van Royen, R. P., Weisskopf, V. F. 1967. *Nuovo Cimento A* 50:617
73. Barbieri, R., Gatto, R., Kögerler, R. 1976. *Phys. Lett. B* 60:183; Barbieri, R., Gatto, R., Remiddi, E. 1976. *Phys. Lett. B* 61:465
74. Chanowitz, M. 1975. *Phys. Rev. D* 12: 918; Okun, L., Voloshin, M. 1976. *Moscow Rep. ITEP-95-1976*; Brodsky, S. J. et al. 1978. *Phys. Lett. B* 73:203
75. Eichten, E. et al. 1976. *Phys. Rev. Lett.* 36:500
76. Isgur, N. 1976. *Phys. Rev. Lett.* 36:1262
77. De Rújula, A., Jaffe, R. 1977. *MIT Rep. CTP-658*; Lane, K. 1976. Presented at Meet. Am. Phys. Soc., Stanford, CA, Dec. 22–24, 1976
78. Giles, R. C., Tye, S.-H. H. 1977. *Phys. Rev. D* 16:1079; Horn, D., Mandula, J. 1978. *Phys. Rev. D* 17:298

79. Wilczek, F. 1977. *Phys. Rev. Lett.* 39:1304
80. Ellis, J., Gaillard, M. K., Nanopoulos, D. V. 1976. *Nucl. Phys. B* 106:292
81. Harari, H. 1976. *Phys. Lett. B* 64:469
82. Jackson, J. D. 1977. *Rep. TH 2305-CERN*; 1978. *Proc. Rencontre Moriond, 12th*, Mar. 1977. In press
83. Duncan, A. 1976. *Phys. Rev. D* 13:2866; Appelquist, T. 1975. *Caltech Rep. CALT-68-499*
84. De Rújula, A., Georgi, H., Glashow, S. L. 1976. *Phys. Rev. D* 12:147
85. Johnson, K. 1978. Private communication. *MIT Rep.* In press
86. Celmaster, W., Georgi, H., Machacek, M. 1978. *Phys. Rev. D* 17:879, 886
87. Schnitzer, H. J. 1975. *Phys. Rev. Lett.* 35:1540; 1975. *Phys. Rev. D* 13:74; 1976. *Phys. Lett. B* 65:239
88. Pumplin, J., Repko, W., Sato, A. 1975. *Phys. Rev. Lett.* 35:1538
89. Henriques, A. B., Kellet, B. H., Moorhouse, R. G. 1976. *Phys. Lett. B* 64:85; Chan, L.-H. 1977. *Phys. Lett. B* 71:422; Carlson, C. E., Gross, F. 1978. *Phys. Lett. B* 74:404
90. Rapidis, P. A. et al. 1977. *Phys. Rev. Lett.* 39:526
91. Bacino, W. et al. 1977. *Phys. Rev. Lett.* 40:671
92. Chodos, A. et al. 1974. *Phys. Rev. D* 9:3471
93. Bars, I. 1976. *Phys. Rev. Lett.* 36:1521; Tye, S.-H. H. 1976. *Phys. Rev. D* 13:3416; Giles, R. C., Tye, S.-H. H. 1977. *Phys. Rev. D* 16:1079; Ng, T. J., Tye, S.-H. H. 1977. *Phys. Rev. D* 16:2468
94. Lane, K., Eichten, E. 1976. *Phys. Rev. Lett.* 37:477
95. Kogut, J., Susskind, L. 1975. *Phys. Rev. Lett.* 34:767; 1975. *Phys. Rev. D* 12:2821
96a. Weisskopf, V. F., Wigner, E. P. 1930. *Z. Phys.* 63:54, 65:18
96b. Gottfried, K. 1970. *Brandeis Univ. Summer Inst. Theor. Phys. 1967*, ed. M. Chrétien, S. Schweber, Vol. 2. New York: Gordon & Breach; Dashen, R. F., Healy, J. B., Muzinich, I. J. 1976. *Phys. Rev. D* 14:2773
97. Sakurai, J. J. 1969. *Phys. Rev. Lett.* 22:981
98. Le Yaouanc, A. et al. 1977. *Phys. Lett. B* 71:397
99. Goldhaber, G. et al. 1977. *Phys. Lett. B* 69:503
100. Okun, L. B., Voloshin, M. B. 1976. *Zh. Eksp. Theor. Fiz.* 23:369; 1976. *JETP Lett.* 23:333; De Rújula, A., Georgi, H., Glashow, S. L. 1977. *Phys. Rev. Lett.* 38:317; Bander, M. et al. 1976. *Phys. Rev. Lett.* 36:695; Rosenzweig, C. 1976. *Phys. Rev. Lett.* 36:697
101. Machacek, M., Tomozawa, Y. 1978. *Ann. Phys.* 110:407
102. Quigg, C., Rosner, J. 1977. *Phys. Lett. B* 71:153
103. Cahn, R., Ellis, S. D. 1978. *Phys. Rev. D* 17:2338; Barnett, M. 1977. *Proc. Eur. Conf. Part. Phys., Budapest, Hungary, 4–9 July 1977*, ed. L. Jenik, I. Montvay. Budapest: CRIP. 995 pp. Carlson, C. E., Suaya, R. 1977. *Phys. Rev. Lett.* 39:908; Ellis, J. et al. 1977. *Nucl. Phys. B* 131: 285; Hagiwara, T., Kazama, Y., Takasugi, E. 1978. *Phys. Rev. Lett.* 40:76
104. Eichten, E., Gottfried, K. 1977. *Phys. Lett. B* 66:286
105. Quigg, C., Rosner, J. L. 1978. *Phys. Lett. B* 72:462
106. Gaillard, M. K., Lee, B. W., Rosner, J. L. 1975. *Rev. Mod. Phys.* 47:277
107. Jackson, J. D. 1976. *Proc. Summer Inst. Part. Phys., Aug. 2–13, 1976*, ed. M. C. Zipf, p. 147. SLAC, Stanford Univ., CA
108. Feldman, G. J. 1977. *Rep. SLAC-PUB-2068*
109. Lane, K., Weinberg, S. 1976. *Phys. Rev. Lett.* 37:717; Fritzsch, H. 1976. *Phys. Lett. B* 63:419
110. Deshpande, N. G. et al. 1976. *Phys. Rev. Lett.* 37:1305
111. Coremans-Bertrand, G. et al. 1976. *Phys. Lett. B* 65:480; Hand, L. et al. 1975. *Rep. FNAL-Proposal-382*; Burhop, E. H. S. et al. 1976. *Phys. Lett. B* 65:299; Voyvodic, L. 1976. *Rep. Fermilab-FN-289*; Conversi, M. 1975. *Rep. CERN-NP-75-17*; Dine, M. et al. 1976. *Rep. Fermilab-Proposal-P490*
112. Einhorn, M. B., Quigg, C. 1975. *Phys. Rev. D* 12:2015
113. Ellis, J., Gaillard, M. K., Nanopoulos, D. V. 1975. *Nucl. Phys. B* 100:313
114. Altarelli, G., Cabibbo, N., Maiani, L. 1975. *Nucl. Phys. B* 88:285; 1975. *Phys. Lett. B* 57:277; Kingsley, R. L. et al. 1975. *Phys. Rev. D* 11:1919; 1975. *Phys. Rev. D* 12:106; Pais, A., Rittenberg, V. 1975. *Phys. Rev. Lett.* 34:707
115. Glashow, S. L., Weinberg, S. 1977. *Phys. Rev. D* 15:1958; Paschos, E. A. 1977. *Phys. Rev. D* 15:1966; Gaillard, M. K., Lee, B. W., Rosner, J. L. 1975. *Rev. Mod. Phys.* 47:277; Kingsley, R. L. et al. 1975. *Phys. Rev. D* 11:1919, 12:106; Okun, L. B., Zakharov, V. I., Pontecorvo, B. M. 1975. *Lett. Nuovo*

Cimento 13:218; De Rújula, A., Georgi, H., Glashow, S. L. 1975. *Phys. Rev. Lett.* 35:69

116. Brandelik, R. et al. 1977. *Phys. Lett. B* 70:132
117. Cazzoli, E. G. et al. 1975. *Phys. Rev. Lett.* 34:1125
118. Knapp, B. et al. 1976. *Phys. Rev. Lett.* 37:882
119. Wiss, J. et al. 1976. *Phys. Rev. Lett.* 37:1531
120. Brandelik, R. et al. 1977. *Phys. Lett. B* 70:387
121. Feller, J. M. et al. 1978. *Phys. Rev. Lett.* 40:274
122. Kirkby, J. 1977. See Ref. 32, p. 3
123. Barnett, M. 1975. *Phys. Rev. Lett.* 34:41; 1975. *Phys. Rev. D* 11:3246; 1976. *Phys. Rev. D* 13:671; Fayet, P. 1974. *Nucl. Phys. B* 78:14; Gürsey, F., Sikivie, P. 1976. *Phys. Rev. Lett.* 36:775; Ramond, P. 1976. *Nucl. Phys. B* 110:214
124. De Rújula, A., Georgi, H., Glashow, S. L. 1975. *Phys. Rev. D* 12:3589; Wilczek, F. A. et al. 1975. *Phys. Rev. D* 12:2768; Fritzsch, H., Gell-Mann, M., Minkowski, P. 1975. *Phys. Lett. B* 59:256; Pakvasa, S., Simmons, W. A., Tuan, S. F. 1975. *Phys. Rev. Lett.* 35:702
125. Gürsey, F., Sikivie, P. 1976. *Phys. Rev. Lett.* 36:775; 1977. *Phys. Rev. D* 16: 816; Ramond, P. 1976. *Nucl. Phys. B* 110:214; 1977. *Nucl. Phys. B* 126:509; Gürsey, F., Ramond, P., Sikivie, P. 1976. *Phys. Lett. B* 60:177; 1975. *Phys. Rev. D* 12:2166; Gürsey, F., Serdaroglu, M. 1978. *Nuovo Cimento Lett.* 21:28
126. Horn, D., Ross, G. G. 1977. *Phys. Lett. B* 67:460
127. Adler, S. L. 1970. *Lectures on Elementary Particles and Quantum Field Theory*, ed. S. Deser, M. Grisaru, H. Pendleton. Cambridge, MA: MIT Press; Gross, D. J., Jackiw, R. 1972. *Phys. Rev. D* 6:477
128. Knies, G. 1977. See Ref. 32, p. 93
129. Benvenuti, A. et al. 1976. *Phys. Rev. Lett.* 36:1478; 37:189
130. Barish, B. C. et al. 1977. *Phys. Rev. Lett.* 39:741, 1595; 1977. See Ref. 32, p. 239; Holder, M. et al. 1977. *Phys. Rev. Lett.* 39:433; Steinberger, J. 1977. CERN Rep.; Schultze, K. 1977. See Ref. 32, p. 359; Bosetti, P. C. et al. 1977. *Phys. Lett. B* 70:273; Berge, J. P. et al. 1977. *Phys. Rev. Lett.* 39:382
131. Barish, B. C. et al. 1977. *Phys. Rev. Lett.* 38:577; Benvenuti, A. et al. 1977. *Phys. Rev. Lett.* 38:1110, 1183; 1978. *Phys. Rev. Lett.* 40:488; Holder, M. et al. 1977. *Phys. Lett. B* 70:393
132. Barger, V. et al. 1977. *Phys. Rev. Lett.* 38:1190; 1977. *Phys. Rev. D* 16:2141; Albright, C. H., Smith, J., Vermaseren, J. A. M. 1977. *Phys. Rev. Lett.* 38:1187; 1977. *Phys. Rev. D* 16:3182, 3204; 1978. *Phys. Rev. D* 18:108; Albright, C. H., Shrock, R. E., Smith, J. 1978. *Phys. Rev. D* 17:2383; Zee, A., Wilczek, F., Treiman, S. B. 1977. *Phys. Lett. B* 68:369; Barnett, M., Chang, L. N. 1977. *Phys. Lett. B* 72:233; Barnett, M., Chang, L. N., Weiss, N. 1978. *Phys. Rev. D* 17:2266; Langacker, P., Segre, G. 1977. *Phys. Rev. Lett.* 39:259; Barger, V. et al. 1977. *Phys. Lett. B* 70:329; 1977. *Phys. Rev. D* 16:3170; Cox, P. H., Yildiz, A. 1977. *Phys. Rev. D* 16:2897; Pakvasa, S., Sugawara, H., Suzuki, M. 1977. *Phys. Lett. B* 69:461
133. Barnett, M., Chang, L. N. 1977. *Phys. Lett. B* 72:233; Barnett, M., Chang, L. N., Weiss, N. 1978. *Phys. Rev. D* 17:2266; Langacker, P., Segre, G. 1977. *Phys. Rev. Lett.* 39:259; Albright, C. H., Smith, J., Vermaseren, J. A. M. 1978. *Phys. Rev. D* 18:108; Albright, C. H., Shrock, R. E., Smith, J. 1978. *Phys. Rev. D* 17:2383
134. Bletzacker, F., Nieh, H. T., Soni, A. 1977. *Phys. Rev. Lett.* 38:1241; Soni, A. 1977. *Phys. Lett. B* 71:435; Goldberg, H. 1977. *Phys. Rev. Lett.* 39: 1598; Young, B.-L., Walsh, T. F., Yang, T. C. 1978. *Phys. Lett. B* 74: 111; see Ref. 133, Barnett & Chang, Barnett, Chang & Weiss, Albright, Smith & Vermaseren
135. Smith, J., Vermaseren, J. A. M. 1978. *Phys. Rev. D* 17:2288; Barnett, M., Chang, L. N., Weiss, N. 1978. *Phys. Rev. D* 17:2266; Barger, V., Gottschalk, T., Phillips, R. J. N. 1978. *Phys. Rev. D* 17:2284
136. Barnett, M. 1977. *Deeper Pathways in High Energy Physics—Orbis Scientiae 1977*, ed. A. Perlmutter, L. F. Scott, p. 389. New York: Plenum
137. Kobayashi, M., Maskawa, K. 1973. *Prog. Theor. Phys.* 49:652; Harari, H. 1975. *Phys. Lett. B* 57:265; 1975. *Ann. Phys.* 94:391
138. Golowich, E., Holstein, B. R. 1975. *Phys. Rev. Lett.* 35:831; 1977. *Phys. Rev. D* 15:3472; Branco, G., Mohapatra, R. N. 1976. *Phys. Rev. Lett.* 36:926; Wilczek, F. A. et al. 1975. *Phys. Rev. D* 12:2768; Branco, G., Mohapatra, R. N., Hagiwara, T.

1976. *Phys. Rev. D* 13:680
139. Pati, J., Salam, A. 1974. *Phys. Rev. D* 10:275; Fritzsch, H., Minkowski, P. 1976. *Nucl. Phys. B* 103:61; Mohapatra, R. N., Sidhu, D. P. 1977. *Phys. Rev. Lett.* 38:667; De Rújula, A., Georgi, H., Glashow, S. L. 1977. *Ann. Phys.* 109:258; Beg, M. A. B., Zee, A. 1973. *Phys. Rev. Lett.* 30:675; 1973. *Phys. Rev. D* 8:1460; Beg, M. A. B. et al. 1977. *Phys. Rev. Lett.* 38: 1254; Beg, M. A. B. et al. 1977. *Phys. Rev. Lett.* 39:1054
140. Georgi, H., Weinberg, S. 1978. *Phys. Rev. D* 17:275
141. Segre, G., Weyers, J. 1976. *Phys. Lett. B* 65:243; Lee, B. W., Weinberg, S. 1977. *Phys. Rev. Lett.* 38:1237; Lee, B. W., Shrock, R. 1978. *Phys. Rev. D* 17:2410; Barnett, M., Chang, L. N. 1977. *Phys. Lett. B* 72:233; *SLAC Rep.* In preparation; Barnett, M., Chang, L. N., Weiss, N. 1978. *Phys. Rev. D* 17:2266; Langacker, P., Segre, G. 1977. *Phys. Rev. Lett.* 39:259; Langacker, P., Segre, G., Golshani, M. 1978. *Phys. Rev. D* 17:1402
142. Bjorken, J. D., Lane, K. 1978. *SLAC Rep.*
143. Pati, J. C., Salam, A. 1975. *Phys. Lett. B* 58:333; 1977. *Proc. Neutrino 76, Aachen, June 1976*, ed. H. Faissner, H. Reithler, P. Zerwas. Braunschweig: Vieweg
144. Han, M.-Y., Nambu, Y. 1965. *Phys. Rev.* 139:B1006
145. Barnett, M. 1977. *Proc. Part. Fields 1976*, ed. H. Gordon, R. F. Peierls, p. D77. Upton, NY: Brookhaven Nat. Lab.
146. Holder, M. et al. 1977. *Phys. Lett. B* 69:377
147. Altarelli, G., Parisi, G., Petronzio, R. 1976. *Phys. Lett. B* 63:183; Altarelli, G., Parisi, G. 1977. *Nucl. Phys. B* 126: 298; Barnett, M., Georgi, H., Politzer, H. D. 1976. *Phys. Rev. Lett.* 37:1313; Kaplan, J., Martin, F. 1976. *Nucl. Phys. B* 115:333; Barnett, M., Martin, F. 1977. *Phys. Rev. D* 16:2765; Zakharov, V. I. 1977. *Proc. Int. Conf. High Energy Phys., 18th, Tbilisi, July 1976*, ed. N. N. Bogoliubov et al, Vol. II, p. B69. Dubna, USSR: J. Inst. Nucl. Res.; Zee, A., Wilczek, F., Treiman, S. B. 1974. *Phys. Rev. D* 10:2881; Buras, A. J., Gaemers, K. J. F. 1977. *Phys. Lett. B* 71:106; Buras, A. J. 1977. *Nucl. Phys. B* 125:125; Buras, A. J. et al. 1977. *Nucl. Phys. B* 131:308; Fox, G. C. 1978. *Nucl. Phys. B* 134:269; Baluni, V., Eichten, E. 1976. *Phys. Rev. Lett.* 37:1181
148. Chang, L. N., Derman, E., Ng, J. N. 1975. *Phys. Rev. Lett.* 35:6, 1252; 1975. *Phys. Rev. D* 12:3539; Chang, L. N., Ng, J. N. 1977. *Phys. Rev. D* 16:3157; Pais, A., Treiman, S. B. 1975. *Phys. Rev. D* 12:3539; Chang, Derman, E. 1976. *Nucl. Phys. B* 110:40
149. Baltay, C. et al. 1977. *Phys. Rev. Lett.* 39:62
150. von Krogh, J. et al. 1976. *Phys. Rev. Lett.* 36:710; Schultze, K. 1977. See Ref. 32, p. 359
151. Ellis, J. et al. 1977. *Nucl. Phys. B* 131: 285; Ellis, J., Gaillard, M. K., Nanopoulos, D. V. 1976. *Nucl. Phys. B* 109:213
152. Cahn, R. N., Ellis, S. D. 1977. *Phys. Rev. D* 16:1484; Ali, A. 1977. *Rep. CERN-TH-2411*
153. Aubert, J. J. et al. 1975. *Nucl. Phys. B* 89:1; Blanar, G. J. et al. 1975. *Phys. Rev. Lett.* 35:346; Knapp, B. et al. 1975. *Phys. Rev. Lett.* 34:1044; Büsser, F. W. et al. 1975. *Phys. Lett. B* 56:482; Anderson, K. J. et l. 1976. *Phys. Rev. Lett.* 36:237; Antipov, Y. M. et al. 1976. *Phys. Lett. B* 60:309; Snyder, H. D. et al. 1976. *Phys. Rev. Lett.* 36:1415; Branson, J. G. et al. 1977. *Phys. Rev. Lett.* 38:1334; Cordon, M. J. et al. 1977. *Phys. Lett. B* 68:96; Bushnin, Y. B. et al. 1977. *Phys. Lett. B* 72:269; Cobb, J. H. et al. 1977. *Phys. Lett. B* 68:101; Amaldi, E., et al. 1977. *Lett. Nuovo Cimento* 19:152; Bamberger, A. et al. 1978. *Nucl. Phys. B* 134:1, 1978
154. Sivers, D. 1976. *Nucl. Phys. B* 106:95; Barnett, M., Silverman, D. 1975. *Phys. Rev. D* 12:2037; Gunion, J. 1975. *Phys. Rev. D* 12:1345; Green, M. B., Jacob, M., Landshoff, P. 1975. *Nuovo Cimento* 29:123; Donnachie, A., Landshoff, P. 1976. *Nucl. Phys. B* 112:233
155. Binkley, M. et al. 1976. *Phys. Rev. Lett.* 37:578
156a. Ellis, S., Einhorn, M., Quigg, C. 1976. *Phys. Rev. Lett.* 36:1263; Carlson, C. E., Suaya, R. 1976. *Phys. Rev. D* 14:3115; 1977. *Phys. Rev. D* 15:1416; 1978. *Phys. Rev. D* 18:760; Ellis, S., Einhorn, M. 1975. *Phys. Rev. D* 12: 2007
156b. Cahn, R., Ellis, S. 1977. *Phys. Rev. D* 16:1484
157. Cordon, M. J. et al. 1977. *Phys. Lett. B* 68:96
158. Cobb, J. H. et al. 1978. *Phys. Lett. B* 72:497

159a. Fritzsch, H. 1977. *Phys. Lett. B* 67: 217; Halzen, F. 1977. *Phys. Lett. B* 69:105; Gaisser, T. K., Halzen, F., Paschos, E. 1977. *Phys. Rev. D* 15: 2577; Gluck, M., Owens, J. F., Reya, E. 1978. *Phys. Rev. D* 17:2324
159b. Ellis, J. et al. 1977. *Nucl. Phys. B* 131:285; Jones, L., Wyld, H. 1978. *Phys. Rev. D* 17:2332; Owens, J., Reya, E. 1978. *Phys. Rev. D* 17:3003
160. Drell, S., Yan, T.-M. 1970. *Phys. Rev. Lett.* 25:316
161. Carlson, C. E. Private communication
162. Bjorken, J. D. 1977. See Ref. 32, p. 960; Nieh, H. T. 1977. *Stony Brook Rep. ITP-SB-77-64*
163. Sivers, D. 1976. *Nucl. Phys. B* 106:96
164. Coremans-Bertrand, G. et al. 1976. *Phys. Lett. B* 65:480
165. Babcock, J., Sivers, D., Wolfram, S. 1978. *Phys. Rev. D* 18:162; Gaisser, T. K., Halzen, F., Kajantie, K. 1975. *Phys. Rev. D* 12:1968; Pilachowski, L., Tuan, S. F. 1975. *Phys. Rev. D* 11:3148; Barnett, M. 1975. *Phys. Rev. D* 12:3441; Gaisser, T. K., Halzen, F. 1975. *Phys. Rev. D* 11:3157; 1976. *Phys. Rev. D* 13:171; Hinchliffe, I., Llewellyn Smith, C. H. 1976. *Phys. Lett. B* 61:472; 1976. *Nucl. Phys. B* 114:45; Bourquin, M., Gaillard, J. M. 1976. *Nucl. Phys. B* 114:334; McKay, D. W., Young, B.-L. 1977. *Phys. Rev. D* 15:1282; Cox, P. H., Park, S. Y., Yildiz, A. 1977. *Phys. Lett. B* 70:317; Jones, L. M., Wyld, H. W. 1978. *Phys. Rev. D* 17:1782; Gustafson, G., Peterson, C. 1977. *Phys. Lett. B* 67:81; Halzen, F., Matsuda, S. 1978. *Phys. Rev. D* 17:1344
166. Barger, V., Phillips, R. 1975. *Phys. Rev. D* 12:2623; Field, R. D., Quigg, C. 1975. *Rep. Fermilab-75/15-THY*
167. Carlson, C. E., Freund, P. G. O. 1972. *Phys. Lett. B* 39:349
168. Sivers, D., Townsend, J., West, G. 1976. *Phys. Rev. D* 13:1234; Walsh, T. 1975. *Nuovo Cimento Lett.* 14:290; Boreskov, K. G., Ioffe, B. L. 1976. *Moscow Rep. ITEP-102-1976*; Ioffe, B. L., Okun, L. B., Zakharov, V. I. 1975. *Moscow-ITEP Rep.*; Aviv, R. et al. 1975. *Phys. Rev. D* 12:2862; Pumplin, J., Repko, W. 1975. *Phys. Rev. D* 12:1376; Horn, D. 1975. *Phys. Lett. B* 58:323; Humpert, B., Wright, A. C. D. 1977. *Phys. Rev. D* 15:2503; Humpert, B. 1977. *Phys. Lett. B* 68:66
169. Camerini, U. et al. 1975. *Phys. Rev. Lett.* 35:483
170. Anderson, R. L. et al. 1977. *Phys. Rev. Lett.* 38:263
171. Gittleman, B. et al. 1975. *Phys. Rev. Lett.* 35:1616; Knapp, B. et al. 1975. *Phys. Rev. Lett.* 34:1040; Nash, T. et al. 1976. *Phys. Rev. Lett.* 36:1233
172. Babcock, J., Sivers, D., Wolfram, S. 1978. *Phys. Rev. D* 18:162; Chen, M.-S., Kane, G. L., Yao, Y.-P. 1976. *Mich. Rep. UM HE 76-17*; Novikov, V. A. et al. 1978. *Nucl. Phys. B* 136:125; Jones, L. M., Wyld, H. W. 1978. *Phys. Rev. D* 17:759
173. Shifman, M. A., Vainshtein, A. I., Zakharov, V. I. 1976. *Phys. Lett. B* 65:255
174. Feldman, G. J. et al. 1977. *Phys. Rev. Lett* 38:1313
175. Blietschau, J. et al. 1977. *Nucl. Phys. B* 118:218; Benvenuti, A. et al. 1976. *Phys. Rev. Lett.* 37:1039; Merritt, F. S. et al. 1978. *Phys. Rev. D* 17:2199; Holder, M. et al. 1977. *Phys. Lett. B* 71:222, 72:254; Schultze, K. 1977. See Ref. 32, p. 359
176. Cline, D. et al. 1976. *Phys. Rev. Lett.* 37:252, 648; Lee, W. et al. 1976. *Phys. Rev. Lett.* 37:186; Sulak, L. R. et al. 1977. See Ref. 143, 1977; Pohl, M. et al. 1978. *Phys. Lett. B* 72:489
177. Kluttig, H., Morfin, J. G., Van Doninck, W. 1977. *Phys. Lett. B* 71: 446
178. Hasert, F. J. et al. 1973. *Phys. Lett. B* 46:121; Blietschau, J. et al. 1976. *Nucl. Phys. B* 114:189; Reines, F. et al. 1976. *Phys. Rev. Lett.* 37:315; Reithler, H. 1977. See Ref. 32, p. 343
179. Sehgal, L. M. 1977. *Phys. Lett. B* 71:99; Hung, P. Q., Sakurai, J. J. 1977. *Phys. Lett. B* 72:208; Abbott, L., Barnett, M. 1978. *Phys. Rev. Lett.* 40:1303; Barnett, M. 1976. *Phys. Rev. D* 14:2990; Albright, C. H. et al. 1976. *Phys. Rev. D* 14:1780; Barger, V., Nanopoulos, D. V. 1977. *Nucl. Phys. B* 124:426; Sidhu, D. P. 1976. *Phys. Rev. D* 14:2235; Bernabeu, J., Jarlskog, C. 1977. *Phys. Lett. B* 69:71; Langacker, P., Sidhu, D. P. 1978. *Phys. Lett. B* 74:233
180. Baird, P. E. G. et al. 1976. *Nature* 264:528; Sandars, P. G. H. See Ref. 32, p. 343
181. Feinberg, G. 1977. *Columbia Rep. CU-TP-111*; 1978. *Proc. Ben Lee Mem. Int. Conf., Batavia, Ill.*, Oct. 20–22, 1977. In press
182. Cahn, R. N., Kane, G. L. 1977. *Phys. Lett. B* 71:348; Marciano, W. J., Sanda, A. I. 1978. *Phys. Rev. D* 17: 3055
183. Sachs, R. G. 1963. *Ann. Phys.* 22:239;

Wolfenstein, L. 1964. *Phys. Rev. Lett.* 13:562; Mohapatra, R. N. 1972. *Phys. Rev. D* 6:2023; Pais, A. 1973. *Phys. Rev. D* 8:625; 1972. *Phys. Rev. Lett.* 29:1719; 1973. *Phys. Rev. Lett.* 30:114; Lee, T. D. 1973. *Phys. Rev. D* 8:1226; 1974. *Phys. Rep. C* 9:143; Mohapatra, R. N., Pati, J. C., Wolfenstein, L. 1975. *Phys. Rev. D* 11:3319
184. Weinberg, S. 1976. *Phys. Rev. Lett.* 37:657
185. Sikivie, P. 1976. *Phys. Lett. B* 65:141
186. Maiani, L. 1976. *Phys. Lett. C* 62:183; Pakvasa, S., Sugawara, H. 1976. *Phys. Rev. D* 14:305
187. Geweniger, C. et al. 1974. *Phys. Lett. B* 48:487; Messner, R. et al. 1973. *Phys. Rev. Lett.* 30:876
188. Bjorken, J. D., Weinberg, S. 1977. *Phys. Rev. Lett.* 38:622
189. Depommier, P. et al. 1977. *Phys. Rev. Lett.* 39:1113
190. Bryman, D. A. et al. 1972. *Phys. Rev. Lett.* 28:1469
191. Cheng, T.-P., Li, L.-F. 1977. *Phys. Rev. Lett.* 38:381; 1977. *Phys. Rev. D* 16:1425; Bilenkii, S. M., Petkov, S. T., Pontecorvo, B. 1977. *Phys. Lett. B* 67:309
192. Bjorken, J. D., Lane, K., Weinberg, S. 1977. *Phys. Rev. D* 16:1474
193. Korenchenko, S. M. et al. 1976. *Sov. Phys. JETP* 43:1
194. Fitch, V. L. et al. 1967. *Phys. Rev.* 164:1711
195. Glashow, S. L. 1977. *Harvard Rep. HUTP-77/A008*
196. Fritzsch, H. 1977. *Phys. Lett. B* 67:451
197. Appelquist, T., Carazzone, J. 1975. *Phys. Rev. D* 11:2856
198. Georgi, H., Quinn, H., Weinberg, S. 1974. *Phys. Rev. Lett.* 33:451
199. Buras, A. J. et al. 1978. *Nucl. Phys. B* 135:66
200. Georgi, H., Glashow, S. L. 1974. *Phys. Rev. Lett.* 32:438
201. Fritzsch, H., Minkowski, P. 1975. *Ann. Phys.* 93:193; 1976. *Nucl. Phys. B* 103:61
202. Gürsey, F., Serdaroglu, M. 1978. *Nuovo Cimento Lett.* 21:28
203. Ramond, P. 1976. *Nucl. Phys. B* 110:214
204. Reines, F., Crouch, M. F. 1974. *Phys. Rev. Lett.* 32:493
205. Gell-Mann, M., Ramond, P., Slansky, R. 1977. *Los Alamos Rep. LA-UR-77-2059*
206. Weinberg, S. 1977. *Harvard Rep. HUTP-77/A057*
207. Gel'fand, Y., Likhtman, E. P. 1971. *Zh. Eskp. Teor. Fiz. Pis'ma Red.* 13:452 (in English 13:323); Wess, J., Zumino, B. 1974. *Phys. Lett. B* 49:54
208. Freedman, D., van Nieuwenhuizen, P., Ferrara, S. 1976. *Phys. Rev. D* 13:3214; Deser, S., Zumino, B. 1976. *Phys. Lett. B* 62:335
209. Grisaru, M., van Nieuwenhuizen, P., Vermaseren, J. 1976. *Phys. Rev. Lett.* 37:1662

Ann. Rev. Nucl. Part. Sci. 1980. 30: 337–81

4

CHARMED MESONS PRODUCED IN e^+e^- ANNIHILATION

Gerson Goldhaber
Department of Physics and Lawrence Berkeley Laboratory,
University of California, Berkeley, California 94720

James E. Wiss
Department of Physics, University of Illinois, Urbana, Illinois 61801

CONTENTS

1 INTRODUCTION

The study of high energy electron-positron annihilation has revealed a rich spectrum of new phenomena and led to the discovery of new, hadronically stable particles—the charmed mesons. These particles are particularly interesting because they are the lightest particles containing the fourth quark; thus they serve to test in a novel way many theoretical ideas about quarks and their interactions.

Before the existence of charm was generally accepted, elementary particles were assumed to be constructed from three constituents or quarks: up (u), down (d), and strange (s). Their quantum numbers are summarized in Table 1. This three-quark model for elementary particles, first proposed independently by Gell-Mann and Zweig in 1964 (Gell-Mann 1964, Zweig 1964) proved enormously successful in explaining the spectroscopy of the known hadrons, as well as many of the features of their interactions. Until the early 1970s there was little experimental need for a fourth quark. But then in 1970 Glashow, Illiopoulos & Maiani (GIM) demonstrated that the inclusion of a new quark, the charmed quark (c), would solve a problem with the Weinberg-Salam model of the weak interaction—the nonobservation of the strangeness-changing neutral current (Aronson et al 1970, Carithers et al 1973, Clark et al 1971, Klems et al 1970). The existence of a fourth quark, of course, implied the existence of numerous new particles. Using the quark model convention of constructing mesons from quark-antiquark pairs, one could anticipate the existence of $c\bar{u}$, $c\bar{d}$, $c\bar{s}$, and $c\bar{c}$ mesons. The lowest-lying states of these first three mesons, dubbed the D^0 ($c\bar{u}$), D^+ ($c\bar{d}$), and F^+ ($c\bar{s}$), are the subject of this review. The family of mesons consisting of $c\bar{c}$ are known as the psions and their discovery in 1974 in conjunction with the discovery of charmed mesons in 1976 gave the first compelling evidence for the validity of the charm theory. (See, for example, Chinowsky 1977.)

We begin our review by summarizing the first experimental indications for the existence of charm as obtained from experiments in e^+e^- annihilation. This includes a brief discussion of the role of charm in the understanding of the ψ mesons, as well as the unraveling of the intricate structure present in the e^+e^- total hadronic cross section. Next we discuss the discovery of the D^0 and D^+, and detail those properties crucial to their identification as charmed particles. We then review the properties of the D^0 and D^+ learned through studies at the $\psi(3770)$ resonance. We summarize

Table 1 Quark quantum numbers

	u	d	s	c
Baryon number	1/3	1/3	1/3	1/3
Spin	1/2	1/2	1/2	1/2
Charge	+2/3	−1/3	−1/3	+2/3
Isospin	1/2	1/2	0	0
I_3	+1/2	−1/2	0	0
Strangeness	0	0	−1	0
Charm	0	0	0	1

the compelling evidence that indicates that this state decays nearly exclusively into $D\bar{D}$, which thus makes it particularly useful in establishing inclusive and exclusive D branching fractions.

Our discussion of branching fractions includes two particularly important D decay modes, $D^0 \to \pi^+\pi^-$ and $D^0 \to K^+K^-$. These processes are suppressed relative to $D^0 \to K^-\pi^+$ in the standard charm model, and thus serve as a critical test of that theory. This is followed by a discussion of the D semileptonic decay modes, which provide useful information on the D^0 and D^+ lifetimes.

Turning our attention to the data collected beyond the $\psi(3770)$, we discuss the properties and production mechanisms of the excited charmed mesons, the D^{*0} and the D^{*+} · D production just above the $\psi(3770)$ appears to be dominated by the three quasi-two-body processes, $e^+e^- \to D\bar{D}$, $D^*\bar{D}+\bar{D}^*D$, and $D^*\bar{D}^*$, in accordance with early theoretical predictions. The relative yield of each process, on the other hand, are somewhat surprising, and have led to considerable theoretical speculation. Finally, we summarize evidence for the existence of the F meson, which is as yet not on as solid a footing as the D^0, D^+ isodoublet.

There has been a great deal of theoretical and experimental work on charmed particles within the last decade. Because of the limitations of space and time, we confine our review to the experimental aspects of charmed meson production through e^+e^- annihilation. Hence we do not review the many important results on production of charmed mesons by neutrino, photon, and hadron beams, nor discuss the important results on charmed baryons. Our emphasis is entirely on experimental matters. The theoretical aspects of charm were recently reviewed in this series by Appelquist, Barnett & Lane (1978). When we found it necessary to reference theoretical works, we did so in the spirit of illustration and made no attempt to be exhaustive or judgmental.

2 THE e^+e^- ANNIHILATION PROCESS

2.1 *QED and Nonresonant Annihilation*

The electron-positron annihilation process has been studied using the colliding-beam technique since the early 1960s (Schwitters & Strauch 1976). The early motivation for such experiments was first the study of QED processes, such as $e^+e^- \to \mu^+\mu^-$, and later the study of $e^+e^- \to$ hadrons. Both processes are assumed to be dominated by the s-channel exchange of a virtual photon as illustrated in Figure 1. The cross section for the process $e^+e^- \to \mu^+\mu^-$ (neglecting the mass of the muon and radiative corrections) is $\sigma_{QED} = 4\pi\alpha^2/3s$ where $s = E_{cm}^2$. In analogy with Figure 1*a* we expect, as indicated schematically in Figure 1*c*, that the total hadronic cross section at high energies is

$$\sigma_{\text{had}} = \frac{4\pi\alpha^2}{3s} \sum Q_i^2$$

where Q_i are the contributing quark charges, and i ranges over those flavors with quark masses $<\frac{1}{2}E_{\text{cm}}$ and the three quark colors.

Below a center-of-mass energy of about 1.5 GeV, hadronic production by e^+e^- annihilation is dominated by the decays of vector mesons as shown in Figure 1*b* with sizable contributions from the ρ, ω, and φ. Many of the beautiful results obtained in this region are summarized by Perez-y-Jorba (1969).

Early data collected beyond the ρ, ω, and φ resonance region but below 3 GeV appeared to be consistent with the predictions of the naive point-like parton model, as shown in Figure 1*c*. The σ_{had} should scale in the limit of negligible quark masses (and small QCD corrections) as s^{-1}. The

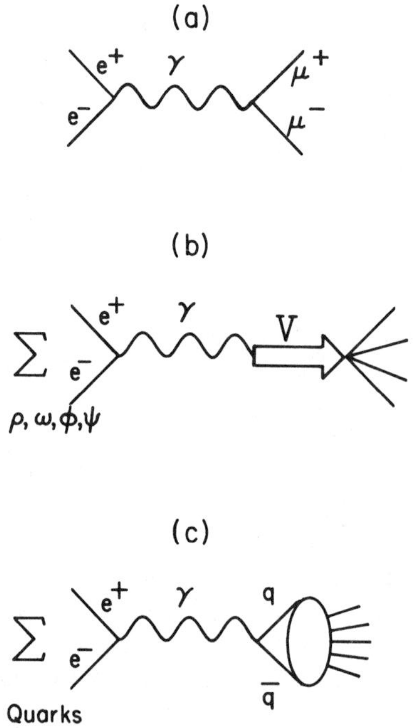

Figure 1 (*a*) The 1γ diagram for the QED process $e^+e^- \to \mu^+\mu^-$. Two types of 1γ contribution to $e^+e^- \to$ hadrons: (*b*) Hadronic decays of vector mesons. This should dominate when $s^{1/2}$ is near the mass of the vector mesons. (*c*) Hadronic contribution from the point-like production of quarks.

traditional way of expressing this scaling behavior evokes the famous ratio

$$R \equiv \frac{\sigma_{e^+e^- \to \text{hadrons}}}{\sigma_{\text{QED}}}.$$

As illustrated in Figure 1 the naive quark model prediction for this ratio is $R = \sum Q_i^2$.

Experimental measurements in the range $1.5 \lesssim E_{\text{cm}} \lesssim 3$ GeV taken prior to 1974 were consistent with values of R ranging from 2 to 3 (Richter 1974). Very recent data taken in this energy region by ADONE and DCI show consistency with $R = 2$ with considerably smaller error bars (summarized by Feldman 1978).

The value of $R = 2$ is precisely what one would expect in the original three-quark model (quark charges of 2/3, $-1/3$, and $-1/3$) with the added color degree of freedom, whereby each quark is present in three colors.

As early as 1973, data accumulated at CEA indicated a rise in R to a value of 4.7 ± 1.1 nb at $s^{1/2} = 4$ GeV (Litke et al 1973, Tarnopolsky et al 1974). This rise in R was later corroborated by the SLAC-LBL collaboration working at SPEAR (Richter 1974). Although initial theoretical response was varied (Ellis 1974), one hypothesis for the rise in R was the opening up of new degrees of freedom such as the threshold of a new quark, or the production of a new heavy lepton (Perl 1980). We now believe there were indeed contributions to R from both effects.

2.2 *The Discovery of the J/ψ*

The J/ψ was simultaneously observed in p-Be collisions (Aubert 1974) and e^+e^- annihilations (Augustin et al 1974). In the hadronic production experiment the J appeared as a narrow enhancement in the electron pair invariant mass distribution in the reaction p Be $\to e^+e^-X$. This enhancement had a width of 20 MeV, which was consistent with experimental resolution. The ψ was observed in e^+e^- annihilations as an enhancement in the total hadronic and lepton pair cross section for center-of-mass energies near the ψ mass of 3.095 ± 0.002 GeV, with a measured width compatible with the e^+e^- experimental resolution of 2 MeV.

Because the ψ appeared as an s-channel resonance in e^+e^- annihilation, one could make a precision width measurement using $\int \sigma(E)\, dE$ (where E is the center-of-mass energy and σ is the total resonant cross section) and the measured muon pair branching ratio of $\sim 7\%$, even though the total width was much smaller than the center-of-mass resolution of a storage ring. This analysis yielded the remarkable result $\Gamma_\psi = 69 \pm 15$ keV. The properties of known vector mesons are summarized in Tables 2 and 2a. Although the ψ is over three times as massive as the conventional ρ, ω, and φ vector mesons, it is narrower by about two orders of magnitude.

Table 2 Resonance parameters for vector mesons[a]

State	Mass (MeV)	Total width Γ (MeV)	Partial width Γ_e (keV)	Branching fraction B_e	Reference
ρ	776 ± 3	155 ± 3	6.7 ± 0.8	$(4.3 \pm 0.5) \times 10^{-5}$	b
ω	782.6 ± 0.3	10.1 ± 0.3	0.76 ± 0.17	$(7.6 \pm 1.7) \times 10^{-5}$	b
φ	1019.6 ± 0.2	4.1 ± 0.2	1.31 ± 0.10	$(31 \pm 1) \times 10^{-5}$	b
ψ	3095 ± 4	0.069 ± 0.015	4.8 ± 0.6	$(69 \pm 9) \times 10^{-3}$	SLAC-LBL Mark I
ψ'	3684 ± 5	0.228 ± 0.056	2.1 ± 0.3	$(9.3 \pm 1.6) \times 10^{-3}$	SLAC-LBL Mark I
	3772 ± 6	28 ± 5	0.35 ± 0.09	$(1.2 \pm 0.3) \times 10^{-5}$	LGW
ψ''	3770 ± 6	24 ± 5	0.18 ± 0.06	$(0.7 \pm 0.2) \times 10^{-5}$	DELCO
	3764 ± 5	24 ± 5	0.28 ± 0.05	$(1.2 \pm 0.2) \times 10^{-5}$	Mark II
4.04[c]	4040 ± 10	52 ± 10	0.75 ± 0.10	$(1.4 \pm 0.4) \times 10^{-5}$	DASP
4.16[c]	4159 ± 20	78 ± 10	0.77 ± 0.20	$(0.9 \pm 0.3) \times 10^{-5}$	DASP
4.41	4414 ± 7	33 ± 10	0.44 ± 0.14	$(1.3 \pm 0.3) \times 10^{-5}$	SLAC-LBL Mark I

[a] Other states have been reported between the φ and the ψ by experiments at Frascati and Orsay; we do not include them here.
[b] World averages compiled by the Particle Data Group (Bricman et al 1978).
[c] The SLAC-LBL and DELCO data do not separate this region into two states.

Table 2a New results on the Υ resonances

State	Mass (MeV)	Total width Γ (MeV)	Partial width Γ_e (keV)	Branching fraction B_μ	Reference
$\Upsilon(9.4)$	9460 ± 10	~ 0.05	1.28 ± 0.27	$(2.1 \pm 1.4) \times 10^{-2}$	DORIS[a]
	9433.1 ± 0.4[b]	—	1.15 ± 0.08	—	CLEO[c]
	9434.5 ± 0.4[b]	—	—	—	CUSB[d]
	9461.6 ± 0.5[e] }	$0.039^{+0.013}_{-0.008}$	1.31 ± 0.12	$(4.0 \pm 1.4) \times 10^{-2}$	LENA[f,g]
	9463.0 ± 0.5[e] }		1.35 ± 0.11	$(2.9 \pm 1.3) \times 10^{-2}$	DASP 2[g]
$\Upsilon'(10.0)$	10015 ± 20	—	0.33 ± 0.14	—	DORIS[a]
	$M(\Upsilon) + 560.7 \pm 0.8$[h]	—	(0.23 ± 0.08)[i]	—	CLEO[j]
	9993.0 ± 1.0[b]	—	(0.39 ± 0.06)[i]	—	CUSB[d]
	$M(\Upsilon) + 553.6 \pm 1.7$[e]	—	(0.49 ± 0.09)[i]	—	LENA[g]
	$M(\Upsilon) + 557 \pm 2$[e]	—	(0.45 ± 0.05)[i]	—	DASP 2[g]
$\Upsilon''(10.3)$	$M(\Upsilon) + 891.1 \pm 0.7$[k]	—	(0.31 ± 0.09)[i]	—	CLEO[j]
	10323.2 ± 0.7[b]	—	(0.32 ± 0.04)[i]	—	CUSB[d]
$\Upsilon'''(10.6)$	$M(\Upsilon) + 1112 \pm 2$[l]	21.5 ± 5.7[m,n]	(0.21 ± 0.06)[i]	—	CLEO[c]
	$M(\Upsilon) + 1114 \pm 2$[l]	10.8 ± 0.9[n,o]	(0.25 ± 0.07)[i]	—	CUSB[p]

[a] Values for Υ are averages of PLUTO, DASP 2, and DESY-Heidelberg; and for Υ', DASP 2, and DESY-Heidelberg, as quoted by Meyer (1979).
[b] There is a ± 30 MeV/c^2 systematic error including a 15 MeV/c^2 uncertainty due to the preliminary machine calibration at the Cornell Electron Storage Ring (CESR).
[c] Results from CESR quoted at the 1980 Experimental Meson Spectroscopy Conference at Brookhaven National Laboratory (Bebek 1980).
[d] Columbia University–Stony Brook experiment at CESR (Böhringer et al 1980).
[e] There is a ± 10 MeV uncertainty in the DORIS ring energy calibration.
[f] LENA is a new NaI-Pb glass detector at DORIS.
[g] Results from DORIS quoted at the 1980 Experimental Meson Spectroscopy Conference at Brookhaven National Laboratory (Schröder 1980).
[h] In addition to ± 3 MeV/c^2 systematic error.
[i] In units of $\Gamma_{ee}(\Upsilon)$.
[j] CLEO experiment at CESR (Andrews et al 1980).
[k] In addition to a ± 5 MeV/c^2 systematic error.
[l] In addition to a ± 4 MeV/c^2 systematic error.
[m] The CESR beam spread has been unfolded from this width using a Breit-Wigner form for the resonance.
[n] This is a finite width resonance, and could well be the counterpart for the "B mesons" of what the ψ'' is for charmed mesons.
[o] The CESR beam spread has been folded from this width using a Gaussian form for the resonance.
[p] Results from CESR quoted at the 1980 Experimental Meson Spectroscopy Conference at Brookhaven National Laboratory (Lee-Franzini 1980, Finocchiaro 1980).

Because of the anomalous ψ width, it was historically important to establish that the ψ was indeed a vector meson as implied by its s-channel production in e^+e^- annihilation. This prejudice was partially borne out through the observations of psi-photon interference in the process $e^+e^- \rightarrow \mu^+\mu^-$, which established the ψ quantum numbers as $J^{PC}(\psi) = 1^{--}$ (Boyarski et al 1975). In addition, data on the photoproduction of the ψ (Knapp et al 1975, Camerini et al 1975, Gittleman et al 1975) demonstrated that the ψ-nucleon total cross section was ~ 1 mb—the same order of magnitude as for the conventional vector mesons.

A search for additional narrow vector mesons formed in e^+e^- annihilation found the $\psi(3684)$ or ψ' with a width of 228 keV, the $\psi(4400)$ with a width of 33 ± 10 MeV, and, considerably later, the $\psi(3770)$ with a width of 25 MeV.

As seen in Figure 2 these resonances occur in the region where R makes a transition from a value of ~ 2.5 to a plateau around 4 to 5.

2.2.1 THE OZI RULE AND THE ψ The discovery of the new narrow vector meson states encouraged the proponents of a new quark. In fact such states were being predicted by charm enthusiasts concurrently with the experimental observation of the ψ (Appelquist et al 1978). Within the charm picture the new vector mesons were assumed to be constructed from ^3S or ^3D states of $c\bar{c}$. Many such vector mesons can be constructed in the theory by placing the new quarks in various levels of radial excitation. The narrow width of the ψ could be explained as a manifestation of the phenomenological OZI rule (Okubo 1963, Zweig 1964, Iizuka 1966), which was previously proposed to explain the suppression of certain strong interaction processes that can only occur via "disconnected" quark diagrams.

A disconnected diagram allows at least one external particle to be isolated by making a cut that does not intersect any quark line. For example, Figure 3 demonstrates that $\varphi \rightarrow \pi^+\pi^-\pi^0$ proceeds via a disconnected or OZI-suppressed diagram, whereas $\varphi \rightarrow K^+K^-$ does not. Experimentally one does find that the decay $\varphi \rightarrow K\bar{K}$ is preferred over $\varphi \rightarrow \pi^+\pi^-\pi^0$, even though the former process has a much smaller phase space than the latter.

Using this rule one could explain the narrow widths of the ψ and ψ' by asserting that they had masses below the threshold for the pair production of the lowest-lying states (presumably D's) containing the new quark. If this were true, all decay diagrams for the ψ and ψ' would be disconnected and hence suppressed.

The dramatic increase in the width of the $\psi(4400)$ and, as determined much later, the $\psi(3770)$, would then represent the opening up of the $D\bar{D}$ decay channel. Hence one would expect the relation

$$\tfrac{1}{2}M_{\psi'} < M_D < \tfrac{1}{2}M_{\psi(3770)}$$

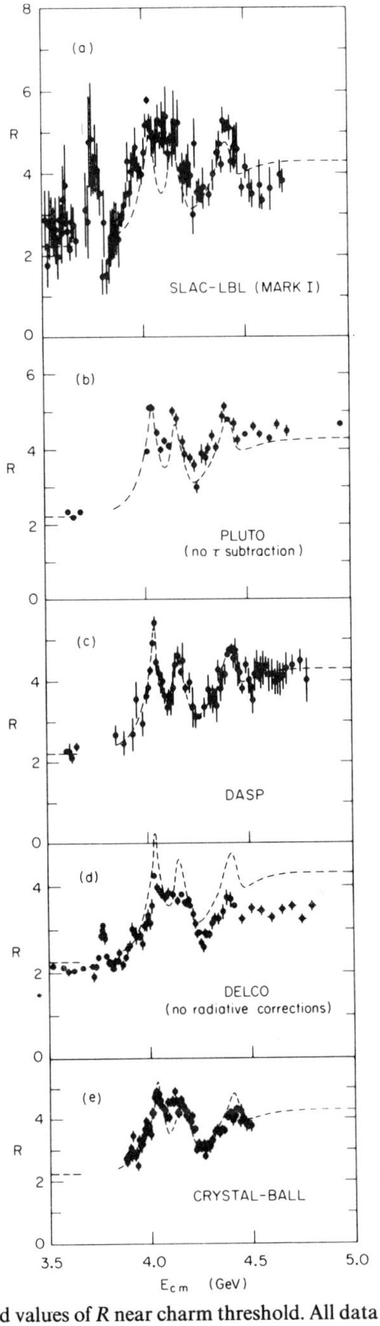

Figure 2 The observed values of R near charm threshold. All data are corrected for radiative effects and contamination from the τ, unless otherwise indicated. The hand-drawn curve, which follows the DASP points, has been applied to the other measurements in order to facilitate a comparison. Systematic errors, which range from 10 to 15%, are not included in the error bars. (Summary by Kirkby 1979.)

or $1.842 < M_D < 1.885$ GeV, with the upper bound coming through considerable hindsight.

2.2.2 PROPERTIES OF THE ψ If the psion family were comprised of $q\bar{q}$ states of a new quark, one would expect the presence of additional relatively narrow but strongly decaying states with $J^{PC} \neq 1^{--}$. These states would be constructed from p-wave or other excitations of the new quark. Several such additional states have been observed via radiative decays from the ψ'. The beautiful phenomenology associated with the many $c\bar{c}$ states below the charm threshold was reviewed by Chinowsky (1977) and hence is not discussed here. Suffice it to say that it is indeed difficult to account for the multitude of new narrow states produced in e^+e^- without accepting the presence of a new quark.

Additional evidence for this interpretation of the new mesons comes from a study of the ψ decay modes. We have previously noted that the OZI explanation for the narrow width of the ψ and ψ' implies the existence of a fourth quark. The large mass of the ψ suggests that this fourth quark is much heavier than the conventional three quarks, and hence if the new states consisted of quark-antiquark combinations of the new quark they must be isosinglet states. A study of pion multiplicities in the decays of the ψ demonstrated that it has odd G-parity and thus (in light of its odd C-parity) the ψ has even isospin (Jean-Marie et al 1976). Even isospin, coupled with the observation of a substantial $\psi \to p\bar{p}$ decay mode, proves that the ψ is an isosinglet as expected. Corroborating evidence comes from the observation of a $\Lambda\bar{\Lambda}$ decay mode (Peruzzi et al 1978).

To summarize, the observation and properties of the new mesons discovered in e^+e^- annihilation provide a vast body of data which is easily accommodated into the framework of a new, heavy quark. However, there is little to link this quark to the charm quark of the GIM model from a study of the ψ family alone. The definitive proof of the charm hypothesis comes from the observation of states of nonzero charm—the D^0 and D^+. We turn now to a brief discussion of the properties one would expect in the GIM model for these mesons.

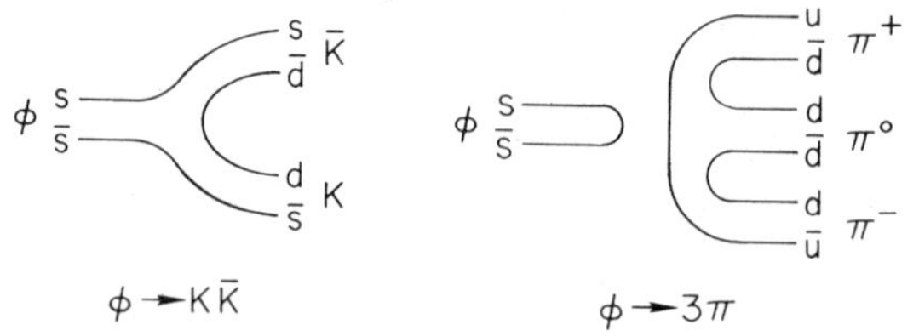

Figure 3 Illustration of OZI-allowed ($\varphi \to K\bar{K}$) and -forbidden ($\varphi \to \pi\pi\pi$) decays of the φ meson.

3 CHARM

3.1 *The GIM Model*

Several excellent and comprehensive reviews of the theoretical aspects of charm have been written since the seminal review of Gaillard, Lee & Rosner (1975). In this section we outline those aspects of the theory necessary for the understanding of the experimental material to follow.

As previously stated, the charmed quark was partly motivated to provide a means of cancelling first-order strangeness-changing neutral current effects such as $K_L \to \mu^+\mu^-$ and a large weak $K_S - K_L$ mass difference. In Figure 4 we show a diagram for these processes within the three-quark theory. Although these diagrams appear to contribute to the second-order weak interaction, the loop integral enhances their strength, making them comparable to a first-order diagram. The factors appearing at the diagram vertices follow from the form of the precharm weak hadronic current:

$$J_h^\mu = \bar{u}\gamma^\mu(1-\gamma^5)(d \cos \theta_c + s \sin \theta_c), \qquad 1.$$

where θ_c is the Cabibbo angle, introduced in 1963 to relate the strength of

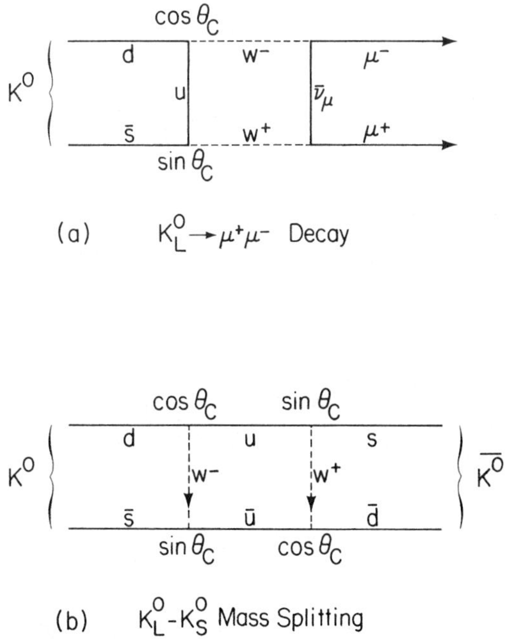

Figure 4 Quark diagrams for the decay $K_L^0 \to \mu^+\mu^-$ and the K_L^0/K_S^0 mass splitting with only u, d, and s quarks.

strangeness-conserving weak current to the strangeness-changing weak current. Present measurements indicate that the Cabibbo angle is small ($\theta_c = 13.2 \pm 0.5°$).

Figure 5 shows alternative diagrams for the processes employing the charmed quark in place of the up quark. These diagrams will tend to cancel the diagrams of Figure 4 provided the new vertex factors are as shown —that is, if Equation 1 is modified to

$$J_h^\mu = (\bar{u}\bar{c})\gamma^\mu(1-\gamma^5)\begin{pmatrix} \cos\theta_c & \sin\theta_c \\ -\sin\theta_c & \cos\theta_c \end{pmatrix}\begin{pmatrix} d \\ s \end{pmatrix}. \qquad 2.$$

Equation 2, employing two quark doublets and a unitary mixing matrix, is the hadronic current of the GIM model (Glashow et al 1970).

3.2 *Predicted Decays of Charmed Mesons*

Using Equation 2 and the bilinear weak Lagrangian $L = J^{\dagger\mu}J_\mu + J^\mu J_\mu^\dagger$, one obtains the results listed in Tables 3 and 4 for nonleptonic and semileptonic decays of charmed mesons. As illustrated in the tables, one expects a wide

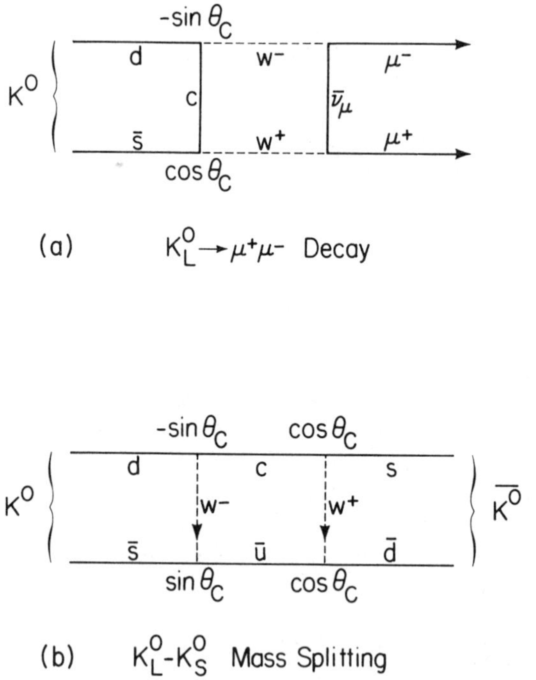

Figure 5 Cancelling quark diagrams for the decay $K_L^0 \to \mu^+\mu^-$ and the K_L^0/K_S^0 mass splitting in the GIM model.

Table 3 Nonleptonic decays of charmed mesons in the GIM model

Cabibbo factors	Flavor structure	Selection rules	D decay examples	F decay examples
$\cos^4\theta_c \cong 0.90$	csud	$\Delta c = \Delta s$	$D \to K^- m\pi$	$F^+ \to m\pi, K^+K^-\pi$
$\sin^2\theta_c \cos^2\theta_c \cong 0.05$	cdud and csus	$\Delta s = 0$	$D \to m\pi, K^-K^+m\pi$	$F^+ \to K^+m\pi$
$\sin^4\theta_c \cong 0.0025$	cdus	$\Delta c = -\Delta s$	$D \to K^+m\pi$	$F^+ \to K^+K^+m\pi$

Table 4 Semileptonic decays of charmed mesons

Cabibbo factors	Flavor structure	Selection rules	D examples	F^+ examples
$\cos^2\theta_c \cong 0.95$	$\bar{c}s$	$\Delta c = \Delta s = \Delta q$[a]	$D \to K^-e^+\nu m\pi$	$F^+ \to m\pi e^+\nu$, $\to K^+K^-e^+\nu$
$\sin^2\theta_c \cong 0.05$	$\bar{c}d$	$\Delta c = \Delta q, \Delta s = 0$	$D \to m\pi e^+\nu$	$F^+ \to Ke^+\nu$

[a] The term Δq is the change in charge of the hadronic system

disparity in rates between Cabibbo-favored decays such as $D^0 \rightarrow K^-\pi^+$ or $D^0 \rightarrow K^- e\nu$, singly suppressed decays such as $D^0 \rightarrow \pi^+\pi^-$ or $D^0 \rightarrow \pi e\nu$, and doubly suppressed decays such as $D^0 \rightarrow K^+\pi^-$, owing to the smallness of the Cabibbo angle. We compare the predictions of Tables 3 and 4 to the data in Section 4.

3.3 *Beyond Charm*

Since the 1977 discovery of the Υ by Lederman and collaborators (Herb et al 1977), considerable indirect evidence has accumulated for the existence of another quark, the b quark, which is much more massive ($\sim$5 GeV) than the c quark. The simplest extension of the Weinberg-Salam-GIM model involves the introduction of a new quark doublet (t, b) where the b quark is the lighter quark with charge $-1/3$ responsible for the Υ, and the t quark is the heavier quark of charge 2/3 whose presence is still speculation.

The model of Kobayashi & Maskawa (1973) incorporates this new doublet into the weak current by proposing that the u, c, t quarks mix with the d, s, b quarks via a general 3×3 unitary mixing matrix. Hence

$$J_h^\mu = (\bar{u}\bar{c}\bar{t})\gamma^\mu(1-\gamma^5)M \begin{pmatrix} d \\ s \\ b \end{pmatrix},$$

where

$$M = \begin{pmatrix} C_1 & S_1C_3 & S_1S_3 \\ -S_1C_2 & C_1C_2C_3 - S_2S_3e^{i\delta} & C_1C_2S_3 + S_2C_3e^{i\delta} \\ S_1S_2 & -C_1S_2C_3 - C_2S_3e^{i\delta} & -C_1S_2S_3 + C_2C_3e^{i\delta} \end{pmatrix}$$

and

$$C_i = \cos\,\theta_i, \quad S_i = \sin\,\theta_i, \quad i = 1, 2, 3.$$

Within this picture the single Cabibbo angle present in Equation 2 (now called θ_1) is augmented by two new angles θ_2 and θ_3 and a phase factor ($e^{i\delta}$). We note that the original GIM predictions for the weak transitions between the u, d, s, and c quarks are recovered in the limit $\theta_2, \theta_3 \rightarrow 0$.

Nonzero values for the new mixing angles will affect the weak decays of conventional particles as well as the decays of charmed particles. In particular, we see that the Cabibbo-suppressed processes $D^0 \rightarrow \pi^+\pi^-$ and $D^0 \rightarrow K^+K^-$ will no longer have identical mixing-angle factors as was the case in the GIM model. Within the sector of older phenomena, we see that the predictions for processes involving $u \rightleftarrows d$ transitions such as neutron beta decay still involve only the original Cabibbo angle θ_1. We can thus retain the value $\theta_1 = 13.2 \pm 0.5°$ as measured by comparing the rate of $n \rightarrow pe\bar{\nu}$ to that for $\mu \rightarrow e\nu\bar{\nu}$ (Nagels et al 1976).

The amplitude for u $\rightleftarrows$ s transitions, however, acquires a new mixing factor to become $\sin\theta_1 \cos\theta_3$. The success of the original Cabibbo model applied to processes such as $\Lambda \to pe\bar{\nu}$ and $K \to \pi e\nu$ thus limits the Kobayashi-Maskawa correction to $|\cos\theta_3| > 0.87$ (Schrock & Wang 1978). Less direct theoretical arguments based on the possible contributions of the t quark to the $K_L - K_S$ mass difference set a limit of $\theta_2 < 30°$ (Harari 1977). These limits, when applied to the D decays could cause deviations from the GIM predictions by factors of two, although the basic pattern of enhanced vs suppressed decays would be expected to hold.

4 D MESONS

4.1 *Discovery of D Mesons*

D mesons were first observed by the SLAC-LBL Mark I collaboration at SPEAR in 1976 using data collected at center-of-mass energies ranging from 3.9 to 4.6 GeV (Goldhaber et al 1976, Peruzzi et al 1976). The first D decay modes observed included $D^0(1863) \to K^-\pi^+$, $K^-\pi^+\pi^+\pi^-$ and $D^+(1868) \to K^-\pi^+\pi^+$. A substantial body of evidence soon accumulated linking the narrow enhancements in the $K^-\pi^+$, $K^-\pi^+\pi^+\pi^-$, and $K^-\pi^+\pi^+$ invariant mass distributions to the D^0, D^+ charmed isodoublet. We outline the evidence below.

4.1.1 EVIDENCE FOR ASSOCIATED PRODUCTION Both the $D^0(1863)$ and $D^+(1868)$ are produced in final states containing a $D\bar{D}$ pair, as one would expect for particles containing a quantum number conserved by the electromagnetic interaction. This is evidenced by two observations: (*a*) No D's are observed in e^+e^- annihilation at either the ψ or ψ', although a substantial amount of Mark I data was collected at these resonances. The ψ' in particular is located just below $D\bar{D}$ threshold. (*b*) No evidence is seen in recoil mass spectra against the D^0 or D^+ system for events with recoil masses smaller than the D candidate mass of 1863 MeV.

4.1.2 EVIDENCE FOR WEAK HADRONIC DECAYS Particles carrying a quantum number conserved in the strong or electromagnetic interaction must decay weakly. This is evidenced by five observations:

Narrow width All reported sightings of the D^0 and D^+ into inclusive decay modes report an observed width that is consistent with experimental mass resolutions. The data with the best mass resolution set a limit $\Gamma_{D^0,D^+} < 2$ MeV.

Parity violation The observation of parity violation in the decays $D^0 \to K^-\pi^+$ and $D^+ \to K^-\pi^+\pi^+$ is reminiscent of the $\theta - \tau$ problem for K decays of the 1950s, which led to the hypothesis of a parity-violating weak interaction. Because of the small mass difference between the $D^0(1863)$ and $D^+(1868)$ it is natural to assume that they are members of the same isodoublet and hence have the same spin and parity. The $D^0 \rightarrow K\pi$ decay final state must have a natural spin parity of $0^+, 1^-, 2^+, \ldots$. A study of the $D^+ \rightarrow K^-\pi^+\pi^+$ Dalitz plot (Wiss et al 1976) rules out D^+ final state spin-parity assignments of $J^P = 1^-$ and 2^+, while 0^+ is forbidden by angular momenta considerations for three pseudoscalars. Hence, neglecting possible higher spin assignments for the D system, one is left with a contradiction most naturally resolved by assuming that the D^0 and D^+ decay through the parity-violating weak decay.

An exotic final state Because of the $\Delta s = \Delta c$ selection rule for the Cabibbo-favored hadronic decays of charmed particles, the D^+ must always decay hadronically into final states of positive charge and negative strangeness. Within the context of the conventional three-quark model such final states are labeled as exotic because they cannot be constructed from quark-antiquark pairs using the u, d, or s quarks. If the $K^-\pi^+\pi^+$ enhancement at 1863 MeV that we implicitly associate with the weak decay $D^+ \rightarrow K^-\pi^+\pi^+$ actually represented the strong decay of a noncharmed meson, that meson would be exotic and would be the first compelling observation of such a state. The observation of the $I_z = 3/2$ $K^-\pi^+\pi^+$ enhancement when combined with the nonobservation of an enhancement in $K^-\pi^+\pi^-$, the $I_z = -1/2$ brother, rules out such an interpretation, however.

Semileptonic decay The observation of an appreciable semileptonic branching ratio, as discussed below, again suggests that D's do not decay strongly.

Evidence for a GIM pattern of decays The Cabibbo-suppressed decay modes $D^0 \rightarrow K^-K^+$ and $\pi^+\pi^-$ have recently been observed in the SLAC-LBL Mark II detector at SPEAR. The dominant two-body decay mode, however, is $D^0 \rightarrow K^-\pi^+$. Hence the D^0 is observed to decay into both strange and nonstrange final states, which implies that the decay mechanisms do not conserve strangeness and are thus weak. This pattern of decay modes is characteristic of charm in the GIM model.

Much of the information presented above linking the early $(K\pi)^0$, $(K3\pi)^0$, and $(K2\pi)^+$ signals to the D^0D^+ charmed doublet was performed at the Mark I detector using data collected from 3.9–4.6 GeV with an emphasis on the 4.028-GeV resonance region. Although charmed mesons are copiously produced near 4.028 GeV, the considerable structure in the total cross section near this enhancement precludes a clean Breit-Wigner fit to determine the cross section beneath the peak. Such information would have been useful, for, as we discuss below, it provides a means of measuring the absolute branching fractions for the D decay modes.

4.2 *Hadronic Decays of D Mesons*

4.2.1 THE $\psi(3770)$ RESONANCE After the Mark I detector at SPEAR stopped taking data, the $\psi(3770)$ or ψ'' resonance was discovered in the lead glass wall (LGW) and direct electron counter (DELCO) experiments (Rapidis et al 1977, Bacino et al 1978). Comparing the width of the ψ' ($\Gamma = 228$ keV) with that of the ψ'' ($\Gamma = 25$ MeV) we note that the effect of the OZI suppression at the ψ' is no longer present at the ψ'', since it lies above $D\bar{D}$ threshold. Furthermore the ψ'' lies below DD^* threshold; it can only decay into D's via the process $D^0\bar{D}^0$ or D^+D^-.

The ψ'' has been studied extensively in the LGW and DELCO experiments and recently again in the SLAC-LBL Mark II experiments (Lüth 1979, Schindler et al 1980). Figure 6 shows the *R* distribution

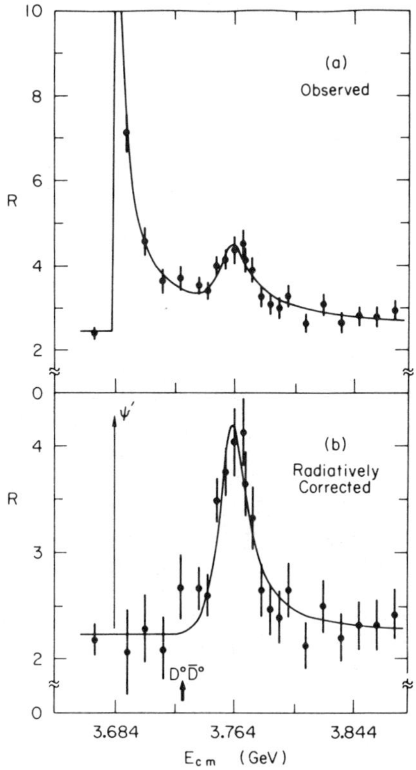

Figure 6 The value of $R \equiv \sigma_{\text{had}}/\sigma_{\text{QED}}$ in the vicinity of the ψ'' obtained by the Mark II collaboration (Schindler et al 1980); (*a*) before and (*b*) after radiative correction using the technique of Jackson & Scharre (1975). The curve is a fit of the data to Equation 3.

observed in the Mark II experiment, where R is the ratio of the observed hadronic cross section to the theoretical QED μ pair cross section σ_{QED}. The latter is obtained from calibration against observed Bhabha pairs. Figure 6*a* gives the R distribution where the $\tau^+\tau^-$ cross section has been subtracted. Figure 6*b* gives the R distribution after the radiative tails from the J/ψ and ψ' have been subtracted as well. The errors shown are statistical. The resonance is fitted to a p-wave Breit-Wigner expression (Barbaro-Galtieri 1968) with an energy-dependent total width $\Gamma_{\mathrm{tot}}(E_{\mathrm{cm}})$ that takes account of closeness to the different $D^0\bar{D}^0$ and D^+D^- thresholds. The explicit fitting function employed is

$$R(E_{\mathrm{cm}}) = \frac{1}{\sigma_{\mathrm{QED}}}\frac{3\pi}{M^2}\frac{\Gamma_{\mathrm{ee}}\Gamma_{\mathrm{tot}}(E_{\mathrm{cm}})}{(E_{\mathrm{cm}}-M)^2+\frac{1}{4}\Gamma^2_{\mathrm{tot}}(E_{\mathrm{cm}})} \qquad 3.$$

and

$$\Gamma_{\mathrm{tot}}(E_{\mathrm{cm}}) \propto \frac{p_+^3}{1+(rp_+)^2}+\frac{p_0^3}{1+(rp_0)^2}$$

where p_+ (p_0) is the momentum of the pair-produced D^+ (D^0) and r the interaction radius. The quantities M (the resonance mass) and Γ_{ee} (the partial width to electrons) were additional free parameters. The fit is not sensitive to r, which was fixed at 2.5 Fermi. Table 5 summarizes the results of this fit for the various experiments. We note that the Mark II results are consistent with those of DELCO and the LGW except for a shift in the central mass that is 6–8 MeV lower than previous values. In addition the Mark II value for the width of the decay into e^+e^- of 276 ± 50 eV lies between the earlier reported values.

From theoretical arguments (Eichten et al 1975, Lane & Eichten 1976, Gottfried 1978) the ψ'' is believed to be a $^3D_1$ state of charmonium which is, however, mixed with the 2^3S_1 state, dominant in the ψ'. The relatively large Γ_{ee} value gives an estimate for this mixing angle of $20.3 \pm 2.8°$.

4.2.2 CHARMED-MESON BRANCHING RATIOS Without knowledge of the total D production cross section it is difficult to measure the branching ratio

Table 5 Measurements of the $\psi(3770)$ resonance parameters

Experiment	Mass (MeV/c^2)	Γ_{tot} (MeV)	Γ_{ee} (eV)	ΔM[a] (MeV/c^2)
DELCO	3770 ± 6	24 ± 5	180 ± 60	86 ± 2
LGW	3772 ± 6	28 ± 5	345 ± 85	88 ± 3
Mark II	3764 ± 5	24 ± 5	276 ± 50	80 ± 2

[a] ΔM is the mass difference between the $\psi(3684)$ and $\psi(3770)$.

for a given exclusive final state. One can, however, readily measure the product $\sigma \cdot \mathrm{B}$ by counting the number of events observed in a given channel and dividing by the acceptance and luminosity. Charm production at the ψ'' offers the considerable advantage that $\sigma(\mathrm{D\bar{D}})$ can be determined if one is willing to assume two things: (*a*) The ψ'' is a state of definite isospin (0 or 1); this allows a prediction of the $\mathrm{D}^0/\mathrm{D}^+$ production ratio, namely $\sigma(\mathrm{D}^0)/\sigma(\mathrm{D}^+) \simeq p_0^3/p_+^3$ as expected for p-wave production. This reflects the D^0 and D^+ mass difference (1863.3 MeV and 1868.3 MeV respectively). (*b*) The ψ'' decays nearly entirely into $\mathrm{D\bar{D}}$ ($\sim 99\%$). This is based on the $\Gamma_{\mathrm{tot}}(\psi')$ to $\Gamma_{\mathrm{tot}}(\psi'')$ ratio ($\sim 1/100$); i.e., the OZI-suppressed portion of the ψ'' decay width is of the same magnitude as the $\Gamma_{\mathrm{tot}}(\psi')$.

Assumption (*b*) has been tested with limited statistical accuracy using events where both members of a $\mathrm{D\bar{D}}$ pair are observed decaying into exclusive final states. The D branching ratio obtained by this technique can be used to compute the absolute D production cross section at the ψ'' (Schindler 1979).

Using these assumptions and their fit to the radiatively corrected ψ'' resonance shape, the LGW collaboration determined that the $\mathrm{D}^0\mathrm{\bar{D}}^0$ and $\mathrm{D}^+\mathrm{D}^-$ cross sections averaged over their particular set of ψ'' running energies were 11.5 ± 2.5 nb and 9.1 ± 2.0 nb, respectively (Peruzzi et al 1977). Most of this running was done within 2 MeV of their nominal ψ'' mass. The Mark II collaboration, on the other hand, collected 49,000 hadronic events at an energy 7 MeV above their nominal ψ'' mass. They find that the total $\mathrm{D}^0\mathrm{\bar{D}}^0$ and $\mathrm{D}^+\mathrm{D}^-$ cross sections for their running conditions are 8.0 ± 1.0 (± 1.2) nb and 6.0 ± 0.7 (± 1.0) nb, respectively (Schindler 1979), where the numbers in parentheses are the systematic error estimates.

The beam-constrained mass distributions obtained for several D^0 and D^+ final states are shown in Figure 7 for the LGW data and Figures 8–10 for the Mark II data. The beam-constrained mass M_b is calculated using the relationship

$$M_\mathrm{b} = (E_\mathrm{b}^2 - p^2)^{1/2},$$

where E_b is the storage-ring single-beam energy and p is the momentum of the D^0, D^+ candidate as determined by the magnetic detector. Such a technique implicitly assumes D^0's and D^+'s are pair produced at the ψ'' and hence have exactly half the total center-of-mass energy. In the spirit of this assumption, some background is eliminated by histograming only events with a detector-measured energy within 50 MeV of the beam energy. Use of the beam-constrained mass offers unparalleled resolution and background rejection.

Fits to the signals shown in the above figures have been used to determine $\sigma \cdot B$ and B for various decay modes. This information is summarized in Table 6.

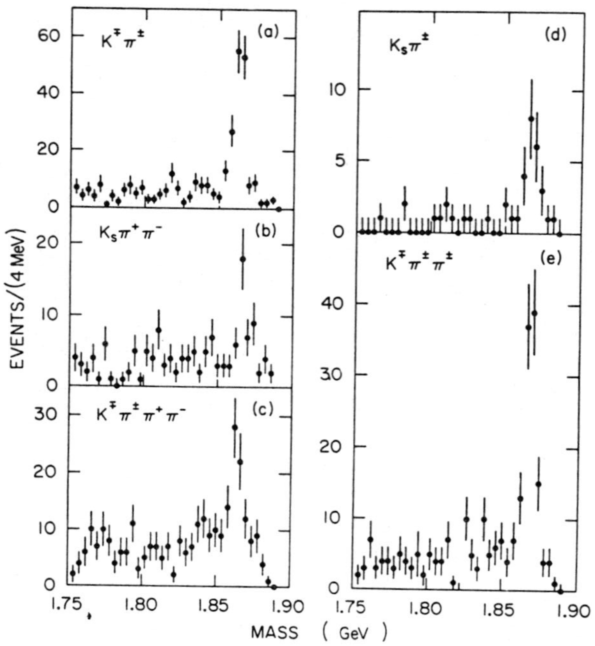

Figure 7 The beam-energy-constrained mass distributions for the indicated D^0, D^+ decay modes obtained by the LGW collaboration at the ψ''.

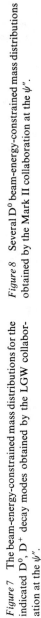

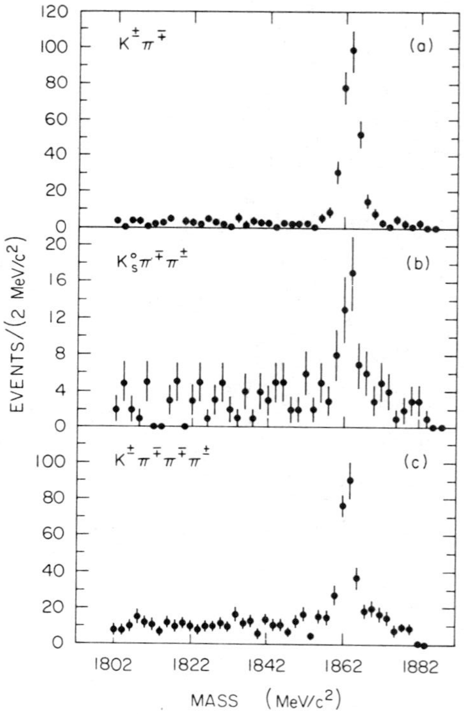

Figure 8 Several D^0 beam-energy-constrained mass distributions obtained by the Mark II collaboration at the ψ''.

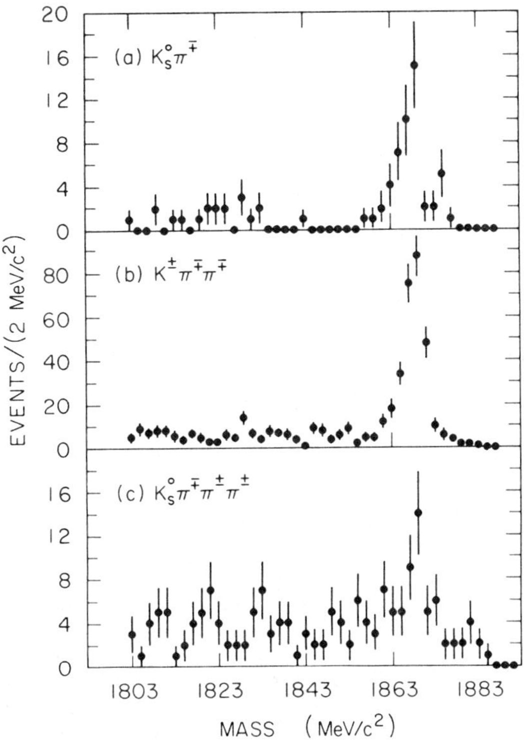

Figure 9 Several D^+ beam-energy-constrained mass distributions obtained by the Mark II collaboration at the ψ''.

Table 6 D meson branching ratios

Mode	LGW B (%)	Mark II B (%)
$K^-\pi^+$	2.2 ± 0.6	3.0 ± 0.6
$\bar{K}^0\pi^0$	—	2.2 ± 1.1
$\bar{K}^0\pi^+\pi^-$	4.0 ± 1.3	3.8 ± 1.2
$K^-\pi^+\pi^0$	12.0 ± 6.0	8.5 ± 3.2
$K^-\pi^+\pi^+\pi^-$	3.2 ± 1.1	8.5 ± 2.1
$\pi^+\pi^-$	—	0.09 ± 0.04
K^+K^-	—	0.31 ± 0.09
$\bar{K}^0\pi^+$	1.5 ± 0.6	2.3 ± 0.7
$K^-\pi^+\pi^+$	3.9 ± 1.0	6.3 ± 1.5
$\bar{K}^0\pi^+\pi^0$	—	12.9 ± 8.4
$\bar{K}^0\pi^+\pi^+\pi^-$	—	8.4 ± 3.5
$K^-\pi^+\pi^+\pi^+\pi^-$	—	<4.1[a]
$\bar{K}^0K^+$	—	0.5 ± 0.27

[a] 90% confidence limit.

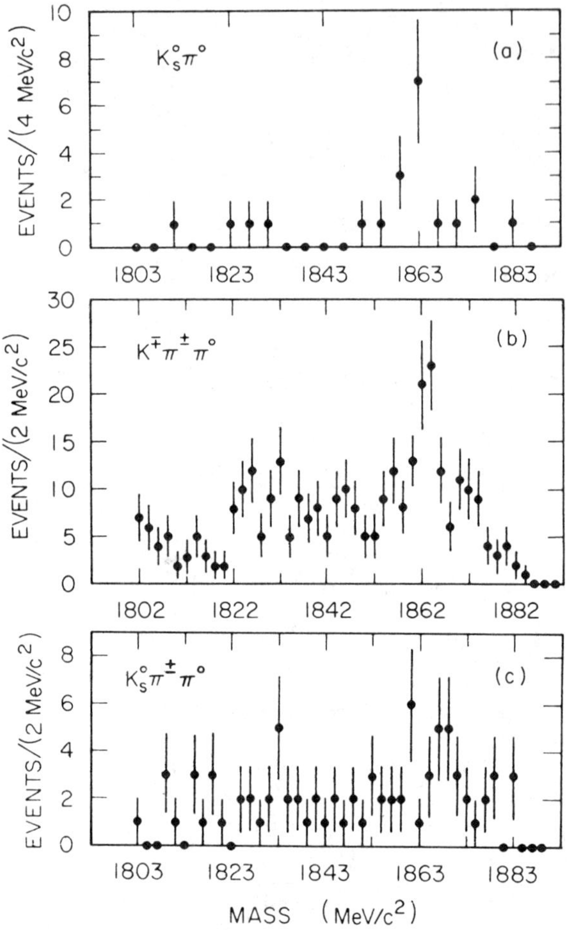

Figure 10 Several D beam-energy-constrained mass distributions for decays involving π^0's obtained by the Mark II collaboration at the ψ''.

4.2.3 CABIBBO-SUPPRESSED DECAY MODES An intrinsic part of the GIM mechanism for charm is the prediction that in addition to the principal (Cabibbo-favored) D decay modes, which lead to K^- or $\bar{K}^0$ in the final states, there are also Cabibbo-suppressed modes producing zero strangeness final states. The Cabibbo-favored and -suppressed modes for D^0 two-particle final states are illustrated by the quark diagrams in Figure 11, where θ_A is the familiar Cabibbo angle and θ_B is a new angle that in the four-quark model is associated with the flavor mixing of charmed quarks. The

GIM assumption is that $\theta_A = \theta_B$. Experimentally one can independently measure these angles using

$$\tan^2\theta_A = \frac{\Gamma(D^0 \to K^-K^+)}{\Gamma(D^0 \to K^-\pi^+)} \quad \text{and} \quad \tan^2\theta_B = \frac{\Gamma(D^0 \to \pi^-\pi^+)}{\Gamma(D^0 \to K^-\pi^+)}.$$

The above expressions neglect the phase space corrections due to the K, π mass difference, which will raise the $\pi^+\pi^-$ rate by 7% and lower the K^+K^- rate by 8%.

Figure 12 shows the Mark II $\pi^-\pi^+$, $K^-\pi^+$, and K^-K^+ invariant mass distributions for two-particle combinations with momentum within 30 MeV/c of the expected D pair momentum at the ψ''. Aside from the signals in the three channels at the D mass one notes kinematic reflections shifted by about ± 120 MeV/c^2 from the D mass due to $\pi \leftrightarrow K$ misidentifications. A fit to the data yields 235 ± 16 $K^\mp\pi^\pm$ events, 22 ± 5 K^+K^- events, and 9 ± 3.9 $\pi^+\pi^-$ events. After correcting for the relative efficiencies one obtains

$$\frac{\Gamma(D^0 \to K^-K^+)}{\Gamma(D^0 \to K^-\pi^+)} = 0.113 \pm 0.03 \qquad 4.$$

and

$$\frac{\Gamma(D^0 \to \pi^-\pi^+)}{\Gamma(D^0 \to K^-\pi^+)} = 0.033 \pm 0.015, \qquad 5.$$

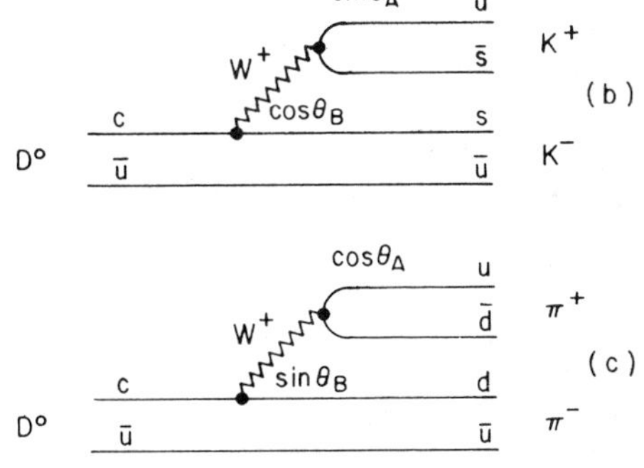

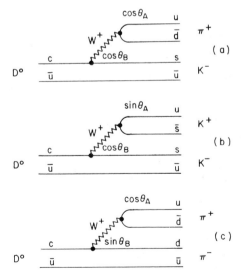

Figure 11 Quark diagrams for D^0 decays into two charged particles.

where the quoted errors include systematic effects. The results clearly demonstrate the existence of the Cabibbo-suppressed decay modes of roughly the expected magnitude, $\tan^2\theta \simeq 0.05$, although the $\pi\pi$ ratio is lower by about one standard deviation and the KK ratio is higher by about two standard deviations.

We note that the discrepancy between Equation 4 and the pre-charm measurements of the Cabibbo angle cannot be explained by the presence of additional mixing angles in the Kobayashi-Maskawa model. Hence the discrepancy (if not statistical in origin) implies a violation of SU(3) invariance due to unknown dynamical effects. It is thus premature to use Equation 5 as a measure of the "new" Cabibbo angle (Abrams et al 1979).

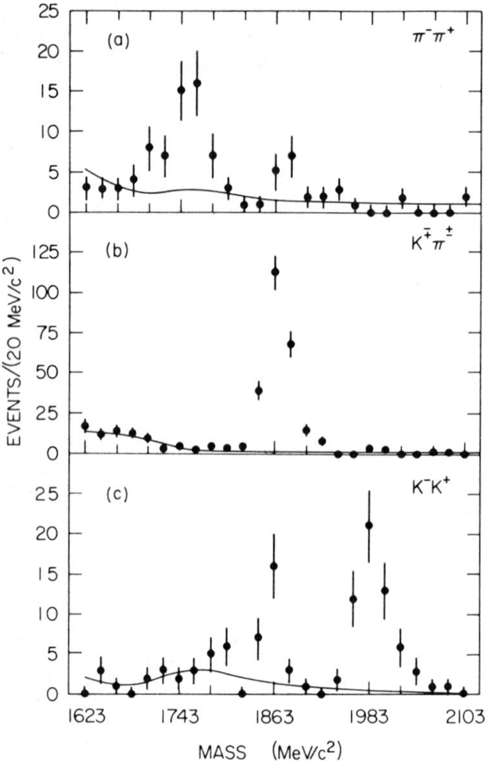

Figure 12 Evidence for two-body Cabibbo-suppressed decay modes obtained by the Mark II collaboration at the ψ''. Candidates are required to have momenta within 30 MeV of that expected for D pair production. Both (*a*) and (*c*) show prominent reflection peaks not centered at the D^0 mass because of π/K misidentification by the time-of-flight system. The solid curves are background estimates.

4.3 *Semileptonic Decay Modes*

The prompt leptons created in the semileptonic decay of charmed mesons have been used to estimate the rate for hadronic charm production, as well as to trigger detectors looking for the decay of charmed particles into exclusive hadronic final states. Because it is possible to estimate theoretically the D semileptonic width [i.e. $\Gamma(D \to Ke\nu)$], a measurement of the semileptonic branching ratio can be used to estimate the D lifetime. At the time of this writing, emulsion as well as precision bubble and streamer chamber lifetime studies are being performed. These experiments can observe D's traveling finite distances before decaying, directly measure their lifetimes, and thus check current theoretical ideas on D semileptonic decays (Voyvodic 1979).

Because D semileptonic decays produce a neutrino in the final state, their study in e^+e^- annihilation invariably requires the measurement of inclusive final state electron or muon rates at center-of-mass energies where charmed mesons are known to be copiously produced. Inclusive electrons rather than muons are generally studied since the charged lepton in $D \to \bar{K}e^+\nu$ or $D \to \bar{K}\mu^+\nu$ has a momentum spectrum that peaks near 500 MeV for D's produced near threshold. Hence the muons would be produced with momenta too low to be separated clearly from pions using conventional hadron filters.

4.3.1 THE AVERAGE SEMILEPTONIC BRANCHING RATIO Several basic techniques for extracting the D semileptonic branching ratio have been discussed in the literature. The simplest method (experimentally) estimates the branching fraction $B(D \to eX) \equiv \Gamma(D \to eX)/\Gamma(D \to \text{all})$ from the ratio of the total electron to the total charmed meson inclusive cross section:

$$B(D \to eX) = \frac{\sigma(e^+e^- \to e^\pm + \text{hadrons})}{\sigma(e^+e^- \to DX)}. \qquad 6.$$

This method, the first historically (Braunschweig et al 1976, Burmester et al 1976), continues to provide the best statistical information on the D semileptonic branching ratio. It suffers from inevitable systematic problems, however. In evaluating the numerator, one must be careful to exclude (or correct for) contributions from heavy lepton decays or electromagnetic processes. The most common way of minimizing this contamination is to demand that the electrons be accompanied by two or more charged hadrons. Since the τ heavy lepton is known to decay predominantly into final states containing leptons and single hadrons, the multihadronic background due to $e^+e^- \to \tau^+\tau^-$ production are expected to lie at about the 25% level (Perl 1980, Barbaro-Galtieri 1978) and can be subtracted using the measured τ branching fraction and computable production cross sections.

Several techniques, all subject to various systematic uncertainties, have been used to estimate the denominator of Equation 6. Extraction of $\sigma(e^+e^- \rightarrow DX)$ is relatively straightforward at the ψ'' since, as discussed earlier, it is natural to assume that this resonance decays exclusively into D's via

$$\psi'' \rightarrow (56 \pm 3)\% \ D^0\bar{D}^0 \rightarrow (44 \pm 3)\% \ D^+D^-. \qquad 7.$$

The fractions used in Equation 7 follow from the assumption of equal D^0, D^+ production corrected for p-wave threshold factors. Since it is impossible to separate the semileptonic decays of the D^0 from those of the D^+ without using tagged events or measuring the dielectron rate as well as the single electron rate, one in effect measures a weighted average of the D^0 and D^+ semileptonic branching ratios with weights given by Equation 7.

At center-of-mass energies above the ψ'' it is sometimes possible to extract $\sigma(e^+e^- \rightarrow DX)$ by counting hadronic D decays into exclusive final states such as $D^0 \rightarrow K^-\pi^+$, $D^+ \rightarrow K^-\pi^+\pi^+$ and dividing by the hadronic branching ratios measured for these states at the ψ''. When this is not possible because of statistical limitations, or in data predating the ψ'', the D inclusive cross section can be estimated at a given center-of-mass energy via

$$\sigma(e^+e^- \rightarrow DX) = (R - R_{\text{old}}) \times \sigma_{\text{QED}},$$

where R is the τ-lepton-corrected ratio of hadrons to μ pairs, and R_{old} is the value of R below the $\psi(3095)$, which presumably represents the cross-section contribution of the "old" u, d, and s quarks. Being cognizant of the possible contributions to both the numerator and denominator of Equation 6 from other charmed objects such as the F^+ and charmed baryons, some authors refer to the semileptonic branching ratios obtained through this technique as the average "charm" semileptonic branching ratio rather than the D semileptonic branching ratio.

Table 7 summarizes the average semileptonic branching ratios derived through measurements of the single electron rate by various groups. Figures 13–15 give the average semileptonic branching ratio vs E_{cm} for the DASP, LGW, and DELCO data (note there is a factor of two difference between the quantities plotted in Figure 15 and the other two). All data sets show a remarkable constancy in the value for the average branching ratio even though the relative contribution of the D^0, D^+, F^+, and Λ_c must be changing as a function of energy. At $E_{\text{cm}} = 4.028$ GeV, for example, measurements of exclusive final states show that $70 \pm 10\%$ of D's are neutral compared to the $56 \pm 3\%$ neutral-D fraction assumed at the ψ'' (Rapidis 1979).

Table 7 The branching ratio for D → $e\nu X$

Experiment	E_{cm} (GeV)	Branching ratio (%)
DASP Wiik & Wolf 1978	3.99 → 4.08	8.0 ± 2.0
LGW Feller et al 1978	ψ''	7.2 ± 2.8
DELCO Kirkby 1979	ψ''	8.0 ± 1.5
Mark II Lüth 1979	ψ''	9.8 ± 3.0
	Average	8.0 ± 1.1

4.3.2 THE INCLUSIVE ELECTRON MOMENTUM SPECTRUM Aside from studying the rate for charm-associated inclusive electron production, it is interesting to study the momentum distribution. Figure 16 shows the inclusive momentum distribution obtained at the ψ'' by DELCO for events with ≧2 additional charmed particles. As the curves of the figure show, the momentum spectrum is consistent with the distribution expected for a mixture of D → $Ke\nu$, $K\pi e\nu$, and $\pi e\nu$ semileptonic decays. The $Ke\nu$ and $K\pi e\nu$ contributions dominate over the Cabibbo-suppressed $\pi e\nu$ mode and

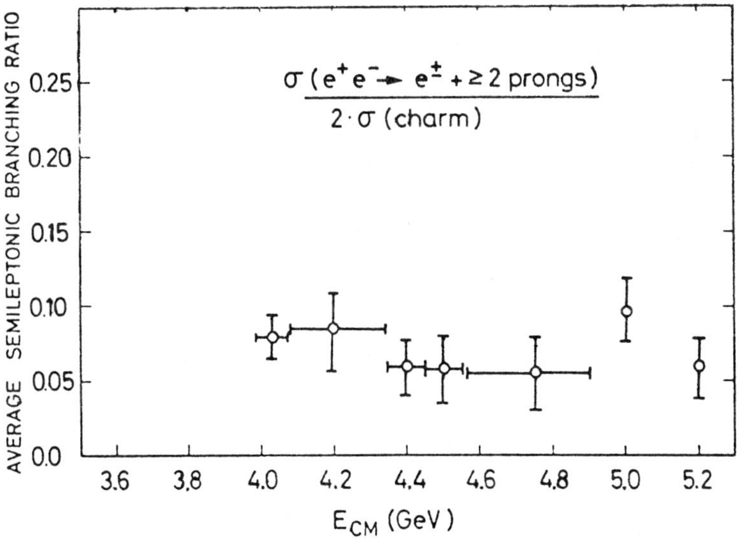

Figure 13 The average semileptonic branching ratio for charmed hadrons as a function of energy. The error bars are statistical only (from DASP data).

appear to be roughly equal. The exact ratio of $K\pi e\nu$ to $Ke\nu$ depends sensitively on how much of the $K\pi$ contribution comes from primary K*(890) formation.

4.3.3 TAGGED EVENTS A second, potentially much cleaner, technique for extracting the D semileptonic branching ratio involves counting the number of events containing an electron recoiling against a "tagging"

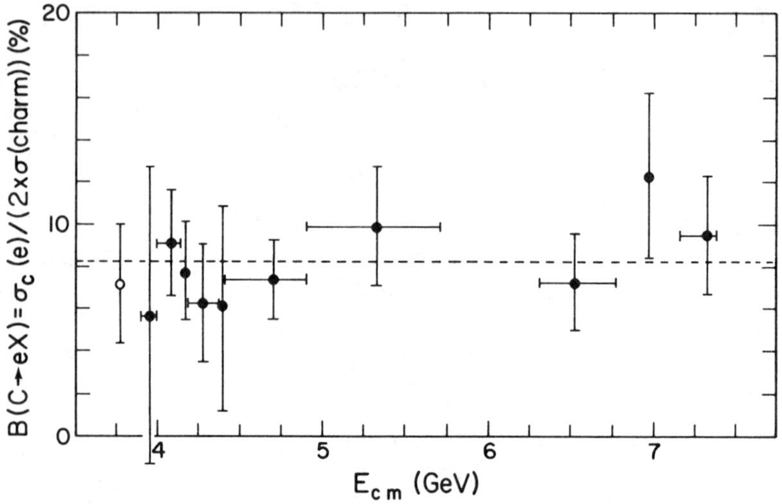

Figure 14 The branching fraction for charmed-particle decay into an electron plus additional particles as a function of energy (LGW collaboration) The dashed line indicates the average value of the ratio for $3.9 < E_{cm} < 7.4$ GeV.

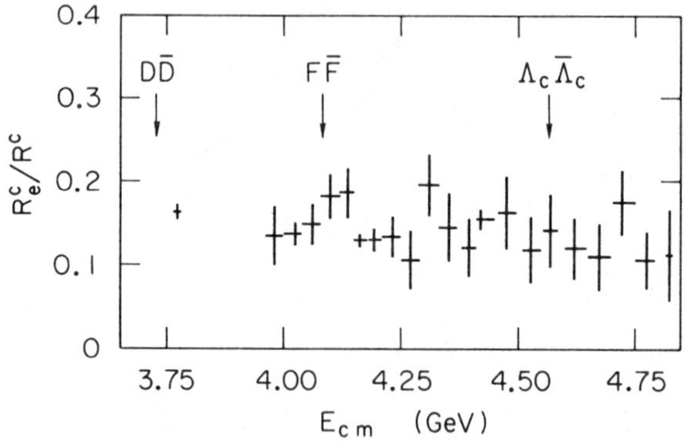

Figure 15 The ratio of R_e^c to the total charm hadronic cross section, R^c, measured by DELCO. The ratio R_e^c/R^c is equal to $2B_e(1-B_e)$, where the branching ratio $B_e = B$(charm $\rightarrow e\nu X$) (Kirkby 1979).

$D^0 \to K^-\pi^+$, $K^-\pi^+\pi^+\pi^-$ or $D^+ \to K^-\pi^+\pi^+$ candidate produced at the ψ''. Because the ψ'' cannot decay into final states containing a D*, one is guaranteed that D^-'s are always produced against tagged D^+'s and $\bar{D}^0$'s are always produced against tagged D^0's. Hence the tagging technique allows one to measure separately the D^+ and D^0 semileptonic branching ratio. Owing to the smallness of tagging branching ratios, however, the number of tagged electron events is much smaller than the number of inclusive electron events, and such studies suffer from larger statistical errors.

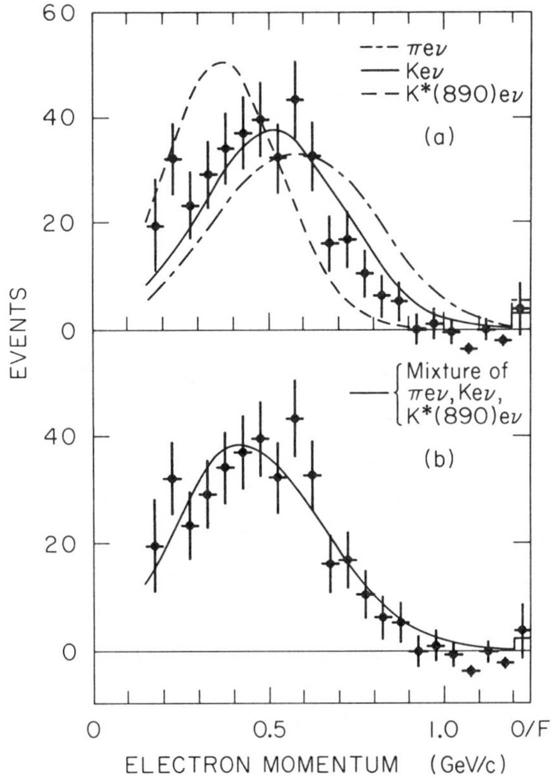

Figure 16 The electron momentum spectrum from D decays at the ψ'', measured by DELCO. The curves have been fitted to the data below 1 GeV/*c* and correspond to the following hypotheses:

(*a*) $D \to \pi e\nu$ (*dot-dashed curve*, χ^2/dof = 80.9/16), $D \to Ke\nu$ (*solid curve*, χ^2/dof = 23.4/16), $D \to K^*(890)\,e\nu$ (*dashed curve*, χ^2/dof = 53.8/16).

(*b*) Contributions from $D \to Ke\nu$ (55%), $D \to K^*e\nu$ (39%), and $D \to \pi e\nu$ (6%) with χ^2/dof = 11.2/15 (Kirkby 1979).

Table 8 summarizes the Mark II collaboration's information on the D^0 and D^+ semileptonic branching ratio (Lüth 1979). Wrong-sign e-tag events were used to compute the background level due to an $\sim 6\%$ π/e misidentification. We see from Table 8 that the D^0 and D^+ semileptonic branching ratios are unequal at about a two-standard-deviation level of significance.

We should also mention that the use of tagged events at the ψ'' has provided important information on the kaon content of D decays, and the D decay final state multiplicity distribution (Feller 1979, Schindler 1979, Lüth 1979, Vuillemin et al 1978).

4.3.4 TWO-ELECTRON FINAL STATES The above result is corroborated by the DELCO collaboration's analysis of the two-electron vs one-electron rate at the ψ'' (Kirkby 1979). In the limit of perfect acceptance, the single (N_1) and double (N_2) electron event rates due to D decays are related to the neutral (B_0) and charged (B_+) semileptonic branching ratios by

$$\begin{aligned} N_1 &= 2N_0B_0(1-B_0)+2N_+B_+(1-B_+) \\ N_2 &= N_0B_0^2+N_+B_+^2, \end{aligned} \qquad 8.$$

where N_0 and N_+ are the number of $D^0\bar{D}^0$ and D^+D^- events produced. One sees from Equation 8 that in the limit of small branching ratios a measurement of N_1 determines essentially a line in the B_0 vs B_+ plane whereas a measurement of N_2 determines an elliptical arc. One thus expects that a simultaneous measurement of N_2 and N_1 will lead to two ambiguous solutions for B_0 and B_+. Figure 17 indicates the experimental regions in the B_+ vs B_0 plane that are consistent to within one standard deviation with the data presented in Table 9. The data have been corrected for electron detection efficiency using the two extreme models that $D \to Ke\nu$ (Figure 17*a*) or $D \to K^*e\nu$ (Figure 17*b*). Under either assumption it appears that $B_0 \gg B_+$ or $B_+ \gg B_0$. DELCO uses the K_S content in the two-electron events to distinguish between these two possibilities. Although both D^0's

Table 8 Semileptonic decays of D^+ and D^0, Mark II data at the ψ''

Decay mode	Tags (no.)	Electrons (no.)	Background	$B(\%)$
$D^+ \to e^+$	295 ± 18	38	15 ± 1	16.8 ± 6.4
$\to e^-$		4	3.9 ± 0.5	
$D^+ \to e^+$	480 ± 23	36	19 ± 1	5.5 ± 3.7
$\to e^-$		19	12 ± 1	

Table 9 The DELCO multiprong electron data sample at the ψ''

Event description	Event topology		
	1 electron	2 electrons	2 electrons + "V" (K_s^0)
Observed	1416	21	8
Background	692	4.6	1.8
Charm signal	724	16.4	6.2

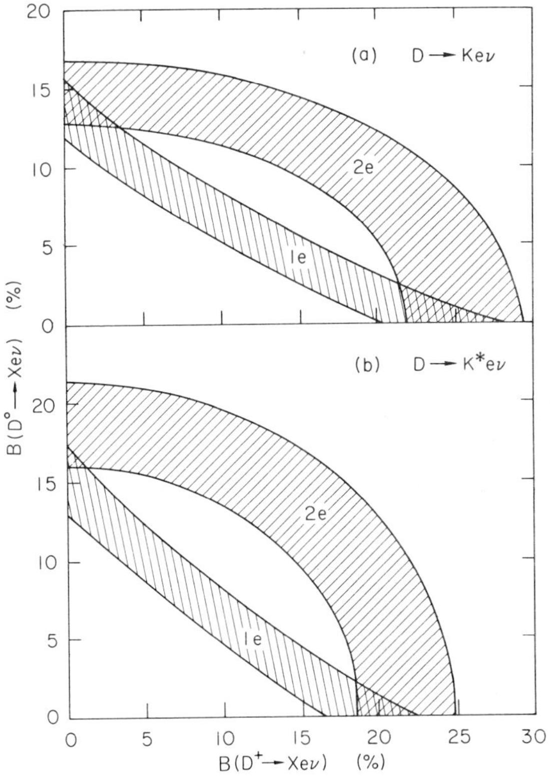

Figure 17 The allowed solutions for the D^0 and D^+ semileptonic branching ratios in the DELCO 1e and 2e multiprong data at the ψ''. The shaded regions, which correspond to $\pm 1\sigma$ limits, are plotted for two extreme assumptions of the detection efficiencies: (*a*) all $D \rightarrow K e \nu$ and (*b*) all $D \rightarrow K^* e \nu$ (Kirkby 1979).

and D^+'s can decay into K_S, the D^0 must do so via the decay sequence $D^0 \to (K\pi)^- e^+ \nu$ where the isodoublet $K\pi$ system decays into a charged kaon 2/3 of the time ($I = 1/2$) and a neutral kaon only 1/3 of the time. The D^+, on the other hand, can produce K_S via both $D^+ \to K_S e^+ \nu$ and $K_S \pi^0 e^+ \nu$. In fact, a rather large fraction of the two-electron events (8 of 16.4) have a K_S, which suggests the solution $B_+ \gg B_0$. Combining the information on the single-electron and two-electron rate at the ψ'' with the K_S content of the two-electron events, DELCO finds the D^+ and D^0 semileptonic branching ratios to be $24 \pm 4\%$ and $< 5\%$ (95% confidence level), respectively, in agreement with the trend and values of the Mark II data of Table 8.

4.3.5 THE D^0, D^+ LIFETIMES Because of the isosinglet character of the Lagrangian responsible for the Cabibbo-favored D semileptonic decay, one expects that $\Gamma(D^+ \to Km\pi e^+ \nu) = \Gamma(D^0 \to Km\pi e^+ \nu)$ for any number m of pions (see for example Pais & Treiman 1977). Hence, neglecting the Cabibbo-suppressed decay modes one has the relation $B_0 \Gamma(D^0 \to \text{all}) = B_+ \Gamma(D^+ \to \text{all}) = \Gamma(D \to KeX)$. In terms of lifetimes, one thus obtains $\tau(D^+)/\tau(D^0) > 4$ (95% confidence level) for the DELCO data, and $\tau(D^+)/\tau(D^0) = 3.1^{+4.1}_{-1.3}$ for the Mark II. One can estimate the order of magnitude of the D lifetime using the theoretical calculation for the $D \to Ke\nu$ width of about 10^{11} sec^{-1} (see for example Fakirov & Stech 1978) and the ratio $\Gamma(D \to Ke\nu)/\Gamma(D \to eX) = 45 \pm 24\%$ obtained through the DELCO fit to the ψ'' electron momentum spectrum. These results suggest that the D^+ lifetime is on the order of 10^{-12} sec while the D^0 is three to five times shorter lived.

Owing to the large phase space and number of hadronic decay modes available to both charmed mesons, it appears surprising that hadronic decays of the D^0 can be enhanced by at least a factor of three to five relative to those of the D^+, as suggested by the recent data. We note that the average charm semileptonic branching ratio reported by all groups appears to be remarkably constant as a function of center-of-mass energy. Within the limits of the present statistical and systematic uncertainties these measurements are not inconsistent with the observation of different lifetimes; however, one should eventually be able to measure significant variations in the semileptonic branching ratio in data taken at different center-of-mass energies.

5 THE EXCITED STATES OF CHARM

5.1 *Observation of $D^{*+} \to \pi^+ D^0$*

Structure present in the early D meson recoil spectra obtained by the SLAC-LBL Mark I collaboration suggested the presence of D*'s or heavier, new charmed mesons produced against the D in e^+e^- annihilation.

Direct evidence for the D^{*+} was obtained by this group (Feldman et al 1977) in data collected from 5 to 7.8 GeV, the energy limit of SPEAR. In an effort to observe the pion cascade process $D^{*+} \rightarrow D^0\pi^+$, $D^0 \rightarrow K^-\pi^+$ candidates with masses in the D region from 1.820 to 1.910 GeV and momenta exceeding 1.5 GeV were paired with extra pions of the appropriate charge. Figure 18 shows the resulting mass difference, $M(D^0\pi^+) - M(D^0)$, distribution. Relatively large momenta D^0's were required in order that the pions produced in the process $D^{*+} \rightarrow \pi^+ D^0$ have sufficient momenta to be observed in their detector.

A clear, nearly background-free signal is seen in the $D^0\pi^+$, $\bar{D}^0\pi^-$ distribution at a mass difference of 145.3 ± 0.5 MeV. The narrow width of this peak sets the limit $\Gamma_{D^{*+}} < 2\ \mathrm{MeV}/c^2$ (90% confidence level). The slight enhancement at the same mass difference for $D^0\pi^- + \bar{D}^0\pi^+$ events can be explained by π/K misidentification by the time-of-flight system. The relative smallness of this peak compared to the peak of Figure 18*a* sets a

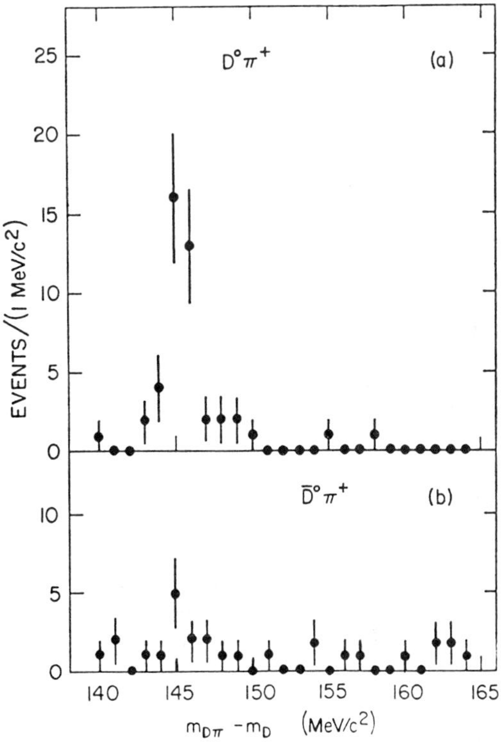

Figure 18 The $D^0\pi^+$ (*a*) and $\bar{D}^0\pi^+$ (*b*) mass difference distributions for $D^0 \rightarrow K^-\pi^+$ candidates lying within 45 MeV of the D^0 mass obtained by the Mark I collaboration. The slight signal in (*b*) is consistent with that expected from double time-of-flight misidentification. The charge conjugate reactions are included.

limit on the conjectured $D^0-\bar{D}^0$ mixing process: less than 16% of produced D^0's mix into $\bar{D}^0$'s within the D^0 lifetime. Finally it was found that a substantial (i.e. 25±9%) fraction of D^0's produced in their data sample with momenta exceeding 1.5 GeV are from the D^{*+} pionic decay.

5.2 *Evidence for the D^{*0}*

A natural extrapolation of this D^{*+} observation is that there exists an excited state of the D^0, as well. Thus one would have an excited (D^{*0}, D^{*+}) isodoublet which might, for example, be a spin excitation of the (D^0, D^+)

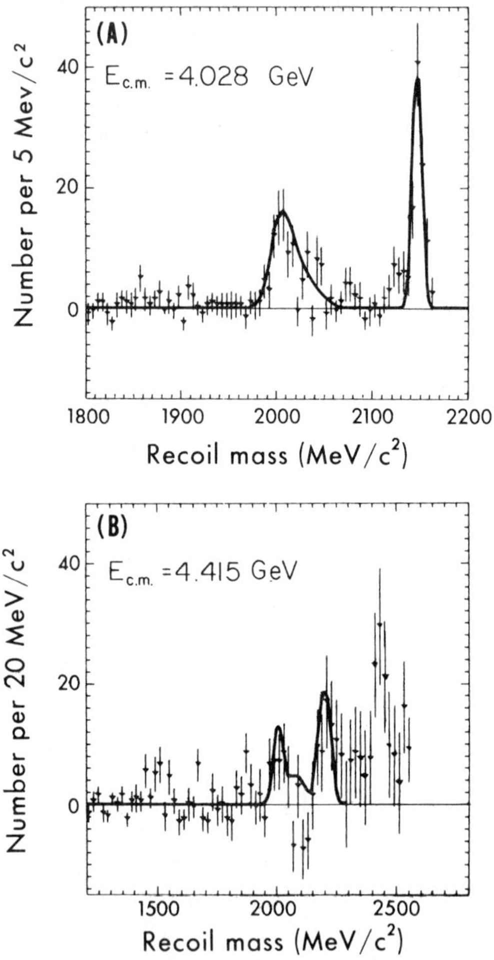

Figure 19 The D^0-subtracted recoil spectra obtained by the Mark I collaboration at two fixed energies. Please note the scale changes.

system. Figure 19*a*, which shows the recoil mass distribution against $D^0 \rightarrow K^-\pi^+$ candidates collected at a center-of-mass energy of 4.028 GeV, provides evidence for the existence of the D^{*0} (Goldhaber et al 1977). This distribution has background $K^{\mp}\pi^{\pm}$ combinations with masses straddling the D^0 signal subtracted out. The two peaks present in Figure 19*b* are attributed to the process $e^+e^- \rightarrow D^*\bar{D}$ or $\bar{D}^*D$, which produces a peak near 2.01 GeV, and $e^+e^- \rightarrow D^*\bar{D}^*$; $D^* \rightarrow \pi^0 D^0$, which produces a narrow reflection peak near 2.15 GeV. We note the relative smallness of the reaction $e^+e^- \rightarrow D\bar{D}$, which would produce a peak near 1.863 GeV.

The narrowness of the reflection peak near 2.15 GeV must follow from the small Q value for the pionic cascade process as well as the improved recoil mass resolution for D*'s produced nearly at threshold. This small Q value allows the radiative D* decay process, $D^{*0} \rightarrow \gamma D^0$, to compete favorably with the pionic cascade, as we demonstrate later. Processes such as $e^+e^- \rightarrow \bar{D}^0_1$, D^{*0}; $D^{*0} \rightarrow D^0_2\pi^0$ where D^0_2 rather than D^0_1 is observed decaying into $K^-\pi^+$ give rise to the trailing tail of the peak near 2.15 GeV.

The solid curves superimposed on Figure 19*a* give the expected shapes for the $D^*\bar{D}$ and $D^*\bar{D}^*$ contributions to the D^0 recoil spectrum at $E_{cm} = 4.028$ GeV. These curves neglect any $D^* \rightarrow \gamma D$ contributions, however. Figure 19*b* shows the D^0 spectrum obtained for data collected at $E_{cm} = 4.415$ GeV overplotted with the expected shape of contributions from these same two processes.

Comparison of the second peak in Figure 19*a* and *b* shows that this peak moves and broadens in the manner expected for a kinematic reflection. The origin of the peak near 2.44 GeV in the $E_{cm} = 4.415$ GeV is as yet unclear—it may arise from multibody final states such as $D^{*0}\bar{D}^{*0}\pi^0$ or it may be possible evidence for a D**.

5.3 *The Masses and Branching Ratios of the D*'s*

In order to analyze quantitatively D production at $E_{cm} = 4.028$ GeV, a fit was performed to the joint D^0 and D^+ momenta spectra. Owing to the kinematic simplicity of the symmetric annihilation process, the D momentum can be trivially related to the value of the recoil mass against the D. The momentum variable does, however, offer an advantage because the D momentum resolution is a relatively insensitive function of momentum near threshold. The decay mode $D^{*+} \rightarrow \pi^+ D^0$ couples the charged and neutral D momentum spectra, thus necessitating a single fit to both.

Figure 20*a* illustrates eight contributions of the D^0 momentum spectrum at $E_{cm} = 4.028$ GeV in terms of the three basic processes:

(i) $e^+e^- \rightarrow D\bar{D}$
(ii) $e^+e^- \rightarrow D\bar{D}^* + \bar{D}D^*$
(iii) $e^+e^- \rightarrow D^*\bar{D}^*$

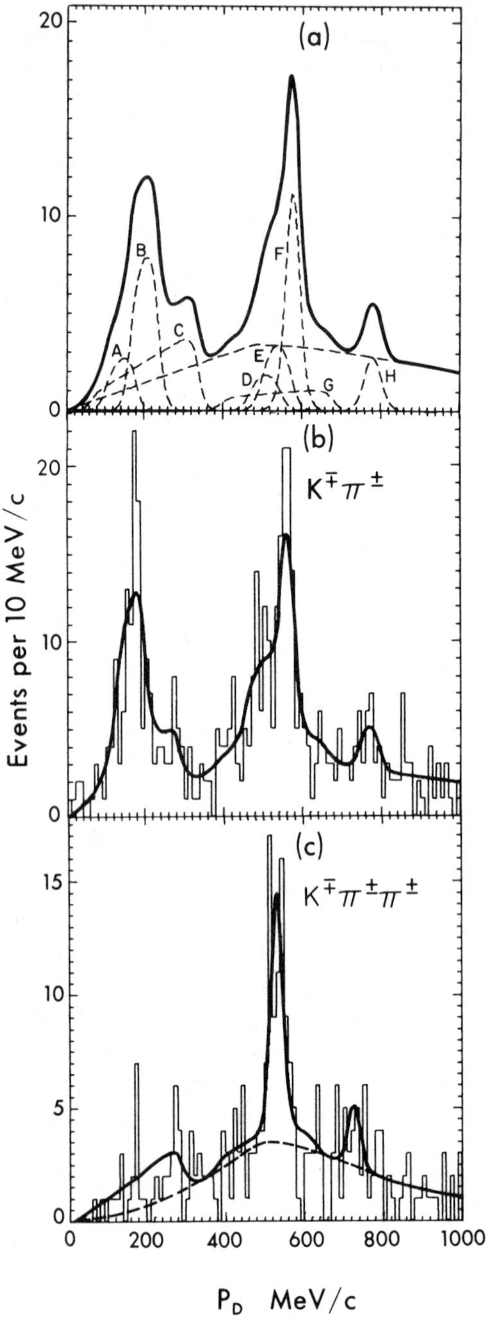

Figure 20 The D momentum spectrum at 4.03 GeV from Goldhaber et al (1977).

(*a*) Contribution to the expected D^0 momentum spectrum from:

A: $e^+e^- \rightarrow D^{*+}D^{*-}, D^{*+} \rightarrow \pi^+D^0$

B: $\rightarrow D^{*0}\bar{D}^{*0}, D^{*0} \rightarrow \pi^0D^0$

C: $\rightarrow D^{*0}\bar{D}^{*0}, D^{*0} \rightarrow \gamma D^0$

D: $\rightarrow D^{*+}D^-, D^{*+} \rightarrow \pi^+D^0$

E: $\rightarrow D^{*0}\bar{D}^0, D^{*0} \rightarrow \pi^0D^0$

F: $\rightarrow \bar{D}^{*0}D^0$, direct D^0

G: $\rightarrow D^{*0}\bar{D}^0, D^{*0} \rightarrow \gamma D^0$

H: $\rightarrow D^0\bar{D}^0$, direct D^0

(*b*) $D^0 \rightarrow K^-\pi^+$ momentum spectrum, the curve is the result of the fit;

(*c*) $D^+ \rightarrow K^-\pi^+\pi^+$ momentum spectrum where the curve is the result of the fit and the dashed line is the background.

where D*'s decay ultimately into D's via the reactions:

(iv) $D^{*+} \rightarrow \pi^+ D^0$
(v) $D^{*+} \rightarrow \pi^0 D^+$
(vi) $D^{*+} \rightarrow \gamma D^+$
(vii) $D^{*0} \rightarrow \pi^0 D^0$
(viii) $D^{*0} \rightarrow \gamma D^0$.

The fit does indeed show that D production at this energy is overwhelmingly dominated by the two-body processes (i) through (iii). Less than 10% of the D^0's were found to arise from the three-body process $D^0\bar{D}^0\pi^0$. The positions and shapes of these contributions depend sensitively on the D* masses and D* − D mass differences. The relative areas of these contributions are functions of the rates for processes (i) through (iii) and the various D* branching ratios.

Table 10 Results from simultaneous fits to the D^0, D^+ momentum spectra at $E_{cm} = 4.028$ GeV (from Goldhaber et al 1977)

	Fit parameter	Normal fit	Isospin constrained fit	Estimated values
Masses in MeV/c^2	M_{D^0}	1864 (1.5)[a]	1862 (0.5)[a]	1863 ± 3[b]
	M_{D^+}	1874 (2.5)	1873 (2.0)	1874 ± 5
	$M_{D^{*0}}$	2006 (0.5)	2007 (0.5)	2006 ± 1.5
	$M_{D^{*+}}$	2009 (1.5)	2007 (0.5)	2008 ± 3
Branching ratios	$B(D^{*0} \rightarrow \gamma D^0)$	0.45 (0.08)	0.75 (0.05)	0.55 ± 0.15
	$B(D^{*+} \rightarrow \pi^+ D^0)$[c]	—	0.60 ± 0.15	—
	$\frac{B(D^+ \rightarrow K^-\pi^+\pi^+)}{B(D^0 \rightarrow K^-\pi^+)}$[c]	—	1.60 ± 0.60	—
D^0 source fractions	$D^0\bar{D}^0$	0.05 (0.03)	0.05 (0.02)	0.05 ± 0.03
	$D^0\bar{D}^{*0} + \bar{D}^0 D^{*0}$	0.42 (0.04)	0.34 (0.04)	0.38 ± 0.08
	$D^{*0}\bar{D}^{*0}$	0.47 (0.05)	0.32 (0.05)	0.40 ± 0.10
	$D^{*+}D^-$; $D^{*+} \rightarrow \pi^+ D^0$	0.03 (0.02)	0.09 (0.04)	0.06 ± 0.05
	$D^{*+}D^{*-}$; $D^{*+} \rightarrow \pi^+ D^0$	0.03 (0.03)	0.20 (0.07)	0.11 ± 0.10
D^+ source fractions	D^+D^-	0.09 (0.05)	0.09 (0.05)	0.09 ± 0.05
	$D^{*+}D^- + D^{*-}D^+$	0.65 (0.07)	0.58 (0.06)	0.62 ± 0.09
	$D^{*+}D^{*-}$	0.26 (0.08)	0.33 (0.08)	0.29 ± 0.10

[a] Quantities in parentheses are typical statistical errors for a single fit.
[b] Errors quoted include estimated systematic uncertainty.
[c] These values can only be obtained under the assumptions of the isospin constrained fit. The quoted errors do not reflect possible breakdown of these assumptions.

Table 10, reprinted from Goldhaber et al (1977), summarizes the information obtained from two fits to the joint $D^0 \rightarrow K^-\pi^+$, and $D^+ \rightarrow K^-\pi^+\pi^+$ momentum spectra. The detailed assumptions for both fits are described in this reference, but a few words are in order. Both fits embody the constraint $M_{D^{*+}} - M_{D^0} = 145.2$ MeV/c^2. The "normal" fit treats the two D^{*+} contributions to the D^0 spectrum, via process (iv), as independent parameters, whereas the "isospin-constrained" fit relates the number of D^0 arising from $D^{*+}D^{*-}$ production to that from $D^{*+}D^-$ production via a universal $D^{*+} \rightarrow \pi^+D^0$ branching ratio and the assumption that, apart from p^3 threshold factors, the rates for the charged versions of processes (i) through (iii) equal the rates for the neutral versions.

Both fits match the experimental momentum spectra reasonably well. In addition, the D masses and ratio of D^+/D^0 branching fractions obtained in this fit agree well with the results of later work by the LGW and Mark II collaborations obtained at the ψ''. These results are compared in Table 11. Figure 21 summarizes the mass relationships between the D and D* systems.

We see from Table 10 that D^0 production at $E_{em} = 4.028$ GeV is dominated by nearly equal contributions from reactions (ii) and (iii). This is notable in light of the 16-MeV Q value for reaction (iii) compared to a Q of 159 MeV and 312 MeV for reactions (ii) and (i) respectively. One expects some enhancement of reaction (iii) due to the larger number of available final state spins, but the spin factor is considerably smaller than the enhancement implied by the data when corrected by the expected p^3 threshold factors. Various explanations for this enhancement have been offered, such as the possibility that the 4.028 resonance is a "molecular" state, or that the radial nodes of the $c\bar{c}$ wave function dictate the relative rates of reactions (i) to (iii) (see Appelquist et al 1978 for a discussion and references to the theoretical papers).

Table 11 Comparison of D results

	SLAC-LBL (Mark I)	LGW	Mark II
M_{D^+} (MeV/c^2)	1874 ± 5	1868.3 ± 0.9[a]	1868.4 ± 0.5[a]
M_{D^0} (MeV/c^2)	1863 ± 3	1863.3 ± 0.9[a]	1863.8 ± 0.5[a]
$\dfrac{B(D^+ \rightarrow K^-\pi^+\pi^+)}{B(D^0 \rightarrow K^-\pi^+)}$	1.6 ± 0.6	$\dfrac{3.9 \pm 1}{1.8 \pm 0.5} = 2.2 \pm 0.8$	$\dfrac{6.3 \pm 1.5}{3.0 \pm 0.6} = 2.1 \pm 0.7$

[a] These masses are still dependent on the absolute SPEAR calibration. They are thus defined relative to the ψ mass. The mass difference $\Delta = M_{D^+} - M_{D^0}$ is known more precisely as some errors cancel. The values are: $\Delta = 5.0 \pm 0.8$ MeV/c^2 for the LGW and $\Delta = 4.7 \pm 0.3$ MeV/c^2 for the Mark II experiments respectively.

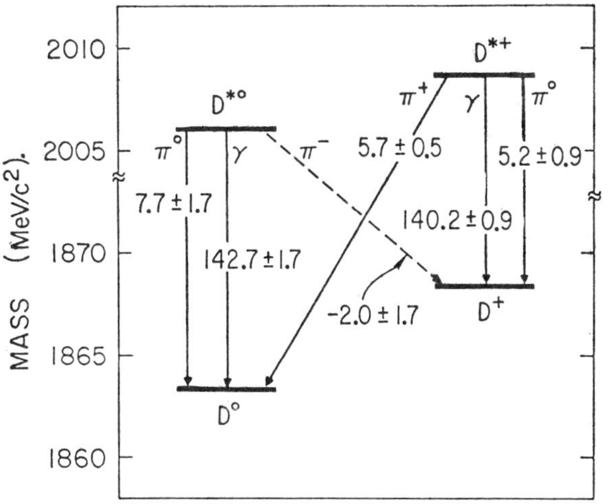

Figure 21 Mass level diagram giving the best current values for D* and D states. The arrows represent different decay modes of the D*; the numbers across the lines represent the Q for each decay expressed in MeV. The decay $D^{*0} \to D^+\pi^-$ is kinematically forbidden. The masses are (D^0) 1863.3 ± 0.9, (D^+) 1868.3 ± 0.9, (D^{*0}) 2006.0 ± 1.5, and (D^{*+}) 2008.6 ± 1.0 MeV/c^2.

5.4 *Spins of the D, D* Mesons*

Because the D and D* are the two lightest charmed particles, it is a priori probable that the D is a pseudoscalar residing in the SU(4) multiplet of the pion, while the heavier D* is a vector residing in the SU(4) multiplet of the $\rho(770)$. The mass difference between particles in the vector and pseudoscalar multiplets ($^3S_1$ and $^1S_1$) is presumably due to the hyperfine interaction between the quarks. Because of the large mass splitting within these SU(4) multiplets, and the presumably reduced strength of the quark hyperfine splitting for states containing a heavy, charmed quark, these conclusions may not necessarily hold. Present data are insufficient to establish the unique spin-parity assignments of the low-lying charmed mesons, although there is experimental information from SPEAR.

Some information is available from the study of the D* decay modes discussed earlier. Observation of the decay $D^* \to \pi D$ along with the reaction $e^+e^- \to D\bar{D}^*$ demonstrates that the D and D* are not both spinless.[1] Evidence for the radiative decay $D^* \to \gamma D$ corroborates this conclusion.

The LGW collaboration (Peruzzi et al 1977) obtained information on the angular distribution (in θ) of the D momentum vector with respect to the e^+e^- annihilation axis for the reaction $e^+e^- \to D\bar{D}$ obtained in data

[1] If the D and D* were both spinless they would require even relative parity to couple to a photon via $e^+e^- \to DD^*$. However, then the decay $D^* \to \pi D$ would fail to conserve parity.

collected at the ψ''. Figure 22 shows the background-subtracted $\cos\theta$ distributions for $D^0(D^0) \rightarrow K^{\mp}\pi^{\pm}$ and $D^{\pm} \rightarrow K^{\mp}\pi^{\pm}\pi^{\mp}$ events. Fits of these distributions to the form

$$\frac{dn}{d\cos\theta} \propto 1 + \alpha\cos^2\theta$$

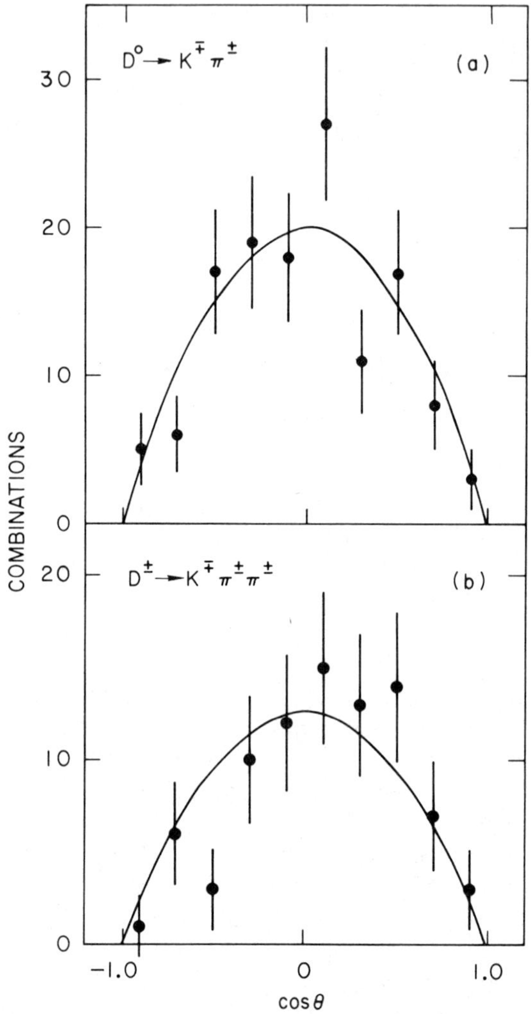

Figure 22 Distribution of events in the cosine of the angle between the incident e^+ beam and the D momentum for (*a*) $D^0(\bar{D}^0) \rightarrow K^{\mp}\pi^{\pm}$ and (*b*) $D^{\pm} \rightarrow K^{\mp}\pi^{\pm}\pi^{\pm}$, after background subtraction. The curves represent $\sin^2\theta$, the required distribution for the production of spinless D mesons.

yield the values $\alpha = -1.04 \pm 0.10$ for the $D^{\pm}$ and $\alpha = -1.00 \pm 0.09$ for the $D^0(\bar{D}^0)$, which are remarkably close to the value $\alpha = -1$ required for production of spinless particles. Their result suggests that the D is spinless as expected; however, two D's of higher spin could couple fortuitously to give a $\sin^2 \theta$ polar distribution as well.

Spin information on the D and D* is available from SLAC-LBL collaboration data on D's produced in e^+e^- annihilations at a center-of-mass energy of 4.028 GeV. At this energy D^0's are primarily produced via the processes $e^+e^- \rightarrow D^*\bar{D}^*$ and $D\bar{D}^*$. One can obtain a relatively pure sample of D's from either process by applying an appropriate cut on the measured recoil mass against the D^0.

The distribution for the angle between the D* momentum and the annihilation axis, obtained for $D^*\bar{D}^*$ production, was fit to the form

$$\frac{\mathrm{d}n}{\mathrm{d}\cos\theta} \propto 1 + \alpha \cos^2 \theta.$$

The measured value $\alpha = -0.3 \pm 0.3$ tends to rule out spinless D*'s at the two-standard-deviation level.

A study of the joint production and decay angular distribution for the process $e^+e^- \rightarrow \bar{D}^{*0}D^0$, $D^0 \rightarrow K^-\pi^+$ provides additional information on the charmed-meson spins. One can uniquely predict this distribution for the case of spin-0 D and spin-1 D* and vice versa if one assumes that the two mesons have even relative parity (as evidenced under these spin assignments by the observation of the reaction $D^* \rightarrow \pi D$). In particular there would be a considerable anisotropy in the $D^0 \rightarrow K\pi$ decay if a vector D^0 was produced against a pseudoscalar D*. Such an anisotropy is inconsistent with the SLAC-LBL data at about the three-standard-deviation level. Their data are fully consistent with the expected distribution for a pseudoscalar D and vector D* (Nguyen et al 1977).

In summary, all known data on charmed-meson spin is consistent with the expected vector character of the D* and pseudosclar character of the D. The two mesons cannot both be spinless, and the case of vector D's and pseudosclar D*'s is explicitly excluded. Nothing as of yet is known about the possibilities where the sum of the D and D* spins exceeds one.

6 THE F MESON

Little is known experimentally about the third charmed meson—the F^+. This isosinglet meson is constructed from a c and $\bar{s}$ quark and should decay predominantly into final states of zero strangeness according to the GIM model. Since the many possible multipion F^+ decay modes are expected to have huge backgrounds, investigators have tended to search for F^+

decaying into less common particles such as $K^+K^-\pi^+$, K^+K_S, or η and multipions.

At the time of this writing the only published observation of the F^+ comes from the DASP collaboration (Brandelik et al 1979), who looked for the process $e^+e^- \to F\bar{F}^*$ where the $F \to \pi\eta$ and $\bar{F}^* \to \gamma\bar{F}$ (the $F^* \to \pi F$ decay does not conserve isospin).

Figure 23 shows the $\gamma\gamma$ invariant mass distribution for hadronic events collected at the indicated center-of-mass energies. The most pronounced η signal occurs near $E_{cm} = 4.42$ (Siegrist et al 1976), which is at the location of a 33-MeV wide resonance in R. Requiring the presence of an additional soft γ ($E_\gamma < 140$ MeV) in the event, as would be the case for F*F or F*F* production, appears to enhance the η signal relative to the background in the 4.42-GeV data (see Figure 24). These observations are thus suggestive of a substantial F* contribution to η's produced at $E_{cm} = 4.42$ GeV.

In order to observe F^+'s decaying into a specific final state, events collected at 4.42 GeV were fitted to the hypothesis $e^+e^- \to FF^* \to \pi\eta F\gamma$ where the $\eta \to \gamma\gamma$ candidates were constrained to the precise η mass, and the $\pi\eta$ system was constrained to the mass of the missing F. A scatter plot of the $\pi\eta$ mass versus its recoil mass is shown in Figure 25 for events with an acceptable χ^2. There is a clustering of six events present in this plot that are F, F* candidates, where $M_F = 2.04 \pm 0.01$ GeV/c^2 and $M_{F^*} = 2.15 \pm 0.04$ GeV/c^2. These six events also tend to fit the hypothesis $e^+e^- \to F^*\bar{F}^*$, where $M_F = 2.00 \pm 0.04$ and $M_{F^*} = 2.1 \pm 0.02$ GeV. Here the DASP collaboration take average values and quote $M(F) = 2.03 \pm 0.06$ GeV and $M(F^*) = 2.14 \pm 0.06$ GeV.

The six events imply $\sigma \cdot B(F^+ \to \eta\pi^+) = 0.41 \pm 0.18$ nb at the 4.42 resonance, a value close to the Mark II 95% confidence level upper limit of 0.26 nb at $E_{cm} = 4.42$ GeV and 0.33 nb at $E_{cm} = 4.16$ GeV. Clearly further work must be done to clarify the physics of the F.

ACKNOWLEDGMENTS

We want to thank Mrs. C. Frank-Dieterle for her help and meticulous care in preparing and compiling this manuscript.

This work was supported primarily by the US Department of Energy under Contracts No. W-7405-ENG-48 (Lawrence Berkeley Laboratory) and DE-AC02-76ER01195 (University of Illinois).

We wish to thank the Staff of the Aspen Center for Physics for the hospitality extended to us during the Summer of 1979 while this review was being prepared in part.

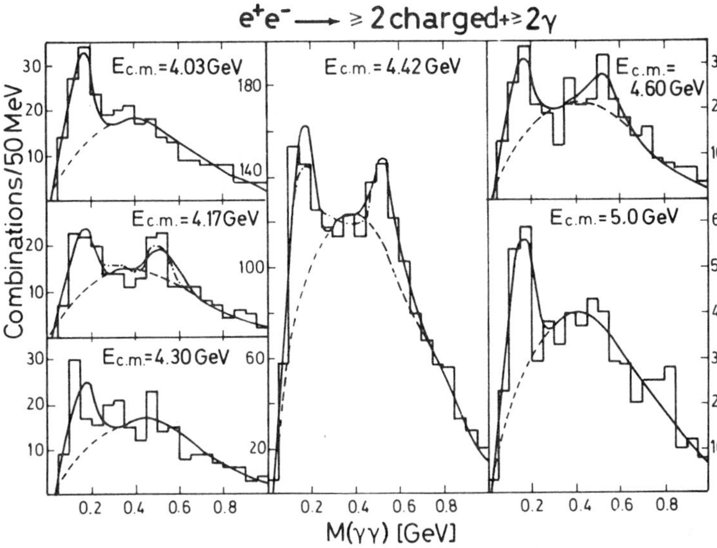

Figure 23 DASP data on inclusive η production. Note that the η signal is cleanest in the vicinity of the $\psi(4.42)$.

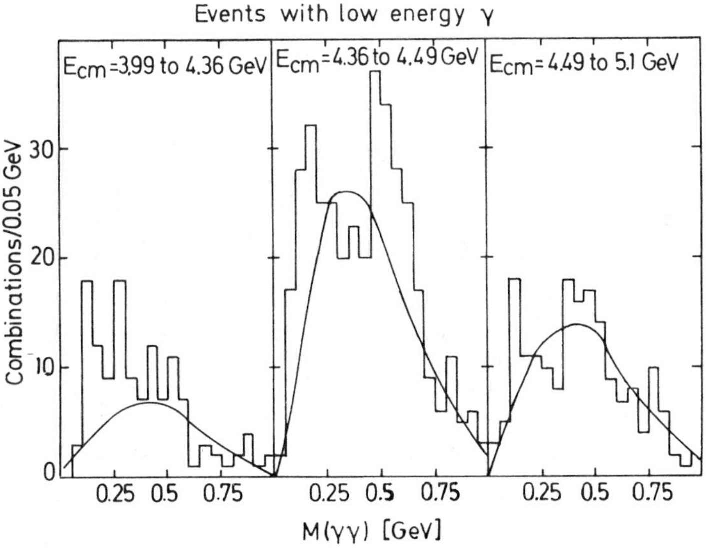

Figure 24 DASP data on η production for events with low energy γ rays. Note the η signal in the energy region near the $\psi(4.42)$ is improved if one demands the presence of an additional soft γ ($E_\gamma < 140$ MeV). Such a γ could result from the process $F^* \rightarrow \gamma F$.

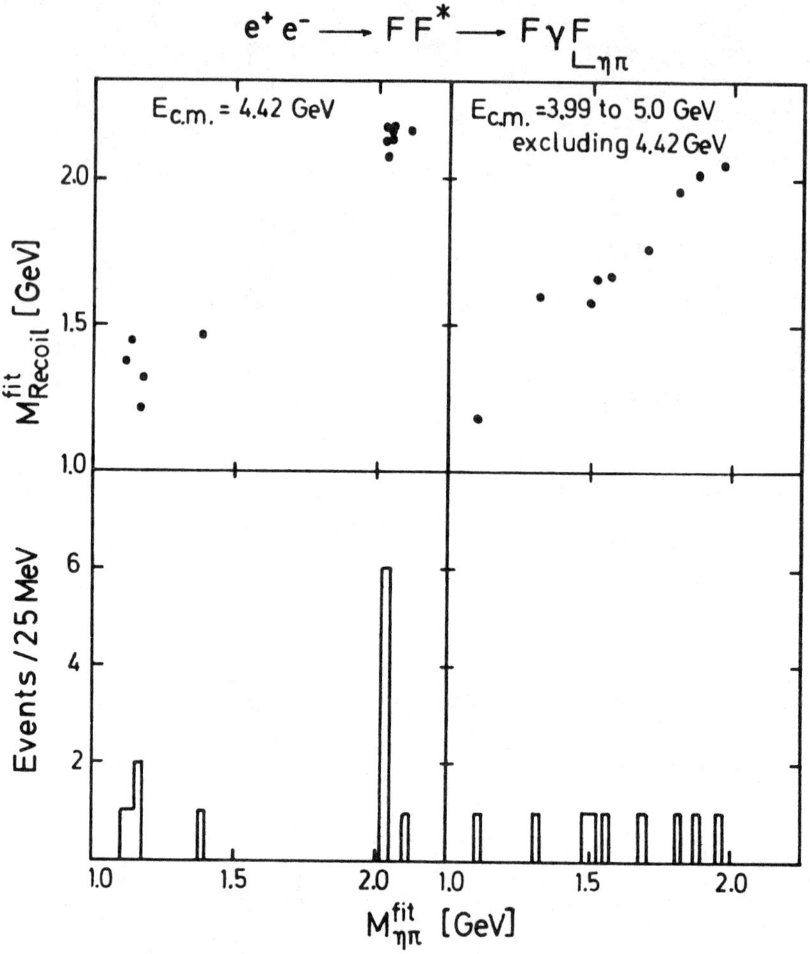

Figure 25 DASP data showing evidence for F production. A clustering of six events is seen in the $M_{\eta\pi}$ vs M_{Rec} scatter plot near $M_{\eta\pi} = 2.04$ GeV/c^2.

Literature Cited

Abrams, G. S. et al. 1979. *Phys. Rev. Lett.* 43:481

Andrews, D. et al. 1980. *Phys. Rev. Lett.* 44:1108

Appelquist, T., Barnett, R. M., Lane, K. 1978. *Ann. Rev. Nucl. Part. Sci.* 28:387

Aronson, S. H. et al. 1970. *Phys. Rev. Lett.* 25:1057

Aubert, J. J. 1974. *Phys. Rev. Lett.* 33:1404

Augustin, J.-E. et al. 1974. *Phys. Rev. Lett.* 33:1406

Bacino, W. et al. 1978. *Phys. Rev. Lett.* 40:671

Barbaro-Galtieri, A. 1968. *Advances in Particle Physics*, Vol. D, ed. R. Cool, R. Marshak, p. 193. New York: Interscience

Barbaro-Galtieri, A. 1978. *Production and Decay of Charm Particles in e^+e^- Collisions*, Lawrence Berkeley Lab., Rep. LBL-8537, LBL-9247 (1979)

Bebek, C. 1980. *Exp. Meson Spectrosc. Conf. EMS-80*. Brookhaven Natl. Lab.

Böhringer, T. et al. 1980. *Phys. Rev. Lett.* 44:1111
Boyarski, A. M. et al. 1975. *Phys. Rev. Lett.* 34:1357
Brandelik, R. et al. 1979. *Z. Phys. C* 1:233
Braunschweig, W. et al. 1976. *Phys. Lett. B* 63:471
Bricman, C. et al. 1978. *Phys. Lett. B* 75:1
Burmester, J. et al. 1976. *Phys. Lett. B* 64:369
Camerini, U. et al. 1975. *Phys. Rev. Lett.* 35:483
Carithers, W. A. et al. 1973. *Phys. Rev. Lett.* 31:1025
Chinowsky, W. 1977. *Ann. Rev. Nucl. Sci.* 27:393
Clark, A. R. et al. 1971. *Phys. Rev. Lett.* 26:1667
Eichten, E. et al. 1975. *Phys. Rev. Lett.* 34:369
Ellis, J. 1974. *Proc. XVII Int. Conf. High Energy Phys., London*, p. IV–20
Fakirov, D., Stech, B. 1978. *Nucl. Phys. B* 133:315
Feldman, G. J. et al. 1977. *Phys. Rev. Lett.* 38:1313
Feldman, G. J. 1978. *Proc. XIX Int. Conf. High Energy Physics, Tokyo*, p. 777
Feller, J. M. 1979. PhD thesis. Univ. Calif., Berkeley. *LBL-9017*
Feller, J. M. et al. 1978. *Phys. Rev. Lett.* 40:1677
Finocchiaro, G. 1980. See Bebek 1980
Gaillard, M. K., Lee, B. W., Rosner, J. 1975. *Rev. Mod. Phys.* 47:277
Gell-Mann, M. 1964. *Phys. Lett.* 8:214
Gittleman, B. et al. 1975. *Phys. Rev. Lett.* 35:1616
Glashow, S. L., Illiopoulos, J., Maiani, L. 1970. *Phys. Rev. D* 2:1285
Goldhaber, G. et al. 1976. *Phys. Rev. Lett.* 37:255
Goldhaber, G. et al. 1977. *Phys. Lett. B* 69:503
Gottfried, K. 1978. *Phys. Rev. Lett.* 40:598
Harari, H. 1977. Proc. 1977 Summer Inst. on Part. Phys., Stanford Linear Accel. Cent. *SLAC-204*, p. 1
Herb, S. W. et al. 1977. *Phys. Rev. Lett.* 39:252
Iizuka, J. 1966. *Suppl. Prog. Theor. Phys.* 37:21
Jackson, J. D., Scharre, D. L. 1975. *Nucl. Instrum. Methods* 128:13
Jean-Marie, B. et al. 1976. *Phys. Rev. Lett.* 36:291
Kirkby, J. 1979. In *Proc. 9th Int. Symp. Lepton and Photon Interactions at High Energies*, p. 107. Batavia, Ill.: Fermilab
Klems, J. H., Hildebrand, R. H., Stiening, R. 1970. *Phys. Rev. Lett.* 24:1086
Knapp, B. et al. 1975. *Phys. Rev. Lett.* 34:1040
Kobayashi, M., Maskawa, T. 1973. *Prog. Theor. Phys.* 49:652
Lane, K., Eichten, E. 1976. *Phys. Rev. Lett.* 37:477
Lee-Franzini, J. 1980. See Bebek 1980
Litke, A. et al. 1973. *Phys. Rev. Lett.* 30:1189
Lüth, V. 1979. In *Proc. 9th Int. Symp. Lepton and Photon Interactions at High Energies*, p. 78. Batavia, Ill.: Fermilab
Meyer, H. 1979. In Proc. 9th Int. Symp. Lepton and Photon Interactions at High Energies, p. 214. Batavia, Ill.: Fermilab
Nagels, M. M. et al. 1976. *Nucl. Phys. B* 109:1
Nguyen, H. K. et al. 1977. *Phys. Rev. Lett.* 39:262
Okubo, S. 1963. *Phys. Rev. Lett.* 5:165
Pais, A., Treiman, S. B. 1977. *Phys. Rev. D* 15:2529
Perez-y-Jorba, J. 1969. *Proc. IV Int. Symp. on Electron and Photon Interactions at High Energies, Liverpool*, p. 213
Perl, M. 1980. *Ann. Rev. Nucl. Part. Sci.* 30: 299–335
Peruzzi, I. et al. 1976. *Phys. Rev. Lett.* 37: 569
Peruzzi, I. et al. 1977. *Phys. Rev. Lett.* 39:1301
Peruzzi, I. et al. 1978. *Phys. Rev. D* 17:2901
Rapidis, P. A. et al. 1977. *Phys. Rev. Lett.* 39:526
Rapidis, P. A. 1979. PhD thesis. Stanford Univ. *SLAC-220*
Richter, B. 1974. *Proc. XVII Int. Conf. High Energy Phys., London*, p. IV–37
Schindler, R. H. 1979. PhD thesis. Stanford Univ. *SLAC-219*
Schindler, R. H. et al. 1980. *Phys. Rev. D*. In press
Schrock, R. E., Wang, L. L. 1978. *Phys. Rev. Lett.* 41:1692
Schröder, H. 1980. See Bebek 1980
Schwitters, R. F., Strauch, K. 1976. *Ann. Rev. Nucl. Sci.* 26:89
Siegrist, J. L. et al. 1976. *Phys. Rev. Lett.* 36:700
Tarnopolsky, G. et al. 1974. *Phys. Rev. Lett.* 32:432
Voyvodic, L. 1979. In *Proc. 9th Int. Symp. Lepton and Photon Interactions at High Energies*, p. 569. Batavia, Ill.: Fermilab
Vuillemin, V. et al. 1978. *Phys. Rev. Lett.* 41:1149
Wiik, B. H., Wolf, G. 1978. DESY-78/23 (Unpublished report); also Balian, R., Llewellyn-Smith, C. H., eds. 1977. *Ecole d'été de physique théorique, 29th session, Les Houches, 1976*. Amsterdam: North-Holland
Wiss, J. E. et al. 1976. *Phys. Rev. Lett.* 37:1531
Zweig, G. 1964. *CERN-TH-401*

Ann. Rev. Nucl. Part. Sci. 1980. 30: 299–335

5

THE TAU LEPTON[1]

Martin L. Perl
Stanford Linear Accelerator Center, Stanford University, Stanford, California 94305

CONTENTS

1 INTRODUCTION

In the last few years there has been an important addition to the known elementary particles—the tau (τ) lepton. It is an important addition first because, to the best of our knowledge, the tau is a fundamental particle.

[1] Work supported by the Department of Energy, contract DE-AC03-76SF00515.

That is, unlike most of the so-called elementary particles such as the proton or pion, the tau is not made up of simpler particles or constituents (see Section 1.1). Second, the tau is important because all its measured properties agree with its designation as a lepton. Hence it joins the very small lepton family of particles; a family that prior to the tau's discovery had only four members: the electron (e), its associated neutrino (ν_e), the muon (μ), and its associated neutrino (ν_μ). Third, the tau is important because, along with the discovery of the fifth quark, it appears to confirm some general theoretical ideas about the connection between leptons and quarks. A brief discussion on quarks and the lepton-quark connection is presented in Sections 1.1 and 2.2.

This article reviews the experimental work done on the tau, why we believe it is a lepton, the measured properties of the tau, and the experimental work still to be done on the tau. I present just enough theory to provide a framework for discussing the experimental results. Correspondingly, I present a full set of experimental references (up to November 1979), but only a few general theoretical references.

The history of the discovery of the tau was reviewed by Feldman (1978a); I only outline it here. It is an old idea to look for leptons with masses greater than that of the electron or muon—the so-called heavy leptons. The first searches for a heavy charged lepton using electron-positron collisions were carried out by Bernardini et al (1973) and by Orioto et al (1974) at the ADONE e^+e^- storage ring. They looked for the electromagnetic production process

$$e^+ + e^- \rightarrow l^+ + l^-, \qquad 1.$$

where l represents the new lepton. The ADONE storage ring did not have enough energy to produce the tau.

The first evidence for the tau was obtained (Perl 1975, Perl et al 1975) in 1974 at the SPEAR e^+e^- storage ring by using the reaction in Equation 1. Subsequent experiments at SPEAR, which is at the Stanford Linear Accelerator Center (SLAC), and at the DORIS e^+e^- storage ring at the Deutsches Elektronen-Synchrotron (DESY) confirmed this discovery and measured the properties of the tau. Recent reviews have been given by Kirkby (1979), Flügge (1979), Feldman (1978a), and Tsai (1980).

1.1 *The Definition of a Lepton*

The definition of a lepton is based upon our experience with the electron, muon, and their associated neutrinos. We use these criteria to classify a particle as a lepton (Perl 1978):

1. The lepton does not interact through the strong interaction. Thus the lepton is differentiated from the hadron particle family, such as the pion, proton, and ψ/J.

2. The lepton interacts through the weak interactions and, if charged, through the electromagnetic interaction.

3. The lepton has no internal structure or constituents. I shall call a particle without internal structure or constituents a point particle. This is, of course, always a provisional definition since going to higher energy may reveal the internal structure or the constituents of a particle. However, the requirement on a lepton is to be understood in contrast to the properties of hadrons. That is, hadronic properties such as electromagnetic form factors are explained in terms of the hadron's internal constituents—the quarks. The form factor concept provides a quantitative test of whether or not a particle has internal structure (Section 3.2).

The leptons that we now know share two additional properties that may not be central to the definition of a lepton (Perl 1978):

4. The known leptons have spin 1/2. We can, however, conceive of particles that have properties 1–3 listed above and yet have other spins, zero for example (Farrar & Fayet 1980).

5. All the known leptons obey a lepton conservation law. This is defined formally in Section 2.1. I will give an intuitive definition here. A lepton, such as the e^-, possesses an intrinsic property called lepton number, which cannot disappear. This property can either be transferred to an associated neutrino (transferred from e^- to ν_e) or it can be cancelled by combining the lepton with its antilepton (e^- combined with e^+). As with the property of spin-1/2 we do not know if this is intrinsic to all leptons or only an accidental property of the known leptons.

1.2 *The Tau Lepton*

The tau lepton has the crucial lepton defining properties 1–3 listed in the previous section. It also has property 4, namely spin 1/2; and very probably it has property 5, lepton conservation. It is easiest to get a general picture of the properties of the τ by comparing it with the e and μ (Table 1).

The astonishing property of the tau is its large mass of about 1782 MeV/c^2, 3600 times the electron mass and 17 times the muon mass. Until the tau was discovered many physicists held the vague idea that the simplicity and lack of structure of the leptons were associated with their relatively small mass. The masses of the electron, muon, and their associated neutrinos are all smaller than the mass of the lightest hadron, the neutral pion, which has a mass of 135 MeV/c^2. Indeed, the word lepton

Table 1 The known leptons

Properties	Electron	Muon	Tau
Charged lepton symbol	e^-, e^+	μ^-, μ^+	τ^-, τ^+
Charged lepton mass (MeV/c^2)	0.51	105.7	1782^{+3}_{-4} [a]
Charged lepton lifetime (s)	stable	2.20×10^{-6}	$<2.3 \times 10^{-12}$ [a]
Charged lepton spin	1/2	1/2	1/2
Associated neutrino	$\nu_e, \bar{\nu}_e$	$\nu_\mu, \bar{\nu}_\mu$	$\nu_\tau, \bar{\nu}_\tau$ [b]
Associated neutrino mass	<60 eV/c^2 [c]	<0.57 MeV/c^2 [c]	<250 MeV/c^2 [a,c]

[a] The detailed discussion of these measurements appears in Section 4.
[b] All measured properties of the τ are consistent with it having a unique neutrino, ν_τ; however, more experimental work needs to be done on these properties, as discussed in Section 4.4.
[c] Could be 0.

comes from the Greek lepto meaning fine, small, thin, or light. However, this is certainly not descriptive of the tau whose mass is greater than that of many hadrons, almost twice the proton mass for example. Nevertheless the term lepton has been kept for the tau; often the oxymoron heavy lepton is used.

The relatively large mass of the tau allows it to decay to a variety of final states. Some of the decay modes that have been measured (Section 5) are

$$\begin{aligned}
&\tau^- \rightarrow \nu_\tau + e^- + \bar{\nu}_e \\
&\tau^- \rightarrow \nu_\tau + \mu^- + \bar{\nu}_\mu \\
&\tau^- \rightarrow \nu_\tau + \pi^- \\
&\tau^- \rightarrow \nu_\tau + \rho^- \\
&\tau^- \rightarrow \nu_\tau + \pi^- + \pi^+ + \pi^- \\
&\tau^- \rightarrow \nu_\tau + \pi^- + \pi^+ + \pi^- + \pi^0
\end{aligned} \qquad 2.$$

An analogous set of decay modes occurs for the τ^+. Note that in all measured τ decays a neutrino is produced, which indicates that the τ obeys a lepton conservation rule.

2 THEORETICAL FRAMEWORK

By definition, a lepton interacts through the weak and electromagnetic forces but not through the strong interactions, and it has no internal structure or constituents. To proceed further, we need a more restrictive theoretical framework. I shall impose on the lepton conventional weak interaction theory (Bailin 1977, Zipf 1978) and some sort of lepton conservation rule, since these are restrictions that the τ obeys. However, the reader should keep in mind that there may exist leptons that do not obey these restrictions.

2.1 *Weak Interactions and Lepton Conservation*

Consider a charged and neutral lepton pair (L^-, L^0) with the same lepton number $n_L = +1$. Their antiparticles (L^+, $\bar{L}^0$) have lepton number $n_L = -1$. Lepton conservation means that in all reactions the sum of the n_L's of all the particles remains unchanged.

Assuming (*a*) conventional weak interaction theory, (*b*) that the L^- is heavier than the L^0, and (*c*) that there is sufficient mass difference between the L^- and the L^0, the following sorts of decays will occur (Figure 1):

$$L^- \to L^0 + e^- + \bar{\nu}_e \tag{3}$$

$$L^- \to L^0 + \mu^- + \bar{\nu}_\mu \tag{4}$$

$$L^- \to L^0 + (\text{hadrons})^-. \tag{5}$$

In Figure 1*c* the quark-antiquark pair $d\bar{u}$ replaces the lepton-neutrino pair, and the quarks convert to hadrons. If the L^0 is heavier than the L^- the reverse decays

$$L^0 \to L^- + e^+ + \bar{\nu}_e \tag{6}$$

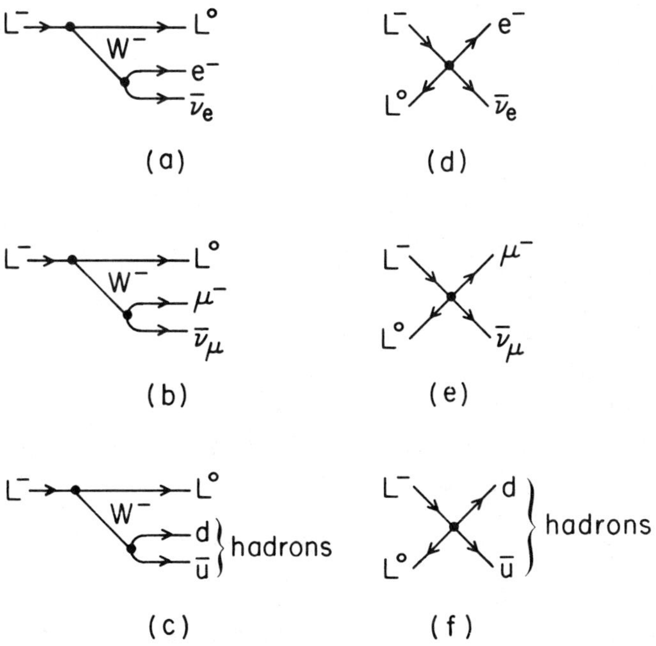

Figure 1 Diagrams for the decay of the τ into quarks or leptons: (*a–c*) intermediate boson view, (*d–f*) four-fermion coupling view.

$$L^0 \to L^- + (\text{hadrons})^+ \qquad 7.$$

will occur. We shall assume that lepton-W and quark-W vertices have conventional Weinberg-Salam theory couplings (Bailin 1977, Zipf 1978).

The W propagator diagrams of Figures 1*a–c* have become the conventional way to represent weak decays. However, when a lepton, such as the τ, has a mass much smaller than the proposed W mass (1.8 GeV/c^2 compared to roughly 100 GeV/c^2) the W propagator has no observable effect. Therefore for some of the discussions of the decays of the τ I shall use Figures 1*d*, *e*, which diagram the old four-fermion coupling of Fermi weak interaction theory (Bjorken & Drell 1964).

2.2 *Simple Models for New Charged Leptons*

2.2.1 SEQUENTIAL LEPTON MODEL In this model (Perl & Rapidis 1974) a sequence of charged leptons of increasing mass is assumed, each lepton type having a separately conserved lepton number and a unique associated neutrino of smaller, but not necessarily zero, mass. That is, there is a particle (and antiparticle) sequence:

$$\begin{array}{cc} \text{charged lepton} & \text{associated neutrino} \\ e^-(e^+) & \nu_e(\bar{\nu}_e) \\ \mu^-(\mu^+) & \nu_\mu(\bar{\nu}_\mu) \\ l^-(l^+) & \nu_l(\bar{\nu}_l) \\ \vdots & \vdots \end{array} \qquad 8.$$

Decays of the $l^\pm$ through the electromagnetic interaction such as $l^\pm \to e^\pm + \gamma$ or $l^\pm \to e^\pm + \gamma$ are forbidden. The $l^\pm$ can only decay through the weak interaction as described in Section 2.1, namely

$$\begin{array}{l} l^- \to \nu_l + e^- + \bar{\nu}_e \\ l^- \to \nu_l + \mu^- + \bar{\nu}_\mu \\ \vdots \\ l^- \to \nu_l + (\text{hadrons})^-. \end{array} \qquad 9.$$

The vertical dots in Equation 9 indicate decays to all associated lepton-neutrino pairs of sufficently small masses. In Equation 9 and in the remainder of this paper we list only the decay modes of the negatively charged lepton; the decay mode of the positively charged lepton is obtained by changing every particle to its antiparticle. The neutrinos in this model are stable because their associated charged lepton has a larger mass and their lepton number is conserved. The search for the τ was based on this model (Perl 1975) and to the best of our knowledge the τ conforms to this model.

2.2.2 ORTHOLEPTON MODEL In this model (Llewellyn Smith 1977) the new charged lepton, l^-, has the same lepton number as a smaller mass, same-sign, charged lepton, such as the e^-. Let us use the e^- example. We expect the dominant decay to occur through the electromagnetic interaction

$$l^- \to e^- + \gamma. \tag{10}$$

However, current conservation forbids the l-γ-e vertex from having the usual form $\bar{\psi}_l \gamma^\mu \psi_e$ (Low 1965, Perl 1978), and this decay mode might be suppressed. Therefore decays through the weak interactions such as

$$\begin{aligned} &l^- \to e^- + e^+ + e^- \\ &l^- \to \nu_e + e^- + \bar{\nu}_e \\ &l^- \to e^- + (\text{hadrons})^0 \\ &l^- \to \nu_e + (\text{hadrons})^- \end{aligned} \tag{11}$$

could in principle be detected.

2.2.3 PARALEPTON MODEL In this model (Llewellyn Smith 1977, Rosen 1978) the l^- has the same lepton number as a smaller mass, opposite-sign, charged lepton, such as the e^+. Electromagnetic decays such as $l^- \to e^- + \gamma$ are now forbidden. The decay modes through the weak interaction are

$$l^- \to \bar{\nu}_e + e^- + \bar{\nu}_e \tag{12}$$

$$l^- \to \bar{\nu}_e + \mu^- + \bar{\nu}_\mu \tag{13}$$

$$l^- \to \bar{\nu}_e + (\text{hadrons})^- \tag{14}$$

In this illustration the l^- has the same lepton number as the e^+ and hence as the $\bar{\nu}_e$.

2.3 *$e-\mu-\tau$ Universality*

A special case falling within the sequential heavy lepton model is the model in which the e, μ, and τ differ only by having (*a*) different masses and (*b*) different and separately conserved lepton numbers. In this model the e, μ, and τ have the same spin (1/2), the same electromagnetic interactions, and the same weak interactions. They are all point particles and they are all associated with different massless, spin-1/2 neutrinos. Thus the comparative properties of the charged lepton depend only on the masses being different. We call this $e-\mu-\tau$ universality.

2.4 *Leptons and Quarks*

The Weinberg-Salam theory of the unification of weak and electromagnetic interactions (Bailin 1977, Zipf 1978) provides a quantitative model for new leptons that is related to the sequential lepton model. In its current form,

Weinberg-Salam theory classifies the leptons and quarks into left-handed doublets, containing at least

$$\text{leptons} = \begin{pmatrix} \nu_e \\ e^- \end{pmatrix}_L, \begin{pmatrix} \nu_\mu \\ \mu^- \end{pmatrix}_L, \begin{pmatrix} \nu_\tau \\ \tau^- \end{pmatrix}_L$$
$$\text{quarks} = \begin{pmatrix} u \\ d' \end{pmatrix}_L, \begin{pmatrix} c \\ s' \end{pmatrix}_L, \begin{pmatrix} t \\ b \end{pmatrix}_L, \qquad 15.$$

and right-handed singlets, containing at least

$$\text{leptons} = e_R, \mu_R, \tau_R$$
$$\text{quarks} = u_R, d'_R, c_R, s'_R, t_R, b_R. \qquad 16.$$

This classification assumes that the t quark exists and that the ν_τ is unique. The weak-electromagnetic interaction only connects particles to themselves or to the other member of the doublet. For leptons this is equivalent to lepton conservation. For the first two quark doublets there is only approximate conservation because the d′ and s′ quarks are mixed by the Cabibbo angle θ_c. That is

$$d' = d \cos \theta_c + s \sin \theta_c$$
$$s' = -d \sin \theta_c + s \cos \theta_c \qquad 17.$$

where *d* and *c* are pure quark states.

The τ plays an important role in this model, because with the τ there are three sets (usually called generations) of leptons and three sets of quarks (assuming the t quark exists). This theory does not require equal numbers of generations of leptons and quarks. But if it happens that the numbers of generations are equal, that is certainly very significant with respect to the connection between leptons and quarks.

Our immediate need for this theory is, however, more mundane. The theory predicts that the weak interactions between the members of a doublet are the same for all doublets. Hence from the $e - \nu_e$ or $\mu - \nu_\mu$ weak interactions we can predict the $\tau - \nu_\tau$ weak interactions if this theory is correct. Specifically, it predicts that (*a*) the $\tau - \nu_\tau$ coupling will be V − A and (*b*) the coupling constant will be the universal Fermi weak interaction constant $G_F \approx 1.02 \times 10^{-5}/M^2_{\text{proton}}$. We discuss these predictions in Sections 4.2 and 4.4.

3 THE IDENTIFICATION OF THE TAU AS A LEPTON

The identification of the tau as a lepton is intertwined with all the properties of the tau. Therefore in a general sense the subject of this entire review is the demonstration that the tau is a lepton. However, it is useful to summarize this demonstration in this section.

3.1 *Decay Process Signatures*

In this discussion the tau is treated as a sequential lepton. The tau may be an electron-associated ortholepton with the decay $\tau^- \to e^- + \gamma$ strongly suppressed compared to the weak interaction decay modes (see Section 4.4), but this possibility does not alter the present discussion.

A crucial signature for identification of a particle as a sequential lepton is that it decays only via the weak interaction and that the various decay branching ratios are explained by the weak interactions. We can roughly calculate the weak interaction predictions for the τ decay by using Figure 1 and replacing the L, L^0 pair by the τ, ν_τ pair. Because the quark decay mode (Figure 1*c*) occurs in the different colors, there are five diagrams of equal weight. Therefore, we expect that the leptonic decays $\nu_\tau e^- \bar{\nu}_e$ or $\nu_\tau \mu^- \bar{\nu}_\mu$ will each occur 20% of the time and the semileptonic decays via the quark mode will occur 60% of the time.

A more precise calculation of the branching ratios uses (*a*) conventional Weinberg-Salam theory; (*b*) the masses of the τ and the final-state particles; (*c*) some theoretical concepts like the conserved vector current (CVC); and (*d*) some specific experimental parameters, for example, the pion lifetime, required to calculate the decay rate for $\tau^- \to \pi^- \nu_\tau$. Many of these

Table 2 Predictions for τ^- branching ratios

Mode	Branching ratio (%)	Additional input to calculation
$e^- \bar{\nu}_e \nu_\tau$	16.4–18.0	none
$\mu^- \nu_\mu \nu_\tau$	16.0–17.5	none
$\pi^- \nu_\tau$	9.8–10.6	π^- lifetime
$K^- \nu_\tau$	≈ 0.5	θ_{Cabibbo}
$\rho^- \nu_\tau$	20–23	CVC plus e^+e^- annihilation cross sections
$K^{*-} \nu_\tau$	0.8–1.5	Das-Mathur-Okubo sum rules plus θ_{Cabibbo}
$A_1^- \nu_\tau$	8–10	Weinberg sum rules
(2 or more π's, K's)$^- \nu_\tau$	23–25	quark model, CVC, $\sigma(e^+e^- \to \text{hadrons})$, etc
$A_1^- \nu_\tau$ + (2 or more π's, K's)$^- \nu_\tau$ subtotal[a]	31–35	

[a] Does not include the ρ, K^*, or A_1 modes.

calculations were first made by Thacker & Sakurai (1971) and by Tsai (1971). Table 2 gives the branching ratios, based on these references and on the work of Gilman & Miller (1978), Kawamoto & Sanda (1978), Pham, Roiesnel & Truong (1978), and Tsai (1980). We assume a massless ν_τ, spin 1/2 for the τ and ν_τ, V$-$A coupling, and the Weinberg-Salam weak interaction theory; and we use the additional inputs listed in the third column of Table 2. Two of the branching ratios are uncertain. The decay rate of the mode with three or more π's or K's, the multihadron decay mode, is difficult to calculate precisely (Section 5.3). The calculation of the A_1 decay mode depends upon knowing for certain that the A_1 exists, and on knowing the properties of the A_1 (Section 5.2). Since the total of the branching fractions must be one, any change in these decay rates will change all the branching fractions. In addition some of the calculations are uncertain because they depend on experimental data such as the total cross section for $e^+ + e^- \rightarrow$ hadrons. Therefore a range of theoretical predictions is given for some of the branching fractions in Table 2.

Note that the crude prediction using Figure 1 is quite good; the individual leptonic branching ratios are calculated to be 16–18% rather than 20%, so that the total semileptonic branching ratio prediction increases from 60% to 64–68%.

All of the decay modes listed in Table 2, except $\tau^- \rightarrow \nu_\tau + K^-$, have been seen and their measured branching ratios agree with the calculations (Section 5). Of equal importance, τ decays modes that would occur through the strong or electromagnetic interactions have not been found (Section 5.3). The tau's decay processes are thus consistent with it being a lepton and inconsistent with it being a hadron.

G. Feldman has remarked that in the W exchange model of τ decays (Figure 1*a*–*c*) all the decay modes of the τ are decay modes of the W if the ν_τ is excluded. Hence the consistency of the measured with the predicted branching ratios may be thought of as repeated proof that the τ acts as a lepton in the $\tau - W - \nu_\tau$ vertex. The measurements themselves can be viewed as a study of the decay modes of a virtual W.

3.2 e^+e^- *Production Process Signatures*

3.2.1 THEORY There are four general observations we can make about tau production in e^+e^- annihilation.

1. Taus should be produced in pairs via the one-photon exchange process (Figure 2*a*)

$$e^+ + e^- \rightarrow \gamma_{\text{virtual}} \rightarrow \tau^+ + \tau^- \qquad 18.$$

once the total energy, E_{cm}, is greater than twice the τ mass (m_τ).

2. For spin 0 or 1/2, the production cross section for point particles is

known precisely from quantum electrodynamics:

spin 0: $$\sigma_{\tau\tau} = \frac{\pi\alpha^2\beta^3}{3s} \qquad 19.$$

spin 1/2: $$\sigma_{\tau\tau} = \frac{4\pi\alpha^2}{3s}\frac{\beta(3-\beta^2)}{2} \qquad 20.$$

where $s = E^2_{\text{cm}}$; $\beta = v/c$, v being the cms velocity of the τ and c being the velocity of light; and α is the fine structure constant. The $\sigma_{\tau\tau}$ for higher spins was discussed by Tsai (1978), Kane & Raby (1980), and Alles (1979). I restrict further discussion in this section to spin 1/2, which is appropriate to the τ. It has become customary in e^+e^- annihilation physics to remove the $1/s$ dependence of cross sections (Equations 19 and 20) by defining

$$R = \sigma/\sigma_{e^+e^-\to\mu^+\mu^-}, \qquad 21.$$

where

$$\sigma_{e^+e^-\to\mu^+\mu^-} = \frac{4\pi\alpha^2}{3s} \qquad 22.$$

Then for spin 1/2 we expect

$$R_\tau = \frac{\beta(3-\beta^2)}{2}, \qquad 23.$$

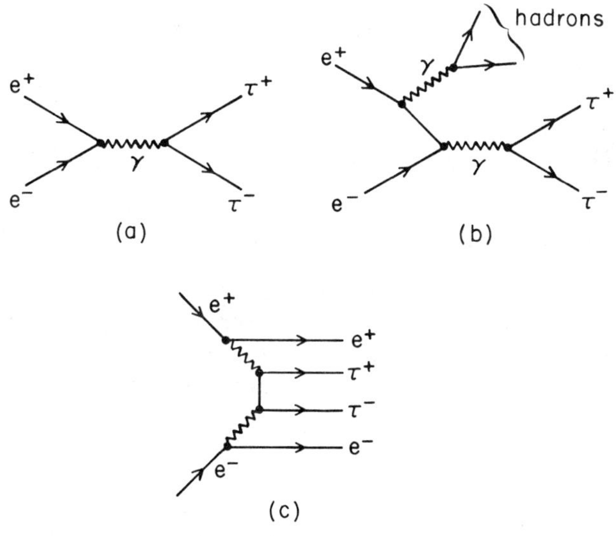

Figure 2 Diagrams for (*a*) $e^+e^- \to \tau^+\tau^-$ via one-photon exchange, (*b*) higher order production of a τ pair with hadrons, and (*c*) $e^+e^- \to e^+e^-\tau^+\tau^-$ via a two-virtual-photon process.

which has the simple property that $R_\tau \to 1$ as $\beta \to 1$; that is, as $E_{\rm cm}$ rises above the τ threshold. If the τ has internal structure, then Equation 20 is modified by a form factor $F(s)$

$$\sigma_{\tau\tau} = \frac{4\pi\alpha^2}{3s} \frac{\beta(3-\beta^2)}{2} |F(s)|^2. \tag{24.}$$

We expect that the internal structure will cause

$$|F(s)| \ll 1, \quad \text{when } E_{\rm cm} \gg 2m_\tau. \tag{25.}$$

This is what happens in pair production of hadrons such as $e^+e^- \to \pi^+\pi^-$ or $e^+e^- \to p\bar{p}$. A point particle has $F(s) = 1$ for all s.

3. The production process

$$e^+ + e^- \to \tau^+ + \tau^- + \text{hadrons} \tag{26.}$$

should be very small compared to $e^+ + e^- \to \tau^+ + \tau^-$. This is because for a lepton the reaction in Equation 26 can only occur in a higher order process such as the one in Figure 2*b*, where two extra powers of α will appear in the cross section. On the other hand, for hadrons the reaction in Equation 26 is the common one. For example: in the region of several GeV the cross section for $e^+ + e^- \to K^+ + K^- + \text{hadrons}$ is much larger than the cross section for $e^+ + e^- \to K^+ + K^-$.

4. At sufficiently high energy, tau pairs should be produced in higher order electromagnetic processes (Figure 2*c*) such as

$$e^+ + e^- \to \tau^+ + \tau^- + e^+ + e^- \tag{27.}$$

and

$$e^+ + e^- \to \tau^+ + \tau^- + \gamma + \gamma \tag{28.}$$

These production processes are discussed in Section 6 (future studies of the τ) because there are not yet any published data on them.

3.2.2 EXPERIMENTAL RESULTS BELOW 8 GeV Since the τ decays before detection, all production cross section measurements depend upon detection of some set of τ decay modes. Two sets have been used.

1. $e^\pm\mu^\mp$ events: the production and decay sequence

$$\begin{aligned} &e^+ + e^- \to \tau^+ + \tau^- \\ &\tau^+ \to e^+ + \nu_e + \bar{\nu}_\tau \\ &\tau^- \to \mu^- + \bar{\nu}_\mu + \nu_\tau \end{aligned} \tag{29.}$$

leads to $e^\pm\mu^\mp$ pairs being the only detected particles in the event. These two-prong, total charge zero, $e\mu$ events constitute a very distinctive signature, and thereby led to the discovery of the tau (Perl 1975, Perl et al 1975). The SPEAR data (Perl 1977a) on the energy dependence of the production of such events are shown in Figures 3 and 4. Figure 3 shows $R_{e\mu,\mathrm{observed}}$, defined as

$$R_{e\mu,\mathrm{observed}} = 2R_\tau B(\tau \to e\nu\text{'s})B(\tau \to \mu\nu\text{'s})A_{e\mu}, \qquad 30.$$

where the B's are branching fractions and $A_{e\mu}$ is the acceptance and efficiency of the apparatus. In the apparatus used for these data (the SLAC-LBL Mark I magnetic detector) $A_{e\mu}$ was almost independent of E_{cm}; hence the $R_{e\mu,\mathrm{observed}}$ values are proportional to R_τ. Note that the sharp threshold at about 3.7 GeV and the leveling out of the R value as E_{cm} increases above 5 GeV are in agreement with Equation 23. This is shown explicitly in Figure 4 (Perl 1977a) where R_τ was calculated from Equation 30, and is in agreement with Equation 23—the theoretical curve.

2. $e^\pm$ hadron$^\mp$ and $\mu^\pm$ hadron$^\mp$ events: the production and decay

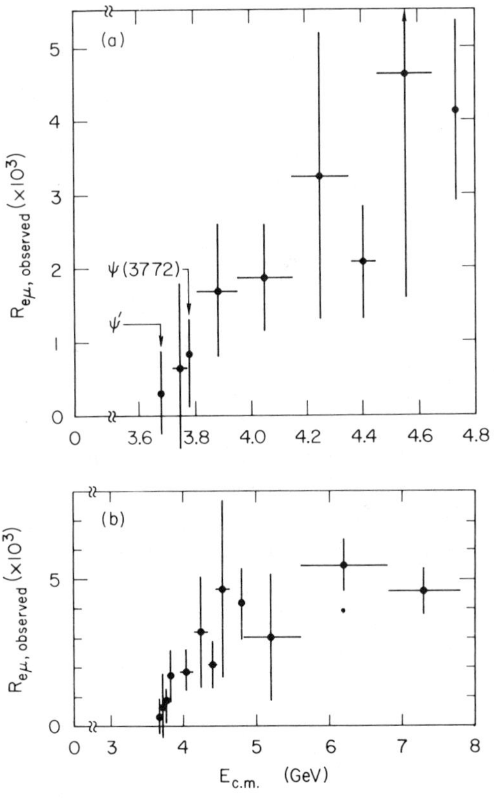

Figure 3 $R_{e\mu,\mathrm{observed}}$ for (*a*) $3.6 \leqq E_{\mathrm{cm}} \leqq 4.8$ GeV and (*b*) $3.6 \leqq E_{\mathrm{cm}} \leqq 7.8$ GeV (Perl 1977a).

sequence

$$\begin{aligned} &e^+ + e^- \to \tau^+ + \tau^- \\ &\tau^+ \to e^+ + \nu_e + \bar{\nu}_\tau \\ &\tau^- \to \text{hadron}^- + \nu_\tau, \quad (\text{hadron} = \pi \text{ or } K) \end{aligned} \qquad 31.$$

leads to two-prong events consisting of one hadron and one opposite-sign electron. The restriction to events with one hadron is necessary to reduce the background (Brandelik et al 1978, Kirkby 1979) from charmed-particle production and decay processes such as

$$\begin{aligned} &e^+ + e^- \to D^+ + D^- \\ &D^+ \to e^+ + \nu_e + K^0 \\ &D^- \to \text{hadrons} \end{aligned}$$

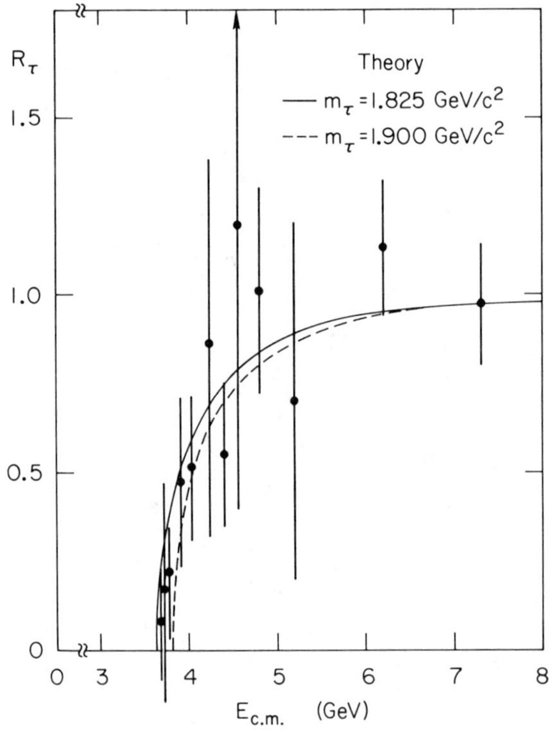

Figure 4 R_τ compared to theoretical curves for point-like, spin-1/2 particles for two τ masses (Perl 1977a). These older data indicated a τ mass between 1825 and 1900 MeV/c^2; the presently accepted value is 1782 MeV/c^2.

since charm-related events are predominantly multihadronic. Figure 5 shows early results from the DASP detector at DORIS (Brandelik et al 1978); more recent results from the DELCO detector at SPEAR are given in Figure 6 (Kirkby 1979). The data in both figures are consistent with Equation 23.

If the τ decays to μ + neutrinos instead of e + neutrinos in Equation 31, then $\mu^{\pm}$ hadron$^{\mp}$ events are produced. These μ-single-hadron events can also be separated from charmed-particle-related μ-multihadron events, and

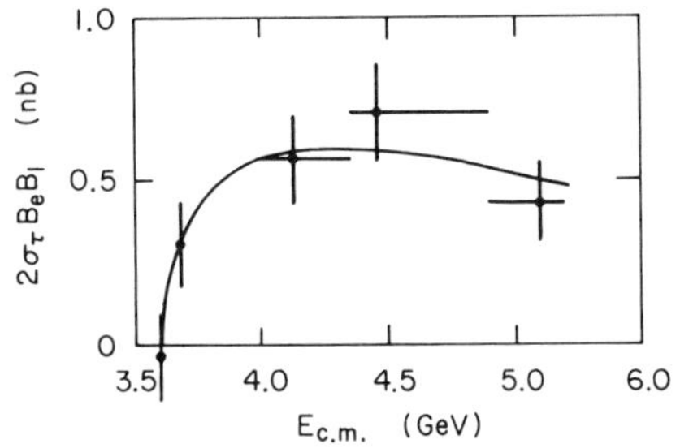

Figure 5 Cross section for e-hadron events from DASP (Brandelik et al 1978) compared to theoretical curve for point-like, spin-1/2 particles.

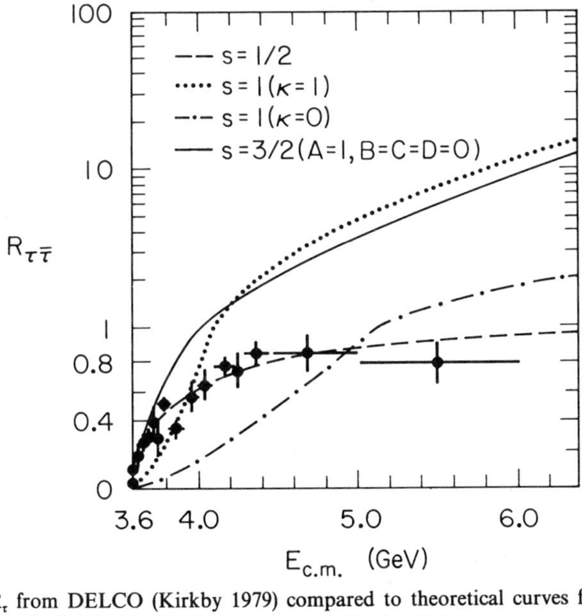

Figure 6 R_τ from DELCO (Kirkby 1979) compared to theoretical curves for point-like particles with spin 0, 1/2, 1, or 3/2. The constants A, B, C, and D are vertex parameters (Tsai 1978).

were very important in the early work on the τ (Cavalli-Sforza et al 1976, Feldman et al 1976, Burmester et al 1977a,b).

The eμ, e-hadron, and μ-hadron data discussed above, and all other published data on e-hadron and μ-hadron events (e.g. Barbaro-Galtieri et al 1977, Bartel et al 1978), are consistent with Equation 23 and hence with Equation 20. Therefore all the production data are consistent with the τ being a point particle with spin 1/2, required properties for a lepton (Section 1.1). We can discuss this more quantitatively by examining how much the form factor $F_\tau(s)$ (Equation 24) is allowed to deviate from one. There are two possibilities: (*a*) We expect that this deviation will be largest at high E_{cm}. Looking at the higher energy data in Figure 4 we see that R_τ is consistent with $F(s) = 1$ within about $\pm 20\%$. Therefore these data allow a maximum deviation of $F(s)$ from one of about $\pm 10\%$ at these energies. (*b*) Alternatively, we can use a model for $F(s)$; the usual choice is (Hofstadter 1975, Barber et al 1979)

$$F(s) = 1 \mp \frac{s}{s - \Lambda_\pm^2}. \tag{32}$$

Note that the larger Λ is, the less $F(s)$ deviates from one. This model has recently been applied to very high energy data from the PETRA e^+e^- colliding-beams facility at DESY.

3.2.3 EXPERIMENTAL RESULTS ABOVE 8 GeV As E_{cm} increases in the PETRA and PEP energy range, $\sim$10–40 GeV, τ pair events become increasingly distinctive for two reasons. (*a*) The increased energy of each of the τ's causes their respective decay products to move in opposite and roughly colinear directions. (*b*) Since the τ decays predominantly to one or three charged particles, the total charged multiplicity of τ events is small compared to the average charged multiplicity of hadronic events (≈ 12) (Wolf 1979). Figure 7 from the PLUTO group at PETRA is an example. A particularly striking signature (Barber et al 1979) is

$$\begin{aligned} &e^+ + e^- \to \tau^+ + \tau^- \\ &\tau^+ \to \mu^+ + \text{neutrinos} \\ &\tau^- \to \text{hadrons}^- \text{ or } \nu^- \text{ or } e^- + \text{neutrinos.} \end{aligned} \tag{33}$$

The restriction to single hadron decays used in Section 3.2.2 is no longer necessary.

Figure 8 from the Mark-J Collaboration at PETRA (Barber et al 1979) shows the production cross section for τ pairs using the signature in Equation 33. The application of Equation 32 to these data yields, with 95% confidence (Barber et al 1979),

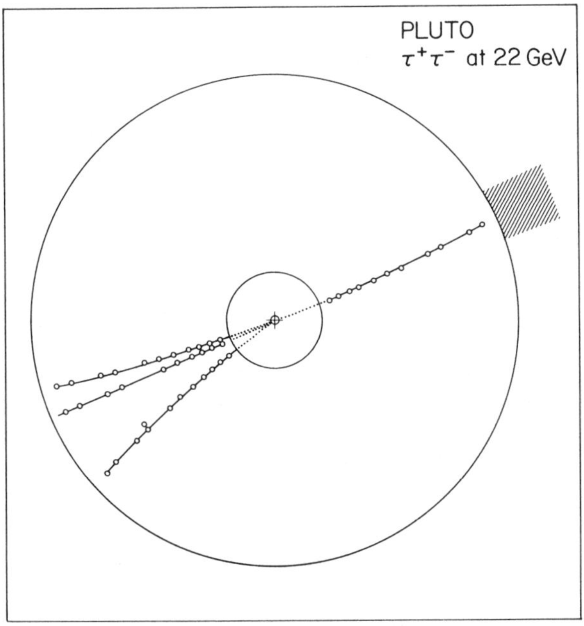

Figure 7 An example of a τ pair event at $E_{\mathrm{cm}} = 22$ GeV obtained by the PLUTO group at PETRA. The single track is an electron, and the three tracks going in the other direction are pions.

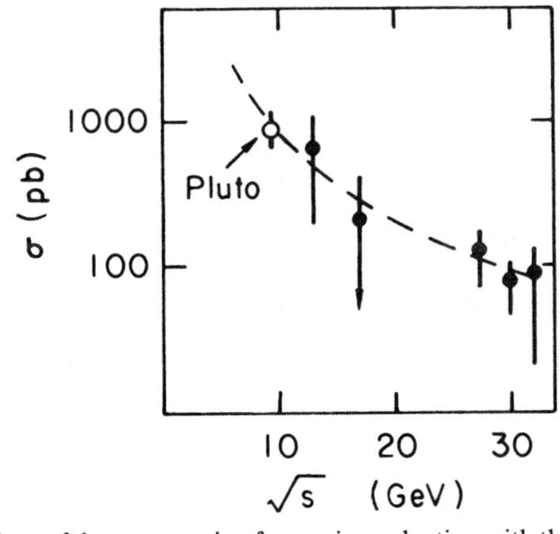

Figure 8 Comparison of the cross section for τ pair production with the theoretical curve for a point-like spin-1/2 particle. The data are from the Mark J detector at PETRA (Barber et al 1979), except for the point marked PLUTO obtained by the PLUTO group at DORIS.

$$\Lambda_- > 53 \text{ GeV}, \quad \Lambda_+ > 47 \text{ GeV}. \tag{34}$$

These lower limits on Λ_τ mean that the data show no deviation from $F(s) = 1$, as we can see directly from Figure 8. Hence these new PETRA data are consistent with the τ being a point particle. Very recent results from the TASSO Collaboration at PETRA (Brandelik et al 1980) are also consistent with the τ being a point particle.

4 PROPERTIES OF THE TAU

4.1 *Mass*

The τ mass (m_τ) is best determined by measuring the threshold for $e^+ + e^- \rightarrow \tau^+ + \tau^-$. Table 3 summarizes three recent measurements, all of which are consistent. The DELCO measurement is based on the largest statistics and I use that value: 1782^{+3}_{-4} MeV/c^2.

4.2 *Spins, $\tau - \nu_\tau$ Coupling, and ν_τ Mass*

The general form for the $\tau - \nu_\tau$ weak interaction four-current is

$$J^\lambda_{\tau\nu_\tau} = g_\tau(\psi^\dagger_{\nu_\tau}\mathcal{O}\psi_\tau)^\lambda, \quad \lambda = 0, 1, 2, 3 \tag{35}$$

where g is a coupling constant, ψ_{ν_τ} and ψ_τ are spin functions, and $\mathcal{O}$ is an operator. The decay of the τ is dependent on the current. For example, in the decay $\tau^- \rightarrow \nu_\tau + e^- + \bar{\nu}_e$ the four-particle matrix element (Figure 1*d*) is

$$T = J^\lambda_{\tau\nu_\tau} j_{\lambda,e\bar{\nu}_e}, \tag{36}$$

where

$$j_{\lambda,e\bar{\nu}_e} = g\bar{u}_{\nu_e}\gamma_\lambda(1-\gamma_5)u_e. \tag{37}$$

Here the u's are the usual Dirac spinors and the γ's are the usual Dirac γ matrices (Bjorken & Drell 1964). Equation 37 is of course the conventional V−A weak interaction current (Bjorken & Drell 1964). Also

$$\sqrt{2}g^2 = G_F = 1.02 \times 10^{-5}/M^2_{\text{proton}}. \tag{38}$$

Hence the determination of $J^\lambda_{\tau\nu_\tau}$ involves the simultaneous determination of

Table 3 Measurements of the τ mass using the production threshold

Experiment	Mass (MeV/c^2)	Figure	Reference
DELCO	1782^{+3}_{-4}	9	Bacino et al 1978, Kirkby 1979
DASP	1807 ± 20	5	Brandelik et al 1978
DESY-HEIDELBERG	1787^{+10}_{-18}	10	Bartel et al 1978

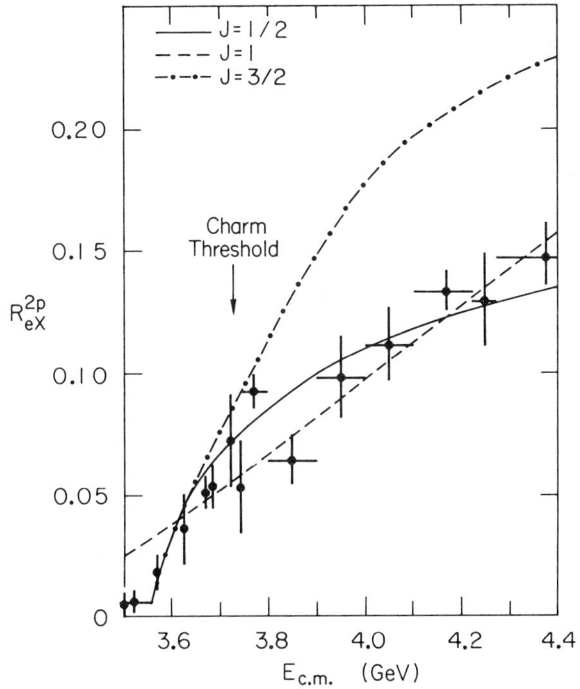

Figure 9 The threshold behavior of R_{ex} for e-hadron and eμ events from DELCO (Kirkby 1979).

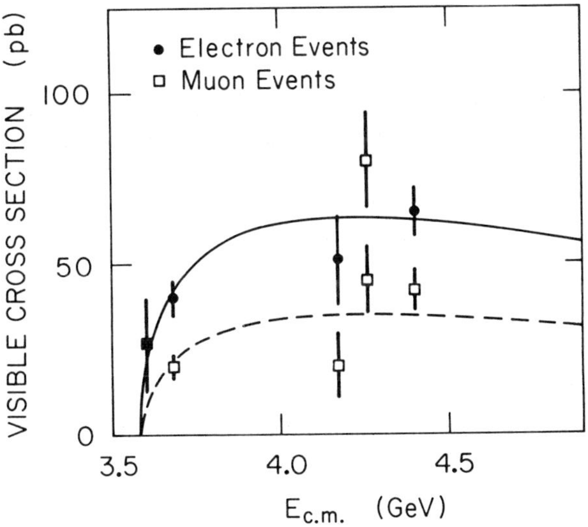

Figure 10 The threshold behavior of the cross section for e-hadron and μ-hadron events measured by the DESY-Heidelberg group (Bartel et al 1978). The theoretical curve is for a point-like, spin-1/2 particle.

g_τ, the τ spin, the ν_τ spin, the form of the operator $\mathcal{O}$, and even the determination of the ν_τ mass.

Fortunately, as discussed in Section 3.2.1, the τ pair production cross section depends upon the τ spin; hence we can independently determine that quantity. Equation 19 for spin 0 predicts a maximum R_τ value of 0.25; however, the measured maximum value is 1, hence spin 0 is excluded. Spin 1 and higher integral spins predict a β^3 factor (Tsai 1978). The behavior of $\sigma_{\tau\tau}$ near threshold (Figure 9) excludes a β^3 behavior and hence spin 1 or higher integral spins.

If the τ is not a point particle, spin 3/2 or higher half integral spins cannot at present be excluded. One can always select the arbitrary, energy-dependent parameters that occur in the τ-γ-τ vertex for spin 3/2, 5/2, . . . , so that $\sigma_{\tau\tau}$ mimics the spin-1/2 case (Kane & Raby 1980). Such a model must also explain the $\tau^- \rightarrow \nu_\tau + \pi^-$ branching ratio, which requires spin 1/2 for a point particle (Alles 1979, Kirkby 1979). Further tests of the higher half integral spin proposal can be made; however, my strong instincts are to accept the spin-1/2 assignment.

The ν_τ spin is limited to half integral values by the existence of decay modes such as $\tau^- \rightarrow \nu_\tau + \pi^-$ and $\tau^- \rightarrow \nu_\tau + e^- + \bar{\nu}_e$. Furthermore, the measured value of the decay rate for $\tau^- \rightarrow \nu_\tau + \pi^-$ excludes spin 3/2 for a point-like τ or a point-like ν_τ (Alles 1979, Kirkby 1979). Higher half integral spins for the ν_τ have not been analyzed. Once again I use Ockham's razor and assign to the ν_τ a spin of 1/2 as the simplest hypothesis consistent with all observations on the ν_τ and τ.

Equation 35 now reduces to

$$J^\lambda_{\tau\nu_\tau} = g_\tau \bar{u}_{\nu_\tau} \mathcal{O}^\lambda u_\tau \qquad 39.$$

where the u's are Dirac spinors and $\mathcal{O}$ is some combination of scalar, pseudoscalar, vector, axial vector, and tensor current operators (Bjorken & Drell 1964, Rosen 1978). Insertion of Equation 39 into the matrix element (Equation 36) for

$$\tau^- \rightarrow \nu_\tau + e^- + \bar{\nu}_e$$

or 40.

$$\tau^- \rightarrow \nu_\tau + \mu^- + \bar{\nu}_\mu$$

shows that the e^- or μ^- momentum spectrum will depend on the form of $\mathcal{O}$. A similar situation occurred in the determination of the matrix element in μ decay

$$\mu^- \rightarrow \nu_\mu + e^- + \bar{\nu}_e \qquad 41.$$

(Marshak, Riazuddin & Ryan, 1969); except in that case the electron polarization was also measured, whereas in the τ case (Equation 40) the e^-

or μ^- polarization has not been measured. As discussed by Marshak, Riazuddin & Ryan (1969) the measurement of the e^- or μ^- momentum spectrum does not completely determine $\mathcal{O}$ in Equation 39. This dilemma has been resolved by all experimenters who have worked on the τ by (*a*) assuming that only vector (V) and axial vector (A) currents occur in Equation 39, as in all other weak interactions; and (*b*) allowing the relative strengths of the V and A currents to be fixed by measurement.

With these assumptions, Equation 39 reduces to

$$J^{\lambda}_{\tau\nu_\tau} = g_\tau \bar{u}_{\nu_\tau}\gamma^{\lambda}(v - a\gamma_5)u_\tau;\ v, a \text{ real}, \quad v^2 + a^2 = 1. \qquad 42.$$

Table 4 gives values of v and a for special choices of the current. In the τ rest system the normalized momentum distribution of the e or μ in Equation 40 is

$$\mathrm{d}P/\mathrm{d}y = y^2[6(v-a)^2(1-y) + (v+a)^2(3-2y)] \qquad 43.$$

where y = e or μ momentum/maximum e or μ momentum, and the e or μ mass and all neutrino masses are set to zero (Bjorken & Drell 1964). Of course, the e or μ momentum is measured in the laboratory system where the τ is in motion, and Equation 43 must be transformed properly.

Early studies of the v and a parameters were carried out using eμ events (Perl et al 1976, Barbaro-Galtieri et al 1977), e-hadron events (Yamada 1977, Brandelik et al 1978, Barbaro-Galtieri et al 1977), and μ-hadron events (Feldman et al 1976, Burmester et al 1977a,b). All these studies are consistent with V−A; Figure 11 from the PLUTO group is an example. Where the statistics are sufficient V+A is excluded (Figure 12).

The most definitive study was carried out by the DELCO group (Kirkby 1979) (Figure 13), who determined the ρ Michel parameter (Michel 1950). With this parameter Equation 43 becomes

$$\mathrm{d}P/\mathrm{d}y = 4y^2[3(1-y) + (2\rho/3)(4y-3)] \qquad 44.$$

and

$$\rho = 3(v+a)^2/8. \qquad 45.$$

Table 4 Parameters for the $\tau - \nu_\tau$ current

Type	v	a	ρ
V−A	$\frac{1}{\sqrt{2}}$	$\frac{1}{\sqrt{2}}$	3/4
pure V	1	0	3/8
pure A	0	1	3/8
V+A	$\frac{1}{\sqrt{2}}$	$\frac{-1}{\sqrt{2}}$	0

They find $\rho = 0.72 \pm 0.15$, assuming the ν_τ mass is zero, which is in excellent agreement with V−A and in disagreement with V, A, or V+A (Table 4). Kirkby (1979) presents other evidence for V−A using $\langle p_e \rangle / E_{cm}$.

The effect of a nonzero ν_τ mass on the e or μ momentum spectrum is shown in Figure 12; the larger the ν_τ mass, the fewer the events with very energetic e's or μ's. Thus the ν_τ mass determination interacts with the V−A test. It has been shown that V+A is excluded for any value of the ν_τ mass (Bacino et al 1978, Kirkby 1979). However, the limited statistics of all the momentum spectrum measurements allow some deviation from V−A combined with some deviation of the ν_τ mass from zero. No study of the combined deviations has been done. Indeed, it has become conventional to assume that the $\tau - \nu_\tau$ current is precisely V−A and then to set an upper limit on the ν_τ mass. Table 5 gives three such determinations in historical order; all are consistent with a zero mass, although the limits are still quite

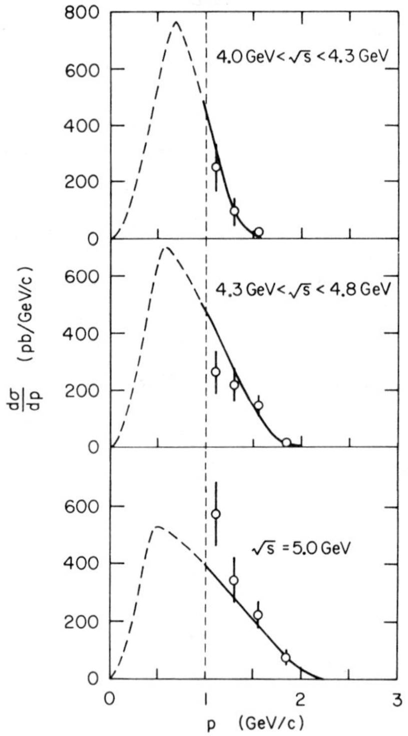

Figure 11 The momentum spectrum for muons in μ-hadron events obtained by the PLUTO group (Burmester et al 1977a). Muons with momenta less than the vertical dashed line could not be identified in the detector. The theoretical curve is for the decay $\tau^- \rightarrow \nu_\tau + \mu^- + \bar{\nu}_\mu$ with V−A coupling, all spins 1/2, and a massless ν_τ.

large because the spectra involve the ratio of the squares of the neutrino and tau masses.

To summarize this section: all measurements are consistent with

$$\begin{aligned} \tau \text{ spin} &= 1/2 \\ \nu_\tau \text{ spin} &= 1/2 \\ \tau - \nu_\tau \text{ current} &= \text{V} - \text{A} \\ \nu_\tau \text{ mass} &= 0 \end{aligned} \qquad 46.$$

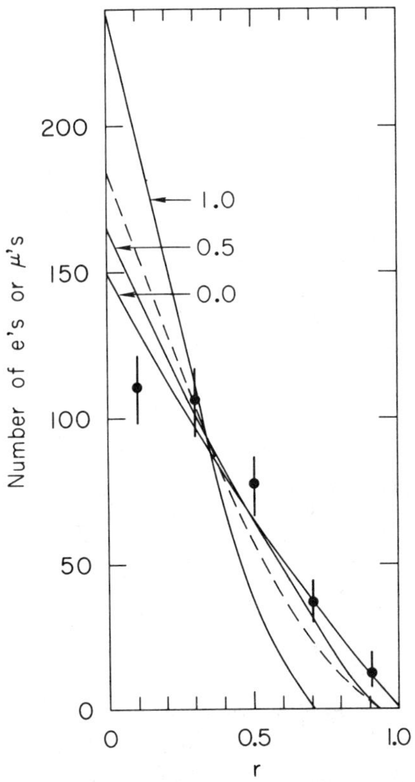

Figure 12 The momentum spectrum of e's or μ's in eμ events from the SLAC-LBL magnetic detector group (Perl et al 1976): $r = (p - 0.65)/(p_{\max} - 0.65)$, where p is the momentum of the e or μ in GeV/c and $p_{\max}$ is its maximum value. The 0.65 constant is the lowest momentum at which e's and μ's could be identified in this detector. The solid theoretical curves are for the decays $\tau^- \to \nu_\tau + e^- + \bar{\nu}_e$ or $\tau^- \to \nu_\tau + \mu^- + \bar{\nu}_\mu$ with V − A coupling, all spins 1/2, and a mass for the ν_τ indicated by the attached number in GeV/c^2. The dashed theoretical curve is for V + A coupling and a massless ν_τ.

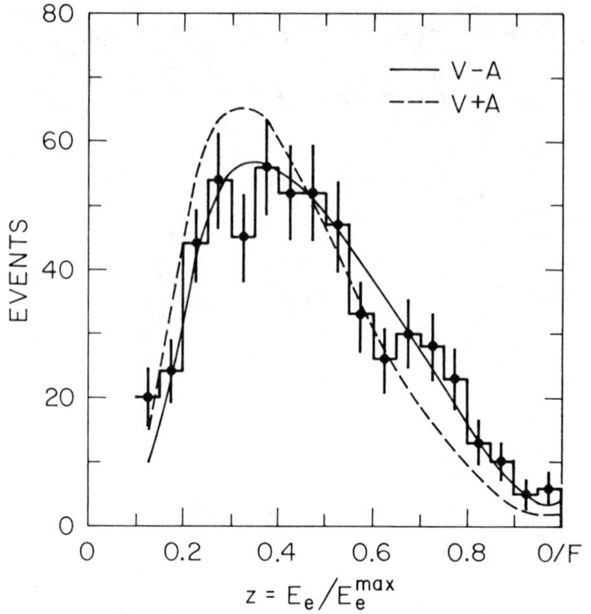

Figure 13 The normalized momentum spectrum for electrons from e-hadron and eμ events from DELCO (Kirkby 1979). The theoretical curves are for the decays $\tau^- \to \nu_\tau + e^- + \bar{\nu}_e$ with all spins 1/2, a massless neutrino, and the indicated coupling. The fits to V − A (*solid*) and V + A (*dashed*) give χ^2 per degrees of freedom of 15.9/17 and 53.7/17, respectively.

Table 5 Upper limits on ν_τ mass

Group	Upper limit on ν_τ mass (MeV/c^2)	Confidence (%)	Reference
SLAC-LBL	600	95	Perl et al 1977b
PLUTO	540	90	Knies 1977
DELCO	250	95	Kirkby 1979

Those are the standard set of parameters of the τ (Tsai 1980). They are, of course, compatible with e-μ-τ universality. As I have noted in this section, there has been a liberal use of Ockham's razor; measurements should continue on questions such as better limits on the ν_τ mass.

4.3 *Lifetime*

We have reduced the $\tau - \nu_\tau$ current to

$$J^\lambda_{\tau\nu_\tau} = g_\tau \bar{u}_{\nu_\tau} \gamma^\lambda (1 - \gamma_5) u_\tau \qquad 47.$$

but the value of g_τ is still undetermined. It cannot be determined by measurement of branching ratios or decay mode distributions because it appears as a constant multiplier in all decay mode matrix elements. One way to determine it is to measure the τ lifetime T_τ; other ways are discussed in Section 6. A specific example is helpful here. The decay rate for $\tau^- \to \nu_\tau + e^- + \bar{\nu}_e$ (Tsai 1971, Thacker & Sakurai 1971) is

$$\Gamma(\nu_\tau e^- \bar{\nu}_e) = \frac{2g_e^2 g_\tau^2 m_\tau^5}{192\pi^3}. \qquad 48.$$

Note the analogy to the $\mu \to \nu_\mu + e^- + \bar{\nu}_e$ decay rate

$$\Gamma(\nu_\mu e^- \bar{\nu}_e) = \frac{G_F^2 m_\mu^5}{192\pi^3}, \qquad 49.$$

where $G_F^2 = 2g^4$. Then

$$T_\tau = \frac{B(\nu_\tau e^- \bar{\nu}_e)}{\Gamma(\nu_\tau e^- \bar{\nu}_e)} = B(\nu_\tau e^- \bar{\nu}_e)\left(\frac{g^2}{g_\tau^2}\right)\left(\frac{m_\mu^5}{m_\tau^5}\right)T_\mu, \qquad 50.$$

where $B(\nu_\tau e^- \bar{\nu}_e)$ is the branching ratio for $\tau^- \to \nu_\tau + e^- + \bar{\nu}_e$ and T_μ is the μ lifetime. Using $B(\nu_\tau e^- \bar{\nu}_e) = 0.17$ (Section 5.1), Equation 50 predicts

$$T_\tau = 2.7 \times 10^{-13} \text{ s, for } g_\tau = g. \qquad 51.$$

The measurement of T_τ requires a study of the average decay length of the τ's produced in $e^+ + e^- \to \tau^+ + \tau^-$; so far experiments have only been able to put an upper limit on this decay length. The smallest upper limit (Kirkby 1979, Bacino et al 1979a) is

$$T_\tau < 2.3 \times 10^{-12} \text{ s, (95\% confidence limit).} \qquad 52.$$

Hence from Equation 50

$$g_\tau^2/g^2 > 0.12 \text{ (95\% confidence limit),} \qquad 53.$$

or defining $G_\tau = \sqrt{2}\, g_\tau^2$,

$$G_\tau/G_F > 0.12 \text{ (95\% confidence limit).} \qquad 54.$$

This is consistent with $G_\tau = G_F$, and hence with e-μ-τ universality and with Weinberg-Salam theory; but $G_\tau = G_F$ has not yet been proven.

4.4 *Lepton Type and Associated Neutrino*

Is the τ a sequential lepton with a unique lepton number, or does it share lepton number with the μ or e? Consider the μ first. If the τ were a μ-related ortholepton (Section 2.2.2) the reaction

$$\nu_\mu + \text{nucleus} \to \tau^- + \text{anything} \qquad 55.$$

should occur. Analogously if the τ were a μ-related paralepton (Section 2.2.3)

$$\nu_\mu + \text{nucleus} \rightarrow \tau^+ + \text{anything} \qquad 56.$$

would occur. Cnops et al (1978) looked for the reactions in Equations 55 and 56 by using a neon-hydrogen-filled bubble chamber exposed to a ν_μ beam. They found no events; their upper limit on $G_{\tau-\nu_\mu}$ was $G_{\tau-\nu_\mu}/G_F < 0.025$ with 90% confidence. However, in these μ-related models $G_{\tau-\nu_\mu}$ is identical with what I call G_τ in Equation 54, and must have the value $G_{\tau-\nu_\mu}/G_F > 0.12$. This excludes the μ-related ortholepton and paralepton models.

The equivalent tests for e-related models have not been done because ν_e beams have not been built. Heile et al (1978) excluded the e-related paralepton model using an argument of Ali & Yang (1976). If the τ were an e-related paralepton its leptonic decay modes would be

$$\begin{aligned} &\tau^- \rightarrow \bar{\nu}_e + e^- + \bar{\nu}_e \\ &\tau^- \rightarrow \bar{\nu}_e + \mu^- + \bar{\nu}_\mu. \end{aligned} \qquad 57.$$

The two identical $\bar{\nu}_e$'s in Equation 57 constructively interfere leading to the decay width ratio (Ali & Yang 1976)

$$\Gamma(\tau^- \rightarrow e^- + \text{neutrinos})/\Gamma(\tau^- \rightarrow \mu^- + \text{neutrinos}) \approx 2. \qquad 58.$$

The measurements of Heile et al (1978) show this ratio to be close to one; hence this model is wrong.

There is no way, using present data, to exclude the e-related ortholepton model for the τ. That is, it is possible that the τ^- and e^- have the same lepton number. The decay mode

$$\tau^- \rightarrow e^- + \gamma \qquad 59.$$

is then allowed. It has not been seen; the upper limit is 2.6% (Section 5.3). However, this does not exclude this model because the τ-γ-e vertex can be suppressed (Section 2.2.2).

More complicated models (Altarelli et al 1977, Horn & Ross 1977) such as the τ decaying into a mixture of ν_e and ν_μ have been considered and excluded. They are reviewed by Gilman (1978) and Feldman (1978b). However, one can always devise a complicated model that cannot be excluded. For example, suppose the τ shares its neutrino with a heavier τ'. That model cannot be excluded using present data, and if true the τ would not be a sequential lepton.

As we have done before, we select the simplest model consistent with all the data; that is the sequential model. It is, of course, consistent with e-μ-τ universality and with Weinberg-Salam theory.

5 DECAY MODES OF THE TAU

5.1 *Purely Leptonic Decay Modes*

Conventional weak interaction theory (Bjorken & Drell 1964) predicts that the decay width for

$$\tau^- \rightarrow \nu_\tau + l^- + \bar{\nu}_l, \quad l = e \text{ or } \mu$$

is

$$\Gamma(\nu_\tau l^- \bar{\nu}_l) = \frac{G_F^2 m_\tau^5}{192\pi^3}, \qquad 60.$$

where the mass of the l is neglected. This is a very basic and simple calculation and the only parameter is G_F; hence, the measurement of the purely leptonic branching fractions is crucial.

Feldman (1978b) reviewed all the data on B_e and B_μ, the branching fractions for

$$\tau^- \rightarrow \nu_\tau + e^- + \bar{\nu}_e$$

and 61.

$$\tau^- \rightarrow \nu_\tau + \mu^- + \bar{\nu}_\mu$$

respectively. All measurements are consistent and their average values (Table 6) agree with the theoretical predictions from Table 2. Table 6 gives two sets of values for B_e and B_μ. One set assumes they are unrelated; the other set assumes they are connected via

$$\Gamma(\nu_\tau \mu^- \bar{\nu}_\mu)/\Gamma(\nu_\tau e^- \bar{\nu}_e) = 1 - 8y + 8y^3 - y^4 - 12y^2 \ln y = 0.972, \qquad 62.$$

where $y = (m_\mu/m_\tau)^2$ (Tsai 1971).

5.2 *Single Hadron or Hadronic Resonance Decay Modes*

In this section I consider the decay modes

$$\tau^- \rightarrow \nu_\tau + \pi^-$$

$$\tau^- \rightarrow \nu_\tau + K^-$$

$$\tau^- \rightarrow \nu_\tau + \rho^-$$

$$\tau^- \rightarrow \nu_\tau + K^*(890)^-$$

$$\tau^- \rightarrow \nu_\tau + A_1^-.$$

The first mode, $\tau^- \rightarrow \nu_\tau + \pi^-$, has the decay width (Tsai 1971)

$$\Gamma(\nu_\tau \pi^-) = \frac{G_F^2 f_\pi^2 \cos^2\theta_c m_\tau^3}{16\pi}\left(1 - \frac{m_\pi^2}{m_\tau^2}\right)^2. \qquad 63.$$

Table 6 Purely leptonic decay mode branching fractions (in %) (Feldman 1978b)

	B_e and B_μ free	$B_\mu = 0.972B_e$	Theory
B_e	16.5 ± 1.5	17.5 ± 1.2	16.4–18.0
B_μ	18.6 ± 1.9	17.1 ± 1.2	16.0–17.5

Here in comparison to Equation 60 two more parameters appear: θ_c, the Cabibbo angle, and f_π, the coupling constant that appears in the π decay width (Tsai 1971)

$$\Gamma(\pi^- \to \bar{\nu}_\mu + \mu^-) = \frac{G_F^2 f_\pi^2 \cos^2 \theta_c}{8\pi} m_\pi m_\mu^2 \left(1 - \frac{m_\mu^2}{m_\pi^2}\right)^2. \qquad 64.$$

However, $f_\pi^2 \cos^2 \theta_c$ can be evaluated experimentally from Equation 64, so the calculation of $\Gamma(\nu_\tau \pi^-)$ is firm.

Table 7 gives four published branching ratios for this decay mode. They are consistent with each other and with the theoretical prediction of 9.8–10.6% (Table 2). Dorfan (1979) recently gave a preliminary value of 10.7 $\pm 2.1\%$ based on a new analysis of SLAC-LBL Mark II data.

The second mode, $\tau^- \to \nu_\tau + K^-$, is Cabibbo suppressed, and weak interaction theory predicts (Tsai 1971)

$$\Gamma(\nu_\tau K^-)/\Gamma(\nu_\tau \pi^-) = \tan^2 \theta_c \frac{(1 - m_K^2/m_\tau^2)^2}{(1 - m_\pi^2/m_\tau^2)^2}. \qquad 65.$$

The smallness of this branching fraction, $\tan^2 \theta_c \approx 0.05$, and the difficulty of separating K's from the much larger π background has so far prevented the measurement of this mode.

Table 7 Branching fractions for $\tau^- \to \nu_\tau + \pi$

Experiment	Mode	Branching fraction[a] (%)	Reference
SLAC-LBL	$x\pi$	$9.3 \pm 1.0 \pm 3.8$	Feldman 1978b
PLUTO	$x\pi$	$9.0 \pm 2.9 \pm 2.5$	Alexander et al 1978a
DELCO	$e\pi$	$8.0 \pm 3.2 \pm 1.3$	Bacino et al 1979b
SLAC-LBL	$x\pi$	$8.0 \pm 1.1 \pm 1.5$	Hitlin 1978
	$e\pi$	$8.2 \pm 2.0 \pm 1.5$	
Average		8.3 ± 1.4	

[a] The first error is statistical, the second systematic.

The two measurements of $\tau^- \to \nu_\tau + \rho^-$ give

$$B(\tau^- \to \nu_\tau + \rho^-) = 20.5 \pm 4.1\% \quad \text{(Abrams et al 1979, Dorfan 1979)} \qquad 66.$$

$$B(\tau^- \to \nu_\tau + \rho^-) = 24 \pm 9\% \quad \text{(Brandelik et al 1979).}$$

These values are consistent with theoretical predictions of 20–23% (Table 2). Figure 14 shows that the ρ momentum spectrum is consistent with the flat spectrum expected for this two-body decay.

The decay mode $\tau^- \to \nu_\tau + K^*(890)^-$, like $\tau^- \to \nu_\tau + K^-$, is Cabibbo suppressed and will be suppressed relative to $\tau^- \to \nu_\tau + \rho^-$ by a factor of $\tan^2 \theta_c \approx 0.05$. Hence a branching fraction of about 1% is expected. The first measurement of this mode was reported by Dorfan (1979) using SLAC-LBL Mark II data:

$$B[\tau^- \to \nu_\tau + K^*(890)^-] = 1.3 \pm 0.4 \pm 0.3\%, \qquad 67.$$

in good agreement with theory. In Equation 67 the first error is statistical and the second is systematic.

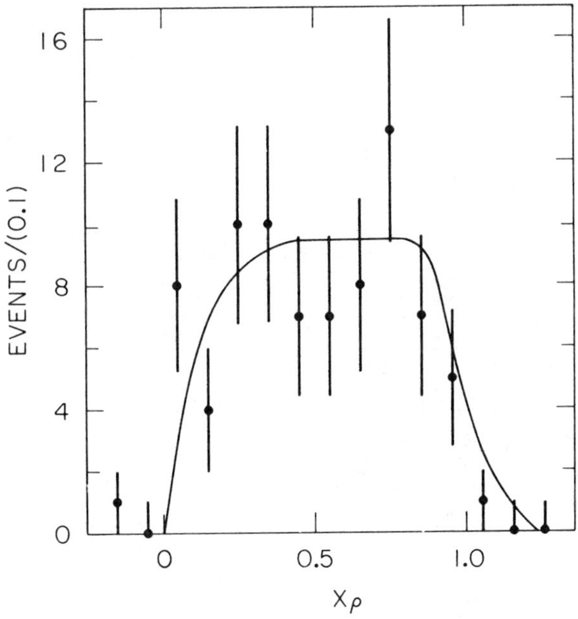

Figure 14 The distribution $X_\rho = (E_\rho - E_{min})/(E_{max} - E_{min})$ where E_ρ is the ρ energy and E_{max} and E_{min} are its maximum and minimum values from the SLAC-LBL magnetic detector (Abrams et al 1979). The theoretical curve is for $\tau^- \to \nu_\tau + \rho^-$ corrected for detector acceptance and measurement errors. Without these corrections the curve would be flat for $X_\rho = 0$ to 1 and zero above $X_\rho = 1$.

The search for the decay mode $\tau^- \to \nu_\tau + A_1^-$ is important (Sakurai 1975) because it may be the only way to establish the existence of the A_1 resonance, which is assumed to have mass ≈ 1100 MeV/c^2, $I^G = 1^-$, $J^P = 1^+$ (Bricman et al 1978). It is because of this importance that I consider the present data suggestive of the A_1's existence but not yet conclusive. The experimental analysis is difficult for two reasons: (*a*) The mode is found via

$$\tau^- \to \nu_\tau + A_1^-, \qquad A_1^- \to \pi^- + \pi^+ + \pi^- \qquad 68.$$

and this has to be separated from higher multiplicity $\pi^\pm$ and π^0 decay modes in which some π's are undetected; (*b*) After the $\pi^-\pi^+\pi^-$ mode is separated out, one has to show that it contains a resonance with the expected properties (Basdevant & Berger 1978).

I only summarize the experimental situation here. The PLUTO group published two papers giving evidence for the A_1 (Alexander et al 1978b, Wagner et al 1980). Figure 15 shows their recent analysis (Wagner et al 1980) using

$$\tau^- \to \nu_\tau + \rho^0 + \pi^- \to \nu_\tau + \pi^- + \pi^+ + \pi^-. \qquad 69.$$

The SLAC-LBL group has also studied this decay mode (Jaros et al 1978); Figure 16 shows their data on Equation 68.

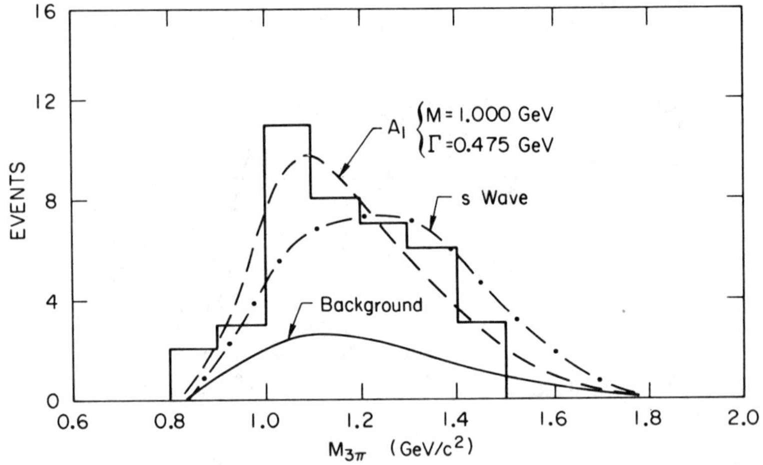

Figure 15 Invariant $\rho^0\pi$ mass distribution from the PLUTO group (Wagner et al 1980), compared to an s-wave without and with an imposed A_1 resonance of mass $= 1.0$ GeV/c^2 and width $= 0.475$ GeV/c (*dashed curve*), added to the expected background (*solid curve*).

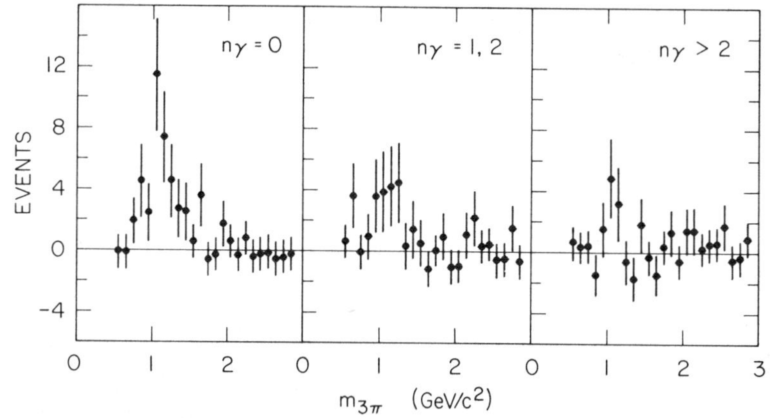

Figure 16 Invariant $(3\pi)^{\pm}$ mass distribution from four-prong events $\mu^{\mp}(3\pi)^{\pm}$ obtained by the SLAC-LBL magnetic detector (Jaros et al 1978). The number of photons accompanying the event is n_{γ}; hence (*a*) contains the possible $A^{\pm} \rightarrow (3\pi)^{\pm}$ signal, and (*b*) and (*c*) are measures of the possible feeddown background from events with π^0's.

5.3 *Multihadron Decay Modes*

The multihadron decay modes are

$$\tau^- \rightarrow \nu_\tau + (n_\pi \pi + n_K K)^-$$
$$n_\pi + n_K \geqq 2 \qquad 70.$$

where the hadrons do not come from a single resonance such as the ρ or A_1. There is no general method for calculating the decay width $\Gamma(\nu_\tau + n_\pi \pi + n_K K)$ because that would require us to know how a W with a small virtual mass converts into hadrons, and of course we do not yet know this. Instead, several special methods have been used. Decay modes with even numbers of pions were considered by Gilman & Miller (1978). They used CVC to connect the vector hadronic weak interaction current to the vector hadronic electromagnetic current. We have experimental information on the latter since we know the multihadron cross sections for

$$e^+ + e^- \rightarrow \gamma_{\text{virtual}} \rightarrow \text{hadrons}. \qquad 71.$$

That is, Equation 71 is used to predict some of the rates for

$$W_{\text{virtual}} \rightarrow \text{hadrons} \qquad 72.$$

Pham, Roiesnel & Truong (1978) used current algebra and PCAC to calculate rates for the axial vector hadronic current; that is, for the decay modes with odd numbers of pions.

Table 8 Some measurements of multihadron decay modes, including the A_1 resonance if it exists

Mode for τ^-	Branching fraction (%)	Reference
$\nu_\tau + \pi^- + \pi^+ + \pi^- + \geqq 0\pi^0$	18 ± 6.5	Jaros et al 1978
$\nu_\tau + \pi^- + \pi^+ + \pi^-$	7 ± 5	Jaros et al 1978
$\nu_\tau + \pi^- + \pi^+ + \pi^- + \pi^0$	11 ± 7	Jaros et al 1978
$\nu_\tau + \rho^0 + \pi^-$	5.4 ± 1.7	Wagner et al 1980
$\nu_\tau + \geqq 3$ charged particles $\pm \geqq 0\pi^0$	28 ± 6	Kirkby 1979

Unfortunately we do not yet have the required measurements to compare in detail with these calculations. This is because the experimental separation of the various multihadron modes is obscured (*a*) by the great difficulty of detecting π^0's in multihadron events; and (*b*) by the possibility of some charged π's escaping detection. Therefore, it is premature to attempt a precise comparison of theory and measurement of the multihadron decay modes. Table 8 lists existing measurements. All of them are compatible with calculations, given the large errors in the measurements and the uncertainties in the calculations. The branching fraction of $28 \pm 6\%$ for modes with at least two charged hadrons should be compared with the A_1 + multihadron subtotal in Table 2 of 31–35%. The latter are expected to be larger because they include some modes with one charged π and two or more π^0's.

5.4 *Sequential Model Forbidden Decay Modes*

If the τ lepton number were not perfectly conserved or if the τ, e, and μ lepton numbers were slightly mixed, the τ might decay electromagnetically into modes such as

$$\begin{aligned} &\tau^- \rightarrow e^- + \gamma \\ &\tau^- \rightarrow \mu^- + \gamma \\ &\tau^- \rightarrow e^- + e^+ + e^- \\ &\tau^- \rightarrow e^- + \mu^+ + \mu^- \\ &\tau^- \rightarrow \mu^- + \mu^+ + \mu^- \end{aligned} \qquad 73.$$

None of these decay modes have been found. Branching fraction upper limits are given in Table 9, along with upper limits on some other modes forbidden by the sequential model. The limits in Table 9 were obtained several years ago; increased statistics obtained since will allow searches that are five to ten times more sensitive.

Table 9 Upper limits on branching fractions for decay modes forbidden by the sequential model

Mode	Upper limit on branching fraction (%)	Confidence limit (%)	Experimental group or detector	Reference
$\tau^- \to e^- + \gamma$ $\tau^- \to \mu^- + \gamma$	12	90	PLUTO Group	a
$\tau^- \to e^- + \gamma$	2.6	90	LBL-SLAC lead glass wall	b
$\tau^- \to \mu^- + \gamma$	1.3	90	LBL-SLAC lead glass wall	b
$\tau^- \to$ (3 charged leptons)$^-$	1.0	95	PLUTO Group	a
$\tau^- \to$ (3 charged leptons)$^-$	0.6	90	SLAC-LBL magnetic detector	b
$\tau^- \to$ (3 charged particles)$^-$	1.0	95	PLUTO Group	a
$\tau^- \to \rho^- + \pi^0$	2.4	90	SLAC-LBL magnetic detector	b

[a] Flügge 1977.
[b] Perl 1977a.

6 FUTURE STUDIES OF THE TAU

6.1 *Future Studies of the Properties of the Tau*

As indicated throughout this review, there is more experimental work to be done on the properties of the τ for two reasons:

(*a*) Just as we continue to test the leptonic nature of the e and μ and the validity of the laws of lepton conservation, so we should continue to test these properties of the τ. The relatively large mass of the τ might be associated with deviation from $e-\mu-\tau$ universality. (*b*) Measurement of the properties of the τ can teach us about other areas of elementary particle physics, such as the existence of the A_1.

The main τ studies to be done are the following:

1. high energy, high statistics measurement of $\sigma(e^+e^- \to \tau^+\tau^-)$;
2. better determination of the limits on the ν_τ mass;
3. measurement of the τ lifetime and determination of G_τ;
4. discovery and measurement of $\tau^- \to \nu_\tau + K^-$;
5. elucidation of $\tau^- \to \nu_\tau + A_1$ and study of A_1 properties if it exists;
6. detailed study of multihadron decay modes of τ; and
7. better limits on sequential model forbidden decay modes.

6.2 *The Tau as a Decay Product*

Several known or proposed very heavy hadrons should decay to one or more τ's (Table 10). The measurement of these decay modes provides a separate measurement of G_τ, tests our standard model for the τ, such as the spin being 1/2, and provides information about the heavy hadron.

6.3 *Tau Production in Photon-Hadron Collisions*

We should be able to produce pairs of heavy charged leptons, such as τ pairs, by the very high energy Bethe-Heitler process (Kim & Tsai 1973, Tsai 1974, Smith et al 1977):

$$\gamma + \text{nucleus} \rightarrow \tau^+ + \tau^- + \text{anything}. \qquad 74.$$

We cannot learn much about the τ from this process, if our standard model of the τ is correct. However, this is a test of the model that should be done. An alternative method would use very high energy μ's via

$$\begin{aligned} &\mu + \text{nucleus} \rightarrow \mu + \text{anything} + \gamma_{\text{virtual}} \\ &\gamma_{\text{virtual}} \rightarrow \tau^+ + \tau^-. \end{aligned} \qquad 75.$$

The detection of the τ pairs in the reactions of Equations 74 and 75 is difficult because the τ decay products are immersed in an enormous background of hadrons, e's, and μ's. The most promising signature appears to be $e^\pm\mu^\mp$ pairs.

6.4 *Tau Production in Hadron-Hadron Collisions*

Taus can be produced in hadron-hadron collisions in two types of processes: 1. Virtual photons from quark-antiquark annihiliations can produce $\tau^+\tau^-$ pairs (Chu & Gunion 1975, Bhattacharya et al 1976). 2. Heavy hadrons such as those in Table 10 can be produced, and subsequently decay to $\tau\nu_\tau$ or $\tau^+\tau^-$ pairs.

Table 10 Some predicted decay modes of heavy hadrons to one or two τ's

Hadron	Quark composition	Predicted decay mode	Predicted branching fraction (%)	Reference
F	$c\bar{s}$	$F^+ \rightarrow \tau^+ + \nu_\tau$	≈ 3	Albright & Schrock 1979
B[a]	$b\bar{u}$, $b\bar{d}$	$B^+ \rightarrow \tau^+ + \nu_\tau$	0.5–1.0	Ellis et al 1977
Υ	$b\bar{b}$	$\Upsilon \rightarrow \tau^+ + \tau^-$	≈ 3.5	Eichten & Gottfried 1977, Ellis et al 1977
T[b]	$t\bar{t}$	$T \rightarrow \tau^+ + \tau^-$	≈ 8	Ellis et al 1977

[a] The evidence for the existence of the B is scanty at present.
[b] This is a proposed particle; there is no evidence for its existence at present.

Unfortunately the detection of the τ's is a very difficult task because of the overwhelming background from e, μ, and hadron production. However, the second process can be used to produce ν_τ's in sufficient quantity to permit their detection.

6.5 *Future Studies of the Tau Neutrino*

The production and decay sequence

$$\text{proton} + \text{nucleon} \rightarrow \text{F}^- + \text{hadrons} \tag{76.}$$

$$\text{F}^- \rightarrow \tau^- + \bar{\nu}_\tau \tag{77.}$$

$$\tau^- \rightarrow \nu_\tau + \text{charged particles} \tag{78.}$$

can produce a ν_τ beam (Barger & Phillips 1978, Albright & Schrock 1979, Sciulli 1978). However, the neutrinos from π and K decay would overwhelm the ν_τ signal unless the majority of the π's and K's interact before they decay. Therefore the entire proton beam must be dumped in a thick target. There is still some problem with ν_e's and ν_μ's from D meson and other charmed-particle semileptonic decays, but the detection of the ν_τ appears feasible.

The detection of the ν_τ and the study of its interactions can provide the following information. (*a*) We can verify that the ν_τ is a unique lepton and is not a ν_e, thereby testing the electron-related ortholepton model (Section 4.4). (*b*) More generally we can determine if the ν_τ behaves conventionally with normal weak interactions. (*c*) We can measure the product of the cross section for Reaction 76 and the branching ratio for Reaction 77.

7 SUMMARY

All published studies of the τ are consistent with it being a point-like, spin-1/2, sequential charged lepton. These studies are consistent with $e-\mu-\tau$ universality, and they are consistent with conventional Weinberg-Salam theory. There is more experimental work to be done on measuring the properties of the tau and on using the tau to study other particles.

ACKNOWLEDGMENTS

In the writing of this review I was greatly aided by the reviews done by Gary Feldman, Gustave Flügge, Jasper Kirkby, and Y. S. (Paul) Tsai.

I wish to take this opportunity to thank my collaborators in the SLAC-LBL magnetic detector experiments at SPEAR; it was in these experiments that the first evidence for the existence of the tau was discovered. Two of these collaborators were particularly important to me. One is Gary Feldman who was a constant companion in my work on the tau. The other is Burton Richter; without his knowledge, experimental skill, and leadership SPEAR would not have been built.

Literature Cited

Abrams, G. S. et al. 1979. *Phys. Rev. Lett.* 43:1555

Albright, C. H., Schrock, R. E. 1979. *Phys. Lett. B* 84:123

Alexander, G. et al. 1978a. *Phys. Lett. B* 78:162

Alexander, G. et al. 1978b. *Phys. Lett. B* 73:99

Ali, A., Yang, T. C. 1976. *Phys. Rev. D* 14:3052

Alles, W. 1979. *Lett. Nuovo Cimento* 25:404

Altarelli, G. et al. 1977. *Phys. Lett. B* 67:463

Bacino, W. et al. 1978. *Phys. Rev. Lett.* 41:13

Bacino, W. et al. 1979a. *Phys. Rev. Lett.* 42:749

Bacino, W. et al. 1979b. *Phys. Rev. Lett.* 42:6

Bailin, D. 1977. *Weak Interactions*. London, England: Sussex Univ. Press. 406 pp.

Barbaro-Galtieri, A. et al. 1977. *Phys. Rev. Lett.* 39:1058

Barber, D. P. et al. 1979. *Phys. Rev. Lett.* 43:1915

Barger, V., Phillips, R. J. 1978. *Phys. Lett. B* 74:393

Bartel, W. et al. 1978. *Phys. Lett. B* 77:331

Basdevant, J. L., Berger, E. L. 1978. *Phys. Rev. Lett.* 40:994

Bernardini, M. et al. 1973. *Nuovo Cimento* 17:383

Bhattacharya, R., Smith, J., Soni, A. 1976. *Phys. Rev. D* 13:2150

Bjorken, J. D., Drell, S. D. 1964. *Relativistic Quantum Mechanics,* New York: McGraw Hill. 414 pp.

Brandelik, R. et al. 1978. *Phys. Lett. B* 73:109

Brandelik, R. et al. 1979. *Z. Phys. C* 1:233

Brandelik, R. et al. 1980. *Phys. Lett. B* 92:199

Bricman, C. et al. 1978. *Phys. Lett. B* 75:1. This is the Particle Data Group's review of particle properties.

Burmester, J. et al. 1977a. *Phys. Lett. B* 68:297

Burmester, J. et al. 1977b. *Phys. Lett. B* 68:301

Cavalli-Sforza, M. et al. 1976. *Phys. Rev. Lett.* 36:558

Chu, G., Gunion, J. F. 1975. *Phys. Rev. D* 11:73

Cnops, A. M. et al. 1978. *Phys. Rev. Lett.* 40:144

Dorfan, J. 1979. In *Particles and Fields—1979*, ed. B. Margolis, D. G. Stairs, p. 159. AIP Conf. Proc. 59, Particles and Fields Subseries No. 19. New York: Am. Inst. Phys. (Publ. 1980)

Eichten, E., Gottfried, K. 1977. *Phys. Lett. B* 66:286

Ellis, J., Gaillard, M. K., Nanopoulos, D. V., Rudaz, S. 1977. *Nucl. Phys. B* 131:285

Farrar, G. R., Fayet, P. 1980. *Phys. Lett. B* 89:191

Feldman, G. J. et. al. 1976. *Phys. Lett. B* 63:466

Feldman, G. J. 1978a. *Proc. Int. Meet. Frontier of Physics, Singapore*. In press; also issued as Stanford Linear Accel. Cent. *SLAC-PUB-2230*

Feldman, G. J. 1978b. In *Neutrinos—78*, ed. E. C. Fowler, p. 647. West Lafayette, Ind: Purdue Univ.

Flügge, G. 1977. In *Exp. Meson Spectrosc. 1977*, ed. E. Von Goeler, R. Weinstein, p. 132. Boston: Northeastern Univ.

Flügge, G. 1979. *Z. Phys. C* 1:121

Gilman, F. J. 1978. Stanford Linear Accel. Cent. *SLAC-PUB-2226*

Gilman, F. J., Miller, D. H. 1978. *Phys. Rev. D* 17:1846.

Heile, F. J. et al. 1978. *Nuc. Phys. B* 138:189

Hitlin, D. 1978. Quoted in *Proc. 19th Int. Conf. High Energy Phys.*, ed. S. Homma, M. Kawaguchi, H. Miyazawa, p. 784. Tokyo: Phys. Soc. Jpn.

Hofstadter, R. 1975. In *Proc. 1975 Int. Symp. Lepton Photon Interact. High Energies*, ed. W. T. Kirk, p. 869. Stanford, Calif: SLAC

Horn, D., Ross, G. G. 1977. *Phys. Lett. B* 67:460

Jaros, J. A. et al. 1978. *Phys. Rev. Lett.* 40:1120

Kane, G., Raby, S. 1980. *Phys. Lett. B* 89:203

Kawamoto, N., Sanda, A. I. 1978. *Phys. Lett. B* 76:446

Kim, K. J., Tsai, Y. S. 1973. *Phys. Rev. D* 8:3109

Kirkby, J. 1979. In *Proc. 1979. Int. Symp. Lepton Photon Interact. High Energies*, ed. T. B. W. Kirk, H. D. I. Arbarbanel, p. 107. Batavia, Ill: FNAL

Knies, G. 1977. In *Proc. 1977 Int. Symp. Lepton Photon Interact. High Energies*, ed. F. Gutbrod, p. 93. Hamburg: DESY

Llewellyn Smith, C. H. 1977. *Proc. R. Soc. London Ser. A* 355:585

Low, F. E. 1965. *Phys. Rev. Lett.* 14:238

Marshak, R. E., Riazuddin, Ryan, C. P. 1969. *Theory of Weak Interactions in Particle Physics*, New York: Wiley-Interscience. 761 pp.

Michel, L. 1950. *Proc. Phys. Soc. London Sect. A* 63:514

Orioto, S. et al. 1974. *Phys. Lett. B* 48:165

Perl, M. L., Rapidis, P. 1974. Stanford Linear Accel. Cent. *SLAC-PUB-1499*

Perl, M. L. 1975. In *Proc. Summer Inst. Part. Phys., SLAC-191*, ed. M. C. Zipf, p. 333. Stanford: SLAC

Perl, M. L. et al. 1975. *Phys. Rev. Lett.* 35:1489

Perl, M. L. et al. 1976. *Phys. Lett. B* 63:466
Perl, M. L. 1977a. See Knies 1977, p. 145
Perl, M. L. et al. 1977b. *Phys. Lett. B* 70:487
Perl, M. L. 1978. *New Phenomena in Lepton-Hadron Physics,* ed. D. E. C. Fries, J. Wess, p. 115. New York: Plenum. This reference discusses broader definitions of a lepton.
Pham, T. N., Roiesnel, C., Truong, T. N. 1978. *Phys. Lett. B* 78:623
Rosen, S. P. 1978. *Phys. Rev. Lett.* 40:1057
Sakurai, J. J. 1975. See Hofstadter 1975, p. 353
Sciulli, F. 1978. See Feldman 1978b, p. 863
Smith, J., Soni, A., Vermaseren, J. A. M. 1977. *Phys. Rev. D* 15:648
Thacker, H. B., Sakurai, J. J. 1971. *Phys. Lett. B* 36:103
Tsai, Y. S. 1971. *Phys. Rev. D* 4:2821
Tsai, Y. S. 1974. *Rev. Mod. Phys.* 46:815
Tsai, Y. S. 1978. Stanford Linear Accel. Cent. *SLAC-PUB-2105*
Tsai, Y. S. 1980. Stanford Linear Accel. Cent. *SLAC-PUB-2403*
Wagner, W. et al. 1980. *Z. Phys. C* 3:193
Wolf, G. 1979. See Kirkby 1979, p. 34
Yamada, S. 1977. See Knies 1977, p. 69
Zipf, M. C., ed. 1978. *Proc. Summer Inst. Part. Phys. SLAC-215, Stanford Linear Accel. Cent., Stanford Univ. Calif.*

Ann. Rev. Nucl. Part. Sci. 1983. 33 : 143–97

6

PHYSICS WITH THE CRYSTAL BALL DETECTOR

Elliott D. Bloom

Stanford Linear Accelerator Center, Stanford University, Stanford, California 94305

Charles W. Peck

High Energy Physics, California Institute of Technology, Pasadena, California 91125

CONTENTS

1. INTRODUCTION AND OUTLINE

At the 1974 PEP Summer Study (1), one of the projects was to explore the possibilities and limitations of detectors optimized to measure photons produced in high energy e^+e^- collisions. It was realized that a device that had high detection efficiency over a large solid angle and that could measure the energy of photons in the region above a few tens of MeV with high precision (in the range of a few percent) would provide a unique capability offered by no existing apparatus. Thus it could possibly yield important and otherwise unattainable information about these fundamental interactions. Furthermore, if it also measured the directions of both photons and charged particles well enough, even a nonmagnetic version of such a device would be able to compete with the large general-purpose magnetic spectrometers then in existence in the reconstruction of certain simple, few-particle final states. And finally, a device designed to absorb all the electromagnetic energy in an event would in fact quickly and directly measure a large fraction of its total energy. This prompt information could form the basis for an admirable trigger having very different biases from those used by the magnetic spectrometers. Thus such a device would be an interesting complementary technique for the investigation of e^+e^- physics. In particular an efficient "all-neutral" trigger would be possible.

Although the thrust of the summer's work had been directed toward instrumentation for PEP (an e^+e^- storage ring at the Stanford Linear Accelerator Center, allowing beam energies up to 15 GeV), which was still in the planning stages at that time, a keen interest in the idea developed among a group of people[1] from Caltech, Harvard, Stanford-HEPL, and SLAC and this led to serious work in the Fall of 1974 toward producing a formal proposal for the existing lower energy storage ring, SPEAR. The startling discoveries of the $J/\psi(3100)$ and $\psi'(3700)$ in November and December of that year spurred on these efforts, especially as people realized there was the likely possibility of a rich gamma ray spectroscopy in the range of a few hundred MeV. Eventually this work led the group to submit a proposal for a nonmagnetic, large solid angle detector whose principal component was a spherical shell of NaI(Tl) with a 10-inch inner radius and

[1] A group from Princeton joined the collaboration in 1977.

a 26-inch outer radius. The device was quickly dubbed the "Crystal Ball" and it has been universally called that ever since. The proposal was approved in the Spring of 1975 and the construction of the detector was completed three years later in the Spring of 1978. Section 2 describes the configuration and performance of the detector.

The Ball was installed at SPEAR in the Fall of 1978 and took data there on e^+e^- collisions in the energy region from 3.1 to 7.4 GeV during the 40 months of calendar time until December 1981. SPEAR actually supplied beam during about half of this time. We spent about five months collecting about 2×10^6 hadronic events at each of the two $^3S_1$ states, the $J/\psi(3100)$ and the $\psi'(3700)$. Typical luminosities at these energies were 0.5×10^{30} cm^{-2} s^{-1} and 1.8×10^{30} cm^{-2} s^{-1}, respectively. About one month was spent at the $\psi''(3770)$ collecting 4×10^4 hadronic events and the rest of the time was spent at energies in the continuum, almost all of which were above charm threshold. We obtained a cumulative exposure of 24.0 pb^{-1} in this region. At the highest energy at which we took data, 7.4 GeV, SPEAR provided a peak luminosity of about 2.0×10^{31} cm^{-2} s^{-1}.

In the Spring of 1982 the Ball was moved as an intact experiment to the Deutsches Elektronen Synchrotron (DESY), Hamburg, Germany to run on the DORIS II e^+e^- storage ring in order to make a parallel study of the Υ system. Data taking in the 10-GeV region, in which Doris II is optimized, had just begun as this paper was being prepared.

This brief review surveys all the Crystal Ball physics results that had been completed as of December 1982. The available space does not permit any detailed discussion of either the experimental details or the theoretical framework that provides the proper setting for the experimental findings described here. Also, the historical perspective is that of participants in the Crystal Ball experiment, one of many that has made major contributions to this field. However, the interested reader can find a discussion of many of the theoretical questions in (2) and (3) and in the literature cited therein; appropriate experimental references and only limited theoretical references are given in this review. A general survey of the physics of psionic matter up to 1977 can be found in (3a). Finally, in Section 11 we briefly discuss some of the analysis projects currently in progress as well as our expectations for results from the just begun exposure of the Crystal Ball at DORIS II.

The principal accomplishments of the Crystal Ball experiment have resulted from the study of radiative and certain hadronic transitions involving the charmonium states. Figure 1 shows the energy level diagram of this system and it also indicates the several radiative and hadronic transitions that have been the focus of the Crystal Ball efforts. The refreshing simplicity of this first-known heavy quark spectrum compared to the corresponding situation among the light quarks (u,d,s) has played an important role in the recent development of particle physics. This positronium-like structure gives strong qualitative evidence for the fundamental $c\bar{c}$ interpretation of charmonium and quantitative details about the energies and transition rates can be compared with phenomenological models motivated by quantum chromodynamics (QCD).

When the Crystal Ball experiment began, however, there were several outstanding difficulties with the then-favored, and now well-established, $c\bar{c}$ model. A total of five states had been reported in two-photon cascade transitions between ψ' and the J/ψ and one state had been reported below the J/ψ in the 3γ decay mode. Preferred quantum numbers for three of the intermediate mass states were indirectly inferred (4–6) from their hadronic decay patterns and mass ordering and these caused them to be identified with the three ^3P_J states. However, the experimental situation concerning the candidate $^1S_0$ states seemed to present an insurmountable challenge to the beautiful $c\bar{c}$ interpretation (7–9).

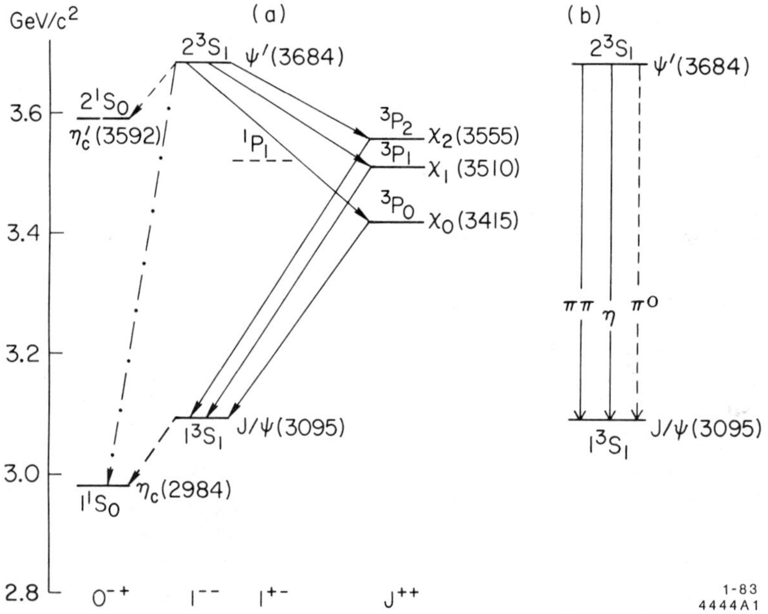

Figure 1 (*a*) The current status of the charmonium spectrum below charm threshold. All the observed photon transitions between these states are shown. Solid lines indicate electric dipole transitions; broken lines indicate allowed magnetic dipole transitions (between states with the same radial wave function); and broken-dotted lines show "hindered" magnetic dipole transitions (between states with different radial wave functions). The $^1P_1$ state is yet to be observed. (*b*) The observed hadronic transitions between the ψ' and the J/ψ. The $\pi\pi$ transitions are allowed, the η transition is $SU(3)_F$ forbidden, and the π^0 transition is $SU(2)_F$ forbidden.

In brief, the $^1S_0$ problem was as follows. In 1977, the DASP collaboration (10, 11) observed a significant signal at 2.83 ± 0.03 GeV/c^2 in the distribution of the highest $\gamma\gamma$ mass from the decay $J/\psi \to \gamma\gamma\gamma$. This state, the X(2830), immediately became a candidate for the 1^1S_0 state, the η_c. Confirming evidence for the state was reported by a Serpukov group using the reaction $\pi^- p \to \gamma\gamma n$ (11a). The measured product branching ratio from the DASP collaboration, $B[J/\psi \to \gamma X(2830)] \cdot B[X(2830) \to \gamma\gamma]$, was $(1.2 \pm 0.5) \times 10^{-4}$. However, no evidence for $J/\psi \to \gamma X(2830)$ was seen in the inclusive γ spectrum from the J/ψ by the SPEAR experiment SP-27 (12), which set an upper limit of 2% for $B[J/\psi \to \gamma X(2830)]$. These results were incompatible with any reasonable $c\bar{c}$ model since this interpretation predicts $B[J/\psi \to \gamma X(2830)]$ to be an order of magnitude larger than the limit set by SP-27. Furthermore, it predicts $B[X(2830) \to \gamma\gamma]$ to be about five times smaller than the lower limit inferred from the DASP and SP-27 results combined. Finally, a hyperfine splitting of 265 MeV is surprisingly large within the $c\bar{c}$ model.

The second serious problem concerned the 2^1S_0 state, the η_c'. Initially, some evidence for an η_c' candidate was reported (13) at a mass of 3455 MeV/c^2 in the cascade process $\psi' \to \gamma 2^1S_0 \to \gamma\gamma J/\psi$ by the Mark I experiment at SPEAR. This observation was not confirmed by a subsequent experiment (14), the DESY-Heidelberg collaboration at DORIS, which independently investigated the radiative cascade process. On the other hand, the DESY-Heidelberg experiment presented evidence for an alternative intermediate state at 3591 MeV/c^2 as a possible η_c' candidate. However, their reported branching ratio for $\psi' \to \gamma\chi(3591) \to \gamma\gamma J/\psi$ was orders of magnitude greater than predicted by the model, if this state were taken to be the η_c'.

One of the first processes measured with the newly commissioned Crystal Ball was $J/\psi \to 3\gamma$. These new observations provided both higher statistics and better resolution than the earlier ones, but they did not confirm the X(2830). A lower limit of 2.2×10^{-5} was initially (15) set for the product branching ratio [this limit was subsequently lowered to 1.6×10^{-5} (90% C.L.); see Section 6.3]. Nor did later data confirm either the $\chi(3455)$ or the $\chi(3591)$ in the radiative cascades from the ψ' to the J/ψ. Thus the experimental status of the two expected $^1S_0$ states was again open.

The first evidence for the η_c in the Crystal Ball came from the inclusive γ spectra observed from the ψ' and, shortly thereafter, that from the J/ψ. Somewhat later, with a doubling of the ψ' data sample, evidence for the η_c' was also found in the ψ' inclusive γ spectrum. The current status of these two charmonium states is discussed in Section 5 below. With these two contributions from the Crystal Ball, there is only one qualitative feature of the expected $c\bar{c}$ spectrum for which no experimental evidence has yet been found, namely, the $^1P_1$ state. Section 7 summarizes our current limits on certain decay modes involving this state.

The radiative transitions involving the three ^3P_J states give rise to the several prominent peaks in the ψ' inclusive γ spectrum shown in Figure 3. So characteristic, in fact, is this spectrum that it has become the logo of the Crystal Ball experiment. The careful study of all the systematics (efficiencies and resolutions) necessary to obtain the branching ratios and natural widths of these states from the inclusive γ spectrum has been recently completed and is discussed in Section 3. The radiative cascade exclusive channels $\psi' \to \gamma\ ^3P_J \to \gamma\gamma e^+e^-$ or $\gamma\gamma\mu^+\mu^-$ were susceptible to more rapid analysis and Section 3 also summarizes our results on product branching ratios, masses, and angular distributions (which strongly support the earlier spin assignments for these states). As a by-product of our study of these exclusive channels, we also made measurements on the three transitions $\psi' \to \pi^0\pi^0 J/\psi$, $\psi' \to \eta J/\psi$, and $\psi' \to \pi^0 J/\psi$. The first simply corroborated the much better results from $\psi' \to \pi^+\pi^- J/\psi$, but the other two yielded significant improvements over earlier work. These hadronic transitions are discussed in Section 4.

Radiative transitions from the J/ψ are especially interesting since their primary mechanism is expected (in the context of QCD) to be $J/\psi \to \gamma gg$ with the two gluons in a singlet state of both color and flavor. Thus any gg bound states that are even under charge conjugation and less massive than the J/ψ are likely to be excited in this decay. At least two candidates for such objects have been observed in the Crystal Ball data. One, with a mass of about 1440 MeV/c^2 was thought to be the 1^{++}, E(1420) meson when it was first found in J/ψ decays by the Mark II experiment. The existence of the state was quickly confirmed by the Crystal Ball. However, only after the partial-wave analysis of twice the initial data sample did the Crystal Ball collaboration find that the 0^{-+} assignment for the state was favored. This state was then named $\iota(1440)$. A second gluonium candidate, the $\vartheta(1640)$, was found by the Crystal Ball in the $\eta\eta$ decay mode and the preferred spin-parity assignment is 2^+. Finally, searching in the channel $J/\psi \to \gamma\eta\pi\pi$, we find no evidence that the 1440-MeV/c^2 state decays into $\eta\pi\pi$ but we do see both the expected signal of $\eta' \to \eta\pi\pi$ and an unexpected very broad enhancement at an $\eta\pi\pi$ mass of 1710 MeV/c^2. The present status of these several interesting possibilities as well as the Crystal Ball's observations on the modes $J/\psi \to \gamma X$, where $X = \pi^0, \eta, \eta'$, f, and f′, is discussed in Section 6.

In addition to the extensive search for the $^1P_1$ charmonium state mentioned earlier, two other searches with negative results have been carried out and are described in Section 7. The first was an attempt to corroborate a strong enhancement in the inclusive η cross section reported by the DASP experiment. This had been interpreted as evidence for $e^+e^- \to F + \cdots \to \eta + \cdots$ where the F is the charmed strange meson. However, no significant enhancement was observed by the Crystal Ball. The second search with negative results was for evidence of the axion. Since there are quite sharp theoretical predictions for radiative decays of the J/ψ and Υ into the axion, we made a detailed investigation of our J/ψ data looking for this decay; nothing was found.

Finally, in addition to the charmonium studies, comprising the bulk of the Crystal Ball results, this experiment has also collected a body of data in the energy region above charm threshold. To date, in addition to the inclusive η cross sections mentioned earlier, we have made total hadronic cross section measurements (R_h) up to the highest energies at SPEAR (Section 8), we have observed production of the f and A_2 by two-photon collisions from which we obtain the decay rates $\Gamma(f \to \gamma\gamma)$ and $\Gamma(A_2 \to \gamma\gamma)$ (Section 9), and, finally, we have made several measurements in the region from charm threshold to 4.5 GeV (Section 10).

2. DESCRIPTION OF THE APPARATUS AND ITS PERFORMANCE

Over the years, many methods have been developed and extensively used for measuring the energy of high energy photons. By the mid-1970s, however, the pioneering work of R. Hofstadter and his colleagues (16) had shown that the technique of total absorption shower counters made of thallium-doped sodium iodide, NaI(Tl), was unsurpassed in the combination of high detection efficiency and energy resolution. Consequently, in spite of the technical difficulties occasioned by the extremely hydroscopic nature of NaI(Tl), this technique, supplemented by fine segmentation of the material, was selected to form the basis of a detector covering nearly the full 4π-sr solid angle about the e^+e^- collision point, the Crystal Ball. The final result of the design was a detector consisting of four main parts. These were a central charged-particle detection system, two hemispherical shells of NaI(Tl), endcaps of tracking chambers followed by sodium iodide covering the beam entry holes into the spherical shell, and a small-angle luminosity monitor. Figure 2 shows the geometric arrangement of the two major components of the detector. Details about the apparatus can be found in (17).

The central tracking system consisted of three concentric cylindrical ionization detectors covering 71%, 83%, and 94% of 4π sr, respectively. The middle detector (18) was a proportional chamber with two gaps, and the other two detectors were magnetostrictive spark chambers. For particles that were detected in both spark chambers, both direction and origin along the beam line could be determined. Those that failed to be detected in both spark chambers were only "tagged," i.e. identified as being charged.

The heart of the detector, of course, was the spherical shell, 16 radiation lengths thick, made of sodium iodide. This thickness is sufficient to contain essentially the entire longitudinal development of electromagnetic showers in our energy range. As shown in Figure 2, the shell is actually a dense packing of truncated triangular pyramids of NaI(Tl). These are optically

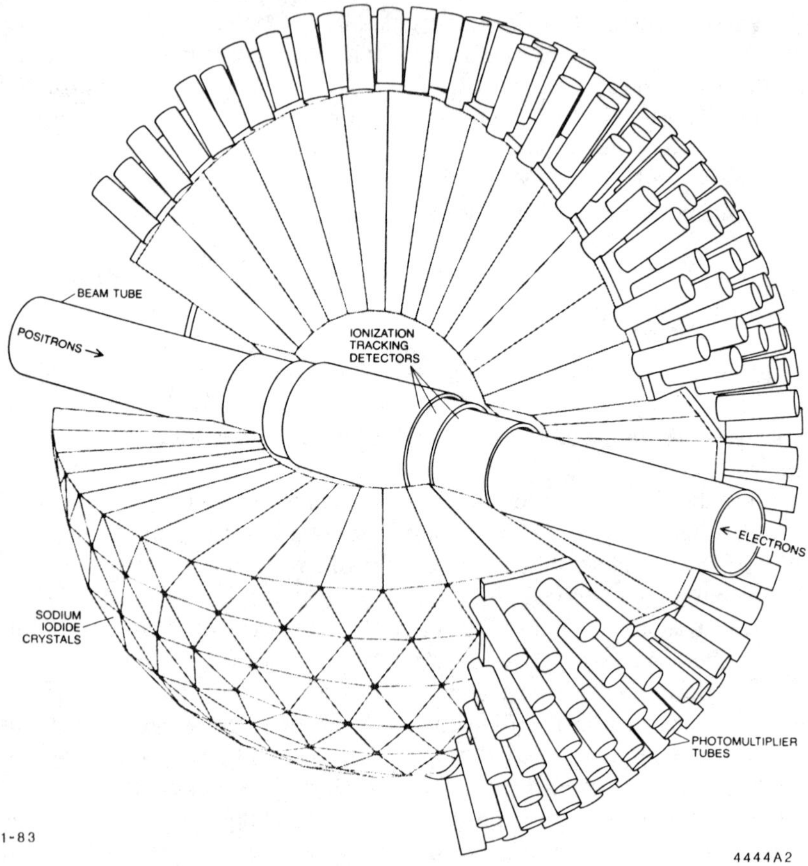

Figure 2 The two principal elements of the Crystal Ball detector, the charged-particle tracking chambers in the 25-cm diameter cavity of the shell, and the NaI(Tl) shell itself. The middle chamber is a continuously sensitive wire proportional chamber and the other two are magnetostrictive spark chambers. The shell itself is segmented into optically separated triangular pyramids in a solidly packed geometry based on an icosahedron. Each pyramid is viewed from the outside by a single photomultiplier. [From *Quarkonium*, by E. D. Bloom and G. J. Feldman. Copyright © 1982 by Scientific American, Inc. All rights reserved.]

isolated one from another, and each is viewed from the outside by a single photomultiplier tube. The only materials separating the individual crystals are thin layers of white paper and aluminum foil (except for the plane separating the two hemispheres). The shell consists of a total of 672 of these crystals and it covers 93% of 4π sr. The missing 7% is due to beam entry holes, but these are almost completely covered by the endcaps.

With this geometric arrangement, we not only measure the amount of energy deposited in the NaI with little loss, but we also obtain information about the transverse structure of this energy deposition. Being minimum ionizing and lacking strong interactions, high energy muons leave simple tracks, with a deposited energy of about 200 MeV distributed over no more than two or three crystals. Electrons and photons with energy greater than about 20 MeV produce electromagnetic showers and deposit all of their energy in a reasonably characteristic pattern covering about 13 crystals. Finally, most hadrons strongly interact in the Ball since it is about one absorption length thick. They thus give rise to somewhat more irregular patterns than electromagnetic showers and the total deposited energy bears little relation to the hadron's energy. This geometric arrangement provides no information about the longitudinal distribution of an energy deposition, but we have found that careful statistical analysis of a transverse pattern is a useful technique for resolving some particle identification ambiguities.

The parameter of particular interest in this detector is its energy resolution for electromagnetic showers. For the energy range of interest, the standard deviation σ_E of this resolution is well approximated by $(0.0255 \pm 0.0013)E^{3/4}$ where E and σ_E are in GeV. Thus, for example, we measure the energy of a 1.55-GeV Bhabha-scattered electron to an accuracy of 36 MeV and that of a 100-MeV photon to an accuracy of 4.6 MeV. An example of the utility of this relatively high resolution is that we can extract the natural widths of the charmonium $^1S_0$ and $^3P_{0,2}$ from our inclusive γ distributions. More generally, the goodness of our photon energy resolution has proved invaluable in allowing us to reliably identify certain reactions by the technique of kinematically constrained statistical fitting, which in turn leads to some of the physics results discussed below. It should be noted, however, that because of the size of electromagnetic showers and the edge effects of the beam holes, the good energy resolution is only available over 85% of 4π sr. A detailed description of tests made on a prototype of the detector and the signal processing method used can be found in (19).

A second parameter of considerable interest is the resolution with which the direction of a photon can be determined. By examining the profile of its shower's energy deposition we can determine the direction of a photon to much better than the size of one module. The limitation on the accuracy of angles determined in this manner is caused by shower fluctuations. The Crystal Ball has achieved a resolution with $\sigma_{\vartheta_\gamma} = 1.5$ to $2°$, where ϑ_γ is the polar angle from the photon's true direction. There is a slight energy dependence in this angular resolution.

An important design goal in this apparatus was to cover as much as possible of the solid angle around the collision point with high efficiency particle detectors. This was achieved by covering the necessary beam holes in the ball with endcaps, 20 radiation lengths thick of individually packaged NaI(Tl) hexagonal prisms covered by two gaps of spark chambers. These brought the total coverage to 98% of 4π sr. Primarily because of edge effects, the energy resolution for photons and electrons going outside the central 85% of 4π sr of the main ball was relatively poor and strongly direction dependent. Consequently, the endcaps were primarily used as veto counters. They allowed us to determine the topology of events with very high confidence, and this was of crucial importance for reducing backgrounds in some of the physics measurements given later.

Finally, for many of our measurements an absolute luminosity determination was necessary. This was provided by a small-angle Bhabha-scattering detector consisting of four counter elements, symmetrically disposed about the beam and centered at a 4° angle to the beam line. Each of the four elements was identical, consisting of three scintillators followed by a shower counter, and covered a solid angle of 4.2×10^{-4} sr. The system provided a counting rate of about 0.7 Hz at the ψ' with our typical luminosity. The accuracy of luminosity determination was better than 3% with this monitor, as checked by using large-angle Bhabha events observed in the full Ball.

The apparatus was triggered and events written on tape when at least one of several overlapping conditions was satisfied. Each of these triggers was based on a coincidence between a beam-crossing signal and the analog sum of signals from the Ball and each required that this sum, proportional to the total energy in the Ball, be greater than some threshold. Generally, a further requirement was also imposed and the more restrictive it was, the lower the total energy threshold. The simplest trigger involved no other requirements and its total energy threshold was normally about 1 GeV. More restrictive triggers involved such event features as charged particles being detected in the proportional chambers, or a requirement on the general pattern of energy deposition in the Ball. In general, the hardware trigger conditions were highly efficient for the classes of events that have been studied with the Crystal Ball, and the Monte Carlo simultations done to determine detection efficiencies have included these hardware trigger conditions.

Data acquisition and general system monitoring were performed by a PDP11/t55 computer. Chestnut et al (20) discuss in detail both the hardware and the flexible complement of software developed for this experiment.

3. THE CHARMONIUM ^3P_J STATES

3.1 *Dominant Features in the Inclusive Photon Spectrum of the ψ'*

After the discoveries of the J/ψ (21, 22) and ψ' (23) in 1974, four experiments measured the inclusive photon spectrum from the ψ' with increasing levels of sensitivity. The first experiment was a two-crystal NaI(Tl) detector (24); it could only place upper limits on radiative transitions to the $^3P_J(\chi_J)$ states. A magnetic detector, measuring converted photons, was able to measure the photon transition to the χ_0 state (13), but was not able to inclusively observe the other transitions. A moderately segmented NaI(Tl) detector (12) finally measured the photon transitions to each of the χ_J states and also inclusively observed the cascade transitions from the χ_2 and χ_1 to the J/ψ. Finally, Figure 3 shows the inclusive spectrum at the ψ' from the Crystal Ball detector, the most sensitive experiment so far. The main spectrum in the figure is from the analysis of approximately 0.9×10^6 ψ' events (the last half of the full sample) obtained at SPEAR. Severe cuts have been made in this spectrum to enhance structure. First, all photons are required to have $|\cos \vartheta_\gamma| < 0.85$, where ϑ_γ is the angle between the photon and the beam direction. The cosine of the angle between each photon and any charged particle is required to be less than 0.9. Pairs of γ's with invariant mass consistent with the mass of the π^0 have been eliminated. Finally, the lateral shower energy deposition in the NaI(Tl) crystals is required to be consistent with a single electromagnetic shower. This "pattern cut" removes most of those minimum ionizing charged particles that were not identified by the tracking-chamber system, many of the spurious energy deposits resulting from interacting charged particles, and some of the high energy π^0's in which the electromagnetic showers from the two photons from the π^0 decay overlap. The pattern cut used for the spectrum in Figure 3, one of many algorithms possible, was designed to optimize the efficiency for photons with energy, E_γ, less than or about 100 MeV.

As is seen in the figure, the photon transitions from the ψ' to the χ_J states and the cascade transitions from the χ_J states to the J/ψ stand out clearly in this inclusive spectrum. Indeed, the strength of these transitions in our detector has allowed frequent checks of the NaI(Tl) energy calibration and resolution over the course of our stay at SPEAR. Typically, two days of reasonable data taking at SPEAR, yielding approximately 2.5×10^4 ψ' decays, allowed an accurate determination of the transition energies to the χ_J states.

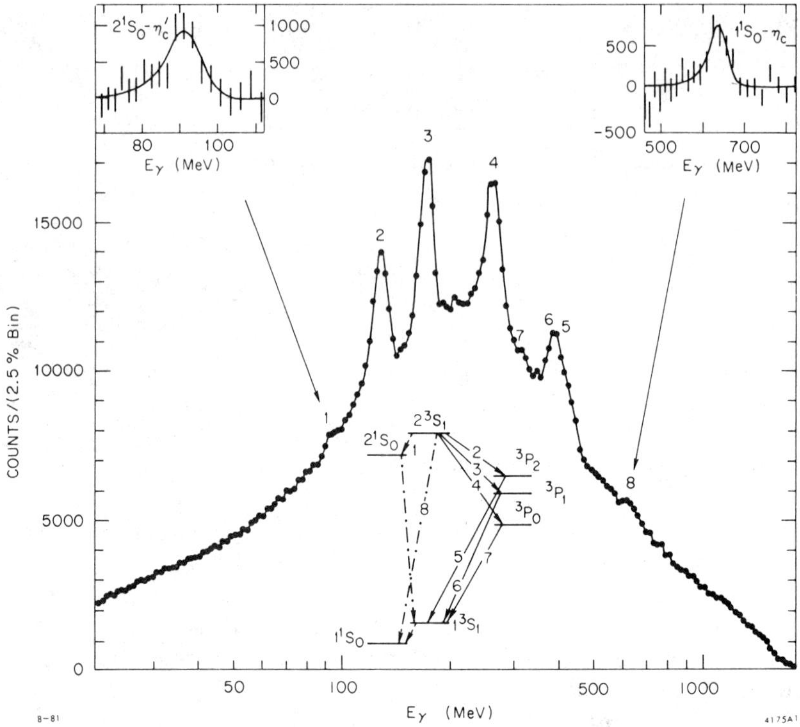

Figure 3 Inclusive γ spectrum at the ψ'. Note that the spectrum is $\Delta N/\Delta(\log E) \cong E\, dN/dE$. The upper inserts show the background-subtracted signals for the η_c and η_c' candidate states. The numbers over the spectrum key the observed spectral features with the expected radiative transitions in the charmonium spectrum inset.

3.2 *The Photon Cascade,* $\psi' \rightarrow \gamma\chi_J \rightarrow \gamma\gamma J/\psi$

A study of the radiative transitions from the ψ' to the χ_J states and the cascade radiative decays from the χ_J states by means of the sequence, $\psi' \rightarrow \gamma\chi_J$, $\chi_J \rightarrow \gamma J/\psi$, $J/\psi \rightarrow \ell^+\ell^-$, where $\ell^+\ell^-$ is e^+e^- or $\mu^+\mu^-$, provides a method for identifying the χ_J states (17) that is almost free of background. Indeed, it was in this reaction sequence that the $\chi_{1,2}$ were first observed (4, 25). Additionally, an analysis of the angular correlations in the cascade final state of Crystal Ball data (17) has permitted a direct measurement of the spin of the $\chi_{1,2}$ states and of the multipole coefficients describing the two individual radiative transitions for each of these states. The decays, $\psi' \rightarrow \eta(\pi^0)J/\psi \rightarrow \gamma\gamma\ell^+\ell^-$ exhibit the same topology as the cascade reactions; these processes have also been studied by the Crystal Ball collaboration (17) in order to separate them from χ_J events as well as for their own sake (cf Section 4.1). Note that the decay, $\psi' \rightarrow \pi^0 J/\psi$, which is forbidden by isospin symmetry, has been observed by the Crystal Ball (17) and the Mark II (26) detectors at SPEAR. The details of the Crystal Ball data analysis for cascade reactions are discussed in (17) and the references cited therein.

Figure 4*a* shows the Dalitz plot of the final event sample (from the first half of the full data set) containing 1206 $\gamma\gamma e^+e^-$ and 1280 $\gamma\gamma\mu^+\mu^-$ decays prior to kinematic fitting. These same events are also shown on the Dalitz plot in Figure 4*b* after they have been kinematically fit to the hypothesis that they arise from $\psi' \to \gamma\gamma J/\psi \to \gamma\gamma\ell^+\ell^-$ (5-C for e^+e^-, 3-C for $\mu^+\mu^-$). The fit kinematics restricts all of the surviving 2234 $\gamma\gamma\ell^+\ell^-$ events to fall within the outer envelope illustrated in Figures 4*a*, *b*; the cuts to the data restrict the events to fall within the inner envelope.

The decay, $\psi' \to \pi^0\pi^0 J/\psi, \pi^0 \to \gamma\gamma, J/\psi \to \ell^+\ell^-$, in which two photons go undetected or have energies less than 20 MeV, is a background of $\sim 5\%$ to

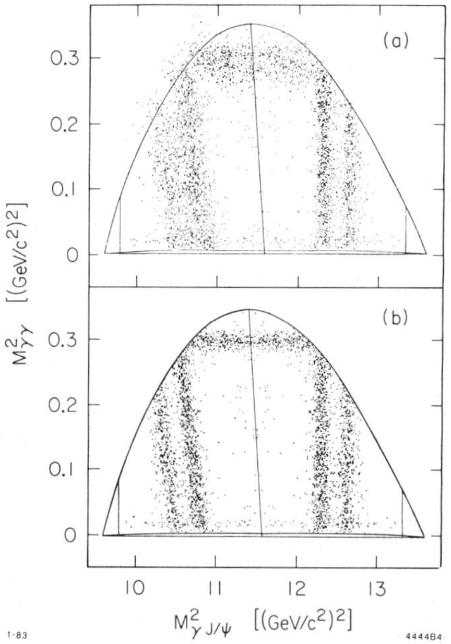

Figure 4 (*a*) Dalitz plot showing events from the two exclusive decays $\psi' \to \gamma\gamma e^+e^-$ and $\psi' \to \gamma\gamma\mu^+\mu^-$. The kinematic boundary is the outer one shown, and inscribed within it are the boundaries imposed by the event selection cuts. Each event appears twice in this plot, once to the right of the almost vertical central dividing line, once to the left. The combination with the lower energy photon is on the right and the clear verticality of the bands shows that the lower energy photon is the first emitted. Horizontal bands corresponding to the η and π^0 are also evident. (*b*) The same as (*a*) but after kinematic fitting. The $\psi' \to \gamma\gamma e^+e^-$ events are subjected to five constraints and $\psi' \to \gamma\gamma\mu^+\mu^-$ to three. The main effect of fitting is to remove background and to improve the energy resolution of the higher energy photon. The latter significantly sharpens the bands on the left and those for the η and the π^0.

the events of Figure 4*a*. This background, as well as all other backgrounds that have been considered (17), is totally negligible in the final fit event sample shown in Figure 4*b*.

In both Figures 4*a* and 4*b* the horizontal band at the top occurs at the η mass; that near the bottom occurs at the π^0 mass. Two strong signals for $\chi_1(3508)$ and $\chi_2(3554)$ appear as vertical bands to the right of the symmetry line shown, which has slope, $d(m_{\gamma\gamma}^2)/d(M_{\gamma J/\psi}^2) = -2$.

The Doppler-shifted bands on the left of the symmetry line (each event is plotted twice, once for each $\gamma J/\psi$ mass combination) are tilted with a slope of -1. The mass resolution for the π^0, η, and the low solution $\gamma J/\psi$ mass is better in Figure 4*b* than in the unfitted plot (Figure 4*a*). This is due to the fact that the kinematic fit reduces the absolute energy error of the higher energy photon to that of the lower energy one.

After separating the η and π^0 bands, the populous states at $\gamma J/\psi$ masses of 3554 and 3508 MeV/c^2 contain 479 and 943 events, respectively. Three of the 20 events associated with $\chi_0(3413)$ are expected to arise from the reaction $\psi' \rightarrow \pi^0\pi^0 J/\psi$. The cuts on the data restrict the $\gamma J/\psi$ mass to the range 3129 to 3644 MeV/c^2; in this region we find no evidence for a fourth χ state.

The branching ratio for a particular $\gamma\gamma\ell^+\ell^-$ decay channel is obtained by taking the number of events observed in the channel, correcting for detection efficiency (from 0.5 to 0.25 for various channels, and typically about 0.4), photon conversion and charged-particle identification efficiencies (0.95 and 0.96, respectively), and dividing by the total number of ψ' produced and the branching ratio for the decay of the J/ψ into dileptons. The J/ψ dilepton branching ratio (27) is the dominant systematic error (13%) in this measurement. The branching ratios obtained by the Crystal Ball are shown in Table 1 in comparison with those obtained from other experiments. There is good agreement for the χ_2 and χ_1 measurements; however, only the Crystal Ball measures a significant χ_0 branching ratio. Only upper limits are given for $\chi(3455)$ and $\chi(3591)$.

Additional information obtained from the Crystal Ball measurements of the photon cascade decays included the spins of the $\chi_{1,2}$ states and the multipolarity of the γ transitions. The particles participating in the cascade sequence $e^+e^- \rightarrow \psi'$, $\psi' \rightarrow \gamma'\chi_J$, $\chi_J \rightarrow \gamma J/\psi$, $J/\psi \rightarrow \ell^+\ell^-$ define the five angles, $\cos\vartheta' = \hat{e}^+ \cdot \hat{\gamma}'$, $\cos\vartheta_{\gamma\gamma} = \hat{\gamma}' \cdot \hat{\gamma}$, $\tan\varphi' = [\hat{e}^+ \cdot (\hat{\gamma}' \times \hat{\gamma})]/\{\hat{e}^+ \cdot [(\hat{\gamma}' \times \hat{\gamma}) \times \hat{\gamma}']\}$, $\cos\vartheta = \hat{\ell}^+ \cdot \hat{\gamma}$, $\tan\varphi = [\hat{\ell}^+ \cdot (\hat{\gamma}' \times \hat{\gamma})]/\{\hat{\ell}^+ \cdot [(\hat{\gamma}' \times \hat{\gamma}) \times \hat{\gamma}]\}$.

The angular distribution function $w(\cos\vartheta', \varphi', \cos\vartheta_{\gamma\gamma}, \cos\vartheta, \varphi, \mathbf{p})$, detailed in (28), which describes the above cascade sequence is a function of the five angles, and of the multipole parameters $\mathbf{p} = (J, a_j', a_j)$, where a_j' and a_j describe the multipole structure for the two radiative transitions.

Table 1 Comparison of Crystal Ball results for $\psi' \to \gamma\gamma J/\psi$ with those from other experiments[a]

State (MeV/c^2)	Crystal Ball	Mark II (26)	Mark I (13)	DESY-Heidelberg (14)
		$B(\psi' \to \gamma\gamma J/\psi)$ (%)		
$\chi(3553.9 \pm 0.5)$*	1.26 ± 0.22	1.1 ± 0.3	1.0 ± 0.6	1.0 ± 0.2
$\chi(3508.4 \pm 0.4)$*	2.38 ± 0.40	2.4 ± 0.6	2.4 ± 0.8	2.5 ± 0.4
$\chi(3412.9 \pm 0.6)$**	0.06 ± 0.02	<0.56	0.2 ± 0.2	0.14 ± 0.09
$\chi(3455)$	<0.02	<0.13	0.8 ± 0.4	<0.25
$\chi(3591)$	<0.04	—	—	0.18 ± 0.06
		$B(\psi' \to mJ/\psi)$ (%)		
η	2.18 ± 0.38	2.5 ± 0.6	4.3 ± 0.8	3.6 ± 0.5
π^0	0.09 ± 0.03	0.15 ± 0.06	—	—

[a] Limits are at the 90% confidence level. Masses as measured by the Crystal Ball are denoted by an asterisk, and those measured by Mark II by a double asterisk. There is an additional 4-MeV/c^2 systematic uncertainty on all the masses.

The multipole coefficients are $a_j(a'_j)$ and they satisfy the relation, $\Gamma(\chi_J \to \gamma J/\psi) \propto \sum_{j=1}^{J+1} |a_j|^2$, and similarly for a'_j. The explicit form of the multipole coefficients is given in (28). Given the standard charmonium model, one expects the electric dipole amplitudes to dominate the transitions. Thus, the coefficients a_3 and a'_3, which are possible in the spin-2 case, can be expected to be very small and they were set to zero.

The data were analyzed by means of a histogram over the five angles. A maximum-likelihood comparison was made to a binned Monte Carlo simulation that was acceptance corrected and constrained to have a total number of events equal to that in the experimental sample. Table 2 contains the results of the likelihood fit.

The multipolarities of the radiative transitions for the $\chi_{1,2}$ are thus found to be predominantly dipole. An earlier analysis (6) also found this to be the case for the χ_1, but only when its spin was assumed to be 1. The data from the Crystal Ball study yield high confidence levels for the spin and multipole values preferred in the standard charmonium models (2).

3.3 *Results from the Full Analysis of the Inclusive Spectrum and Some Comparisons to Exclusive Results*

In this section we focus on the results of a detailed study (29), using the Crystal Ball, of the radiative transitions from ψ' to the χ_J states. Measurements of the natural line widths of the χ_J states will also be discussed briefly. The results are derived from 1.8×10^6 ψ' hadronic decays

Table 2 Results of likelihood fit of data for $\psi' \to \gamma\gamma J/\psi \to \gamma\gamma\ell^+\ell^-$ to correlated angular distributions for various χ spin values[a]

Hypothesis	$-2\ln(L/L_{max})$	a_2'	a_2
$\chi(3508)$ data:			
$J_\chi = 1$	0	$+(0.077^{+0.050}_{-0.045})$	$-(0.002^{+0.020}_{-0.008})$
$J_\chi = 2$	16		
$J_\chi = 0$	162		
$\chi(3554)$ data:			
$J_\chi = 2$	0	$+(0.132^{+0.098}_{-0.075})$	$-(0.333^{+0.292}_{-0.116})$
$J_\chi = 1$	20		
$J_\chi = 0$	40		

[a] The multipole amplitudes have been normalized so that $\sum_{j=1}^{J+1} |a_j|^2 = 1$, and, for spin 2, a_3 has been set to zero.

selected using criteria designed to reject cosmic rays, beam gas, and QED events. These criteria rejected all but a negligible part of the background while maintaining a 94% efficiency for the hadronic events.

The tracks from the hadronic events were selected for the inclusive photon analysis in four different ways. This was done to compare the effects of the different sets of cuts and the resulting different background shapes on the measured photon branching ratios and $\chi_{0,1,2}$ line widths. The following cumulative selection criteria were applied to the data to yield the four ψ' inclusive photon spectra shown in Figure 5:

(*a*) Removal of tracks with $|\cos \vartheta_j| > 0.85$ where ϑ_j is the angle of the track to the positron beam direction. This solid angle restriction ensures that each particle in the spectrum is a fiducial volume of the NaI(Tl) that has a uniform energy resolution and scale. Since both charged and neutral tracks are accepted into this spectrum, an enormous peak at about 200 MeV is observed corresponding to minimum ionizing charged particles passing through the detector. The peak presents a very large background, which dwarfs the χ_J lines. However, these lines are still highly significant and measurable.

(*b*) Removal of charged tracks using tracking chamber information. Most charged particles are removed by this cut, as is evidenced by the great reduction in the relative size of the peak at $\sim$200 MeV; however, the persistence of a remnant bump at the minimum ionization energy indicates some small inefficiency in charged-particle identification.

(*c*) Removal of neutral tracks close to charged tracks, $\cos \vartheta_{ij} < 0.9$, and removal of neutral pairs that reconstruct to a π^0 mass. These last cuts

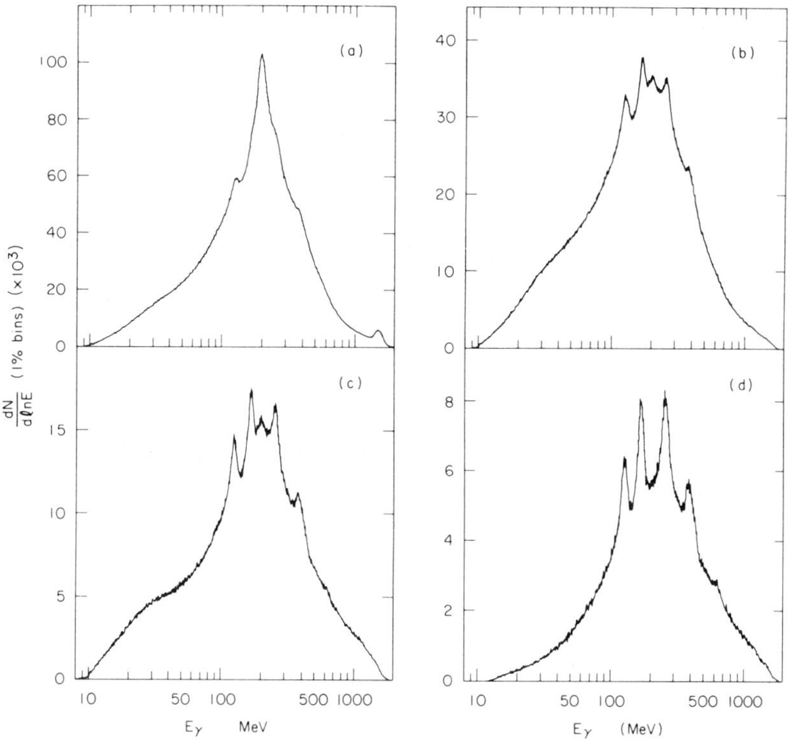

Figure 5 Inclusive γ spectra at the ψ' used in the measurement of $\psi' \to \gamma\chi_J$ and $\psi' \to \gamma\eta_c(2984)$. (*a*) All tracks neutral and charged with $|\cos\,\vartheta| < 0.85$. (*b*) Same as (*a*), except that tracks tagged as charged by the tracking chambers are removed. (*c*) Same as (*b*), except that photons resulting from reconstructed π^0 decays and those near interacting charged particles are removed. (*d*) Same as (*c*), except that each track is required to have a lateral energy deposition pattern consistent with that of an electromagnetically showering particle.

improve the signal-to-noise ratio by about a factor of two while reducing γ detection efficiency by about a factor of 0.7.

(*d*) Removal of tracks identified as minimum ionizing charged particles by their lateral energy deposition in the NaI(Tl) crystals. These charged particles were not rejected in (*b*) because of the charged-particle identification inefficiency of the tracking chambers. In this heavily cut spectrum, the minimum ionizing signal is negligible. The signal-to-noise ratio of the photon transitions has been maximized so that the $\psi' \to \gamma\eta_c$ transition is clearly visible at $E_\gamma \sim 640$ MeV (cf Section 5.1). Note that because of the fine (1%) binning of the data shown in histograms of Figure 5, the signal at $E_\gamma \sim 92$ MeV arising from the transitions $\psi' \to \gamma\eta_c'$ is not clearly visible (cf Section 5.3).

The signals corresponding to the χ_J radiative transitions were obtained from fits to the spectra of Figures 5 (29). The results from the fits are summarized in Figure 6 after corrections for photon detection efficiency, photon conversion probability, and the photon angular distributions arising from the different spins of the χ_J states have been made.

By comparing the branching ratios $B(\psi' \rightarrow \gamma\chi_J)$ extracted from the four spectra, one is able to assess the magnitude of the systematic errors contributing to the measurement. As is seen in Figure 6, the variation among the four branching ratio values for each line is consistent within the statistical errors of the measurements. The fact that consistent results are obtained with such widely different looking spectra gives one confidence in the finally extracted branching ratios.

A second check is the comparison of the cascade branching ratios $B(\psi' \rightarrow \gamma\chi_{1,2}) \cdot B(\chi_{1,2} \rightarrow \gamma J/\psi)$ as measured using the Doppler-broadened secondary transition lines seen in the inclusive photon spectra of Figure 5 with the values obtained from the exclusive events discussed in Section 3.2.

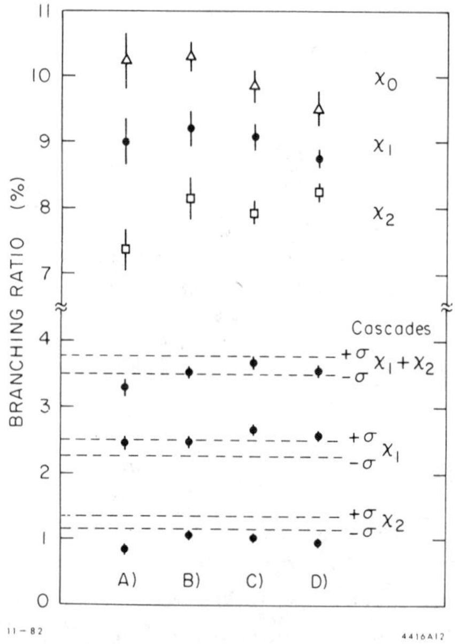

Figure 6 The upper part of the figure shows the observed values of $B(\psi' \rightarrow \gamma\chi_J)$ as obtained from independent analysis of each of the four spectra shown in Figure 5. The lower part compares the cascade product branching ratios $B(\psi' \rightarrow \gamma\chi_{1,2}) \cdot B(\chi_{1,2} \rightarrow \gamma J/\psi)$ from the four spectra (*dots*) with the direct measurements of these products from our analysis of the exclusive events $\psi' \rightarrow \gamma\gamma\ell^+\ell^-$ (*dashed bands*). Since separation of the overlapping lines from the two photons $\chi_1 \rightarrow \gamma J/\psi$ and $\chi_2 \rightarrow \gamma J/\psi$ in the inclusive spectra is difficult, the comparison with the sum is also shown.

The results of the cascade measurement are shown as the points on the bottom part of Figure 6. The dashed bands show the exclusive measurement of the same transitions given in Table 1. As is seen in this figure, the inclusive measurement for the χ_2 is somewhat lower than the exclusive measurement, while for χ_1 it is somewhat higher. However, the sum of the χ_1 and χ_2 branching ratios yields good agreement between the inclusive and exclusive measurements. This effect has also been reproduced in Monte Carlo calculations. It is due to the overlap of the two transitions in the inclusive spectra. That the sum of the inclusive lines is in good agreement with the sum of the exclusive measurements allows an uncertainty in the absolute normalization of the inclusive result of less than 16%, the absolute error in the exclusive measurement [remember that this error is dominated by the uncertainty in $B(\mathrm{J}/\psi \to \ell^+\ell^-)$].

The final results of the analysis, including branching ratios and the values for the natural line widths of the χ_J states, are shown in Table 3. For the branching ratios the first error is dominated by the statistical uncertainty and point-to-point errors in the photon detection efficiency. The second error is an estimate of the overall normalization error due mainly to a $\pm 5\%$ uncertainty in both the hadronic event selection efficiency and the overall photon detection efficiency.

Agreement between these Crystal Ball branching ratio measurements and those of the lower statistics experiment of Biddick et al (12) is within the experimental errors. However, our branching ratios to the χ_J states are consistently higher, and within the point-to-point errors of our measurements there is an indication for an increase in rate from χ_2 to χ_0 transitions. In nonrelativistic models, $\Gamma(\psi' \to \chi_J) \propto (2J+1)E_\gamma^3$, and thus we expect $\Gamma'_0 : \Gamma'_1 : \Gamma'_2 = 1:1:1$ where $\Gamma'_J = \Gamma(\psi' \to \gamma\chi_J)/[(2J+1)E_\gamma^3(\chi_J)]$. As shown in Table 3, we obtain $1:1.07 \pm 0.08 : 1.39 \pm 0.11$, in reasonable agreement

Table 3 Results from the $\psi' \to \gamma\chi_J$[a]

Datum	χ_0	χ_1	χ_2
E_γ (MeV)	$258.4 \pm 0.4 \pm 4$	$169.6 \pm 0.3 \pm 4$	$126.0 \pm 0.2 \pm 4$
$\Gamma(\chi_J)$ (MeV)	(13.5–20.4)	<3.8	(0.85–4.9)
$B(\psi' \to \gamma\chi_J)$ (%)	$9.9 \pm 0.5 \pm 0.8$	$9.0 \pm 0.5 \pm 0.7$	$8.0 \pm 0.5 \pm 0.7$
Ratio $\left\{\dfrac{B(\psi' \to \gamma\chi_J)}{E_\gamma^3(2J+1)}\right\}$	1:	1.07 ± 0.08:	1.39 ± 0.11
$B(\chi_J \to \gamma \mathrm{J}/\psi)$ (%)	0.60 ± 0.17	28.4 ± 2.1	12.4 ± 1.5

[a] When two errors are given, the first error is statistical and the second is systematic. Ranges and upper limits are at 90% confidence levels.

with the simple theory. However, our absolute branching ratios are lower by a factor of two to three than the predictions of the simple nonrelativistic charmonium models (29). Models that include relativistic corrections, variations of the 2S and ^3P_J wave function shapes resulting from higher order corrections, and coupled channels achieve better agreement with the data.

The measurement of the natural line widths of the χ_J states is a tricky one since the Crystal Ball's photon energy resolution is comparable to or greater than these widths. It does appear, however, that the χ_0 is much broader than predicted by QCD, while the χ_1 and χ_2 widths are in good agreement with QCD within errors (29, 30).

4. HADRONIC TRANSITIONS FROM THE ψ' TO THE J/ψ

Figure 1*b* shows the hadronic transitions that have been observed between the ψ' and J/ψ. All of these transitions were observed by at least two experiments and the $\pi\pi$ and η transitions have been observed by many experiments. As the $\pi\pi$ transitions can easily be observed in the charge mode [$B(\psi' \rightarrow \pi^+\pi^- J/\psi) = 33 \pm 2\%$ (27)], excellent measurements of this mode have been made by other detectors stressing charged-particle detection. The Crystal Ball has measured the neutral $\pi^0\pi^0$ mode (31), as a check on measurements of the η and π^0 transitions. Comparison of the neutral $\pi^0\pi^0$ to the charged (32) $\pi^+\pi^-$ mass distributions show the shapes of the two distributions to be the same within error, as is expected from isospin symmetry.

4.1 *The Transitions* $\psi' \rightarrow \eta(\pi^0)J/\psi$

The study of these processes is related to that of the ^3P_J state cascades and so is detailed in (17) (cf Section 3.2). The $m_{\gamma\gamma}$ distribution for all fitted events is shown in Figure 7*a*. Of the events in this figure, 412 candidates for the η events are separated from χ_J and π^0 events by using the cut $m_{\gamma\gamma} >$ 525 MeV/c^2. This cut loses no η events, but does admit some χ_1 events into the η sample.

Monte Carlo calculations determined that 21 χ_1 events and 5 $\pi^0\pi^0$ events are expected in the η sample. The resulting η mode branching ratio is compared in Table 1 with other measurements. The Crystal Ball and Mark II results (26) are in good agreement, while the other measurements shown are larger than our measurement by about a factor of two.

Existence of the transition $\psi' \rightarrow \pi^0 J/\psi$ is apparent in the Dalitz plots of

Figures 4*a*,*b*. A π^0 signal is observed in the diphoton mass plot by removing the dominant background from cascade photons using a cut on the $\gamma J/\psi$ masses. A subtraction of events from the $m_{\gamma\gamma}$ plot of Figure 7*a* with $(M_{\gamma J/\psi})_{\text{high}}$ in the ranges 3410 ± 5 and 3530 ± 60 MeV/c^2, and $m_{\gamma\gamma} >$ 525 MeV/c^2, results in the distribution shown in Figure 7*b*. These data have been fitted to a Gaussian peak with a quadratic background distribution. The fit yields 23 events above background having $m_{\gamma\gamma} < 200$ MeV/c^2. The resulting π^0 mode branching ratio is compared in Table 1 with another measurement from the Mark II (26). The two measurements are in good agreement. This decay violates isospin symmetry. A review of the theoretical literature relevant to our measurement can be found in (17).

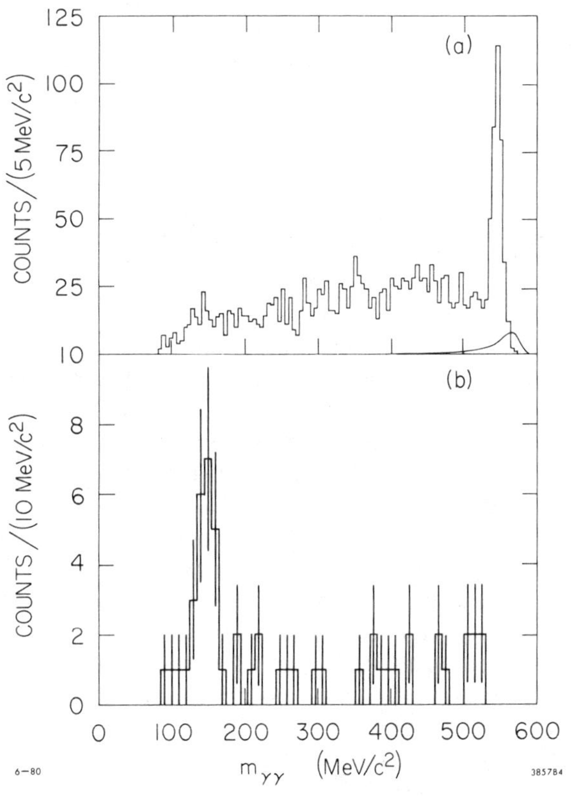

Figure 7 Diphoton masses of fitted events for $\psi' \to \gamma\gamma J/\psi \to \gamma\gamma\ell^+\ell^-$. (*a*) The peak due to the η; the smooth line is a 10-times-magnified calculated curve for the expected contamination from $\psi' \to \pi^0\pi^0 J/\psi$. (*b*) The same as (*a*) except that events consistent with $\psi' \to \eta J/\psi$, and $\psi' \to \gamma\chi_J$ have been removed. These cuts allow the π^0 peak to show clearly.

5. THE CHARMONIUM $^1S_0$ STATES

The Crystal Ball discoveries that created the most excitement were the lack of a signal in $J/\psi \to \gamma\gamma\gamma$ at $M_{\gamma\gamma} = M_{X(2830)}$ (15), which had been reported by the DASP collaboration (10, 11) (cf Section 6.4), and the discovery of an η_c candidates state at $M_{\eta_c} = 2984 \pm 4$ MeV/c^2 by means of the radiative transitions from the ψ' (33) and J/ψ (34). Since the original observations were made, the Crystal Ball has doubled both the ψ' and J/ψ data sets to about 2×10^6 hadronic decays each. This increase in data has allowed a more precise determination of the η_c parameters (29). Furthermore, it has also resulted in the discovery of an η_c' candidate at $M_{\eta_c'} = 3592 \pm 5$ MeV/c^2 via a radiative transition from the ψ'.

The two states at 2984 and 3592 MeV/c^2 can be naturally associated with the 1^1S_0 and 2^1S_0 charmonium states, the η_c and η_c'. As we see in this section, their properties fall within the range of theoretical expectations. Thus, with the work described in previous sections we have come from a state of relative confusion and uncertainty concerning the validity of charmonium as a model of the J/ψ system to one of good agreement between theory and experiment in most cases.

5.1 *Evidence for the 1^1S_0 in Inclusive γ Spectra of the ψ' and the J/ψ*

The analysis of the inclusive photon spectra from the 1.8×10^6 ψ' and 2.2×10^6 J/ψ decays when studying the $\eta_c(2984)$ is very similar to that described in Section 3.3 and is detailed in (29). However, not only were the four spectra from the ψ' shown in Figure 5 and the corresponding four from the J/ψ (not shown) used, but a fifth spectrum from both the J/ψ and ψ' was also included in the analysis. The pattern cuts for this fifth spectrum were designed to improve the efficiency for detection of low energy photons, at the expense of reduced efficiency for removing minimum ionizing charged particles. It is shown for the J/ψ in Figure 8. The inserts on the upper left of Figures 3 and 8 show the result of one of the simultaneous fits made to correspondingly cut J/ψ and ψ' inclusive photon spectra. For the radiative transition to the η_c, the η_c mass and width are constrained to be the same for both spectra. The results of the fits to each of the five pairs of spectra were compared as a consistency check. An additional check was made by measuring the mass and width in the γ spectrum coming from events containing exactly two observed charged particles.

The results of this analysis (29) are $M_{\eta_c} = 2984 \pm 5$ MeV/c^2, $\Gamma_{\eta_c} = 11.5^{+4.5}_{-4.0}$ MeV, $B(J/\psi \to \gamma\eta_c) = (1.27 \pm 0.36)\%$, and $B(\psi' \to \gamma\eta_c) = (0.28 \pm 0.06)\%$. The errors are dominated by the statistical uncertainties, except for the mass error, which is mainly due to the uncertainty in our absolute

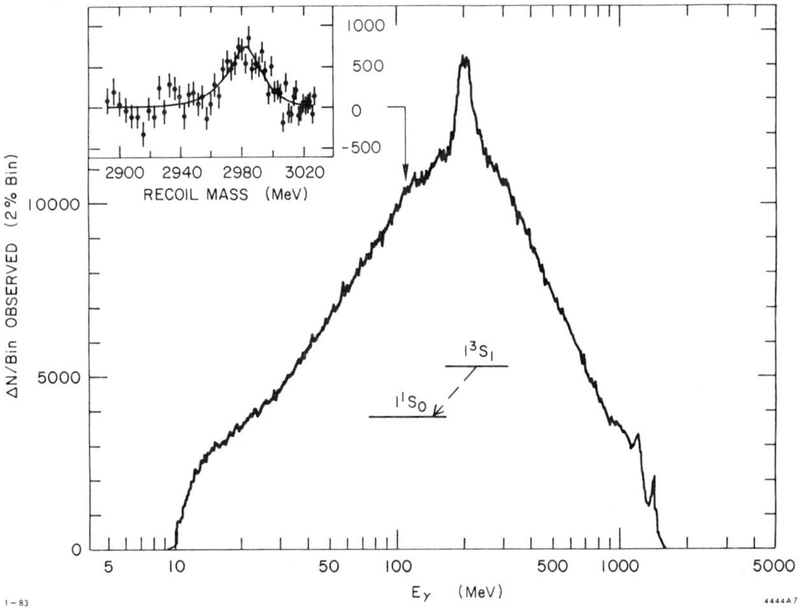

Figure 8 Inclusive γ spectrum at the J/ψ, $\Delta N/\Delta(\log E) \cong E\,dN/dE$. The strong peak at 200 MeV is due to charged particles that were not tagged by the tracking system. The inset shows the background-subtracted signal from the η_c candidate state. The two prominent peaks near the high energy end point of the spectrum are from the monochromatic photons in the reactions $\psi \to \gamma\eta'$ and $\psi \to \gamma\iota(1440)$. The photons from $\psi \to \gamma\eta$ (π^0) were eliminated from this spectrum by a hard QED cut.

energy calibration. These values are in good agreement, within errors, with previously reported Crystal Ball values (34).

5.2 *Hadronic Decays of the 1^1S_0*

Confirmation of the $\eta_c(2984)$ came soon after its discovery in the ψ' and J/ψ inclusive γ spectra with the first reports of the observation of its hadronic decays by the Mark II collaboration at SPEAR (35). A number of decay modes were seen, as is shown in Table 4.

We also have looked for exclusive decays of the $\eta_c(2984)$ into hadrons by performing kinematic fits to exclusive final states with multiple photons and two charged hadrons (34, 36). Remember that the Crystal Ball measures both the energy and angle of electromagnetically showering particles but for charged hadrons (π, K) it measures only the angles well. Secondary interactions of the charged hadrons in the sodium iodide complicate the fitting of some events, but special pattern recognition algorithms have been developed to deal with this effect.

Table 4 η_c branching ratio measurements

Decay mode	Mark II (35) $B(\psi' \to \gamma\eta_c) \cdot B(\eta_c \to X)$	$B(\eta_c \to X)$[a]
$p\bar{p}$	$(8^{+8}_{-4}) \times 10^{-6}$	$(2.9^{+3.0}_{-1.6}) \times 10^{-3}$
$\pi^+\pi^-\pi^+\pi^-$	$(5.7^{+3.9}_{-2.4}) \times 10^{-5}$	$(2.0^{+1.5}_{-0.9}) \times 10^{-2}$
$\pi^+\pi^-K^+K^-$	$(4.0^{+6.0}_{-2.5}) \times 10^{-5}$	$(1.4^{+2.1}_{-0.9}) \times 10^{-2}$
$\pi^+\pi^-p\bar{p}$	$<5 \times 10^{-5}$ (90% C.L.)	$<2.3 \times 10^{-2}$ ($\sim$90% C.L.)
$K^\pm\pi^\mp K^0_s$	$(1.5^{+0.8}_{-0.6}) \times 10^{-4}$	$(5.4^{+3.3}_{-2.4}) \times 10^{-2}$

Decay mode	Crystal Ball $B(J/\psi \to \gamma\eta_c) \cdot B(\eta_c \to X)$	$B(\eta_c \to X)$[b]
$\eta\pi^+\pi^-$	$(3.1 \pm 1.9) \times 10^{-4}$	$(2.6^{+1.8}_{-1.7}) \times 10^{-2}$
$\gamma\gamma$[c]	$<1.6 \times 10^{-5}$ (90% C.L.)	$<1.8 \times 10^{-3}$ ($\sim$90% C.L.)
$K^+K^-\pi^0$	$<1.5 \times 10^{-4}$ (90% C.L.)	$<1.7 \times 10^{-2}$ ($\sim$90% C.L.)

[a] Uses Crystal Ball value for $B(\psi' \to \gamma\eta_c)$.
[b] Uses Crystal Ball value for $B(J/\psi \to \gamma\eta_c)$.
[c] See Section 6.4.

Events with a three-photon, two-charged-particle topology were selected from the sample of J/ψ hadronic decays and subjected to a 3-C kinematic fit to the hypotheses, $J/\psi \to \gamma\eta\pi^+\pi^-$ and $\gamma\eta K^+K^-$, $\eta \to \gamma\gamma$. The energy spectrum for the low energy radiated photon, arising from events that have a χ^2 greater than 0.1 for the $\eta\pi^+\pi^-$ hypothesis, showed a significant signal above background at the $\eta_c(2984)$ mass, within errors. No comparable signal was seen for the ηK^+K^- hypothesis.

The $\gamma\eta\pi^+\pi^-$ data, in principle, contain additional information on the width of the $\eta_c(2984)$. However, given the limited statistics of this measurement, which comes from only half the presently available J/ψ data, we believe the inclusive measurement of the width to be more reliable at this time. From the signal of 18 ± 6 events, we obtain the product branching ratio $B(J/\psi \to \gamma\eta_c) \cdot B(\eta_c \to \eta\pi\pi)$, and branching ratio $B(\eta_c \to \eta\pi\pi)$ given in Table 4. In addition, to compare directly with the Mark II observation of the $K^\pm K^0_s\pi^\mp$ final state of the $\eta_c(2984)$, our upper limit for the $K^+K^-\pi^0$ final state is also given. Note that the Crystal Ball value must be doubled before comparing with the Mark II result, owing to isospin; we assume $I = 0$ for $\eta_c(2984)$. Also, for completeness, the $\gamma\gamma$ final-state branching ratio is given here (cf Section 6.4).

5.3 *Evidence for the 2^1S_0 State in the Inclusive γ Spectrum of the ψ'*

As mentioned in Section 5, a candidate for the 2^1S_0 state or η_c' has been found by the Crystal Ball using inclusive photon decays of the ψ'. In this section we briefly describe our evidence for the state. A more complete description can be found in (37).

The event selection for the analysis as well as the photon selection criteria used are the same as those described in Section 3.3 with two minor changes. First, events with more than 10 charged or more than 10 neutral observed tracks are not considered. Second, a somewhat different lateral shower energy deposition pattern in the NaI(Tl) crystals is used to define photons than in the analysis described in Section 3.3 and 5.1. In this case, an extra premium was placed on good efficiency for $E_\gamma < 100$ MeV. The main spectrum of Figure 3 results from about half the 1.8×10^6 ψ' decays, cut as described in Section 3.1. A signal at 3592 ± 5 MeV is evident in this spectrum. The insert on the upper left of Figure 3 shows the result of performing a fit to the region containing the structure at $E_\gamma \sim 90$ MeV; this insert contains a spectrum obtained from all 1.8×10^6 ψ' decays. A clear signal is obtained with 4.4σ to 6σ significance, depending on how the fit is performed. The properties obtained from the fit for the η'_c candidate state are: $M_{\eta'_c} = 3592 \pm 5$ MeV/c^2, $\Gamma_{\eta'_c} < 8$ MeV (95% C.L.), and $B(\psi' \to \gamma\eta'_c)$, in the range 0.2% to 1.3% with a confidence level of 95%. The confidence interval for the uncertainty in the branching ratio includes the correlation with $\Gamma_{\eta'_c}$.

It should be noted that the DESY-Heidelberg group reported evidence (14) for a state at a mass of 3592 ± 7 MeV/c^2 in the exclusive channel $\psi' \to \gamma\gamma J/\psi$, $J/\psi \to \mu^+\mu^-$. However, as reviewed in Section 3.2, we have looked for evidence of such a state in the cascade decays and found none. If we assume that the object we observe in the inclusive spectrum is the η'_c, then it is expected (37) that $B(\psi' \to \gamma\eta'_c) \cdot B(\eta'_c \to \gamma J/\psi) < 10^{-6}$. This estimate is based on our measured value of $B(\psi' \to \gamma\eta'_c)$ and on theoretical calculations (7) for the η'_c total width and radiative transition rate. The estimate for the hindered magnetic dipole transition $\eta'_c \to \gamma J/\psi$ was based upon our measurement of the similar transition $\psi' \to \gamma\eta_c$, which reduces the estimate's sensitivity to the details of the wave functions. Such a small product of branching ratios has not been accessible to any experiment.

5.4 *Discussion*

Clear signals have been seen for states at $M = 2984 \pm 5$ MeV/c^2, and $M = 3592 \pm 5$ MeV/c^2 by the Crystal Ball detector; the Mark II has confirmed the state at 2984. These states are obvious candidates for the $1^1S_0(\eta_c)$ and $2^1S_0(\eta'_c)$ states of charmonium. What evidence makes these tentative assignments plausible?

First, the η_c is seen to decay into three pseudoscalars and not two. This allows only $0^-, 1^+, \ldots$ assignments for the J^P of the state. As discussed in (3), the measured radiative transition branching ratio $B(J/\psi \to \gamma\eta_c)$ is in reasonable agreement with both the nonrelativistic and QCD sum rule calculations, which assume $J^{PC} = 0^{-+}$ for the observed state. In addition, the branching ratio $B(\psi' \rightarrow \gamma\eta_c)$ was also predicted by theory (38) and these predictions are in agreement with the observation. The mass splitting 1^3S_1–1^0S_1 is predicted by a number of theories, including nonrelativistic models and QCD sum rule calculations (38, 39) and these are again in good agreement with the data. Finally, the width of the η_c is predicted to be 8.3 ± 0.5 MeV using QCD with higher order corrections (40) and the experimental value of $\Gamma_{\eta_c} = 11.5^{+4.5}_{-4.0}$ MeV agrees within the error. The important partial width $\Gamma(\eta_c \rightarrow \gamma\gamma)$ which has been predicted to be 4.2 ± 0.4 keV using the QCD sum rules (41) is well below the Crystal Ball upper limit of $\Gamma(\eta_c \rightarrow \gamma\gamma) < 20$ keV (90% C.L.) (cf Section 6.4). One can thus conclude with some certainty, given the above evidence, that the Crystal Ball state at 2984 ± 4 MeV/c^2 is truly the 1^1S_0 of charmonium.

Unfortunately, the case is not so clear for the η_c' candidate at $M = 3592 \pm 5$ MeV/c^2. Relatively little is known about this state. No exclusive decays have been seen, and only an upper limit exists on its width. Within the limits of uncertainty concerning 2^3S_1–3^3D_1 mixing (42), the agreement between theory and experiment is reasonable for the ψ'–η_c' mass splitting. Also, the observed value for $B(\psi' \rightarrow \gamma\eta_c')$ agrees with nonrelativistic model calculations within the large range allowed by observations. However, confirmation and more information are needed on this state before a firm connection to the 2^1S_0 state of charmonium can be made.

6. RADIATIVE TRANSITIONS FROM THE J/ψ

Other than for J/$\psi \rightarrow \gamma\eta_c$, $\eta_c \rightarrow$ hadrons, interest in radiative decays of the J/ψ first centered on 3γ decays, e.g. $\gamma\eta(\gamma\gamma)$ or $\gamma\eta'(\gamma\gamma)$, and particularly on searches for $\gamma\eta_c(\gamma\gamma)$. However, in recent years the possibility of observing gluonic meson states, particles made up entirely of gluons, has also stimulated much interest in J/ψ radiative decays.

6.1 *The Fundamental Character of Gluonic Mesons in QCD*

The existence of an extensive spectrum of colorless, flavorless bound states of two or more gluons has been firmly predicted by QCD (3). These gluonic bound states were given the name "gluonic mesons" by their inventors, H. Fritzsch and M. Gell-Mann (43). It is expected that the lower mass gluonic meson states are bound states of mostly two gluons; in analogy to quarkonium, a bound state of a quark and antiquark, these systems are called gluonium. It is also expected that, because of their relatively lower masses predicted to lie in the range of 1 to 2 GeV, gluonium states should be by far easier to observe than the higher mass gluonic mesons. Although the existence of gluonium has not yet been experimentally

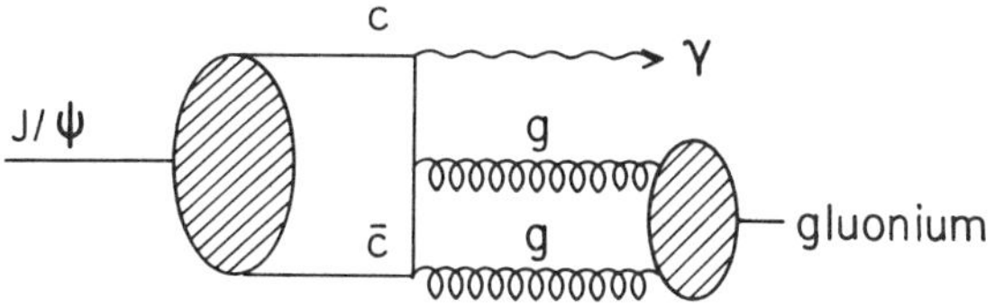

Figure 9 Lowest order QCD diagram for the radiative decay of the J/ψ into a gluonic meson.

established, the interest in this new form of matter has increased considerably since the observation of two new mesons, the $\iota(1440)$ (44, 45) and the $\vartheta(1640)$ (46). These are seen in a reaction thought to be a copious source of gluonic meson states (47), namely, $J/\psi \rightarrow \gamma x$. The mechanism is shown diagramatically in Figure 9. According to lowest order QCD calculations, the hadronic decays of quarkonium $^3S_1$ states, such as the J/ψ, proceed mainly via annihilation of the $q\bar{q}$ system into three gluons. Although this process might seem well suited to the production of gluonium states, it is not since each pair of the three final-state gluons must be in a color octet state. This follows from the fact that the overall state must be a color singlet and each pair recoils against one color octet gluon. However, if a photon is radiated with two gluons in the decay, as shown in Figure 9, the recoiling gluon pair must form a color singlet state, which is even under charge conjugation.

Perturbative QCD indicates (48, 49) a partial width for the process $J/\psi \rightarrow \gamma gg$ of about 8 keV, which is relatively large. Various authors (47) have used duality principles, and other ideas, together with the perturbative result to show that gluonium states should be copiously produced in this process. However, the experimental search for such states has proven to be a difficult and confusing one with a number of guiding theoretical principles losing credibility as the field has matured (3).

6.2 *The "End Point" of the Inclusive γ Spectrum at the J/ψ*

One can qualitatively appreciate the major features of the radiative decays of the J/ψ by viewing the "end point" of the inclusive γ spectrum as measured by the Crystal Ball detector (42), and shown in Figure 10.

Relatively narrow peaks at the $\iota(1440)$ and $\eta'(958)$ are evident, and there is also a broad structure centered at a recoil mass of about 1700 MeV/c^2 (the ϑ has a mass close to 1700 MeV/c^2 but it is not as broad as the structure seen in the inclusive spectrum). The tails of the $\iota(1440)$ structure include the regions where radiative transitions to the f(1270), D(1285), and f′(1515) would appear.

Transitions to the $\eta(549)$ should also be seen, but these are suppressed in this spectrum by the event selection cuts (42). Likewise, even if the $J/\psi \rightarrow \gamma\pi^0$

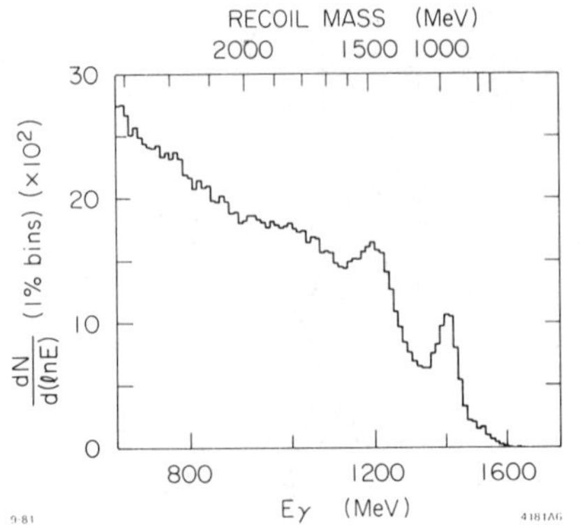

Figure 10 The same as Figure 8 showing details near the high energy end point of the spectrum.

rate were large, no signal would be seen because of these cuts. Up to the time of this writing, very little quantitative analysis of the spectrum in Figure 10 has been done, and so only the above qualitative information can be drawn from it. A strong analysis effort started recently in the Crystal Ball collaboration will, we hope, remedy this situation.

6.3 *Gamma Transitions to Well-Known Particles Using Exclusive Decays*

Experimental measurements have been reported by the Crystal Ball collaboration for the processes $J/\psi \rightarrow \gamma\pi^0, \gamma\eta, \gamma\eta', \gamma f$, and an estimate for $\gamma f'$, $f' \rightarrow \eta\eta$. Crystal Ball measurements for the η and η' have been published (15); however, new measurements by the Crystal Ball collaboration (50) derived from the full data sample of 2×10^6 J/ψ decays, about twice the data of the previously published results, disagree somewhat with the older measurements. The new measurements are in agreement within the errors with three measurements from other experiments (10, 50a, 51).

Table 5 shows the most recent Crystal Ball results for the various decays. Note that the new results on the η' use the $\eta\pi^+\pi^-$, $\eta\pi^0\pi^0$, and $\gamma\rho^0$ decay modes as well as the $\gamma\gamma$ decay mode, which was the only one used in the old result. The Dalitz plot for $J/\psi \rightarrow 3\gamma$ from all our data is shown in Figure 11. Prominent signals are seen for the η and η'. No signal is seen at $M_X = 2830$ MeV/c^2 and $M_{\eta_c} = 2984$ MeV/c^2. Upper limits for these processes are given in Table 5. The direct decay $J/\psi \rightarrow 3\gamma$ as well as the QED process $e^+e^- \rightarrow 3\gamma$ also contribute to the Dalitz plot.

As an example of the analysis of an hadronic final state of the η', Figure 12 shows the signal for $J/\psi \rightarrow \gamma\eta', \eta' \rightarrow \gamma\rho, \rho \rightarrow \pi^+\pi^-$. These events satisfy a 2-C

Table 5 Crystal Ball measurements (except as noted) of $J/\psi \to \gamma\pi^0, \gamma\eta, \gamma\eta', \gamma f, \gamma f'$ [a]

$B(J/\psi \to \gamma\pi^0) = (3.6 \pm 1.1 \pm 0.7) \times 10^{-5}$
$B(J/\psi \to \gamma\eta) = (0.88 \pm 0.08 \pm 0.11) \times 10^{-3}$

η' decay mode	$B(J/\psi \to \gamma\eta') \times 10^{-3}$
$\eta' \to \eta\pi^+\pi^-$	$3.9 \pm 1.0 \pm 1.1$
$\eta' \to \eta\pi^0\pi^0$	$4.2 \pm 0.6 \pm 0.6$
$\eta' \to \gamma\rho^0$	$4.1 \pm 0.4 \pm 0.6$
$\eta' \to \gamma\gamma$	$4.4 \pm 0.9 \pm 0.5$
Average	$4.1 \pm 0.3 \pm 0.6$

$B(J/\psi \to \gamma f) = (1.48 \pm 0.25 \pm 0.30) \times 10^{-3}$
$B(J/\psi \to \gamma f') \cdot B(f' \to \eta\eta) = (0.9 \pm 0.9) \times 10^{-4}$
$B(J/\psi \to \gamma f') \cdot B(f' \to K\bar{K})^{b} = (1.8 \pm 1.0) \times 10^{-4}$
$B(J/\psi \to \gamma X) \cdot B(X \to 2\gamma) < 1.6 \times 10^{-5}$ (90% C.L.),
for $2600 < M_X < 3000$ MeV/c^2 and $\Gamma_X \lesssim 25$ MeV
$B[J/\psi \to 3\gamma \text{ (direct)}] < 5.5 \times 10^{-5}$ (90% C.L.)

[a] Where two errors are given, the first is statistical and the second systematic.
[b] Mark II (64).

fit to the hypothesis $\gamma\gamma\pi^+\pi^-$. They also were subjected to several more constraints:

1. The high energy neutral track was required to satisfy a lateral energy deposition in the NaI(Tl) crystals expected of a high energy photon, rather than two photons from a high energy π^0.
2. Photon pairs forming a π^0 or an η were excluded.
3. The energy of the charged particles had to be less than 1360 MeV.
4. The $\pi\pi$ mass was cut about the ρ mass.

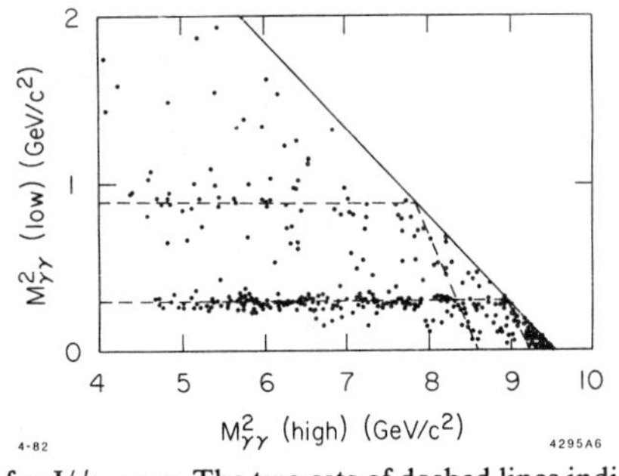

Figure 11 Dalitz plot for $J/\psi \to \gamma\gamma\gamma$. The two sets of dashed lines indicate where events from $J/\psi \to \gamma\eta$ and $J/\psi \to \gamma\eta'$ should be clustered. Except for a QED background, no other signals are seen.

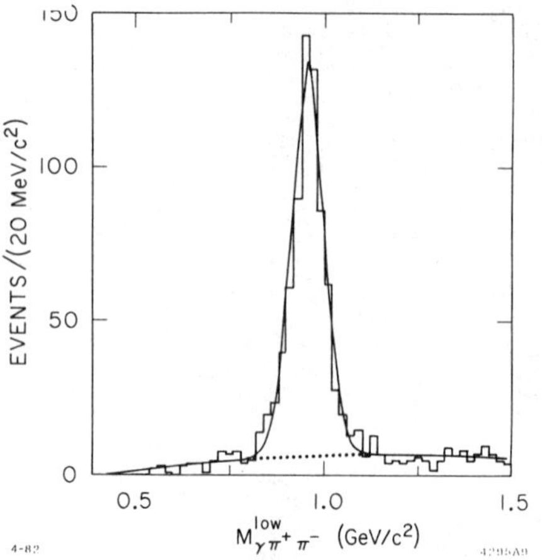

Figure 12 Distribution of the $\gamma_1\pi^+\pi^-$ mass from events satisfying the hypothesis $J/\psi \to \gamma_2\gamma_1\pi^+\pi^-$ where $E_{\gamma_1} < E_{\gamma_2}$ and $m_{\pi\pi}$ has been cut about the ρ^0 mass. The solid curve shows a fit to an η' peak plus a smooth background (*dotted*).

These requirements removed the strong $J/\psi \to \pi\rho$, and $J/\psi \to \gamma\eta'$, $\eta' \to \eta\pi^+\pi^-$ backgrounds. A Monte Carlo calculation gives an efficiency of 24% for the 666 events found and displayed in Figure 12. The ratio, $B(J/\psi \to \gamma\eta')/B(J/\psi \to \gamma\eta)$ has been of theoretical interest with the QCD sum rules (52), and other models (52), yielding values in the range from 3.7 to 4.0. In order to further compare data to theory we calculate from our data, $B(J/\psi \to \gamma\eta')/B(J/\psi \to \gamma\eta) = 4.7 \pm 0.6$.

Table 5 also shows a new Crystal Ball result for $J/\psi \to \gamma\pi^0$ (50). This result is in good agreement with the only other measurement of this quantity by DASP (10).

Although the process $J/\psi \to \gamma f(1270)$ has been well studied in other experiments (53), the analysis of this process in the Crystal Ball (54) provides a useful check on the analysis techniques employed in the ι and ϑ studies (cf Section 6.4). It also provides a check that the Crystal Ball efficiencies are well understood in this complex $\gamma\pi^0\pi^0$ (5γ) final state. In addition, our measurement provides confirmation of previous results, and has also yielded the most precise determination available of the helicity amplitudes for the process $J/\psi \to \gamma f$. Figure 13 shows the $\pi^0\pi^0$ invariant mass distribution; a prominent f(1270) signal is seen with 178 ± 30 events. Figure 14 shows contours of equal probability as a function of x and y, $x = A_1/A_0$ and $y = A_2/A_0$, where A_0, A_1, and A_2 are the f helicity amplitudes (55). The errors on our measurement are small enough that a quantitative comparison with theory can be made. Theoretical predictions for pure M2 and E3 transitions (E1 is offscale), QCD (56), and tensor meson dominance

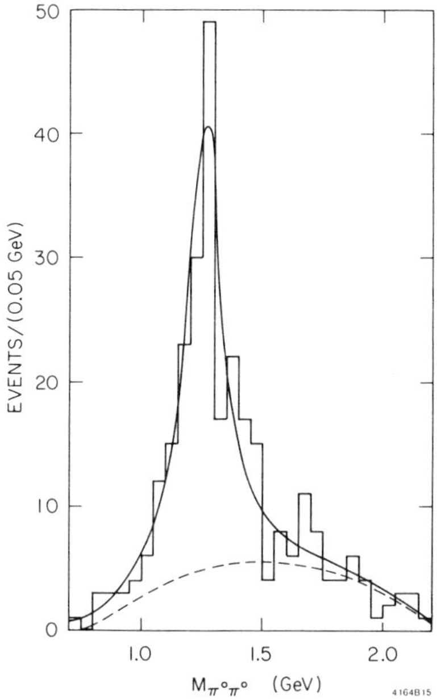

Figure 13 Distribution of the $\pi^0\pi^0$ mass from events that satisfy the 4-C fits to the hypothesis $J/\psi \rightarrow \gamma\pi^0\pi^0$. The solid curve shows a fit to an f peak plus a smooth background. The dashed curve represents the background contribution.

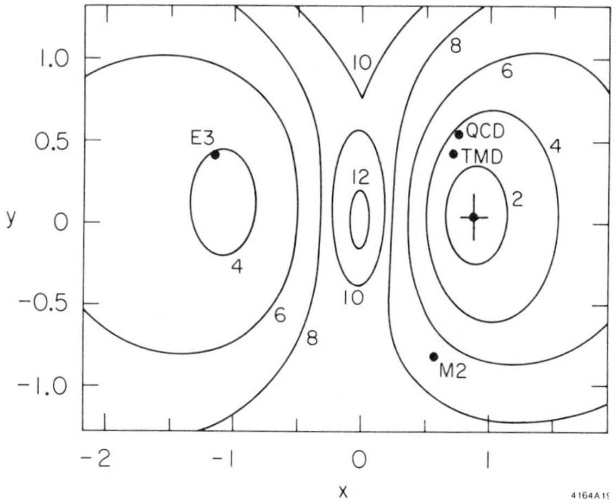

Figure 14 Contours of equal probability as a function of x and y, the ratios of helicity amplitudes in the decay of the f in $J/\psi \rightarrow \gamma f \rightarrow \gamma\pi^0\pi^0$. The data point with error bars represents the measurement and the others are theoretical predictions (see text). Numbers next to the curves are in units of standard deviations.

(TMD) (57) are also shown in Figure 14. All of these predictions are inconsistent with the experimental measurement. In particular, the QCD calculation based on two-gluon exchange (56) is more than three standard deviations from the experimental point.

As discussed in the next section, a by-product of the $\vartheta(1640)$ study has been a rough measurement of $B(J/\psi \to \gamma f') \cdot B(f' \to \eta\eta)$. This result is listed in Table 5, along with the Mark II measurement of $B(J/\psi \to \gamma f') \cdot B(f' \to K\bar{K})$.

6.4 *The Gluonium Candidates, $\iota(1440)$ and $\vartheta(1640)$*

A state at 1440 MeV/c^2 was first seen in the reaction, $J/\psi \to \gamma K^{\pm}K^0_s\pi^{\mp}$, by the Mark II collaboration at SPEAR (44). They tentatively identified it as E(1420), a state with $J^{PC} = 1^{++}$ because their experiment was unable to determine the J^P value. The existence of this state was soon confirmed by the Crystal Ball collaboration at SPEAR (36) using the reaction, $J/\psi \to \gamma K^+K^-\pi^0$. However, much more J/ψ data was needed (2.2×10^6 decays in total) before the Crystal Ball collaboration was able to measure the J^P of the state as 0^- (45).

This 0^{-+} state may have been previously observed in $p\bar{p}$ annihilations (58). The state seen in the $p\bar{p}$ case was named E(1420). However, the 0^{-+} assignment from that experiment was not considered conclusive (59) and so the name "E(1420)" was subsequently assigned to the $J^{PC} = 1^{++}$ state seen in π^-p interactions (60). Thus, the Crystal Ball and Mark II experiments (in collaboration) have named the 0^{-+} state seen in J/ψ radiative decays the $\iota(1440)$ (44).

Figure 15*a* shows the $K^+K^-\pi^0$ invariant mass distribution for events

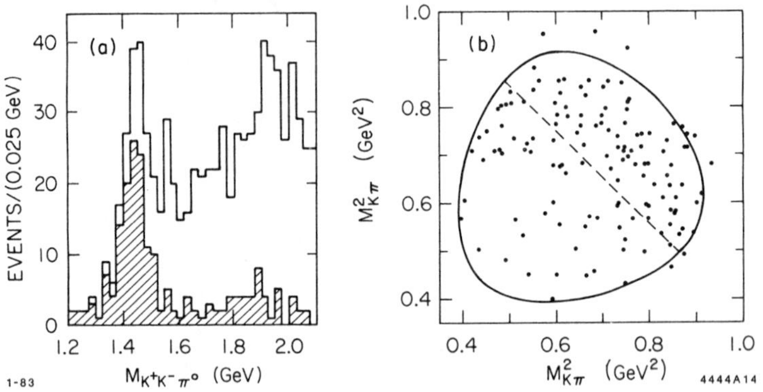

Figure 15 (*a*) Distribution of the $K^+K^-\pi^0$ mass from events consistent with the hypothesis $J/\psi \to \gamma K^+K^-\pi^0$. The events in the shaded region satisfy the further requirement $M_{K^+K^-} < 1125$ MeV/c^2. (*b*) Dalitz plot for $K^+K^-\pi^0$ events from $J/\psi \to \gamma K^+K^-\pi^0$ with $1400 < M_{K\bar{K}\pi} < 1500$ MeV/c^2. The solid curve shows the boundary for $M_{K\bar{K}\pi} = 1450$ MeV/c^2 and the dashed line shows $M_{K\bar{K}} = 1125$ MeV/c^2.

that satisfy 3-C fits to the process $J/\psi \rightarrow \gamma K^+K^-\pi^0$. This analysis is based on 2.2×10^6 produced J/ψ events.

The $K\bar{K}\pi$ Dalitz plot from the Crystal Ball is shown in Figure 15*b*. A low $K\bar{K}$ mass enhancement (in the upper right corner of the plot) is evident. This enhancement has been associated with the $\delta(980)\pi$ decay of the resonance. No evidence for K* bands, which would indicate a $K^*\bar{K}$ + c.c. decay, is observed, although the situation is potentially confusing because of the limited phase space available for the decay and the fact that the K* bands overlap in the region of the δ. The Mark II results are consistent with this. They find the ι to decay primarily into $\delta\pi$.

Before discussing the Crystal Ball spin analysis of the $\iota(1440)$, we review the status of the E(1420). The best estimate of the mass (27) is $M_E = 1418 \pm 10$ MeV/c^2. This is somewhat lower than, but not inconsistent with, the average of the Mark II and Crystal Ball measurements of the ι mass, $M_\iota = 1440 \pm 10$ MeV/c^2. The widths of the E($\Gamma_E = 50 \pm 10$ MeV) and the ι($\Gamma_\iota = 55 \pm 20$ MeV) are also consistent. Thus the mass and width measurements of the ι do not clearly identify it as a state different from the E.

As mentioned previously, the spin of the E was established in an experiment that analyzed the reaction $\pi^- p \rightarrow K_s K^\pm \pi^\mp n$ at 3.95 GeV/c (60). The results of a partial-wave analysis of the $K\bar{K}\pi$ system determined $J^{PC} = 1^{++}$ for the E, thus making it the SU(3) nonet partner of the D(1285) and the A_1. An additional result of the partial-wave analysis of Dionisi et al (60), is that the E decays primarily into $K^*\bar{K}$ + c.c. with $B(E \rightarrow K^*\bar{K} + c.c.)/B[E \rightarrow (K^*\bar{K} + c.c.) \text{ or } \delta\pi] = 0.86 \pm 0.12$.

The spin of the $\iota(1440)$ was determined from a partial-wave analysis of the Crystal Ball data (45). Contributions from five partial waves were included: 1. $K\bar{K}\pi$ flat (phase space); 2. $\delta^0\pi^0$–0^-; 3. $\delta^0\pi^0$–1^+; 4. $K^*\bar{K}$ + c.c.–0^-; 5. $K^*\bar{K}$ + c.c.–1^+. Note that $J^P = 0^+$ is not allowed for a state decaying into three pseudoscalars. $J^P = 1^-$, although allowed for $K^*\bar{K}$ + c.c., would require the Dalitz plot to vanish at the boundaries, which is inconsistent with the data of Figure 15*b*. Amplitudes with $J \geq 2$ were not considered. Contributions from all partial waves except the $K\bar{K}\pi$ phase space contribution were allowed to interfere with arbitrary phase. The $K\bar{K}\pi$ contribution due to phase space was assumed to be incoherent. The full angular decay distributions in each case were included in the amplitudes. The ι and K* helicities were allowed to vary in the fits. The δ and K* parameters were taken to be the standard values (27). In other words, a standard isobar analysis (61) was done here.

The analysis was done for events with $K\bar{K}\pi$ masses between 1300 and 1800 MeV/c^2. The data were divided into five bins of 100 MeV/c^2 each. The standard procedure of eliminating those partial waves that do not contribute significantly to the likelihood was utilized (i.e. the number of events contributed by a given partial wave was required to be larger than the error on that number). The only significant contributions were from $K\bar{K}\pi$ flat, $\delta^0\pi^0$–0^-, and $K^*\bar{K}$+c.c.–1^+. These contributions, corrected for detection efficiency, are shown as a function of $K\bar{K}\pi$ mass in Figure 16. The $K^*\bar{K}$+c.c.–1^+ contribution is relatively small and independent of mass. On the other hand, the $\delta\pi$–0^- contribution shows clear evidence for the resonant structure in the ι signal region ($1400 \leq M_{K\bar{K}\pi} < 1500$ MeV/c^2). This establishes the spin-parity of the ι as 0^-. (The C-parity is required to be even because of the production mechanism.) In addition, contrary to the case of the E(1420), the principal decay of the ι is into $\delta\pi$ and $B(\iota \rightarrow K^*\bar{K}+\text{c.c.})/B[\iota \rightarrow (K^*\bar{K}+\text{c.c.}) \text{ or } \delta\pi] < 0.25$ (90% C.L.).

Since a number of assumptions went into the partial-wave analysis, and, in particular, only a limited number of partial waves were considered, checks were made to show that the results of the analysis were valid. First, maximum likelihood fits were made to the restricted hypothesis that, in each mass interval, only one partial-wave contribution in addition to the flat contribution was allowed. The relative probabilities resulting from fits to the data in the signal region ($1400 \leq M_{K\bar{K}\pi} < 1500$ MeV/c^2) are given in

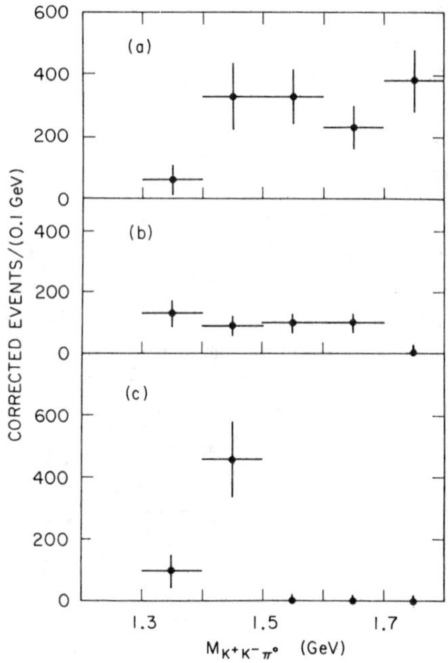

Figure 16 Partial-wave contributions to $J/\psi \rightarrow K^+K^-\pi^0$ as a function of $K\bar{K}\pi$ mass for (*a*) $K\bar{K}\pi$ flat, (*b*) $K^*\bar{K}$+c.c. with $J^P = 1^+$, and (*c*) $\delta\pi$ with $J^P = 0^-$.

Table 6 Relative partial-wave probabilities for various hypotheses for the structure of the $K\bar{K}\pi$ system in $J/\psi \to \gamma K^+K^-\pi^0$ ($1400 \leq M_{K\bar{K}\pi} < 1500$ MeV/c^2)

Partial-wave contribution	Relative probability
flat + $\delta\pi$–0^-	1.0
flat + $\delta\pi$–1^+	0.006
flat + $K^*\bar{K}$ + c.c.–0^-	10^{-7}
flat + $K^*\bar{K}$ + c.c.–1^+	0.01

Table 6. Note that compared to the $\delta\pi$–0^- hypothesis, the next best hypothesis ($K^*\bar{K}$ + c.c.–1^+) has a relative probability of only 1%. This establishes that there is not a strong correlation between the $\delta\pi$ and $K^*\bar{K}$ + c.c. amplitudes. The properties of the ι as measured by the Mark II and Crystal Ball collaboration are shown in Table 7. Also shown is the Crystal Ball upper limit (cf Section 6.5) for $B(J/\psi \to \gamma\iota) \cdot B(\iota \to \eta\pi\pi)$. This upper limit is in mild conflict with the hypothesis that the $K\bar{K}\pi$ decay of the ι is dominated by $\delta\pi$ as hypothesized above (see Table 7), although some theoretical interpretations can avoid this conflict (62). Note that $\delta\pi$ dominance of the ι decay is an important element in our spin-parity analysis of the ι.

The $\vartheta(1640)$ was first observed in the process, $J/\psi \to \gamma\eta\eta$, $\eta \to \gamma\gamma$ by the Crystal Ball collaboration (46). The analysis was based on the full data sample. Figure 17*a* shows the $\eta\eta$ invariant mass distribution for events consistent with $J/\psi \to \gamma\eta\eta$ after a 5-C fit has been performed. Only events with $\chi^2 < 20$ are shown. Because of the limited statistics, it is not possible to establish whether the ϑ peak is one or two peaks (the ϑ and f′). However, it is

Table 7 Parameters for the $\iota(1440)$[a]

Parameter	Crystal Ball	Mark II (44)
M (MeV/c^2)	1440^{+20}_{-15}	1440^{+10}_{-15}
Γ (MeV)	55^{+20}_{-30}	50^{+30}_{-20}
$B(J/\psi \to \gamma\iota) \cdot B(\iota \to K\bar{K}\pi)$[b]	$(4.0 \pm 0.7 \pm 1.0) \times 10^{-3}$	$(4.3 \pm 1.7) \times 10^{-3}$[c]
$B(J/\psi \to \gamma\iota) \cdot B(\iota \to \eta\pi\pi)$[d]	$<2 \times 10^{-3}$ (90% C.L.)	—
C	+	+
J^P	0^-	—

[a] Where two errors are given, the first is statistical and the second systematic.
[b] $I = 0$ is assumed in the isospin correction.
[c] This product branching ratio has been increased by 19% as compared to the value published in (44). This accounts for the differential efficiency correction from the spin-1 to spin-0 case, as discussed in the reference.
[d] Note that one experiment gives $B(\delta \to \eta\pi\pi)/B(\delta \to K\bar{K}) = 1.4 \pm 0.6$ (62a), while $\iota \to \delta\pi$ has been measured as the dominant decay for the $K\bar{K}\pi$ final state.

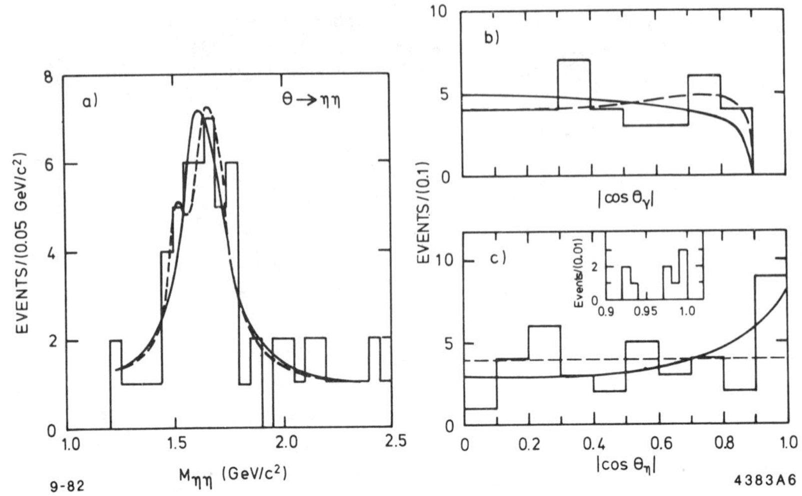

Figure 17 (*a*) The $\eta\eta$ mass distribution from the process $J/\psi \to \gamma\eta\eta$ for $M_{\eta\eta} < 2.5\ \mathrm{GeV}/c^2$. The solid curve is a fit to a flat background plus one Breit-Wigner resonance. The dashed curve is a fit to a flat background plus two Breit-Wigner resonances, one with the mass and width of the f′ but fitted amplitude and the other with all three parameters fitted. A flat background is also included. (*b*) $|\cos\,\vartheta_\gamma|$ and (*c*) $|\cos\,\vartheta_\eta|$ distributions for $J/\psi \to \gamma\vartheta$, $\vartheta \to \eta\eta$. Solid curves are best-fit distributions for a ϑ spin of 2 and the dashed curves are expected distributions for spin 0. The inset shows the $|\cos\,\vartheta_\eta|$ distribution on an expanded scale.

probably most reasonable to assume that the f′ is present and fit for its amplitude. This was not done in (46); however, it was done in (63) and we also use the results from the fit including two resonances. The spin of the ϑ was determined from a maximum likelihood fit to the angular distribution $W(\vartheta_\gamma, \vartheta_\eta, \varphi_\eta)$ for the process $J/\psi \to \gamma\vartheta$, $\vartheta \to \eta\eta$. The parameter ϑ_γ is the polar angle of the γ with respect to the beam axis, and $(\vartheta_\eta, \varphi_\eta)$ are the polar and azimuthal angles of one of the η's with respect to the γ direction in the ϑ rest frame ($\varphi_\eta = 0$ is defined by the electron beam direction). The probability for the spin-0 hypothesis relative to the spin-2 hypothesis is 0.045. Spins greater than 2 were not considered. Note that the $\eta\eta$ decay mode establishes the parity of the state as even. Figures 17*b* and 17*c* show the $|\cos\,\vartheta_\gamma|$ and $|\cos\,\vartheta_\eta|$ distributions, respectively. Although the spin determination depends on information that cannot be displayed in these projections, it is clear that the $|\cos\,\vartheta_\eta|$ distribution plays the major role in the preference for spin 2. This is primarily due to the excess of events with $|\cos\,\vartheta_\eta| > 0.9$. There is no evidence that these events are anomalous.

The Crystal Ball and the Mark II (64) have searched for $J/\psi \to \gamma\vartheta$, $\vartheta \to \pi\pi$. Figure 13 shows the Crystal Ball results for the π^0's from 2.2×10^6 J/ψ decays. The binning in $M_{\pi\pi}$ is 50 (MeV/c^2) per bin. As summarized

Table 8 Parameters for the $\vartheta(1640)$[a]

Parameter	Crystal Ball	Mark II (64)
M (MeV/c^2)	1670 ± 50	1700 ± 30
Γ (MeV)	160 ± 80	156 ± 20
$B(J/\psi \to \gamma\vartheta) \cdot B(\vartheta \to \eta\eta)$	$(3.8 \pm 1.6) \times 10^{-4}$	—
$B(J/\psi \to \gamma\vartheta) \cdot B(\vartheta \to K\bar{K})$[b]	—	$(12.0 \pm 1.8 \pm 5.0) \times 10^{-4}$
$B(J/\psi \to \gamma\vartheta) \cdot B(\vartheta \to \pi\pi)$[b]	$< 6 \times 10^{-4}$ (90% C.L.)	$< 3.2 \times 10^{-4}$ (90% C.L.)
C	+	+
J^P	2^+ (95% C.L.)	2^+ (78% C.L.)

[a] Where two errors are given, the first is statistical and the second systematic.
[b] $I = 0$ is assumed in the isospin correction.

in Table 8, only upper limits were obtained from both the Crystal Ball and Mark II experiments.

The Mark II collaboration has obtained confirming evidence for the ϑ in the process $J/\psi \to \gamma\vartheta$, $\vartheta \to K^+K^-$ (64). They find the spin-parity assignment 2^+ to be favored at the 78% C.L. A summary of the Mark II results on the ϑ is also given in Table 7.

6.5 *Other Radiative Transitions*

The Mark II collaboration (65) reports a signal in the process $J/\psi \to \gamma\rho^0\rho^0$, $\rho^0 \to \pi^+\pi^-$. They interpret their $\rho^0\rho^0$ spectrum in this process as a combination of $\gamma\rho^0\rho^0$ phase space and a Breit-Wigner resonance. A maximum likelihood fit to this hypothesis yields, $M_{\text{res}} = 1650 \pm 50$ MeV/c^2, $\Gamma_{\text{res}} = 200 \pm 100$ MeV. These values are comparable to the mass and width of the ϑ shown in Table 7. Also, they obtain, $B(J/\psi \to \gamma\rho^0\rho^0, M_{\rho^0\rho^0} < 2 \text{ GeV}) = (1.25 \pm 0.35 \pm 0.4) \times 10^{-3}$. Assuming that the $\rho\rho$ is in the decay in an $I = 0$ state, we have $B(J/\psi \to \gamma\rho\rho, M_{\rho\rho} < 2 \text{ GeV}) = (3.75 \pm 1.05 \pm 1.3) \times 10^{-3}$. This branching ratio is approximately equal to the $\iota(1440)$ and η' branching ratios. As a strong note of caution, the Mark II collaboration states that much more data is needed to establish the connection, if any, between the $\rho\rho$ structure and the ϑ meson.

The Crystal Ball collaboration (66) has also found additional structure in the region of the ϑ by examining the process, $J/\psi \to \gamma\eta\pi^+\pi^-$, $\eta \to \gamma\gamma$. Figure 18 shows the $M_{\eta\pi^+\pi^-}$ and $M_{\eta\pi^0\pi^0}$ distributions obtained from the analysis of 2.2×10^6 J/ψ decays. A large signal at $M_{\eta\pi\pi} = M_{\eta'}$ is evident, and in addition, there is a broad enhancement centered at about 1710 MeV/c^2.

Examination of the Dalitz plots for the $\eta\pi^+\pi^-$ events (66) with $1600 < M_{\eta\pi\pi} < 1850$ MeV/c^2 shows no structure. Thus the broad enhancement is not strongly associated with a δ, or any other, resonance in either $\eta\pi^\pm$ or $\pi^+\pi^-$. Three possible interpretations are suggested for this new enhancement.

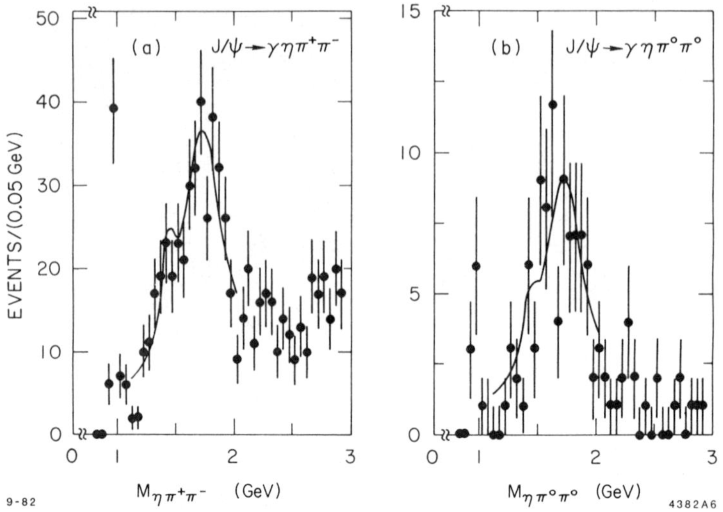

Figure 18 $\eta\pi\pi$ mass spectrum from (*a*) $J/\psi \rightarrow \gamma\eta\pi^+\pi^-$ and (*b*) $J/\psi \rightarrow \gamma\eta\pi^0\pi^0$. The curves are fits including contributions from the $\iota(1440)$ as described in the text.

First, the $\eta\pi\pi$ mass distribution for events with a prompt γ may be quite different from Lorentz invariant phase space. Then the enhancement could arise from the (nonresonant) decay of the J/ψ to a photon plus two gluons. Secondly, the enhancement could be a group of resonances. A third possibility is that it is a single resonance. The data may be fit with a single Breit-Wigner line shape. For the fit, the $\eta\pi^+\pi^-$ and $\eta\pi^0\pi^0$ mass spectra are fit simultaneously with the mass and width parameters constrained to be the same for both channels. A constant background was assumed for the $\eta\pi^0\pi^0$ channel. For $\eta\pi^+\pi^-$, the background was determined by fitting the $\gamma\gamma\pi^+\pi^-$ mass spectrum for events with a $\gamma\gamma$ mass combination in the η sidebands ($320 < M_{\gamma\gamma} \leq 470$ MeV/c^2 or $610 < M_{\gamma\gamma} < 760$ MeV/c^2). The fit has a χ^2 of 66 for 69 degrees of freedom and yields, $M = 1710 \pm 45$ MeV/c^2, $\Gamma = 530 \pm 110$ MeV, where the errors include estimates of the systematic uncertainty.

Using the number of events in the peak, as determined by the fit, and an efficiency obtained from Monte Carlo calculations of 18% (6.6%) for $J/\psi \rightarrow \gamma\eta\pi^+\pi^-(\gamma\eta\pi^0\pi^0)$, one obtains the branching ratios, $B(J/\psi \rightarrow \gamma\eta\pi^+\pi^-) = (3.5 \pm 0.3 \pm 0.7) \times 10^{-3}$, $B(J/\psi \rightarrow \gamma\eta\pi^0\pi^0) = (2.3 \pm 0.3 \pm 0.7) \times 10^{-3}$, where the first error is statistical and the second is systematic. These branching ratios, when added, are comparable to or larger than those for the ι and η'.

The fit shown in Figure 18 also includes a term for the ι, from which the upper limit in Table 7 was obtained. The implications of this low value for $\iota \rightarrow \eta\pi\pi$ are discussed in Section 6.4.

It is of interest to note that if the presently known contributions to radiative decays of the J/ψ in the ϑ region are added together, one obtains, $B[J/\psi \to \gamma\vartheta(\text{region})] \geq B(J/\psi \to \gamma\vartheta + \gamma\rho\rho + \gamma\eta\pi\pi) = (1.1 \pm 0.2) \times 10^{-2}$. This is the second largest branching ratio seen in the J/ψ radiative decays, being just less than that of the $\eta_c(2984)$.

The interpretation of the character of the ι and ϑ and the other new states seen for the first time in radiative transitions from the J/ψ is complex (3). In particular, the discussion of whether some of these states are gluonic mesons is beyond the scope of this review.

7. SEARCHES FOR THE F MESON, THE AXION, AND THE $^1P_1$ STATE

7.1 *The Inclusive η Cross Section*

In 1977, the DASP collaboration reported (67) a strong increase in the inclusive η production in e^+e^- collisions at $E_{CM} \cong 4.4$ GeV (and possibly at 4.17 GeV) relative to the production at 4.03 GeV. They interpreted this as evidence for production of the charmed strange F meson, which is expected to have a strong branching fraction into η's (68). Correlations with electrons and low energy γ's (expected from $F^* \to \gamma F$) strengthened this interpretation. Furthermore, they observed a cluster of events at $E_{CM} = 4.42$ GeV fitting the hypothesis $e^+e^- \to FF^* \to \gamma F\eta\pi$. Based on all of this, they reported that $R(e^+e^- \to F\bar{F}X) \cdot B(F \to \eta x) = 0.46 \pm 0.10$ in the E_{CM} range from 4.36 to 4.49 GeV, where $R(e^+e^- \to F\bar{F}X) = \sigma(e^+e^- \to F\bar{F}X)/\sigma(e^+e^- \to \mu^+\mu^-)$. The final state is written $F\bar{F}X$ to take into account the possibility that F-meson pairs may occur via production of excited F mesons, e.g. $e^+e^- \to F^*\bar{F} \to \gamma F\bar{F}$.

In order to study this interesting phenomenon, the Crystal Ball data were analyzed for inclusive η production. The data sample consisted of hadronic events from six fixed center-of-mass (c.m.) energies and seven c.m. energy bands. The fixed points were the J/ψ, ψ', ψ'', a point at 3.670 GeV in the continuum just below the J/ψ to act as a control from below charm threshold, and two energies above charm threshold (4.028 and 5.200 GeV). The seven energy bands covered a range in E_{CM} from 3.878 to 4.500 GeV. These data were taken in fine scans with steps in E_{CM} of between 2 and 12 MeV. For purposes of measuring $R_\eta = \sigma(e^+e^- \to \eta x)/\sigma(e^+e^- \to \mu^+\mu^-)$, the seven energy bins were chosen to correlate with observed structure in $R(e^+e^- \to \text{hadrons})$ (69).

The method for obtaining the number of produced η mesons at each energy was to study the inclusive $\gamma\gamma$ distribution in the vicinity of the η mass. In all cases, a clear enhancement at the η mass was visible to the naked eye. The number of observed η's was obtained by standard statistical fitting of

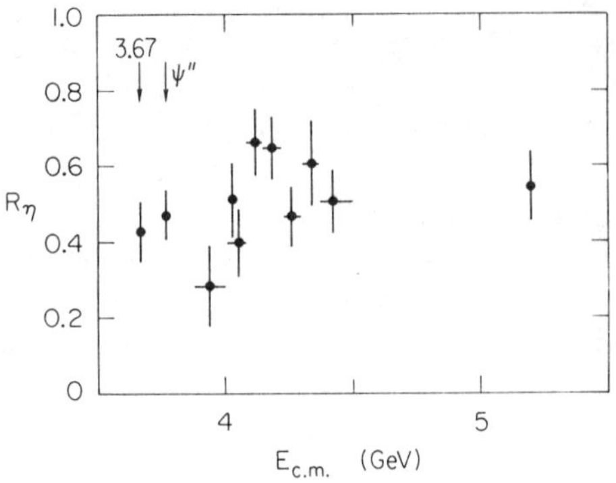

Figure 19 R_η as a function of E_{CM}. The first two points are for $E_{CM} = 3.67$ GeV and the ψ''. The ψ' point is offscale. The error bars include the point-to-point systematic uncertainty, but not the estimated 20% overall systematic uncertainty.

the observed distribution to a smooth background function plus a resolution function of adjustable size centered on the η mass. The observed number was then corrected for the branching ratio for $\eta \to \gamma\gamma$ and the η detection efficiency, which ranged from 38% at the J/ψ to 27% at 5.2 GeV. Uncertainties in the detection efficiencies due to our uncertain knowledge of the details of e^+e^- annihilation physics are included in the limits finally obtained.

Figure 19 shows our results for R_η. The offscale values at the J/ψ and ψ' are excluded[2] and the other points have been corrected for the radiative tails of these two resonances. Although there may be some correlation with the total hadronic cross section, we see that there is no dramatic difference in R_η below and above charm threshold. If we assume that the contribution to R_η due to non-charm physics is constant and that all excess in R_η is due to F decays, we can set limits on $R(e^+e^- \to F\bar{F}X) \cdot B(F \to \eta x)$ by comparing the values for R_η above charm threshold with that below it at 3.67 GeV. The 90% C.L. limits are all below 0.32 and, for the energy band from 4.365 to 4.500 GeV, it is 0.19. This disagrees with the DASP result (67). Most of the disagreement is due to the fact that the earlier experiment saw essentially no η signal at 4.03 GeV whereas the Crystal Ball observed almost the same strength at 4.03 GeV as at other energies, even below charm threshold. At energies above about 4.1 GeV, the cross sections reported by the two experiments are on the average compatible. Up to the time of this writing, the Crystal Ball has found no firm evidence for the elusive charmed strange F meson.

[2] When the results are expressed in terms of f_η, the average number of η's per hadronic event, the two resonances are not special; f_η has a value of about 0.13 and shows little variation over this energy range.

7.2 *The Search for $J/\psi \to \gamma$ Axion*

Because of its exceptionally large solid angle coverage by charged-particle and photon detectors with essentially 100% detection efficiency, and because of its moderately good time resolution (about 3 ns), the Crystal Ball is well adapted to search for certain exotic phenomena, especially those of the class $e^+e^- \to \gamma X$ where X escapes detection for some fundamental reason. An example of such a reaction involving known particles is that in which X is $\nu\bar{\nu}$ resulting from either direct production, or the decay of a light neutral spin-1 gauge boson as suggested by some supersymmetric theories (70), or, at higher energies, the decay of the Z^0. Another possibility, which has been searched for in the Crystal Ball and is reported in (71), is the radiative decay of the J/ψ into an axion. The axion (a) is the Goldstone boson appearing from the breaking of a chiral U(1) symmetry, which has been postulated to avoid large *P*- and *CP*-invariance violations in QCD (72). If the number of quark generations is assumed known, then this theory has only one free parameter, the ratio, x, of the vacuum expectation values of the two Higgs fields present in the theory. However, it does endow the axion with a sufficiently long life and weak enough interactions that it would escape detection in the Ball. The theory reliably predicts that $B(J/\psi \to \gamma a) = (5.7 \pm 1.4) \times 10^{-5}\ x^2$. Positive evidence for an axion or axion-like particle was reported by Faissner et al (73) with a mass $m_\alpha = 250 \pm 25$ keV/c^2 and $x = 3.0 \pm 0.3$ [but these values seem inconsistent with other experiments (74)]. They imply that there should be about 800 events in the Ball with $|\cos \vartheta| < 0.8$ and these events would have the distinctive signature of a single photon with beam energy. No significant numbers of such events are seen.

The dominant background in this search comes from cosmic rays and most of these can be eliminated by restricting attention to the bottom hemisphere of the Ball, which simply reduces the overall detection efficiency to 30%. After making a cosmic ray background subtraction using events out of time with the beam, we obtain a 90% C.L. upper limit of 6.2 events in an energy range from 1.3 GeV to 2 energy-resolution standard deviations above the beam energy. This implies an upper limit of 1.4×10^{-5} (90% C.L.) on the branching fraction. The corresponding upper limit on x is 0.6, in disagreement with the result in (73).

A definitive test of the standard axion model, which eliminates any dependence on x, has been proposed (75) by setting limits on both $J/\psi \to \gamma a$ and $\Upsilon \to \gamma a$. Recent results from the LENA collaboration at DORIS (76) and the CUSB collaboration at CESR (76a) have established that $B(\Upsilon \to \gamma a)$ is less than 9.1×10^{-4} (90% C.L.) and 3.5×10^{-4} (90% C.L.), respectively. These results together with the above Crystal Ball limit on $\psi \to \gamma a$ violate the results of the standard axion model and it now seems necessary to retreat to an even more elusive axion, such as has been proposed in grand unified theories (77).

7.3 *The Search for Decays* $\psi' \to \pi^0\, {}^1P_1$

The only predicted $c\bar{c}$ bound state for which no evidence exists is the ${}^1P_1$ with $J^{PC} = 1^{+-}$. Its mass is expected to be approximately equal to the center of gravity of the ${}^3P_J$ states, or about 3520 MeV (78). Experimental determination of its mass is important since any significant deviation from the above value would suggest a long-range spin-spin term in the quarkonium potential. We have searched extensively in our large ψ' data sample for evidence of this state and have not found it (79).

Single photon transitions between the ψ' and ${}^1P_1$ states are forbidden by C conservation and so one must investigate double photon transitions. Four possibilities are indicated in Figure 20. Estimates based on related measured rates indicate that only the $\psi' \to \pi^0\, {}^1P_1$ process can be reasonably expected to have a branching ratio in the percent range. This process would lead to a monochromatic π^0 in ψ' decay having an energy that is expected to be below about 200 MeV; 165 MeV is favored. Figure 21*a* shows the inclusive π^0 distribution observed in ψ' decays. The evident structures at about 200 MeV and just above 400 MeV are expected backgrounds. They are due to fake π^0's generated with the monochromatic photons from $\psi' \to {}^3P_J$ transitions, and the reactions $\psi' \to \pi^0\pi^0\, J/\psi$ and $\psi' \to \pi^0\, J/\psi$. No other structure is evident and 95% C.L. limits of less than 1.09% have been set for $B(\psi' \to \pi^0\, {}^1P_1)$ for any ${}^1P_1$ mass between 3440 and 3535 MeV/c^2. In particular, at the favored mass of 3520 MeV/c^2, the limit is 0.42%.

In an effort to reduce backgrounds and so increase sensitivity at the expense of a less general result, we have also searched for evidence of the cascade decay $\psi' \to \pi^0\, {}^1P_1 \to \gamma\gamma\gamma\, \eta_c$, where the η_c is constrained in mass but not decay mode. Study of this particular configuration is motivated

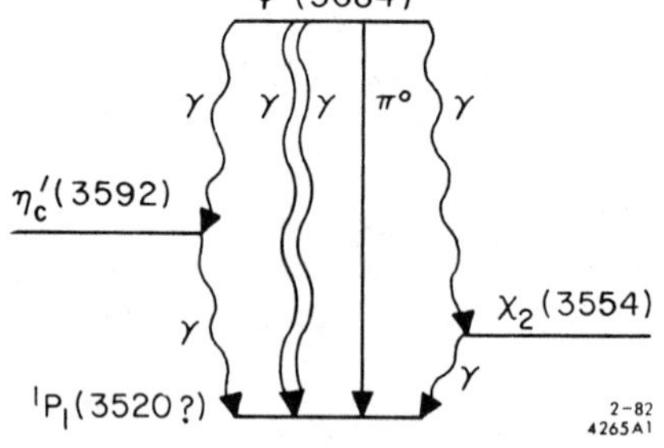

Figure 20 Possible mechanisms contributing to the decay $\psi' \to \gamma\gamma\, {}^1P_1$.

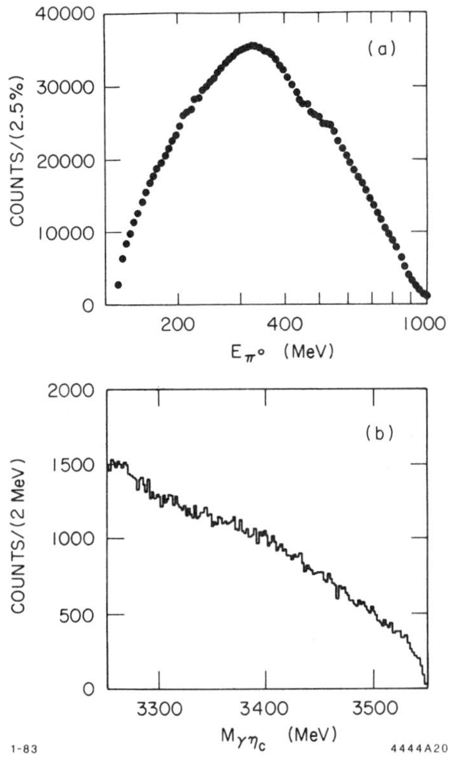

Figure 21 (*a*) The inclusive π^0 energy spectrum from ψ' decays. (*b*) Distribution of the $\gamma\eta_c$ mass for events satisfying the hypothesis $\psi' \rightarrow \pi^0\gamma\eta_c$.

by a reasonable expectation that $B(^1P_1 \rightarrow \gamma\eta_c)$ is in the vicinity of 50%. Figure 21*b* shows the γ(prompt)η_c mass distribution for events fitting the hypothesis $\psi' \rightarrow \pi^0\gamma\eta_c$ and again there is no evidence for the $^1P_1$ state. We have set 95% confidence limits of less than 0.35% for $B(\psi' \rightarrow \gamma\, ^1P_1) \cdot B(^1P_1 \rightarrow \gamma\eta_c)$ for $^1P_1$ masses in the same range as above. At the preferred mass, the limit is 0.14%. Although the $^1P_1$ state has not been found, the limits set are sufficiently low to be theoretically interesting.

8. MEASUREMENTS OF R_h IN THE E_{CM} RANGE OF 5.0 TO 7.4 GeV

One of the most fundamental and difficult measurements that can be done at an e^+e^- storage ring is that of the total hadronic cross section, $\sigma_h = \sigma(e^+e^- \rightarrow \text{hadrons})$. In order to remove the straightforward effects of QED and so reveal the strong interaction effects more clearly, it is customary to normalize the hadronic cross section to the theoretical, lowest order, purely QED cross section $\sigma_{\mu\mu}$ for $e^+e^- \rightarrow \mu^+\mu^-$, which, in the energy range of interest here, is $(4\pi/3)\alpha^2(\hbar c)^2/E_{CM}^2 = 86.8/E_{CM}^2$ (nb, GeV). In the energy range well above charm threshold and below bottom threshold, the prediction of QCD for the normalized cross section, $R_h = \sigma_h/\sigma_{\mu\mu}$, is, to first order, $R_{QCD} = 3\sum_i Q_i^2[1+\alpha_s(s)/\pi+\cdots]$, where $s = E_{CM}^2$, $i = (u, d, s, c)$ is the quark flavor index, Q_i is the charge of the ith quark, and $\alpha_s(s)$ is the QCD running coupling constant (80). The sum over Q_i^2 yields a value of 10/3 and, at $E_{CM} = 6$ GeV, the first order[3] QCD result increases this by about 6% for $\Lambda_{\overline{MS}} = 100$ MeV.

In 1980, Barnett et al (80) made a careful comparison of all available measurements of R_h (81) to the predictions of QCD. They concluded that above 5.5 GeV a potentially serious discrepancy, in the range of 15 to 17%, existed. This was just outside the quoted systematic uncertainties of 10% and there was the exciting possibility that it represented new phenomena or difficulties with QCD. This prompted the Crystal Ball collaboration to undertake a series of new R_h measurements from 5.0 GeV to the top of the SPEAR range, 7.4 GeV (81a). To do this, a total exposure of 12.7 pb^{-1} was distributed over eleven energies in this range.

The two ingredients in the experimental determination of R_h are the integrated luminosity of the exposure and the number of hadronic events produced (and corrected for QED radiative effects). The luminosity was obtained both by the small-angle Bhabha-scattering monitor and by observation of large-angle QED events in the Ball itself. These independent determinations of the integrated luminosity agreed to about 2% and their average was used.

The determination of the radiatively corrected number of hadronic events produced by annihilations involved three steps. First, criteria were developed and applied to efficiently distinguish individual annihilation hadronic events from five classes of backgrounds: cosmic rays, beam-gas collision, QED events, two-photon collisions, and $\tau\bar{\tau}$ events. Next, properly normalized statistical subtractions were made to eliminate the residual backgrounds that slipped through the first step. And finally, the resulting count was corrected for the detection efficiency of both the triggering hardware and the event selection software of the first step. The purely QED radiative effects were also included in this step. Pure samples from each of the five event classes were obtained either experimentally (cosmic rays and beam-gas events) or by Monte Carlo simulation. These were examined and efficient criteria for event selection developed. These criteria were applied to

[3] The next higher order in the QCD calculation using the modified minimal subtraction renormalization scheme is known (80). It contributes only 0.7% to the prediction, well below present experimental precision.

both the data and the five pure samples. The residuals from the backgrounds were then subtracted from the data, and the result was corrected by the detection efficiency determined from the Monte Carlo simulation of the annihilation events. The Monte Carlo calculation incorporated all radiative correction effects in the event generation algorithms and so this last step automatically includes these corrections.

The most important backgrounds were those due to cosmic ray events, beam-gas collisions, and τ decays. Criteria to identify the first two of these were based on the spatial distribution of energy in the Ball and were developed by studying events out of time with the beam and those obtained in runs with the beams separated. The cosmic ray background was reduced to negligible levels by these criteria and the residual beam-gas background, which had to be removed by a statistical subtraction, was typically at the 10 to 12% level. No attempt was made to identify $\tau\bar{\tau}$ events or those arising from $\gamma\gamma$ collisions and so these backgrounds were removed by subtraction only. The pure samples of these two background classes were obtained by Monte Carlo simulation. Together, these two background sources gave a subtraction of about 12%. Finally, the QED contamination was easily reduced to negligible levels by criteria involving leading-particle energy and number of observed particles.

Table 9 Measured values of $R_h = \sigma(e^+e^- \to \text{hadrons})/\sigma(e^+e^- \to \mu^+\mu^-)$[a]

E_{CM} (GeV)	R_h	$(\delta R_h/R_h)_{stat}$	$(\delta R_h/R_h)_{point\ sys}$ (%)
		1981 data	
5.00	3.46	3.3	3.4
5.25	3.60	2.8	1.2
5.50	3.33	2.9	2.4
5.75	3.40	3.1	1.4
6.00	3.25	2.8	2.3
6.25	3.31	2.8	1.4
6.50	3.33	2.8	2.2
6.75	3.38	2.3	1.5
7.00	3.34	2.9	1.5
7.25	3.56	3.0	2.2
7.40	3.32	4.0	2.9
		1980 data	
5.20	3.51	3.5	3.5
6.00	3.43	3.5	3.5
6.75	3.38	3.8	3.5
7.40	3.67	3.5	3.5

[a] The error on R_h consists of three parts, a statistical error $(\delta R_h/R_h)_{stat}$; a systematic error that depends upon E_{CM}, $(\delta R_h/R_h)_{point\ sys}$, and a systematic error that is uniformly applicable to all the 1981 data of 5.3% and to all of the 1980 data of 7.0%.

The important questions of detection efficiency and radiative corrections were answered by subjecting simulated annihilation events generated by the LUND81 (82) Monte Carlo program (including radiative corrections) to the same criteria as the data. The result of this was the product of detection efficiency and the radiative correction factor. This product had a typical value of 1.06.

Table 9 gives the results of the experiment. The point-to-point systematic errors are given in the table and include the effects of uncertainty in the normalization of the several background subtractions and the statistical errors in the several Monte Carlo simulations. In addition, there is an overall systematic scale uncertainty of 5.3% (for the 1981 data) arising from various effects: 3.0% from radiative corrections, 3.0% from the detection efficiency, 2.5% from the luminosity, 1.4% from $\tau\bar{\tau}$ events, 1.3% from beam-gas interactions, and 0.6% from two-photon collisions. Figure 22 compares these results with those of other experiments (82a) and with QCD predictions. Clearly, the trend of the data is completely consistent with that of QCD. The absolute value of the measurements is about 6% lower than

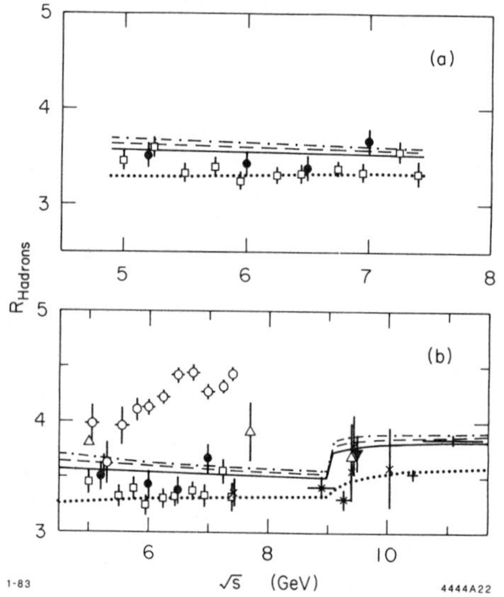

Figure 22 (*a*) Crystal Ball measurements of R_h compared to theoretical prediction. The solid points are from data taken in 1980 and the open squares are from a much larger data sample taken in 1981. The dotted curve is the simple quark-parton model prediction and the others are QCD predictions with $\Lambda_{\overline{MS}} = 100$ MeV (*solid*), 200 MeV (*dashed*), and 300 MeV (*dash-dot*). The error bars do not include the 5.3% (1981 data) or 7.0% (1980 data) overall systematic error. (*b*) Comparison of Crystal Ball results (*solid circles* and *open squares*) with other measurements (82a) (Mark I, *open circles*; PLUTO, *open triangles*; LENA, *crossed dot*; DASP II, *solid triangle*; CUSB, *plus sign*; DESY-Heidelberg, *cross*). The curves are as in (*a*).

the QCD predictions for reasonable values of the QCD parameter $\Lambda_{\overline{MS}}$. This disagreement, of course, is easily accommodated by the systematic scale uncertainty of the data. We conclude that these results rule out the possibility of any new phenomena in this energy range, at least, at the level suggested by the Mark I data.

9. TWO-PHOTON PHYSICS

In addition to studying the physics of e^+e^- annihilations, the Crystal Ball experiment has been able to investigate certain two-photon reactions. More precisely, we can investigate the reaction $e^+e^- \rightarrow e^+e^- +$ hadrons in the configuration that each of the two leptons scatters through a very small angle. Because of the small scattering angle, the outgoing electrons are not detected. To lowest order in QED, then, the hadrons result from the collision of two photons, which though virtual, are very nearly on their mass shell. The Crystal Ball is particularly adapted to study the case that the hadrons decay into only photons and, to date, the work on the four-photon final state has been completed (83). The data for this analysis came from an integrated luminosity of 21 pb^{-1} distributed over an E_{CM} range from 3.9 to 7.0 GeV.

9.1 *Measurement of* $\Gamma(f \rightarrow \gamma\gamma)$ *from* $\sigma(\gamma\gamma \rightarrow f \rightarrow \pi^0\pi^0)$

All neutral events with exactly four energy clusters inside $|\cos \vartheta| < 0.9$, each with more than 20 MeV and with energy deposition patterns consistent with photons, were selected. Furthermore, it was required that the endcaps contain less than 40 MeV, and that the total invariant mass of the event be in the range from 720 MeV to E_{CM}. The resulting sample of events shows a strong peak at zero in the square of the total transverse momentum as expected for 2γ events. Only those in this peak [the cut was at 0.03 $(\mathrm{GeV}/c)^2$] were subsequently used.

In essentially all of the final sample, photon pairings could be made that were consistent with either $\pi^0\pi^0$ or $\pi^0\eta$ being the primary hadrons. Figure 23 (*top*) gives the invariant mass distribution of the $\pi^0\pi^0$ sample, which clearly shows a strong signal near the f mass and no other significant structure. Of special note here is the smallness of the background. This is in contrast to earlier experiments that detected the charged-pion decay mode of the f and tended to be troubled with large nonresonant $\pi^+\pi^-$ and QED $\mu^+\mu^-$ backgrounds (84–87).

To obtain the cross section for $\gamma\gamma \rightarrow \pi^0\pi^0$, the $\pi^0\pi^0$ mass spectrum was corrected for the variation in the $\gamma\gamma$ flux over it (88) and detection efficiency. Figure 23 (*bottom*) gives the resulting cross section with the added restriction that $|\cos \vartheta^*| < 0.7$ where ϑ^* is the angle between the beam and the outoing π^0 direction in the $\pi^0\pi^0$ rest frame. The solid curve shows a fit with three contributions: a relativistic Breit-Wigner function (including slight spreading due to the experimental mass resolution) with mass and width parameters taken from the Particle Data Group compilation (27) for the f; the same for a possible S*(980); and a straight line to describe $\pi^0\pi^0$ nonresonant background. As is clear, the curve does not give a good fit since the data's mass peak is lower than that of the curve by about 40 MeV. The broken curve is the fit obtained when the f mass and width are allowed to be free, the best fit values being 1238 ± 14 MeV/c^2 and 248 ± 38 MeV/c^2, respectively. This mass shift could, in fact, be accommodated within the estimated systematic error of about 2% and the statistical error, but other $\gamma\gamma$ experiments have observed a very similar effect (85, 86), which suggests that

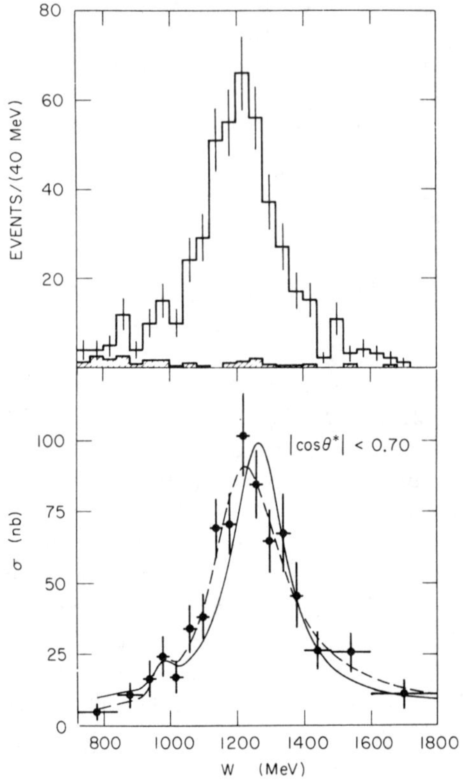

Figure 23 The top part of the figure shows the $\pi^0\pi^0$ mass distribution for 4γ events consistent with $e^+e^- \rightarrow e^+e^-\gamma\gamma \rightarrow e^+e^-\pi^0\pi^0$. The shaded histogram shows the non-$\pi^0\pi^0$ background. The bottom part of the figure gives $\sigma(\gamma\gamma \rightarrow \pi^0\pi^0)$ for $|\cos \vartheta^*| < 0.7$. The curves are described in the text.

the effect is due to some underlying physical mechanism rather than an instrumental artifact. Possible sources of this effect are interference with nonresonant background (89) or the $\varepsilon(1300)$. Finally, there is the interesting possibility that the f may be mixed with a predicted gluonic meson almost degenerate with it (90). In this last case, different production and/or decay channels would yield different resonance shapes, and so f production in the $\gamma\gamma$ channel could possibly give phenomenological resonance parameters different from those found in other hadronic interactions.

Previous determinations of $\Gamma(f \to \gamma\gamma)$ from two-photon collisions have assumed the theoretical prediction that the f is produced predominantly with helicity 2 (91, 92). Because of the negligible backgrounds in this experiment, it is possible to verify this theoretical expectation by observing the ϑ^* angular distribution. This is shown in Figure 24. It is clear that the spin-2 assumption with helicity-2 domination gives a good fit; the other helicity contributions are consistent with zero. If we assume that the mass peak is due to the f and that the decay is purely helicity 2, then we obtain $\Gamma_{f\to\gamma\gamma} = 2.7 \pm 0.2 \pm 0.6$ keV, the first error being statistical and the second systematic. This agrees well with the results from other experiments (84–87).

9.2 *Measurement of* $\Gamma(A_2 \to \gamma\gamma)$ *from* $\sigma(\gamma\gamma \to A_2 \to \pi^0\eta)$

After the $\pi^0\pi^0$ events were removed from the 4γ sample, the resulting events were essentially all $\pi^0\eta$ and their invariant mass peaks at around 1300 MeV/c^2. Identifying this peak with the $A_2(1320)$ and assuming pure helicity 2, we can extract the $\gamma\gamma$ partial width of the A_2. We obtain $\Gamma(A_2 \to \gamma\gamma) = 0.77 \pm 0.18 \pm 0.27$ keV. The naive quark model with ideal mixing predicts a ratio of 9/25 for $\Gamma(A_2 \to \gamma\gamma)/\Gamma(f \to \gamma\gamma)$, which is in agreement with the Crystal Ball observations of $0.29 \pm 0.07 \pm 0.07$ for this ratio.

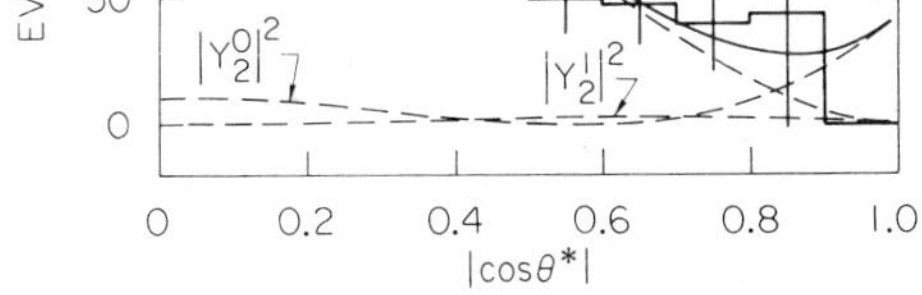

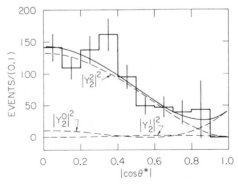

Figure 24 Acceptance-corrected distribution for $|\cos\vartheta^*|$ for $\pi^0\pi^0$ events in the f mass region (1040–1480 MeV/c^2). The solid curve is the best-fit spin-2 distribution and the dashed curves show the contributions from each of the three helicity amplitudes.

9.3 *Other States*

In addition to the two measurements discussed above, we can also set several limits based on the absence of signals. First, as is seen in Figure 23*b*, there is no evidence for $\gamma\gamma \rightarrow S^*(980) \rightarrow \pi^0\pi^0$. From this, we can set the limit $\Gamma(S^* \rightarrow \gamma\gamma) \cdot B(S^* \rightarrow \pi\pi) < 0.8$ keV. This limit is considerably smaller than the value of about 20 keV expected by most theoretical estimates (92), but consistent with a single-quark-exchange calculation (93), which predicts less than 0.4 keV for mesons in the 0^{++} nonet. Finally, no signal for $\gamma\gamma \rightarrow \eta\eta$ was observed. This allows two 95% C.L. limits to be set. First a limit of 0.05 can be put on $\Gamma(f \rightarrow \eta\eta)/\Gamma(f \rightarrow \pi\pi)$. This is consistent with the limit of 0.016 obtained in earlier work (94). And secondly, this absence implies that $\Gamma[\vartheta(1640) \rightarrow \gamma\gamma] \cdot B[\vartheta(1640) \rightarrow \eta\eta]$ is less than 5 keV.

10. MEASUREMENTS IN THE REGION FROM CHARM THRESHOLD TO 4.5 GeV

It has been known for a long time that the energy region above charm threshold, from the ψ'' at 3.77 GeV to about 4.5 GeV is rich in charmed physics phenomena. The ψ'' itself is known to be a "D factory," 4.03 GeV is a "D* factory," F and F* mesons are expected, and, more generally, there should be a rich spectrum of excited D and F mesons produced in e^+e^- collisions in this energy range (95). The strong structure in R_h in this region (81) is ample evidence for this but so far no details have been fully resolved above 4.03 GeV. In order to investigate this potentially interesting area of physics, the Crystal Ball accumulated an exposure at the $\psi''(3770)$ yielding about 1.3×10^4 produced ψ'', and an exposure of 11.3 pb^{-1} distributed over the range 3.8 to 4.5 GeV. Although a great deal of effort has gone into analyzing this data, only the results on R_η discussed above are considered complete. Preliminary reports on some of the other work have been given, however, and we give a short review of them here.

The global structure of the physics in this region is shown by the energy dependence of the normalized hadronic cross section itself, charged and neutral multiplicities, and charged and neutral energy fraction. Preliminary results on some of these from the Crystal Ball data are given in (96). The R_h measurements confirm the structure seen in other experiments (81): clear peaks at 3.77, 4.03, and 4.4 GeV, a broad peak with possibly some substructure around 4.16 GeV followed by a broad "valley" in the 4.2 to 4.3 GeV region. The statistical precision of this set of measurements is very high but much work remains to be done to reduce the point-to-point systematic uncertainties to fully exploit it. Analysis of R_h over the ψ'' excitation curve gives resonance parameters (97) in reasonable agreement with earlier work (98).

The observed neutral energy fraction is quite smooth through the whole region and so seems to be insensitive to the underlying physics. However, the observed neutral multiplicity does show large changes. Together, these two facts suggest that low energy γ's, π^0's, and η's could be useful indicators and these have been stressed in the data analysis. The resonance at $E_{CM} = 4.03$ GeV serves as a source of almost monochromatic low energy π^0's and photons. This results from the combination of a large D* cross section at this energy and low Q values for both the production channels, $D^*\bar{D}^*$ and $D^*\bar{D}+\bar{D}^*D$, and the decay channels $D^* \to \pi^0 D$ and γD. These circumstances give rise to an inclusive photon spectrum at this energy which is quite complex since there are significant contributions from all eight sources (π^0 and γ from both charged and neutral D* decays arising from both $D^*\bar{D}^*$ and $D^*\bar{D}$ + c.c.). Qualitatively, the π^0 and γ contributions are resolvable since the former is peaked around 70 MeV and the latter around 135 MeV (the Doppler broadening gives rise to only a slight overlap). Further information is provided by the π^0 energy spectrum, which shows strong peaking at small kinetic energies.

By contrast to the strong structure in the π^0 and γ inclusive spectra at $E_{CM} = 4.03$ GeV, the corresponding ones below D* threshold at the $\psi''(3770)$ are smooth and featureless. We have used these as background functions in quantifying the effects due to the D* mesons. The spectra allow a new determination of the D^{*0}–D^0 mass difference of $142.2 \pm 0.5 \pm 1.5$ MeV/c^2 in agreement with earlier Mark I results (99). However, because of the relatively large number of cross sections (those for $e^+e^- \to D^{*0}\bar{D}^{*0}$, $D^{*+}D^{*-}$, $D^{*0}\bar{D}^0$ + c.c., $D^{*+}D^-$ + c.c.) and branching ratios (those for $\bar{D}^{*0} \to \pi^0 D^0$ and γD^0, $\bar{D}^{*+} \to \pi^0 D^+$ and γD^+) that are involved, the finite resolution of the apparatus, and the limited statistics, it is only possible to measure directly certain combinations of the physically interesting quantities. Guided by theoretical calculations (100) and Mark I measurements (99), we can make reasonable assumptions about some of these quantities and so obtain $\sigma(D^{*0}\bar{D}^0 + \text{c.c.})/\sigma(D^{*0}\bar{D}^{*0}) \cong 1.6$ and $B(D^{*0} \to \gamma D^0) \cong 0.37$. These results are consistent with those of Mark I (99).

11. SUMMARY AND FUTURE PROSPECTS

Although the people who studied photon detectors for e^+e^- storage rings at the 1974 PEP Summer Study did not know about the soon-to-be-discovered charmed quarks, subsequent events have shown that the practical realization of their ideas has borne rich rewards in understanding this sector of nature. The Crystal Ball detector grew out of that work and, as this brief review has shown, it has proved to be an especially versatile instrument in spite of features (or lack of them) that at first sight would make it seem to be very specialized and restricted in application. Of course, the dominant strength of this detector has been, and always will be, the measurement of monochromatic photons, and it was this capability that allowed the Ball to resolve the old problems with the charmonium interpretation of psionic matter. However, the measurements of R_h, for example, demonstrate the instrument's ability to determine global properties of e^+e^- annihilations. At the other extreme, kinematically constrained fitting to very specific final states that include just two charged particles but are rich in photons has been successfully exploited.

Work is currently in progress on several projects involving our large SPEAR data sample. These include searches for the F meson by means of specific exclusive channels, study of D decays, further work on D* physics, completion of the work on R_h just above charm threshold, measurement of certain interesting exclusive hadronic final states in J/ψ and ψ' decays, and further work on radiative decays of the J/ψ and ψ' and in two-photon physics. However, an increasingly large fraction of the group's efforts is going into new ventures in Υ physics. Within the next few years, we expect to have sufficiently large data samples at the Υ, Υ', and other energies to be able to make contributions toward understanding upsilonic matter comparable to what we have done in the psionic sector. The difficulties are formidable since the rate of data accumulation at the higher energies is considerably smaller than in the J/ψ region and the events are more complex, but the work has begun. Finally, we expect to utilize the higher $\gamma\gamma$ flux in the 10-GeV energy range to explore further questions in two-photon physics.

ACKNOWLEDGMENTS

The work reported here would not have been possible without the dedicated, innovative, and, at times, brilliant efforts of the many members of the Crystal Ball collaboration and their supporting engineers and technicians, the staffs of the accelerators at SLAC, Harshaw Chemical Co., and other people who played important roles in building the detector. The members of the Crystal Ball collaboration were: C. Edwards, R. Partridge, C. Peck, F. Porter (Caltech); D. Antreasyan, Y. Gu, W. Kollmann, M. Richardson, K. Strauch, K. Wacker, A. Weinstein (Harvard); D. Aschman, T. Burnett, M. Cavalli-Sforza, D. Coyne, C. Newman, H. Sadrozinski (Princeton); D. Gelphman, R. Hofstadter, R. Horisberger, I. Kirkbride, H. Kolanoski, K. Königsmann, R. Lee, A. Liberman, J. O'Reilly, A. Osterheld, B. Pollock, J. Tompkins (Stanford-HEPL); E. Bloom, F. Bulos, R.

Chestnut, J. Gaiser, G. Godfrey, C. Keisling, W. Lockman, M. Oreglia, D. Scharre (SLAC). The work of the collaboration was supported in part by the Department of Energy under contracts DE-AC03-76SF00515 (SLAC), DE-AC02-76ER03064 (Harvard), DE-AC03-81ER40050 (Caltech), and DE-AC02-76ER03072 (Princeton); by the National Science Foundation contracts PHY81-07396 (HEPL), PHY79-16461 (Princeton), and PHY75-22980 (Caltech); and by fellowships from the NATO Fellowship, the Chaim Weizmann Fellowship, and the Sloan Foundation, which provided partial support to members of the collaboration.

Literature Cited

1. Bloom, E. D., et al. 1974. *1974 PEP Summer Study Rep., PEP Note 155*; Mast, T., Nelson, J. 1974. *1974 PEP Summer Study Rep. PEP Note 153*. Stanford Univ. Press
2. Bloom, E. D. 1981. *Proc. Summer Inst. Particle Phys.*, Jul. 27–Aug. 7, SLAC-245:1, and references therein
3. Bloom, E. D. 1982. *21st Int. Conf. on High Energy Physics*, Paris, France, July 26–31, C3:407. *J. Phys.* C3, Suppl. 12. (This reference has an extensive review of gluonic mesons and contains references to the relevant theoretical literature.)

3a. Chinowsky, W. 1977. *Ann. Rev. Nucl. Sci.* 27:393

4. Feldman, G. J., et al. 1975. *Phys. Rev. Lett.* 35:821
5. Chanowitz, M. S., Gilman, F. J. 1976. *Phys. Lett.* B63:178
6. Tanenbaum, W., et al. 1978. *Phys. Rev.* D17:1731
7. Appelquist, T., Barnett, R. M., Lane, K. 1978. *Ann. Rev. Nucl. Part. Sci.* 28:387
8. Shifman, M. A. 1978. *Phys. Lett.* B77:80
9. Vainshtein, A., et al. 1978. *Yad. Fiz.* 28:465
10. Braunschweig, W., et al. 1977. *Phys. Lett.* B67:243
11. Wiik, B. H., Wolf, G. 1978. *DESY 78/23*. Hamburg: DESY

11a. Apel, W. D., et al. 1978. *Phys. Lett.* B72:500

12. Biddick, C. J. et al. 1977. *Phys. Rev. Lett.* 38:1324
13. Whitaker, J. S., et al. 1976. *Phys. Rev. Lett.* 37:1596
14. Bartel, W., et al. 1978. *Phys. Lett.* B79:492
15. Partridge, R., et al. 1980. *Phys. Rev. Lett.* 44:712
16. Hughes, E. B., et al. 1972. *IEEE Trans. Nucl. Sci.* 19:126
17. Oreglia, M., et al. 1982. *Phys. Rev.* D25:2259
18. Gaiser, J. E., et al. 1979. *IEEE Trans. Nucl. Sci.* 26:173
19. Chan, Y., et al. 1978. *IEEE Trans. Nucl. Sci.* 25:333
20. Chestnut, R., et al. 1979. *IEEE Trans. Nucl. Sci.* 26:4395
21. Aubert, J. J., et al. 1974. *Phys. Rev. Lett.* 33:1404
22. Augustin, J. E. 1974. *Phys. Rev. Lett.* 33:1406
23. Abrams, G. S., et al. 1974. *Phys. Rev. Lett.* 33:1453
24. Simpson, J. W., et al. 1975. *Phys. Rev. Lett.* 35:699
25. Braunschweig, W., et al. 1975. *Phys. Lett.* B57:407
26. Himel, T. M., et al. 1980. *Phys. Rev. Lett.* 44:920
27. Aguilar-Benitez, M., et al. 1982. *Particle Data Tables*; Particle Data Group, CERN and Univ. Calif. Berkeley, April 1982; *Phys. Lett.* B111:1
28. Karl, G., Meshkov, S., Rosner, J. 1976. *Phys. Rev.* D13:1203
29. Gaiser, J., et al. 1983. *SLAC-PUB-2899*. Stanford: SLAC. To be submitted to *Phys. Rev. Lett.* (A complete discussion of the analysis can be found in Gaiser, J. 1982. PhD thesis, Stanford Univ. SLAC-255. Unpublished.)
30. Bloom, E. D. 1983. In *Physics in Collision: High-Energy ee/ep/pp Interactions*, ed. P. Carlson, W. P. Trower. New York: Plenum
31. Oreglia, M., et al. 1980. *Phys. Rev. Lett.* 45:959
32. Himel, T. M. 1979. PhD thesis, Stanford Univ. SLAC-223 (unpublished)
33. Bloom, E. D. 1979. *Proc. 1979 Int. Symp. on Lepton and Photon Interactions at High Energy*. Batavia, Illinois, August 23–29, p. 92. Batavia: Fermilab

34. Partridge, R., et al. 1980. *Phys. Rev. Lett.* 45:1150
35. Himel, T. M., et al. 1980. *Phys. Rev. Lett.* 45:1146
36. Aschman, D. 1980. *Proc. 15th Rencontre de Moriond*, Les Arcs, France, March 15–21, 2:83. Derux, France: Editions Frontieres
37. Edwards, C., et al. 1982. *Phys. Rev. Lett.* 48:70
38. Eichten, E., et al. 1980. *Phys. Rev.* D21:203
39. Novikov, V. A., et al. 1977. *Phys. Lett.* B67:409; Schifman, M., et al. 1979. *Nucl. Phys.* B147:448
40. Barbieri, R., et al. 1976. *Phys. Lett.* B61:465; Barbieri, R., et al. 1980. *Phys. Lett.* B95:93; Barbieri, R., et al. 1981. *Phys. Lett.* B106:497
41. Reinder, J. L., et al. 1982. *Rutherford Lab. Preprint RL-82-017*
42. Porter, F. C. 1981. *SLAC-PUB-2796*. Stanford: SLAC
43. Fritzsch, H., Gell-Mann, M. 1972. *Proc. 16th Int. Conf. on High Energy Physics*, Chicago, Ill. 2:135. Batavia: Fermilab; and personal cummunication
44. Scharre, D. L., et al. 1980. *Phys. Lett.* B97:329
45. Edwards, C., et al. 1982. *Phys. Rev. Lett.* 49:259
46. Edwards, C., et al. 1982. *Phys. Rev. Lett.* 48:458
47. Brodsky, S., et al. 1978. *Phys. Lett.* B73:203; Koller, K., Walsh, T. 1978. *Nucl. Phys.* B140:449; Bjorken, J. D. 1980. *Proc. Summer Inst. on Particle Physics, SLAC Report No. 224*. Stanford: SLAC
48. Appelquist, T., et al. 1975. *Phys. Rev. Lett.* 34:365
49. Chanowitz, M. S. 1975. *Phys. Rev.* D12:918; Okun, L. G., Voloshin, M. B. 1976. *ITEP Preprint No. ITEP-95-1976*
50. Königsmann, K. C. 1982. *17th Rencontre de Moriond; Workshop on New Spectroscopy*, Les Arcs, France, March 20–26; also *SLAC-PUB-2910 (1980)*. Stanford: SLAC
50a. Scharre, D. L. 1980. Presented at 6th Int. Conf. on Exp. Meson Spectrosc., Upton, N.Y., Apr. 25–26. *Meson Spectroscopy*, p. 329. Brookhaven Natl. Lab., N.Y.
51. Bartel, W., et al. 1977. *Phys. Lett.* B64:483; *Phys. Lett.* B66:489
52. Novikov, V. A., et al. 1979. *Phys. Lett.* B86:347; Novikov, V. A., et al. 1980. *Nucl. Phys.* B165:55; Karl, G. 1977. *Nuovo Cimento* 38:315; Fritzsch, H., Jackson, J. D. 1977. *Phys. Lett.* B66:365; Pham, T. N. 1979. *Phys. Lett.* B87:267
53. Alexander, G., et al. 1978. *Phys. Lett.* B72:493; Brandelik, R., et al. 1978. *Phys. Lett.* B74:292; Alexander, G., et al. 1978. *Phys. Lett.* B76:652
54. Edwards, C., et al. 1982. *Phys. Rev.* D25:3065
55. Kabir, P. K., Hey, A. J. G. 1976. *Phys. Rev.* D13:3161. Note that the first occurrence of $\sin^2 \vartheta_M$ in Eq. (6) of this reference should be replaced by $\sin 2\vartheta_M$.
56. Krammer, M. 1978. *Phys. Lett.* B74:361
57. Gampp, W., Genz, H. 1978. *Phys. Lett.* B76:319
58. Baillon, P., et al. 1967. *Nuovo Cimento* A50:393
59. Montanet, L. 1980. *Proc. 20th Conf. on High Energy Physics*, Madison, Wisconsin, 17–23 July, ed. L. Durand, L. G. Pondrom. New York: Am. Inst. Phys.
60. Dionisi, C., et al. 1980. *Nucl. Phys.* B169:1
61. Herndon, D. J., Söding, P., Cashmore, R. J. 1975. *Phys. Rev.* D11:3165
62. Palmer, W. F., Pinsky, S. S. 1982. *DOE/ER/01545-328, Ohio State Univ. Preprint*
62a. Stanton, R. N., et al. 1979. *Phys. Rev. Lett.* 42:346
63. Scharre, D. L. 1982. *Proc. Orbis Scientiae*, Coral Gables, Fla., ed. A. Perlmutler. New York: Plenum; also *SLAC-PUB-2880*. Stanford: SLAC
64. Franklin, M. E. B. 1982. PhD thesis, Stanford Univ., *SLAC-254* (unpublished)
65. Burke, D. L., et al. 1982. *Phys. Rev. Lett.* 49:632
66. Newman-Holmes, C. 1982. *SLAC-PUB-2971*. Stanford: SLAC
67. Brandelik, R., et al. 1977. *Phys. Lett.* B70:132; Brandelik, R., et al. 1979. *Phys. Lett.* B80:412; Brandelik, R., et al. 1979. *Z. Phys.* C1:233
68. Einhorn, M. B., Quigg, C. 1975. *Phys. Rev.* D12:2015; Ellis, J., Gaillard, M. K., Nanopoulos, D. V. 1975. *Nucl. Phys.* B100:313; Quigg, C., Rosner, J. L. 1978. *Phys. Rev.* D17:239; Fakirov, D., Stech, B. 1978. *Nucl. Phys.* B133:315
69. Partridge, R., et al. 1981. *Phys. Rev. Lett.* 47:760
70. Fayet, P., Mezard, M. 1981. *Phys. Lett.* B104:226
71. Edwards, C., et al. 1982. *Phys. Rev. Lett.* 48:903
72. Pecci, R. D., Quinn, H. R. 1977. *Phys. Rev. Lett.* 38:1440, *Phys. Rev.* D16:1791; Weinberg, S. 1978. *Phys. Rev. Lett.* 40:223; Wilczek, F. 1978. *Phys. Rev. Lett.* 40:279

73. Faissner, H., et al. 1981. *Phys. Lett.* B103:234
74. Goldman, T., Hoffman, C. M. 1978. *Phys. Rev. Lett.* 40:220; Bechis, D. J., et al. 1979. *Phys. Rev. Lett.* 42:1511; Zehnder, A. 1981. *Phys. Lett.* B104:494
75. Porter, F. C., Königsmann, K. C. 1982. *Phys. Rev.* D25:1993, D26:716
76. Niczyporuk, B., et al. 1982. *DESY 82-068.* Hamburg: DESY
76a. Sivertz, M., et al. 1982. *Phys. Rev.* D26:717
77. Dine, M., Fischler, W., Srednicki, M. 1978. *Phys. Lett.* B104:199; Kim, J. 1979. *Phys. Rev. Lett.* 43:193; Wise, M. B., Georgi, H., Glashow, S. I. 1981. *Phys. Rev. Lett.* 47:402; Wilczek, F. 1982. *Phys. Rev. Lett.* 49:1549
78. Eichten, E., et al. 1975. *Phys. Rev. Lett.* 34:369
79. Porter, F. C., et al. 1982. *17th Rencontre de Moriond Workshop on New Flavors,* Les Arcs, France, Jan. 24–30, p. 27
80. Barnett, R. M., Dine, M., McLerran, L. 1980. *Phys. Rev.* D22:594
81. Siegrist, J. 1979. PhD thesis, Stanford Univ., *SLAC Rep. No. SLAC-225.* Stanford: SLAC
81a. Lockman, W., et al. 1983. *SLAC-PUB-3030.* Stanford: SLAC
82. Sjostrand, T. 1982. *LU-TP-82-3*
82a. Siegrist, J. L., et al. 1982. *Phys. Rev.* D26:991 (Mark I); Rice, E., et al. 1982. *Phys. Rev. Lett.* 48:906 (CUSB); Berger, C., et al. 1979. *Phys. Lett.* B81:410 (PLUTO); Bock, P., et al. 1980. *Z. Phys.* C6:125 (DESY-Heidelberg); Albrecht, H., et al. 1982. *Phys. Lett.* B116:383 (DASP II); Niczyporuk, B., et al. 1982. *Z. Phys.* C15:299 (LENA)
83. Edwards, C., et al. 1982. *Phys. Lett.* B110:82
84. Berger, C., et al. 1980. *Phys. Lett.* B94:254
85. Roussarie, A., et al. 1981. *Phys. Lett.* B105:304
86. Brandelik, R., et al. 1981. *Z. Phys.* C10:117
87. Biddick, C. J., et al. 1980. *Phys. Lett.* B97:320
88. Bonneau, G., Gourdin, M., Martin, F. 1973. *Nucl. Phys.* B54:573
89. Brodsky, S. L., Lepage, G. P. 1981. *Phys. Rev.* D24:1808
90. Rosner, J. L. 1981. *Phys. Rev.* D23:2625; Donoghue, J. F., Johnson, K., Li, B. A. 1981. *Phys. Lett.* B99:416
91. Gilman, F. J. 1979. *Int. Conf. on Two-Photon Interactions,* ed. J. F. Gunion, p. 215. Univ. Calif., Davis
92. Greco, M. 1980. *Proc. Int. Workshop on $\gamma\gamma$ Collisions,* Amiens, France, p. 311. *Lect. Notes Phys.,* Vol. 34. Berlin: Springer-Verlag
93. Babcock, J., Rosner, J. 1976. *Phys. Rev.* D14:1286
94. Emms, M. J., et al. 1975. *Nucl. Phys.* B96:155
95. De Rújula, A., Georgi, H., Glashow, S. L. 1976. *Phys. Rev. Lett.* 37:785
96. Tompkins, J. C. 1980. *Quantum Chromodynamics,* ed. A. Mosher, p. 556. Stanford: SLAC
97. Sadrozinski, H. F. W. 1980. *Proc. 20th Int. Conf. on High Energy Physics,* Madison, Wisconsin, July 17–23
98. Rapidis, P., et al. 1977. *Phys. Rev. Lett.* 39:5261; Bacino, W., et al. 1978. *Phys. Rev. Lett.* 40:671
99. Goldhaber, G., et al. 1977. *Phys. Lett.* B69:503
100. Ono, S. 1976. *Phys. Rev. Lett.* 37:655

Ann. Rev. Nucl. Part. Sci. 1983. 33: 1–29

7

UPSILON RESONANCES

Paolo Franzini

Department of Physics, Columbia University, New York, New York 10027

Juliet Lee-Franzini

Department of Physics, SUNY at Stony Brook, Stony Brook, New York 11794

CONTENTS

1. INTRODUCTION

The latest addition to the apparently endless list of strongly interacting particles is a family of vector mesons called the upsilons, Υ. They are the heaviest hadrons known to date, with masses around 10 GeV, and are especially rich in confrontation opportunities between experiments and the continuously developing theories of elementary particles. Some of the salient consequences of the existence of the Υ's are listed in the following. (*a*) Prior to their discovery, all known hadrons were understood as bound or quasibound states of fundamental, point-like constituents, the quarks, which came in four "flavors": u, d, s, c. The discovery of the Υ's required the existence of a fifth quark, the b-flavor quark, b. (*b*) Upsilons are bound states of a $b\bar{b}$ pair whose relative motion is sufficiently slow that relativistic effects are quite negligible, in contrast to light hadrons and even the heavier psions, J/ψ. Schroedinger's equations can be used to accurately describe the level structure and thus we can learn about the interquark forces responsible for the binding of these states, which range in size from 0.2 to 1 fermi. (*c*) The bound upsilons (Υ, Υ', Υ'') decay primarily through the annihilation of the $b\bar{b}$ pair into three "gluons," the quanta of the gauge field coupled to the "color charge" carried by all the quarks. Gluons turn ultimately into hadrons; thus properties of the invisible gluons and of the gluon-quark coupling can be gleaned through the study of the Υ's final states and their decay rates. (*d*) The fourth upsilon is quasibound, i.e. it decays with much higher rate into a pair of b-flavor mesons, the B mesons, each containing a b quark and a light, d or u, quark. B mesons provide means to study the weak interaction of the b quark. (*e*) Finally the massiveness of the b quark implies a nonnegligible coupling to the elusive Higgs scalars, making the Υ's, at present, the best hunting grounds for light Higgs bosons.

This article reviews the experimental work on upsilon spectroscopy, i.e. the study of the level structure, the transitions between states, and the decays of these states. We present only the minimal amounts of theoretical background and results that are necessary to understand the way measurements are performed and the discussion of the results. The reader is referred to the article on the "Sum Rule Approach to Heavy Quark Spectroscopy," by M. A. Shifman in this volume for a detailed theoretical exposition and references to the extensive theoretical reviews.

1.1 *Summary and Plan of the Paper*

The following subsections contain a brief summary of the contents of each section. We first, however, explicitly list some topics of relevance to the subject of this article that will not be reviewed. (*a*) Some B-meson properties are mentioned, but not their decays. (*b*) The fragmentation mechanisms of gluons and quarks, mentioned in the discussion of hadronic decays, are not explained or justified. In particular we do not present data on charged and neutral multiplicities, and production of strange particles and baryons. (*c*) While a great deal of interest and speculation exists concerning the mechanism of hadroproduction of flavored and unflavored heavy particles, this subject is not discussed since, except for the original experiment that discovered the upsilons and similar ones thereafter, there is almost no experimental information on the subject.

1.2 *Discovery of the Υ's and χ_b's*

In 1977, at Fermilab the first three Υ's were discovered, unresolved, as an enhancement in the mass spectrum of muon pairs, produced in collisions of 400-GeV protons on beryllium (1, 2). The first two upsilons (Υ and Υ') were observed fully resolved as resonances in e^+e^- annihilations into hadrons, in 1978 at the DORIS e^+e^- storage ring at the DESY laboratory in Hamburg (3–5), and the third one (Υ'') in 1979 at CESR (Cornell Electron Storage Ring) (6, 7). The fourth Υ (Υ'''), which decays into muon pairs with a branching ratio of the order of 10^{-5}, was discovered in 1980 at CESR as a resonance much broader than the three previous ones (8, 9). The χ_b's (scalar, axial vector and tensor $b\bar{b}$ states) are not produced directly in e^+e^- collisions but are reached via photon emission from the Υ' and Υ''. They were first observed at CESR in 1982–1983 (10–12).

1.3 *Production of the Upsilons in e^+e^- Annihilations*

The masses, leptonic widths, and other properties of the upsilons have been obtained mostly from the study of e^+e^- annihilations into hadrons. The first three Υ's are observed as narrow enhancements in the hadronic cross section with widths determined by the machine energy spread. In the interpretation of the Υ's as $b\bar{b}$ bound states, the mass differences correspond to the excitation energies and the leptonic widths are related to the wave function at the origin. The comparison between measurements and calculations in "potential models," leads to the identification of the four upsilons as the first four triplet S-wave states. The charge of the b quark is determined in this way to be $\frac{1}{3}e$. The Υ''' is a quasibound state lying above the free b-flavor threshold. At energies above the mass of the Υ''', the ratio of the hadronic cross section to the muon pair cross section is observed to increase by $\frac{1}{3}$, which confirms that the b-quark charge is $\frac{1}{3}e$ (13).

1.4 *Transitions and Decays of the Bound Upsilons*

Experimental observation of a three-jet structure in the final state of the lightest Υ is direct evidence that the decay occurs via annihilations of the $b\bar{b}$

pair into three gluons (10, 14–16); this confirms the existence and nonobservability of color and gives tangible proof for the original explanation of the narrowness of the widths of both psions and upsilons. The first two excited Υ's can decay to the ground state, emitting two pions. The partial rates for these transitions compared to those for the ψ' give one of the most convincing proofs of the vector nature of the gluon (17–19). The χ_b's, which are identified as triplet P-wave $b\bar{b}$ states, are reached via electric dipole transition from the excited Υ's. The measured center of gravity of these states gives additional proof of the color independence of the interquark forces (20). Searches for the Weinberg-Wilczek axion in Υ's decays gives strong evidence against the existence of such particles (21).

1.5 *Decay of the Υ'''*

The fourth upsilon has a total width of ~ 20 MeV, whereas the first three have total widths in the range of tens of kilovolts. This suggests that Υ''' can decay into B mesons, a fact confirmed by the observation of a strong signal corresponding to the weak decay of a particle of ~ 5-GeV mass (22, 23), the B meson. Excited B mesons, B*'s, are not produced in Υ''' decays as proved by a search for photons from $B^* \rightarrow B + \gamma$. The B-meson mass is therefore bounded as $5.263 < M_B < 5.278$ GeV (24).

1.6 *Outlook for the Study of Fine and Hyperfine Structure*

The potential models that so successfuly describe the level structure of the $b\bar{b}$ bound states cannot unambiguously predict the fine structure of the ^3P states, nor the hyperfine splitting between $^3S_1$ (Υ's) and $^1S_0$ (η_b's) states. Experimental information is necessary to establish the form of the effective spin-spin and spin-orbit interactions responsible for these effects. The next generation of experiments both at CESR and at DORIS will concentrate in this area, and will require very fine resolution and very high machine intensities.

2. DISCOVERY OF THE UPSILON STATES

2.1 *Discovery of the Υ's*

The upsilons were discovered in 1977 by Lederman's group (Columbia–FNAL–Stony Brook collaboration), in the continuing study of muon-pair production in hadron collisions, at Fermilab, with proton energies up to 400 GeV, thus extending the reachable muon-pair mass to above 15 GeV (1, 2). A plot of the muon-pair cross section versus dimuon mass obtained by this group is shown in Figure 1 (25). A prominent peak, ~ 500 MeV wide (of the order of the experimental mass resolution), is strikingly visible over a sharply falling continuum. The continuum-subtracted signal, shown in

Figure 2, is in fact consistent with the production of two and possibly three states spread over a mass range of about 1 GeV, centered around 10 GeV (25). While the three states were not resolved in the Fermilab experiment, the observed mass spectrum was consistent with the natural width of the states being much smaller than their separation. Their interpretation as bound states of a new flavored heavy quark, of mass around 5 GeV, was almost immediate, after the J/ψ experience of only a couple of years earlier (25a).

The Fermilab discovery prompted the DESY Laboratory to push the maximum energy of the DORIS e^+e^- collider up to 5 GeV per beam, which allowed the observation of the first two upsilons fully resolved. The machine energy spread of DORIS at these energies is around 20 MeV, and the experiments at DORIS obtained the first accurate values for the Υ and Υ' masses as 9.46 and 10.01 GeV, respectively (3–5). The Υ'' was not resolved until the autumn of 1979 (6, 7) when CESR began operation. Shortly thereafter a fourth state, the Υ''', was discovered (8, 9). The Υ'' and Υ''' masses are 10.35 and 10.58 GeV, respectively. This last state is the first Υ with a natural width of tens of MeV, indicative of the fact that it is just above the open flavor threshold. The Υ''' will probably remain for some time the heaviest state in the Υ family to be observed. Figure 3 shows the four resonances as they are seen at CESR.

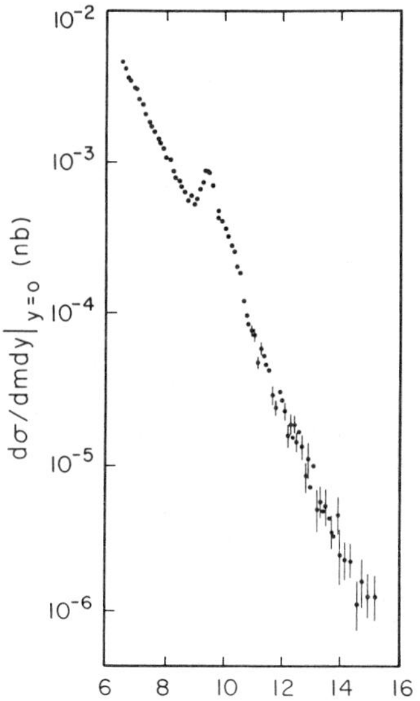

Figure 1 Dimuon mass spectrum observed at Fermilab (25).

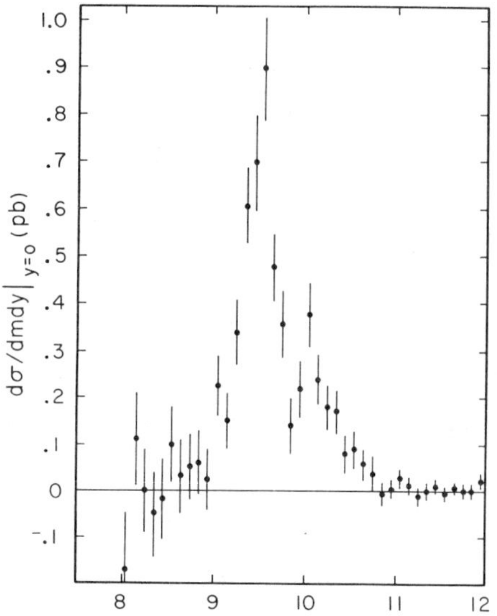

Figure 2 Dimuon mass spectrum after background subtraction (25).

Searches at CESR for additional resonances were extended up to ~11.6 GeV, with negative results (13). However, a step in R ($\equiv \sigma_{\text{had}}/\sigma_{\mu\mu}$) was observed around a total energy $W = 10.58$ GeV, of magnitude approximately one third, which confirmed the crossing of the threshold for the production of a pair of new quarks of charge $\pm\frac{1}{3}e$ (13). The value of R around the Υ''' is shown in an expanded scale in Figure 4.

The four Υ's, copiously produced in e^+e^- hadronic annihilations, are vector mesons with quantum numbers identical to those of the photon. With the aid of potential models these states are identified as S-wave triplet states, with the $\Upsilon(9.46)$ being the ground state and the other three Υ's being the $n = 2, 3, 4$ radially excited states. In addition to the good agreement between measured and calculated excitation energies of the four Υ's, further evidence that they are bound states of the same b quark comes from the observation of hadronic transitions between the various levels. The transition between Υ' and Υ was observed at DORIS and CESR in 1980 (17–19). Transitions from Υ'' to Υ were observed at CESR in 1981 and to Υ' in 1982 (26, 27).

2.2 *Discovery of the χ_b's*

Bound $b\bar{b}$ states other than $^3S_1$ cannot be directly produced in e^+e^- annihilation either because they have wrong J^{PC} values or because the wave function vanishes at the origin. Both hadronic and electromagnetic

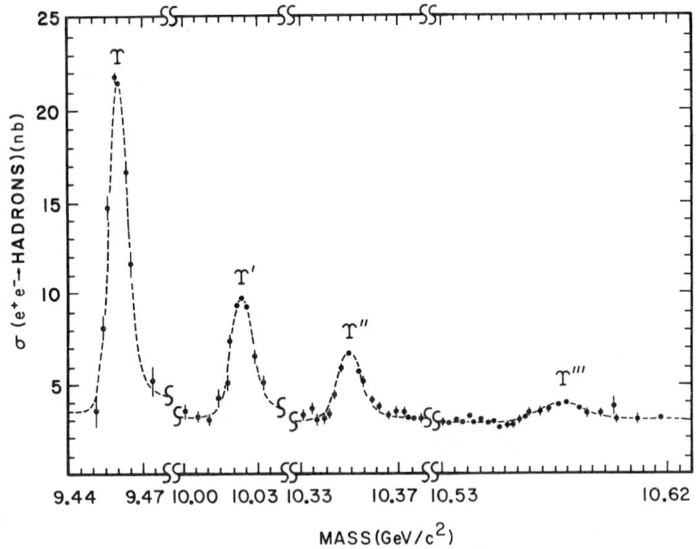

Figure 3 Cross section for e^+e^- annihilations into hadrons at CESR (CUSB data).

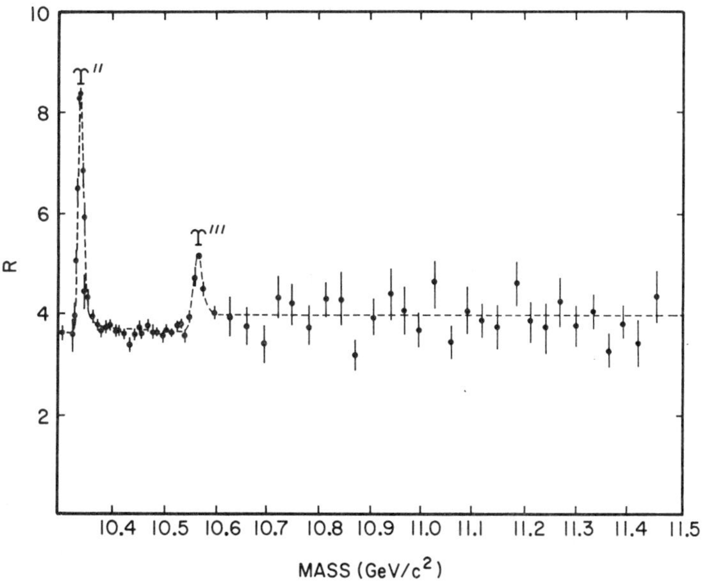

Figure 4 R versus center-of-mass energy around the b-flavor threshold (CUSB data).

transitions can, however, feed-down from the first three Υ's to other states. In particular the ^3P_J states (χ_b's) can be reached via electric dipole (E1) transitions from the ^3S states. Indirect evidence for the decay of ^3S to ^3P states was obtained in 1981 by the CUSB collaboration at CESR (10) (see Section 4.2.2). In 1982 CUSB observed the photons from the transition $\Upsilon'' \rightarrow \chi_b' + \gamma$ in the inclusive Υ'' decay photon spectrum as well as the photons from decays of the χ_b' to Υ' and Υ in exclusive channels. This established for the first time the existence of P-wave bound $b\bar{b}$ states (11, 12).

3. PRODUCTION OF THE UPSILONS IN e^+e^- ANNIHILATIONS

3.1 $\sigma_{\mu\mu}$, σ_{had}, Γ_{ee}, $B_{\mu\mu}$, Γ_{tot}

Positron-electron collisions result in, to lowest order in QED, four processes: (*a*) elastic or Bhabha scattering; (*b*) annihilation into two photons; (*c*) lepton-pair production; and (*d*) hadron production, via production of a quark-antiquark pair. Bhabha scattering is used to measure the "luminosity" L of e^+e^- colliders, defined by $N = TL\sigma$, where T is the time of the measurement, σ the cross section, and N the number of reactions. The two-photon annihilation cross section provides an independent measurement of the luminosity. The cross section for muon-pair production well above the threshold is given by

$$\frac{d\sigma}{d\Omega} = \frac{\alpha^2}{4W^2}(1+\cos^2\theta), \qquad 1.$$

where α is the fine structure constant and θ is the center-of-mass angle between electron and negative muons. The same result applies to the production of any pair of point-like, spin-$\frac{1}{2}$ fermions of charge e. It is therefore assumed to apply for quarks, taking properly into account charge and color. Integrating over the polar angle, we obtain

$$\sigma(e^+e^- \to \mu^+\mu^-) = \frac{4\pi\alpha^2}{3W^2} = \frac{86.8}{W^2}\,\text{nb}$$

(with W in GeV), and for N quark flavors of charge q_i with three colors

$$\sigma(e^+e^- \to \text{hadrons}) = 3(86.8/W^2)\sum_i^N q_i^2\ \text{nb}$$

assuming, according to the quark-parton model, that quarks evolve into hadrons with unit probability. Note that both these cross sections scale as $1/W^2$ as expected for point-like particles. It is convenient to define the dimensionless ratio $R = \sigma_{\text{had}}/\sigma_{\mu\mu}$, which is expected to have no energy dependence except if a new threshold is reached. Below the b-flavor threshold we know of the u, d, s, and c quarks of charges $|q_i| = \frac{2}{3}, \frac{1}{3}, \frac{1}{3}$, and $\frac{2}{3}$ respectively. R is thus predicted to be $\frac{10}{3} = 3.33$. The experimental value is around 3.6. Note that the inclusion of the color factor of 3 is quite necessary for good agreement. The remaining difference is almost entirely accounted for by higher order QCD corrections to the quark-pair cross section (28).

If the total energy W is very close to the mass M of a $J^{PC} = 1^{--}$ bound $q\bar{q}$ state, such as a vector meson V, then the total cross section acquires additional resonant contributions given (29) by

$$\sigma_i = \frac{3\pi}{M^2}\frac{\Gamma_{ee}\Gamma_i}{(W-M)^2+\Gamma_{\text{tot}}^2}, \qquad 2.$$

where σ_i is the cross section to channel i, Γ_{ee} and Γ_i are respectively the partial widths for V decay to e^+e^- and to channel i, and Γ_{tot} is the total decay width of the meson V. Integrating this additional contribution and summing over all channels, one obtains a relation between Γ_{ee} and the total resonant cross section:

$$\int \sigma_{\text{tot,res}}\, dW = \frac{6\pi^2}{M^2}\Gamma_{ee}. \qquad 3.$$

We wish, however, to express the leptonic width in terms of the resonant hadronic cross section only. Defining $B_{\mu\mu} = \Gamma_{\mu\mu}/\Gamma_{\text{tot}}$ and assuming lepton universality, i.e. $B_{ee} = B_{\mu\mu} = B_{\tau\tau}$, we can write

$$\Gamma_{tot} = \Gamma_{had} + 3B_{\mu\mu}\Gamma_{tot} = \frac{\Gamma_{had}}{(1-3B_{\mu\mu})}$$

and thereby obtain

$$\Gamma_{ee} = \frac{M^2}{6\pi^2} \frac{\int \sigma_{had,res}\, dW}{1-3B_{\mu\mu}}. \qquad 4.$$

From measurements of $\sigma_{had}(W)$ and $B_{\mu\mu}$ we can therefore obtain Γ_{ee} and $\Gamma_{tot} = \Gamma_{ee}/B_{\mu\mu}$.

3.2 *Measurements of* R, Γ_{ee}, *and Resonance Masses*

Two types of detectors are used to study the properties of the upsilons: (*a*) magnetic spectrometers such as PLUTO (3), DASP-II (4), and ARGUS at DORIS and CLEO at CESR (30); and (*b*) calorimetric spectrometers such as LENA (15) and the Crystal Ball (see the article by E. D. Bloom and C. W. Peck in this volume) at DORIS and CUSB at CESR (30). Magnetic detectors have tracking in magnetic field, which provides momentum analysis. Calorimetric detectors emphasize precise measurements of electromagnetic energy (e's or γ's) by the use of segmented arrays of active converters such as NaI or lead glass. Tracking is provided with various degrees of accuracy before the electromagnetic calorimeter. These two classes of detectors are complementary in their main aims but can both measure many of the quantities of interest.

An extensive mapping of R versus energy has been obtained in the Υ region. Figure 4 shows the measurements of R obtained with the CUSB detector. The reported values of R are summarized in Table 1. We recall

Table 1 Measurements of R in the Υ region

Energy (MeV)	R[a]	Experiment	
9.4	$3.67 \pm 0.23 \pm 0.29$	PLUTO	(31)
9.5	$3.73 \pm 0.16 \pm 0.28$	DASP-II	(32)
7.4–9.4	$3.37 \pm 0.06 \pm 0.28$	LENA	(33)
10.4–10.5	$3.63 \pm 0.06 \pm 0.37$	CUSB	(13, 34)
10.4–10.5	$3.77 \pm 0.06 \pm 0.26$	CLEO	(35)
ΔR across flavor threshold			
	$0.36 \pm 0.09 \pm 0.03$	CUSB	(13, 34)
	$0.34 \pm 0.09 \pm 0.11$	CLEO	(35)

[a] The first error is statistical and the second is an estimate of the systematic uncertainty.

that above the b-flavor threshold R is expected to be higher by $3(\frac{1}{3})^2 = 0.33$ for $q_b = \frac{1}{3}e$, in good agreement with observation.

At e^+e^- colliders the masses of the vector mesons are determined by the total energy W of the colliding beams at which the resonances are observed. The energy of the stored beams is usually estimated from measurements of the magnets providing the guide field. In this way DORIS obtained for the Υ a mass of 9460 MeV with an estimated uncertainty of 10 MeV (3–5). At CESR a value of 9433 MeV was obtained, with an uncertainty of about 30 MeV (6, 7). Recently at VEPP-4 the $g-2$ depolarizing resonances have been observed at energies close to the Υ mass; this improves better than ten-fold the accuracy with which the Υ mass is determined (36). Their result, $M(\Upsilon) = 9459.7 \pm 0.6$ MeV, in close agreement with the original DORIS value, is therefore used to rescale all the CESR mass measurements and is the only entry for the upsilon mass in the table.

The determination of $B_{\mu\mu}$ is a much harder task, requiring precise knowledge of σ_{had} at the Υ peak and good measurements of the μ-pair yield on and off peak. The first good signal for $\Upsilon \to \mu^+\mu^-$ was obtained by LENA in 1980. CLEO has measured $B_{\mu\mu}$ directly and by comparing the two-pion decay of the Υ' in inclusive and exclusive channels. There is a strong tendency for $B_{\mu\mu}$ to decrease with time. Only these last three values are reported here.

In Table 2 we give the mass, leptonic width, and $B_{\mu\mu}$ for the four resonances. The latest determinations of these parameters are given first, followed by the average values. Systematic uncertainties are not given. They typically are 2 to 5 MeV for masses and $\sim 10\%$ for Γ_{ee} and $B_{\mu\mu}$. Since potential models compute excitation energies and give more accurate predictions for ratios of leptonic widths, these are also given. The fourth upsilon has, at CESR, a width that is clearly wider than the machine energy spread but its shape is not known because of the lack of a detailed scan of the Υ''' peak. This fact is reflected in the difference between the values reported by CLEO and CUSB (34, 35) for mass, Γ_{ee} and Γ_{tot}.

3.3 *Limits on the Production of Other Resonances*

The existence of other narrow resonances, besides the three upsilons, in $\sigma(ee \to \text{hadrons})$ has been suggested in various contexts (see Section 3.4.4). LENA at DORIS has established that in the energy intervals 7.4–7.5 GeV and 8.6–9.4 GeV there are no resonances with leptonic widths larger than 233 eV (41). CUSB (13, 34) and CLEO (35) at CESR have established the following limits:

$10.27 < W < 10.34$ GeV, $\Gamma_{ee} < 20$ eV (CUSB) and $\Gamma_{ee} < 20$ eV (CLEO)
$10.36 < W < 10.55$ GeV, $\Gamma_{ee} < 20$ eV (CUSB) and $\Gamma_{ee} < 40$ eV (CLEO).

Table 2 Measured parameters of the Υ resonances

Parameter	Υ	Υ'	Υ''	Υ'''	Experiment	Reference
M (MeV)	9459.7 ± 0.6				VEPP-4	(36)
		10016.8 ± 1.5			DASP-II	(37)
		10013.6 ± 1.2			LENA	(38)
		10021.2 ± 0.7	10352.1 ± 0.7	10575.0 ± 1.1	CLEO	(35)
		10023.4 ± 0.7	10350.0 ± 0.7	10578.0 ± 3.0	CUSB	(34)
Γ_{ee} (keV)	1.35 ± 0.11	0.61 ± 0.11			DASP-II	(37)
	1.13 ± 0.09	0.53 ± 0.07			LENA	(38)
	1.14 ± 0.05	0.50 ± 0.03	0.35 ± 0.03	0.21 ± 0.05	CUSB[a]	(34)
	1.30 ± 0.05	0.52 ± 0.03	0.42 ± 0.04	0.32 ± 0.03	CLEO	(35)
$B_{\mu\mu}$ (%)	3.8 ± 1.5				LENA	(39)
	3.5 ± 0.8				CLEO	(18)
	2.7 ± 0.3	1.9 ± 1.3	3.3 ± 1.3		CLEO	(40)
			Averages			
M (MeV)	9459.7 ± 0.6	10020.5 ± 0.7	10350.0 ± 0.7	10576.0 ± 3.0		
$M - M_{\Upsilon}$ (MeV)		560.8 ± 0.4	890.3 ± 0.4	1116.5 ± 3.0		
Γ_{ee} (keV)	1.22 ± 0.03	0.52 ± 0.02	0.38 ± 0.02	0.29 ± 0.03		
$\Gamma_{ee}/\Gamma_{ee}(\Upsilon)$		0.42 ± 0.02	0.31 ± 0.02	0.24 ± 0.03		
$B_{\mu\mu}$ (%)	2.83 ± 0.28	1.9 ± 1.3	3.3 ± 1.3	$\sim 10^{-5}$		

[a] From a reanalysis of the data in Reference 34 (P. M. Tuts, paper in preparation).

Higher excitations of the ^{3}S b$\bar{\text{b}}$ states are also expected, with the 5^3S and 6^3S states a few hundred MeV above the Υ''' and with natural widths most likely in the 100-MeV range. Searches at CESR did not have enough sensitivity for the detection of such states. A scan in the energy region $10.6 < W < 11.6$ GeV excludes the existence of resonances with natural widths of 20 MeV and $\Gamma_{ee} > 0.3$ keV (CLEO) and 40 MeV with $\Gamma_{ee} > 0.2$ keV (CUSB).

3.4 *Comparison of Measurements and Theoretical Models*

3.4.1 POTENTIAL MODELS In the potential model description of heavy q$\bar{\text{q}}$ bound states, the Schroedinger equation is solved using an effective central potential $V(r)$. The energy eigenvalues E_n give the excitation energies and, if the wave function can be calculated reliably at the origin, one can obtain the leptonic width Γ_{ee} from the Van Royen & Weisskopf (42) formula:

$$\Gamma_{ee} = 16\pi\alpha^2 q_q^2 \frac{|\Psi(0)|^2}{M_n^2}, \qquad 5.$$

where α is the fine structure constant, q_q is the quark charge, and M_n is the mass. It is more often the case that Equation 5 is solved the other way around to obtain a constraint on the wave function. Equation 5 is valid to zeroth order in quantum chromodynamics, QCD, and in addition one should include relativistic corrections because even for the lowest Υ state $\langle v^2/c^2 \rangle \cong 0.08$. It is reasonable to assume that most of these corrections cancel whenever ratios of widths are considered.

In the last few years, over a dozen potentials have been proposed to simultaneously describe c$\bar{\text{c}}$ and b$\bar{\text{b}}$ system, with considerable success, thus satisfying and proving a fundamental requirement of QCD, namely that the interquark forces be flavor independent. These potentials are virtually indistinguishable at distances of 0.2 to 1 fermi, the region probed by the ψ and the Υ families. For a comparison with data we have chosen four models, each representative of a particular approach in the choice of the functional dependence of V on r and the determination of its parameters. Eichten et al (43) used a linear combination of a confining potential and a Coulomb term, $V(r) = -K/r + r/a^2$, where a and K are determined by the spacing of the first two levels and the quark masses are free parameters. Krasemann & Ono (44) incorporated "asymptotic freedom" in the Bhanot & Rudaz (45) version of the above potential. Defining $\alpha_s = g^2/4\pi$, where g is the strength of the fundamental QCD quark-gluon vertex, asymptotic freedom refers to the fact that α_s vanishes logarithmically for $q^2 \to \infty$. Büchmuller, Grundberg & Tye (46) include higher order QCD corrections in a modified version of the Richardson potential (47). Martin (48) assumes a simple power law potential of the form $V(r) = A + Br^\alpha$. Data from Υ's, ψ's, and ϕ's are used in a global fit to determine the parameters A, B, and α as well as the quark masses m_b, m_c, and m_s. Similar results were obtained by Quigg & Rosner (49), who used $V(r) = \lambda r^v$ as well as the inverse scattering method (50).

3.4.2 QCD SUM RULES This approach to the problem of bound states of heavy quarks does not require a potential but derives the "current" quark mass and the net effect of long-range gluon field fluctuation, as measured by the vacuum expectation value of the gluon field squared $\langle G^2 \rangle$ from experiments. Using these two parameters the masses of other ground states and the leptonic width of the 1^3S state can be calculated (51).

3.4.3 COMPARISON OF CALCULATIONS AND MEASUREMENTS Table 3 lists the calculated quantities for the five examples chosen, and the measured values, for the first four ^{3}S states, labelled in the table as nS.

The agreement between data and calculations is excellent for the first two spacings, which confirms the identification of the Υ' and Υ'' as the 2^3S_1 and 3^3S_1 $b\bar{b}$ states. It is much harder to obtain accurate values for the excitation energy of states above threshold; therefore even an approximate agreement between experiment and calculations for the Υ''' confirms the identification of this last state as the 4^3S_1 $b\bar{b}$ radial excitation. Quigg & Rosner (49) have predicted on general grounds, in the context of potential models, that there should be just three bound $b\bar{b}$ states below the flavor threshold.

3.4.4 OTHER NARROW STATES The ^{3}D states are predicted by potential models to lie some 150 MeV below the corresponding ^{3}S states. While ^{3}D states have $J^{PC} = 1^{--}$ as the photon, they cannot be directly produced in e^+e^- annihilations since their wave function vanishes at the origin. They could, however, be observed if there is significant mixing with a nearby S state. In the $b\bar{b}$ case the second D state could be produced via mixing with the Υ'''. This mixing is, however, expected to be very small (52).

In theories of quark confinement, excitations of the color-flux tube are expected in addition to the potential model levels. The lowest of these modes, referred to as vibrational states, with $J^{PC} = 1^{--}$ has been predicted to exist with excitation energies between 910 and 1000 MeV and a leptonic width between 70 and 270 eV (53). The results of searches at CESR (given in Section 3.3) appear to exclude the existence of these states for the predicted width values, as well as any unexpected production of D states.

3.4.5 THE CHARGE OF THE b QUARK Combining results from potential models with Equation 5, one can in principle obtain the charge of the bound quarks. Quigg & Rosner (49) have given lower bounds for the leptonic widths of the Υ's in terms of those of the ψ's and the quark charges. Comparison of these bounds with the leptonic width of the Υ' establishes

Table 3 Comparison of prediction and measurements for the Υ's

Quantity	Eichten et al (43)	Krasemann & Ono (44)	Büchmuller, Grunberg & Tye (46)	Martin (48)	Voloshin (51)	Experiment
$M(2S)-M(1S)$ (MeV)				565		561
$M(3S)-M(1S)$ (MeV)	898	862	890	900		890
$M(4S)-M(1S)$ (MeV)	1170	1108	1180	1142	—	1116
$\Gamma_{ee}(1S)$ (keV)		1.05	1.07		1.15 ± 0.20	1.22 ± 0.03
$\Gamma_{ee}(2S)/\Gamma_{ee}(1S)$	0.39	0.43	0.44	0.41		0.42 ± 0.02
$\Gamma_{ee}(3S)/\Gamma_{ee}(1S)$	0.27	0.31	0.32	0.35	—	0.31 ± 0.02
$\Gamma_{ee}(4S)/\Gamma_{ee}(1S)$	0.22	0.25	0.26	0.27	—	0.24 ± 0.03

that the b quark has charge $|q_b| < \frac{2}{3}e$. (From the leptonic width of the Υ, one obtains a less stringent bound.) Thus the charge of the b quark is $\frac{1}{3}e$, in agreement with the observed change in R.

4. TRANSITIONS AND DECAYS OF BOUND UPSILONS

4.1 *Decay of the Ground-State* Υ

One of the outstanding properties of the first three Υ's is their width of tens of keV, while their mass is in the 10-GeV range. This fact was first explained in the framework of QCD by Appelquist & Politzer (54) for the J/ψ meson. The decay of a heavy vector meson into hadrons is supposed to proceed mostly via annihilation of the heavy quark antiquark pair into three gluons, in analogy to the decay of triplet positronium into three photons, followed by evolution of the gluons into hadrons with unit probability. Properly accounting for the fact that the three gluons must be in a color singlet state, one obtains (54)

$$\Gamma_{ggg} = \frac{160}{81}\alpha_s^3(\pi^2-9)\frac{|\Psi(0)|^2}{M^2}, \tag{6}$$

which, using current values of α_s and the value of $|\Psi(0)|^2/M^2$ derived from Γ_{ee}, gives $\Gamma_{ggg} \approx 30$ keV in agreement with observation (see Section 4.4).

A direct proof that vector mesons decay into three gluons has been obtained for the first time by observing the three-jet structure in Υ decay. The three-jet structure has been observed at DORIS and CESR. While many global parameters may be employed to characterize the structure of a many-body final state, the one used in various forms by all groups and most amenable to calculations in QCD is the quantity called thrust, defined as

$$T = \mathrm{Max}\left(\sum_i \frac{|\mathbf{p}_i \cdot \mathbf{n}|}{|\mathbf{p}_i|}\right),$$

where $\mathbf{p}_i$ are the particle momenta (energy clusters in a calorimeter) and the unit vector $\mathbf{n}$ is rotated until a maximum is found. For a collinear two-jet final state, $\mathbf{n}$ is the jet axis; for a three-jet event, $\mathbf{n}$ is close to the axis of the most energetic jet. The thrust distribution $\mathrm{d}N/\mathrm{d}T$ can be calculated for two-quark and three-gluon (55) final states.

The thrust axis angular distribution, for three gluons from vector mesons also depends on the gluon spin. The angular distribution for the polar angle of the unit vector $\mathbf{n}$ is given by $1+\rho \cos\theta$, where $\rho = 1.0$ for two-quark jets (hadronic events from continuum e^+e^- annihilations), $\rho = 0.39$ for Υ decays into three spin-1 gluons, and $\rho = -1.0$ for decays into three spin-0 gluons (56). The thrust distributions obtained by PLUTO (14), LENA (15), CLEO (16), and CUSB (10) change significantly between continuum ($q\bar{q}$) and Υ (mostly three gluons), and the data agree well with predictions for three-gluon decays. The parameter ρ has been measured at DORIS (41) and by CLEO (16), who reports the most accurate value of 0.32 ± 0.11. Averaging with the other results gives $\rho = 0.32 \pm 0.09$, in good agreement with the value for decay to three spin-1 gluons.

4.2 *Decays of the Υ' and the Υ''*

Additional decay channels are available to the first two excited Υ's, namely transitions to lower-lying $b\bar{b}$ states. Because of the smallness of the annihilation width, processes with very low Q-value and/or due to electromagnetic interactions can compete favorably with annihilation into three gluons. This is well known for the ψ', whose total width is about three times larger than that of the J/ψ. Half of this width is due to the transition $\psi' \to \psi\pi\pi$ (57) and about 26% to electric dipole (E1) transitions to P-wave $c\bar{c}$ states (58). In the Υ family, these transitions are suppressed with respect to the charm case. For E1 transitions, one expects a suppression of a factor 4 because of the b-quark charge and of a factor ~ 3 from the smaller size of the $b\bar{b}$ system in the approximation that the E1 matrix element $|\langle i|r|f\rangle|^2$ scales like $\langle r^2\rangle$. A similar argument had been advanced by Gottfried (59) for the pion transitions between upsilons, which can be understood in a multipole expansion of the color-electric field. In particular, for spin-1 gluons the lowest contribution to the emission of two pions is a double "E1" transition, thus being suppressed for the upsilon case by a factor of $\langle r^2\rangle^2$ or ~ 10.

4.2.1 TWO-PION TRANSITIONS BETWEEN Υ'S In the upsilon family there are three two-pion transitions between ^3S states: (*a*) $\Upsilon' \to \Upsilon\pi\pi$, (*b*) $\Upsilon'' \to \Upsilon\pi\pi$, and (*c*) $\Upsilon'' \to \Upsilon'\pi\pi$. Reaction (*a*) is the easiest to observe and was detected in 1980 both at DORIS and at CESR. In calorimetric detectors these reactions can only be observed in the decay cascade chain $\Upsilon^n \to \Upsilon^{n-1}\pi\pi$, $\Upsilon^{n-1} \to e^+e^-$ or $\mu^+\mu^-$. For the case of reaction (*a*), the LENA group (19) found 5 $\pi^+\pi^-e^+e^-$ and 2 $\pi^+\pi^-\mu^+\mu^-$ events; the CUSB group (18) found 23 $\pi^+\pi^-e^+e^-$ events in the decay of about 10,000 Υ', and CLEO (18) found 17 $\pi^+\pi^-e^+e^-$ events. These three results combined give $B_{\pi^+\pi^-\ell\bar{\ell}} = 0.0066 \pm 0.0010$. CLEO (18) also detects the two-pion decay of the Υ' to Υ by computing the mass recoiling against any pair of opposite-sign pions. They observe a clear peak at the Υ mass (Figure 5) above a large combinatorial background and obtain $B(\Upsilon' \to \Upsilon\pi^+\pi^-) = 0.191 \pm 0.031$. Combining all results and assuming that $\Gamma(\Upsilon' \to \Upsilon\pi^+\pi^-) = 2\Gamma(\Upsilon' \to \Upsilon\pi^0\pi^0)$, one finally obtains $B_{\Upsilon\pi\pi}(\Upsilon') = 0.29 \pm 0.04$.

The two-pion decays of the Υ'' have much smaller widths; the reaction (*b*) is suppressed because $\Delta n = 2$, whereas the decay to the Υ' is suppressed by phase space because the Q value for the transition is only 47 MeV. Results for the decay to the ground state have been obtained by both CLEO (26) and CUSB (27). A few decays to the Υ' have been observed by CUSB. These results were obtained from the analysis of a sample of about 50,000 Υ''. CLEO (26) also observes the decay to the ground state in the recoiling mass spectrum. In addition, CUSB (27) observed emission of two π^0's by detecting four photons and the rates were consistent with being one half of those for emission of $\pi^+\pi^-$ as expected from isospin invariance. The results reported by the two groups are $B(\Upsilon'' \to \pi^+\pi^-\Upsilon) = 0.049 \pm 0.009$ (CLEO), $B(\Upsilon'' \to \pi^+\pi^-\Upsilon) = 0.039 \pm 0.012$ (CUSB), and $B(\Upsilon'' \to \pi^+\pi^-\Upsilon') = 0.031$

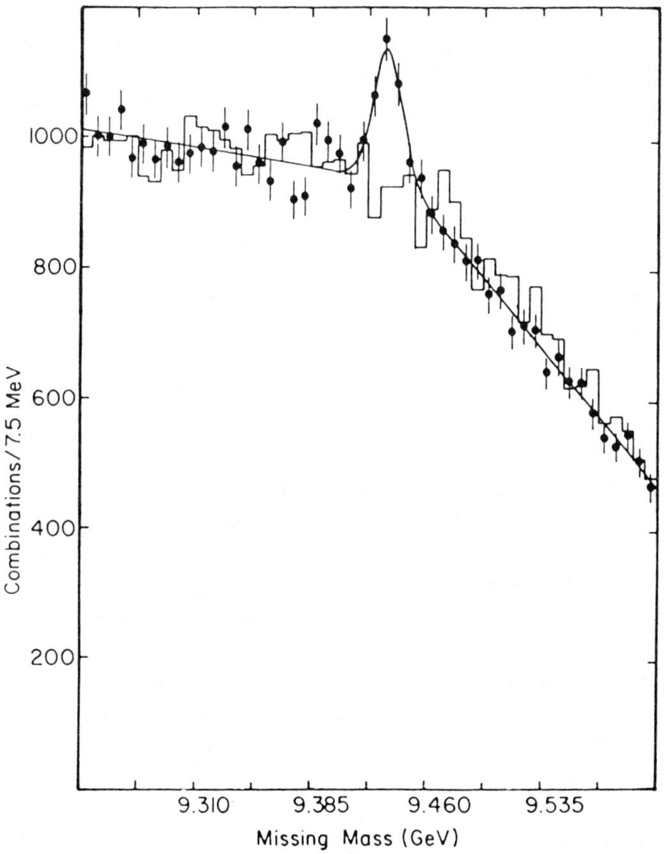

Figure 5 Spectrum of the mass recoiling against any $\pi^+\pi^-$ pair in Υ' decays (*data points*) and against same-sign π pairs (*solid line histogram*) (18).

± 0.020 (CUSB). Again, combining all the results and using isospin invariance, we obtain $B_{\Upsilon\pi\pi}(\Upsilon'') = 0.069 \pm 0.011$ and $B_{\Upsilon'\pi\pi}(\Upsilon'') = 0.046 \pm 0.030$. CLEO has also searched in the recoiling mass spectrum for evidence of two-pion transitions to singlet P-wave $b\bar{b}$ states. They do not observe any signal and establish an upper limit of 3% for this decay (60).

4.2.2 PHOTON TRANSITIONS TO P-WAVE $b\bar{b}$ STATES The study of these transitions in the $b\bar{b}$ system is still in its infancy, partly owing to the low production rates of upsilon resonances thus far. While two million ψ' have been studied in one single experiment at SPEAR (58), the largest equivalent upsilon sample, up to 1982, consisted of about 37,000 Υ''.

Until 1981 the prospects for the search for a signal from E1 transitions were rather dim. Predictions for the branching ratios were in the 5 to 20% range, while the total accumulated samples of Υ' and Υ'' were only 10,000 and 7,000 respectively. A study of the Υ' and Υ'' showed an excess of two-jet event topologies in their decays (10). An explanation of this fact is that a fraction of the Υ's decay via E1 transitions to ^3P_J. These states have $J^{PC} = 2^{++}$, 1^{++}, and 0^{++} and therefore are expected to decay mostly by annihilation of the $b\bar{b}$ pair into two gluons, which leads to a two-jet final state. The experimental observation leads to an estimate of the branching ratio for E1 transition of the order of 10% for the Υ' and 30% for the Υ''. That this branching ratio should be larger for Υ'' is reasonable because most other decay channels are suppressed while the E1 rate is enhanced by the larger value of $\langle r^2 \rangle$ of the Υ''.

Encouraged by these expectations, CESR ran in 1981–1982 with increased luminosity for three months at the Υ'' peak. Some 65,000 hadronic events were observed by CUSB for a total time-integrated luminosity of 14 pb^{-1}, of which about 37,000 were Υ'' decays and the remainder were continuum events. The inclusive photon spectrum from these events shows a clear enhancement centered around 100 MeV (11). This is shown in Figure 6*a* and, after background subtraction, the net excess signal is shown in Figure 6*b*. This signal is attributed to the decay $\Upsilon'' \rightarrow 2^3P_J + \gamma$, where 2 stands for the second P-wave state. The ^{3}P states are expected to show fine structure, with splittings predicted to be in the 15–40-MeV range. The fine structure is not resolved in the spectrum of Figure 6*b*; however, the observed enhancement is significantly broader than the resolution for photons around 100 MeV, which in CUSB is $\sim$17 MeV full width at half maximum (FWHM). The observed spectrum is therefore fitted to three lines with shapes given by the resolution function and with free position and intensity. This fit is shown in Figure 6*b*.

An alternative way to search for E1 transition to P-wave states is to observe the decay chains:

1. $\Upsilon'' \rightarrow 2^3P + \gamma_1 \rightarrow \Upsilon' + \gamma_2 + \gamma_1 \rightarrow \gamma_1 + \gamma_2 + \mu^+\mu^-$ or e^+e^-
2. $\Upsilon'' \rightarrow 2^3P + \gamma_1 \rightarrow \Upsilon + \gamma_2 + \gamma_1 \rightarrow \gamma_1 + \gamma_2 + \mu^+\mu^-$ or e^+e^-
3. $\Upsilon'' \rightarrow 1^3P + \gamma_1 \rightarrow \Upsilon + \gamma_2 + \gamma_1 \rightarrow \gamma_1 + \gamma_2 + \mu^+\mu^-$ or e^+e^-.

The number of events that one can find in this way is very small since it is proportional to the product of three small branching ratios and the fractional solid angle enters at least cubed in the acceptance. However, the background is almost nonexistent and the events can be fitted to satisfy kinematical constraint; this results in better accuracy for the photon energies.

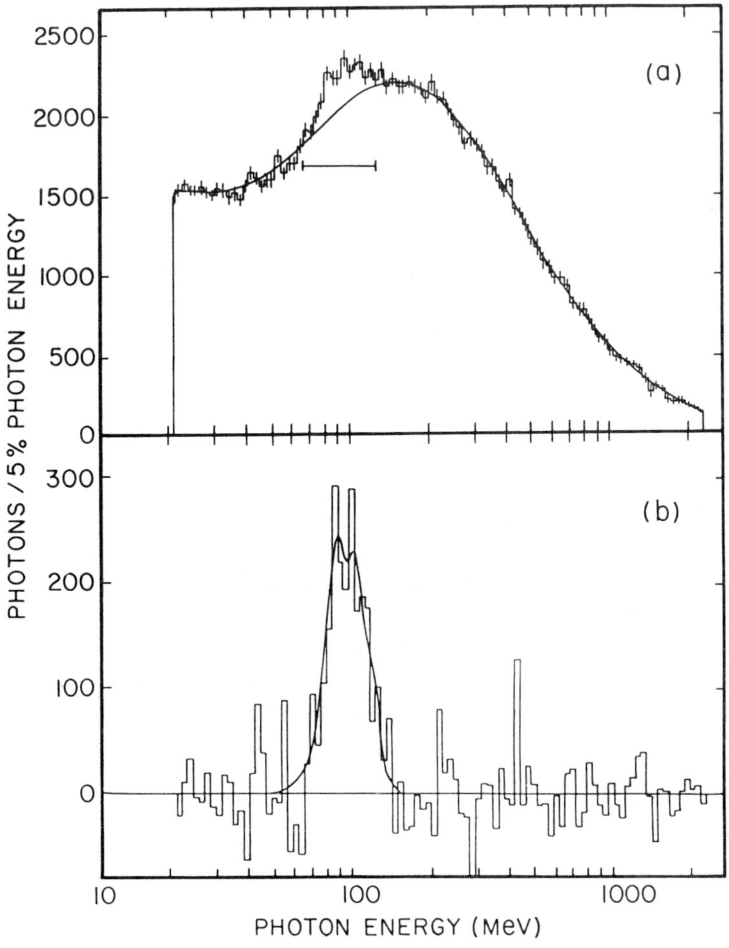

Figure 6 (*a*) the observed inclusive photon spectrum from Υ'' decays, the solid line is a fit to the background; (*b*) the background-subtracted photon signal in Υ'' decays.

Fourteen examples of Reaction 1 and 15 examples of Reaction 2 were observed (12) during the same run described above, in the CUSB detector, with an estimated background of 1.7 and 1.3 events, respectively. One candidate is observed for Reaction 3. The lower energy photons for both Reactions 1 and 2 cluster around 100 MeV, which thus establishes that the 2^3P level lies about 100 MeV below the Υ'', and that the signal observed in the inclusive photon spectrum from Υ'' decays is due to the direct transition to the 2^3P state. The energies of the photons, as extracted from fitting both the inclusive and exclusive spectra, are 84.4 ± 2.0, 99.5 ± 3.2, 117.2 ± 5.0 MeV, and 84.0 ± 3.0, 99.0 ± 2.0, 119.0 ± 5.0 MeV, respectively. The relative intensity of each of the three lines in the inclusive spectrum is consistent with being proportional to $k^3(2j+1)$, as expected for E1 transitions, where k is the photon energy. The overall picture presented by these results is quite a convincing proof that the first excited $b\bar{b}$ P-wave states have indeed been observed.

It is usual to characterize triplet P-wave states by their center of gravity ("cog") and their fine structure. Using the fitted photon energies above, one obtains cog$(2^3\text{P}) \cong M(\Upsilon'') - 93$ MeV $= 10256$ MeV including a correction of ~ 0.5 MeV for recoil. The fine structure splittings are 15 and 18 MeV, with errors of the order of 3 to 7 MeV accounting for systematic uncertainties in addition to the statistical errors. The ratio of these splittings,

$$r = \frac{M(J=2) - M(J=1)}{M(J=1) - M(J=0)},$$

is often used to compare calculations and experiments. The data give $r(2\text{P}) = 0.83 \pm 0.3$. Finally, from the observed enhancement in the inclusive photon spectrum, the CUSB group obtains a branching ratio for $3^3\text{S} \to 2^3\text{P}$ E1 transitions of 0.34 ± 0.03, with an estimated additional systematic uncertainty of 0.03. From the observed numbers of two-photon events, CUSB also obtains the following sums of products of branching ratios for decays of the Υ'' to the Υ' and Υ, via the 2^3P_J states (12):

$$\sum_i B(\Upsilon'' \to \gamma\chi'_{bi})B(\chi'_{bi} \to \gamma\Upsilon') = 0.059 \pm 0.021,$$

$$\sum_i B(\Upsilon'' \to \gamma\chi'_{bi})B(\chi'_{bi} \to \gamma\Upsilon) = 0.036 \pm 0.012;$$

for transitions via the 1^3P_J to the Υ an upper limit is given (12):

$$\sum_i B(\Upsilon'' \to \gamma\chi_{bi})B(\chi_{bi} \to \gamma\Upsilon) < 0.03 \text{ (90\% C.L.)}.$$

Recently, transitions from the Υ' to the first P-wave $b\bar{b}$ level have also been observed by CUSB at CESR, both in the inclusive photon spectrum and in exclusive channels. The cog of the 1^3P_J has a mass of ~ 9901 MeV, the fine structure splittings are 20 and 21 MeV corresponding to a value of $r(1P) \approx 0.9$, and the branching ratio for the decay $\Upsilon' \rightarrow 1^3P_J + \gamma$ is $\sim 0.15 \pm 0.05$ (61).

4.3 *Partial Rates for Υ's Decays*

As described in Section 3.1, a knowledge of $B_{\mu\mu}$ is necessary to obtain Γ_{ee}. The value of $B_{\mu\mu}$ is known at present with reasonable accuracy only for the Υ. For the other two narrow Υ's it is possible, however, to obtain Γ_{tot} indirectly, and therefore $B_{\mu\mu}$, by using the assumption that $\Gamma_{annihilation}/\Gamma_{ee}$ is the same for all three states, together with measurements of the branching ratios for decays without annihilation of the $b\bar{b}$ pair. In addition to the annihilation of the $b\bar{b}$ pair into three gluons, one must also include annihilation into a virtual photon, leading to lepton pairs and quark pairs. For three lepton flavors and by definition of R, the partial width for decays via a virtual photon is given by $(3+R)\Gamma_{ee}$. We can therefore write $\Gamma_{tot} = \Gamma_{ggg} + (3+R)\Gamma_{ee} + \Gamma_{other}$, where Γ_{other} is the decay width for all channels not requiring annihilation (Γ_{ggg} as defined here includes a $\sim 4\%$ (62) contribution from $\Upsilon \rightarrow \gamma gg$). We thus obtain

$$\Gamma^n_{tot} = \frac{\Gamma^n_{ee}}{B_{\mu\mu}(\Upsilon)(1 - B^n_{other})},$$

where n refers to the nth Υ, and B_{other} is the branching ratio for all decays without annihilation. Using the values given in Sections 4.2.1 and 4.2.2, we derive $B_{other}(\Upsilon') = 0.49 \pm 0.06$ and $B_{other}(\Upsilon'') = 0.46 \pm 0.04$. [Searches for other decay modes of the Υ' and Υ'' have given limits of $< 3\%$ (41, 60) for their branching ratios.] The results in Table 4 are derived using the outlined procedure. The directly measured quantities are given first, followed by the derived ones.

4.4 *Comparison with Theory*

4.4.1 HADRONIC TRANSITIONS The two-pion transitions between the Υ's not only confirm the assumptions that these states are composed of the same new quark but also offer insight into the mechanism for such decays, especially in comparison with the ψ' case. The partial width for the decay $\psi' \rightarrow \psi\pi\pi$ is ~ 107 keV (57), while in the Υ' case the width for the $\pi\pi$ transition is only ~ 10 keV, although the Q value for the two decays is very similar. That this should be the case had already been surmised from the fact that production of the Υ' is quite obvious in proton collisions on nuclei, while ψ' production is hardly observed in similar experiments; this implies that the

Table 4 Measured and derived rates and branching ratios for the bound ϒ's

Parameter	ϒ	ϒ′	ϒ″
Inputs			
Γ_{ee} (keV)	1.22 ± 0.03	0.52 ± 0.02	0.38 ± 0.02
$B_{\mu\mu}$ (%)	2.83 ± 0.28		
$B_{\pi\pi}$		0.29 ± 0.04	0.115 ± 0.033
B_{E1}		0.15 ± 0.05	0.34 ± 0.03
Derived			
Γ_{tot} (keV)	43.1 ± 4.3	32.8 ± 4.6	24.6 ± 3.4
Γ_{ggg} (keV)	35.0 ± 4.3	14.9 ± 1.8	10.9 ± 1.3
$\Gamma_{\pi\pi}$ (keV)		9.5 ± 1.4	2.8 ± 0.9
Γ_{E1} (keV)		4.9 ± 1.8	8.4 ± 1.4
$B_{\mu\mu}$ (%)		1.58 ± 0.28	1.54 ± 0.22

$\pi\pi$ feed-down mechanism should be suppressed for the ϒ's (59). Following an original suggestion of Gottfried (59), Yan (63), by a multipole expansion in the "color electric" field, has computed that the rate for $\Upsilon' \rightarrow \Upsilon\pi\pi$ should be $\sim\frac{1}{16}$ the rate for $\psi' \rightarrow \psi\pi\pi$, in good agreement with observation. For spinless gluons the $\pi\pi$ decay width should be proportional to the color charge and therefore identical for ϒ′ and ψ′. The strong suppression of the $\pi\pi$ transition from ϒ′ to ϒ is a particularly beautiful proof that the gluons carry spin 1. Kuang & Yan (64) have also estimated the widths for the $\pi\pi$ transitions of the ϒ″, which, within the present limited statistical accuracy, are consistent with observation. Finally, predictions have been made for the $\pi\pi$ mass spectrum for these transitions. Both CLEO (26) and CUSB (27) results for the ϒ′ are in agreement with the simplest expected shape described by Brown & Cahn (65) and by Yan (63). In the decay $\Upsilon'' \rightarrow \Upsilon\pi\pi$ the enhancement of high mass $\pi\pi$ pairs observed for the ϒ′ does not appear to be present, the spectrum being essentially consistent with phase space.

4.4.2 P-WAVE STATES AND PHOTON TRANSITIONS The discovery of the P-wave $b\bar{b}$ states brings a wealth of new information to be compared with predictions of potential models as well as with results of QCD sum rules and bag model calculations. We can compare three quantities: the center of gravity (cog) of the level masses, the E1 transition rates and the fine structure splitting. The cog's are perhaps easier to predict, because of our good knowledge of the binding potential. The transition rates and splittings are sensitive to relativistic effects [they have not yet been correctly predicted for the ψ′ case (58)] and there is no satisfactory theory of the effective spin dependence of the forces in quarkonium. While all these problems are expected to be less severe for the $b\bar{b}$ case, the predictions for the ratio r for the χ_b' range from 0.48 to 1 and the splittings themselves range from 14 to 24 MeV for the two upper levels and from 17 to 40 MeV for the two lower ones. The spread of the predictions is even larger for the χ_b case. Since r is poorly determined at present, we compare only measurements and calculations for masses and transition rates. Similarly, the products of branching ratios given in Section 4.2.2 are consistent with expectations (12, 64) although the experimental accuracy is very limited. Results of many authors are shown in Table 5 for the ^{3}P level masses and transition rates. Since the rate for E1 transitions is proportional to k^3, where k is the photon energy, we have rescaled all E1 widths using the experimental value $k_{cog}(1P) = 119$ MeV and $k_{cog}(2P) = 93$ MeV.

The agreement between measurements is in fact impressive for potential models, poor for the bag model calculations (70), and extremely poor for the QCD sum rules predictions of the lowest ^{3}P state mass (71, 72).

4.4.3 ^{3}P STATES' CENTER OF GRAVITY AND FLAVOR INDEPENDENCE Quigg & Rosner (49) have pointed out that ratios of level spacings can be directly related to the power v of a potential of the form λr^v. Thus from the ratio $[M(3S)-M(2S)]/[M(2S)-M(1S)]$ they derive $v(\psi) = 0.2$ and $v(\Upsilon) \approx 0.33$. Using the 2S, 1P, and 1S levels gives $v(\psi) \approx 0.15$. From the new information available about the $b\bar{b}$ P-wave states, one obtains $v(\Upsilon) \approx -0.22$ using the 2S, 1P, and 1 S levels (61), and $v(\Upsilon) \approx -0.15$ using the 3S, 2P, and 2S levels (20). The fact that all these values are close to zero confirms the flavor independence of the interquark forces and suggests that the potential becomes more Coulomb-like at small distances.

Table 5 Comparison of predictions and measurements for ^{3}P levels

Author	1^3P cog (MeV)	2^3P cog (MeV)	$\Gamma(2^3S \to 1^3P)^a$ (keV)	$\Gamma(3^3S \to 2^3P)^a$ (keV)
Büchmuller, Grunberg & Tye (46)	9890	10250	4.2	6.1
Büchmuller (66)			4.1	6.8
Eichten et al (43, 64)	9924	10271	4.4	6.1
Martin (48)	9861	10242		
Quigg & Rosner (50)	9888	10244	4.3	7.2
Krasemann (44, 67)	9936	10271	3.1	4.8
Gupta et al (68)	9898	10256		
McClary & Byers (69)	9923	10267	4.4	7.6
Baacke et al (70)	9971	10312	5.2	5.2
Voloshin et al (71)	9835 ± 30			
Bertlmann (72)	9803 ± 10			
Experiment	9901 ± 5	10256 ± 5	4.9 ± 1.8	8.4 ± 1.4

[a] Adjusted using experimental cog's (see text).

4.5 *Search for Rare Decays of the* Υ

Current models of the electroweak interactions require the existence of at least one pseudoscalar Higgs particle (73). The Higgs particle couples to quarks proportionally to the square of the quark mass, which results (74) in the prediction $BR(\Upsilon \to \gamma + \mathrm{H}) = 2.6 \times 10^{-4}(1 - M_{\mathrm{H}}^2/M_{\Upsilon}^2)$. Models with two Higgs fields modify this result by a factor $(v_1/v_2)^2$ for charge-$\frac{1}{3}$ quarks (Υ) while the corresponding ratio for the ψ is multiplied by $(v_2/v_1)^2$, where v_1 and v_2 are the vacuum expectation values of the two Higgs fields (75).

The second model applies to the Weinberg (76) and Wilczek (77) axion, a, which is a light ($M < 1$ MeV), long-lived, and semiweakly interacting pseudoscalar particle, distinguished by its "invisibility." Since v_1/v_2 is not known, one cannot predict the branching ratios for either Υ or $\psi \to \gamma + \mathrm{a}$. However, the product of the two branching ratios is predicted to be 1.6×10^{-8}. The CUSB group found no example of this decay in an equivalent sample of ~60,000 Υ decays, giving an upper limit of 1.2×10^{-4} for the branching ratio. Combining this result with the limit of Edwards et al (78) for $\psi \to \gamma + \mathrm{a}$ of 1.4×10^{-5}, CUSB quote a limit of 0.6×10^{-9} for the product of the two branching ratios at 90% C.L., an unambiguous result arguing against the existence of the axion (21).

In the case of heavier axions the CUSB group has searched for the reaction $\Upsilon \to \mathrm{H} + \gamma$, where the Higgs particle decays into hadrons. While these searches are not at present sensitive to the expected rate for the single Higgs model, they can exclude a large region in the $v_1/v_2, M_{\mathrm{H}}$ plane, the upper boundary of which is defined by values of $(v_1/v_2, M_{\mathrm{H}})$ such as: (2,4), (3,6), (4,7), and (6,8), with M_{H} in GeV (J. Lee-Franzini, personal communication).

5. DECAY OF THE Υ‴

If the mass of the ^{3}S $\mathrm{b\bar{b}}$ state is larger than twice the mass of the lightest bound state of a b quark with a light antiquark such as $\bar{\mathrm{u}}$ or $\bar{\mathrm{d}}$ (B mesons), the Υ‴ can decay into $\mathrm{B\bar{B}}$ pairs with a rate more typical of conventional strong interactions, possibly limited by the available phase space. That the Υ‴ has a natural width of the order of 20 MeV [CLEO gives $\Gamma(\Upsilon''') = 32 \pm 7$ MeV (35) and CUSB reports 20 ± 3 MeV (34); see Section 3.2 for comments] suggests that the free b-flavor threshold has in fact been reached. Support for this assumption was obtained at CESR, from the observation of a sharp increase at the peak of the Υ‴ in the yield of leptons of energies up to ~2.5 GeV as expected from the weak decays of a B meson of mass ~5 GeV (22, 23).

5.1 *b-Flavor Threshold and $B^* \to B+\gamma$*

Of general and practical interest is the knowledge of the B meson mass. An obvious bound is $M(\Upsilon'')/2 < M(\mathrm{B}) < M(\Upsilon''')/2$ since the first state is below threshold and the Υ''' is above. However, if the mass of the B is close to the lower bound, the Υ''' would copiously decay into B*'s (24). From scaling arguments it follows that the B* is about 50 MeV heavier than the B meson, in which case its dominant decay is $\mathrm{B}^* \to \mathrm{B}+\gamma$ (43). Decays of the Υ''' into B* should be detectable by the presence of monochromatic photons around 50 MeV, in the decays of the Υ'''.

A search for these photons has been carried out by the CUSB group at CESR with negative results (24). Their negative result can be expressed as an upper limit for the branching ratio for $\Upsilon''' \to \mathrm{BB}^*$, which they give as $BR < 0.09$, at 90% C.L., for a B*-B mass difference between 45 and 65 MeV. From this limit the same authors (24) establish bounds on the mass of the B meson: $5263 < M_{\mathrm{B}} < 5278$ MeV.

6. OUTLOOK FOR THE STUDY OF SPIN-ORBIT AND SPIN-SPIN INTERACTIONS

6.1 *Fine Structure of the Triplet P-Wave States*

The fine structure of the $2^3\mathrm{P}_J$ states is not at present resolved because their splittings of ~ 15 MeV are very close to the CUSB resolution at ~ 100 MeV. CUSB is in the process of adding an array of bismuth germanate crystals that will improve resolution by a factor of two. The Crystal Ball detector, at DORIS since mid 1982, has approximately 1.5 times better resolution than CUSB. CESR and DORIS have plans for increasing their luminosities, which at present are 550 and 250 nb^{-1} per day, respectively. Good measurement of the fine structure splittings and angular correlations will be performed within the coming year and will allow us to choose among the various ansatz proposed to describe the effective spin-orbit interaction in the $\mathrm{b\bar{b}}$ system.

6.2 *Search for Singlet $b\bar{b}$ States*

The search for the singlet states, $^1\mathrm{P}$ and $^1\mathrm{S}$ or η_{b}, is a much harder proposition. We recall that the lowest $^1\mathrm{P}$ state has not been found in charmonium (79) and that the search for the η_{c}'s had a long and complicated history. The positions of these states are, however, crucial to the understanding of the spin-spin interaction, which is expected to be a short-range effect arising from hard gluon exchange. The singlet-triplet hyperfine splittings for P-wave states are expected to be extremely small

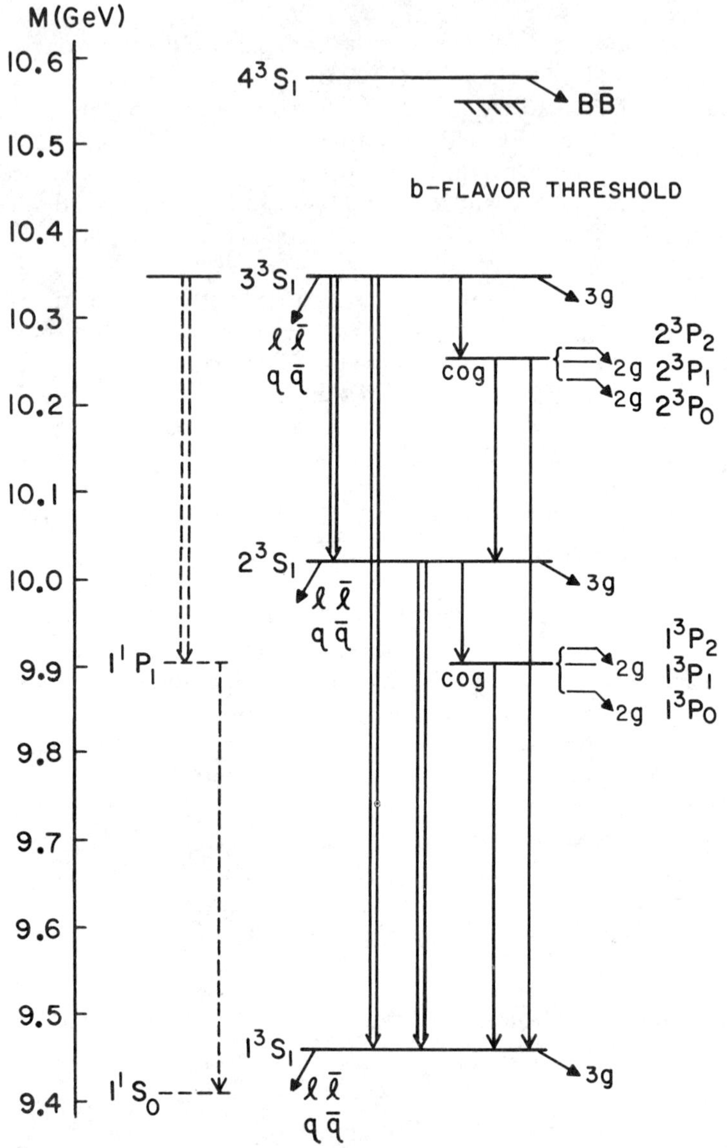

Figure 7 A level diagram of the $b\bar{b}$ bound system; double lines indicate $\pi\pi$ transitions and single lines indicate γ transitions. Solid lines indicate observed levels and transitions (see text).

since P waves vanish at the origin. The hyperfine splittings for S-wave states are also predicted to be small, resulting in branching ratios for the M1 transition $\Upsilon \rightarrow \eta_b + \gamma$ of $\sim 5 \times 10^{-4}$ (43, 67, 80). However, in the Υ system there appears to be an alternate path to reach both the ^{1}P and the η_b that is experimentally promising, as shown on the left in Figure 7. This is the decay chain $\Upsilon'' \rightarrow \pi\pi 1^1$P, 1^1P $\rightarrow \eta_b + \gamma$. Kuang & Yan (64) estimate the first process to have a branching ratio of $\sim 2\%$, and potential models predict for the second transition, an E1 transition with a photon energy of ~ 470 MeV, branching fractions of 0.35 to 0.5. A modus operandi would therefore be to locate the 1^1P state via the two-pion transition, a task mostly for magnetic spectrometers, and to search for a monochromatic photon from the 1^1P $\rightarrow \gamma + \eta_b$, a task for electromagnetic calorimeters. Very good energy resolution and high statistics are required because of the many nearby lines due to Υ's and χ_b's E1 decays. It is estimated that samples of 10^6 Υ'' and Υ' each would yield a definite signal for the η_b in the upgraded CUSB detector.

7 CONCLUSIONS

The present experimental status of upsilon spectroscopy is best summarized by the level diagram of Figure 7, where the observed levels and transitions are illustrated (*solid lines*). We have also reviewed the phenomenal success with which potential models are able to describe most features of the spectrum. We have indicated directions for future forays lying outside the central potential phenomenology. The study of the Υ's will continue to provide us with valuable information about properties of quarks, gluons, and perhaps Higgs particles.

Literature Cited

1. Herb, S. W., et al. 1977. *Phys. Rev. Lett.* 39:252
2. Innes, W. R., et al. 1977. *Phys. Rev. Lett.* 39:1240
3. Berger, Ch., et al. 1978. *Phys. Lett.* 76B:243
4. Dardeen, C. W., et al. 1978. *Phys. Lett.* 76B:246
5. Bienlein, J. K., et al. 1978. *Phys. Lett.* 78B:360
6. Andrews, D., et al. 1980. *Phys. Rev. Lett.* 44:1108
7. Böhringer, T., et al. 1980. *Phys. Rev. Lett.* 44:1111
8. Finocchiaro, G., et al. 1980. *Phys. Rev. Lett.* 45:222
9. Andrews, D., et al. 1980. *Phys. Rev. Lett.* 45:219
10. Peterson, D., et al. 1982. *Phys. Lett.* 114B:277
11. Han, K., et al. 1982. *Phys. Rev. Lett.* 49:1612
12. Eigen, G., et al. 1982. *Phys. Rev. Lett.* 49:1616
13. Rice, E., et al. 1982. *Phys. Rev. Lett.* 48:906
14. Berger, Ch., et al. 1979. *Phys. Lett.* 82B:449; also 1981. *Z. Phys.* C8:101
15. Niczyporuk, B., et al. 1981. *Z. Phys.* C9:1
16. Cabenda, R. C. 1982. PhD thesis, Cornell University (unpublished)
17. Mageras, G., et al. 1981. *Phys. Rev. Lett.* 46:1115
18. Mueller, J., et al. 1981. *Phys. Rev. Lett.* 46:1181
19. Niczyporuk, B., et al. 1981. *Phys. Lett.* 100B:95
20. Lee-Franzini, J. 1983. In *Proc. Summer Institute on Particle Physics*, ed. A.

Mosher, Stanford Univ. Press, Calif. In press
21. Sivertz, M., et al. 1982. *Phys. Rev.* D26:717
22. Bebeck, C., et al. 1981. *Phys. Rev. Lett.* 46:84; also Chadwick, K., et al. 1981. *Phys. Rev. Lett.* 46:88
23. Spencer, L. J., et al. 1981. *Phys. Rev. Lett.* 47:771
24. Schamberger, R. D., et al. 1982. *Phys. Rev.* D26:720
25. Lederman, L. M. 1979. In *Proc. 19th Int. Conf. on High Energy Physics, Tokyo, Japan*, ed. S. Homma, M. Kawaguchi, H. Miyazawa, p. 706. Tokyo: Phys. Soc. Jpn.
25a. Chinowsky, W. 1977. *Ann. Rev. Nucl. Sci.* 27:393
26. Green, J., et al. 1982. *Phys. Rev. Lett.* 49:617
27. Mageras, G., et al. 1982. *Phys. Lett.* 118B:453
28. Barnett, R., Dine, M., McLerran, L. 1980. *Phys. Rev.* D22:594
29. Blatt, J. M., Weisskopf, V. F. 1952. *Theoretical Nuclear Physics*, p. 423. New York: Wiley
30. Franzini, P., Lee-Franzini, J. 1982. *Phys. Rep.* 81:239–291; also Andrews, D., et al. 1982. *Cornell Rep. CLNS 82/538* (unpublished)
31. Berger, Ch., et al. 1979. *Phys. Lett.* 81B:410
32. Weseler, S. 1981. Diplomarbeit, University of Heidelberg (unpublished, in German).
33. Niczyporuk, S., et al. 1983. Z. Phys. C15: In press
34. Rice, E. 1983. PhD thesis, Columbia University (unpublished)
35. Plunket, R. 1983. PhD thesis, Cornell University (unpublished)
36. Artamonov, A. S., et al. 1982. *J. Phys.* 43:C3–789
37. Albrecht, H., et al. 1980. *Phys. Lett.* 93B:500
38. Niczyporuk, B., et al. 1981. *Phys. Lett.* 99B:169
39. Niczyporuk, B., et al. 1981. *Phys. Rev. Lett.* 46:92
40. Andrews, D., et al. 1983. Submitted to *Phys. Rev. Lett.*
41. Bienlein, J. K. 1981. In *Proc. 1981 Int. Symp. on Lepton and Photon Interactions at High Energies*, ed. W. Pfeil, p. 190. Universität Bonn, West Germany
42. Van Royen, R., Weisskopf, V. F. 1967. *Nuovo Cimento* 50A:617
43. Eichten, E., et al. 1980. *Phys. Rev.* D21:203
44. Krasemann, K. H., Ono, S. 1979. *Nucl. Phys.* B154:283
45. Bhanot, G., Rudaz, S. 1978. *Phys. Lett.* 78B:119
46. Büchmuller, W., Grunberg, G., Tye, S.-H. H. 1980. *Phys. Rev. Lett.* 45:103, 587 (E); also Büchmuller, W., Tye, S.-H. H. 1981. *Phys. Rev.* D24:132
47. Richardson, J. L. 1979. *Phys. Lett.* 82B:272
48. Martin, A. 1980. *Phys. Lett.* 93B:338; also 1981. *Phys. Lett.* 100B:511
49. Quigg, C., Rosner, J. L. 1979. *Phys. Rep.* 56:167
50. Quigg, C., Rosner, J. L. 1981. *Phys. Rev.* D23:2625; also Quigg, C., Thacker, H. B., Rosner, J. L. 1980. *Phys. Rev.* D21:234
51. Voloshin, M., Zakharov, V. I. 1980. *Phys. Rev. Lett.* 45:688
52. Eichten, E. 1980. *Phys. Rev.* D22:1819
53. Büchmuller, W., Tye, S.-H. H. 1980. *Phys. Rev. Lett.* 44:850
54. Appelquist, T., Politzer, H. D. 1975. *Phys. Rev. Lett.* 34:43
55. Koller, K., Walsh, T. F. 1978. *Nucl. Phys.* B140:449
56. Koller, K., Krasemann, K. H. 1979. *Phys. Lett.* 88B:119
57. Particle Data Group. 1982. *Phys. Lett.* 111B
58. Gaiser, J. E. 1982. In *Proc. Moriond Workshop on New Flavor*, ed. J. Tran Thanh Van, L. Montanet, p. 11. Gif-sur-Ivette: Edition Frontiere
59. Gottfried, K. 1977. In *Proc. Int. Symp. on Lepton and Photon Interactions at High Energies, Hamburg, Germany*, ed. F. Gutbrod, p. 667. Hamburg: DESY
60. Gilchriese, M. G. D. 1982. In *Proc. 2nd Int. Conf. on Physics in Collision: High Energy ee/ep/pp Interactions, Stockholm.* New York: Plenum
61. Lee-Franzini, J. 1983. In *Proc. 7th Int. Conf. on Experimental Meson Spectroscopy*, ed. S. U. Chung, S. J. Lindembaum. To be published
62. Chanowitz, M. 1975. *Phys. Rev.* D12:918
63. Yan, T.-M. 1980. *Phys. Rev.* D22:1652
64. Kuang, Y.-P., Yan, T.-M. 1981. *Phys. Rev.* D24:2874
65. Brown, L. S., Cahn, R. N. 1975. *Phys. Rev. Lett.* 35:1
66. Büchmuller, W. 1982. See Ref. 58, p. 91
67. Krasemann, K. H. 1981. *CERN Rep. TH.3036*
68. Gupta, S. N., Redford, S. F., Repko, W. W. 1982. *Phys. Rev.* D26:3305
69. McClary, R., Byers, N. 1982. *UCLA Rep. UCLA/82/TEP/12*; also McClary, R. 1982. PhD thesis. Univ. Calif., Los Angeles (unpublished)
70. Baacke, J., Igarashi, Y., Kasperidus, G. 1981. *Dortmund Rep. DO-TH81/10* (unpublished)

71. Voloshin, M., et al. 1980. *ITEP Rep. ITEP-21*; also Voloshin, M., Zakha, V. 1980. *DESY Rep. F15-80/03* (unpublished)
72. Bertlmann, R. A. 1981. *CERN Rep. TH-3192* (unpublished)
73. Ellis, J., Gaillard, M. K., Girardi, G., Sorba, P. 1982. *Ann. Rev. Nucl. Part. Sci.* 32:443
74. Wilczek, F. 1977. *Phys. Rev. Lett.* 39:1304
75. Peccei, R. D., Quinn, H. R. 1977. *Phys. Rev. Lett.* 39:1440
76. Weinberg, S. 1978. *Phys. Rev. Lett.* 40:223
77. Wilczek, F. 1978. *Phys. Rev. Lett.* 40:220
78. Edwards, C., et al. 1982. *Phys. Rev. Lett.* 48:903
79. Porter, F. C., et al. 1982. See Ref. 58, p. 27
80. Barik, N., Jena, S. N. 1982. *Phys. Rev.* D26:618

Ann. Rev. Nucl. Part. Sci. 1983. 33 : 67–104

8

PROGRESS AND PROBLEMS IN PERFORMANCE OF e^+e^- STORAGE RINGS

R. D. Kohaupt and G.-A. Voss

Deutsches Elektronen-Synchrotron DESY, Notkestrasse 85, D-2000 Hamburg 52, West Germany

---

CONTENTS

1. INTRODUCTION

Electron-positron storage rings have been in operation for high-energy physics since 1967. But even before 1967 much pioneering work had been done on the electron-positron ring ADA at Frascati (1), VEPP I at Novosibirsk (2), and the two intersecting Princeton-Stanford electron rings (3). It was the 380-MeV ring VEPP II at Novosibirsk where the first e^+e^- annihilation events were observed in 1966 (4); soon after that events were also observed at the ACO storage ring at Orsay (5).

Since then electron-positron storage rings have played an ever increasing role in elementary particle physics. The simplicity of the initial state of an electron-positron annihilation event, the fact that (given sufficient energy) one can produce any particle together with its antiparticle as long as they have electromagnetic coupling, has made electron-positron storage rings one of the most powerful tools in modern high-energy physics. This is particularly true with storage rings getting into the energy range where weak neutral currents play a significant role.

Figure 1 lists the center-of-mass energies actually obtained for high-energy physics experimentation in all e^+e^- projects built so far. Table 1 gives the operating parameters of those storage rings presently doing e^+e^- physics.

For elementary particle physics certain aspects of electron-positron storage rings are of great importance; these are discussed in the next sections.

1.1 *Energy*

Fifteen years ago it was assumed that the largest energy of practical importance for e^+e^- storage rings would be about 2×3 GeV. This was because form factors for production of strongly interacting particles were throught to decrease with a high power of the center-of-mass energy E_{cms}, and cross sections for QED processes decreasing with E_{cms}^{-2} would also become impractically small at $E_{\text{cms}} > 6$ GeV. In particular, the first point has been shown to be wrong. Production of strongly interacting particles proceeds mainly via production of quarks with different flavors. With the opening of new channels at higher energies, cross sections decrease more slowly than the QED cross sections. Progress in storage ring technology resulted in ever increasing event rates (luminosity) and now makes physics with colliding electron-positron beams with center-of-mass energies up to 200 GeV not only feasible but also very desirable.

The physics of electron storage rings is dominated by energy losses of the circulating beams due to synchrotron radiation. These losses make the installation of large and expensive radiofrequency accelerating systems

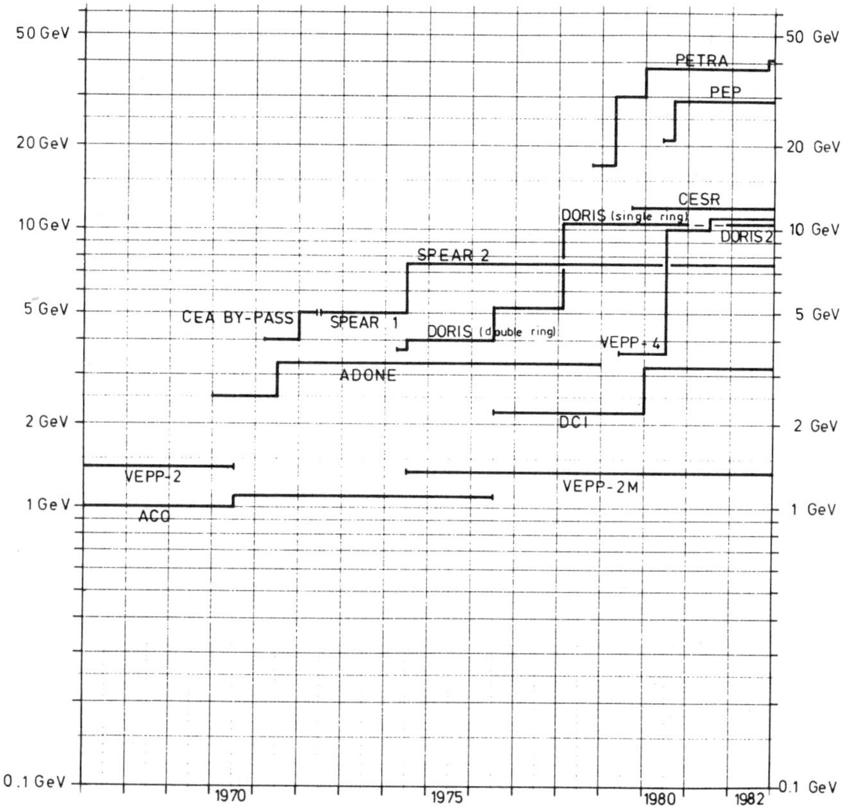

Figure 1 Maximum available center-of-mass energies for e^+e^- colliding-beam physics.

necessary. Cost optimization (6) makes the size and cost of electron and positron storage rings grow quadratically with particle energy. This law sets an upper limit for the energy of electron-positron storage rings at center-of-mass energies around 300 GeV, at which point linear colliders may become more economical (7). Superconducting radiofrequency systems, under development at various laboratories, may push the economical energy limit for electron storage rings somewhat higher; but they do not change the quadratic cost-vs-energy law.

1.2 *Luminosity*

The luminosity L is defined as the event rate n for a particular reaction divided by the cross section σ for that reaction, $L = n/\sigma$ cm^{-2} s^{-1}. (The units cm^{-2} s^{-1} are assumed for luminosity throughout this article.) Peak luminosities in early storage rings were in the 10^{27} to 10^{28} range. Since then

Table 1 Colliding-beam parameters of e^+e^- storage rings presently in operation (December 1982)

	Circumference (m)	Maximum useful collision energy achieved for high-energy physics (GeV)	Maximum luminosity achieved ($cm^{-2}\ s^{-1}$)	Maximum luminosity at the energy (GeV)	Currents for maximum luminosity (mA)
VEPP-2M (Novosibirsk)	17.9	2×0.67	5×10^{30}	2×0.6	2×20
DCI (Orsay)	94.6	2×1.56	1.4×10^{30}	2×1.55	2×125
SPEAR 2 (Stanford)	234	2×3.7	1.6×10^{31}	2×3.7	2×20
DORIS 2 (Hamburg)	288	2×5.1	1.0×10^{31}	2×4.7	2×30
VEPP-4 (Novosibirsk)	366	2×5.5	2×10^{30}	2×4.7	6×6
CESR (Cornell)	768	2×6.0	1.6×10^{31}	2×5.5	2×17
PEP (Stanford)	2200	2×14.5	1.1×10^{31}	2×14.5	6×6
PETRA (Hamburg)	2304	2×20.03	1.7×10^{31}	2×17	4×6

several machines have reached values in the 10^{31} range. How this was achieved is described in the following sections. At the highest energies of a storage ring, luminosities are usually limited by the available rf power, which limits the possible synchrotron radiation power loss by the beams and thereby the beam currents. At low energies, luminosities are limited by the *beam-beam interaction* forces (Section 4). At intermediate energies, luminosities may be limited by the amount of current that can be accumulated and brought to collision. Single-current limitations may be due to *beam instabilities* or *intensity-dependent resonances* (Section 3).

Currents stored in storage rings decay because of the interaction with the residual gas. At average pressures of 10^{-7} torr the typical lifetime due to bremsstrahlung losses is 30 minutes. Ultra high vacua are difficult to obtain in electron-positron storage rings because of gas desorption due to synchrotron radiation (8). Progress in vacuum technologies has resulted in an average pressure smaller than 10^{-8} torr with corresponding lifetimes of several hours for most storage rings now. Since the luminosity is proportional to the product of both currents, it will decrease faster than the currents. For colliding-beam experiments the important number is the *time-averaged luminosity* as given by peak luminosity, luminosity decay time, and the frequency and time needed for refilling the ring with electrons and positrons (Section 2). Average luminosity may range from only 20 up to 90% of peak luminosities, depending on positron production and accumulation rates, on whether injection can take place at the operating energy or if acceleration of newly injected currents to the operating energy is necessary, and on the luminosity decay time.

1.3 *Energy Spread*

Counterrotating currents in electron storage rings are contained in short single bunches. The normal current distribution is Gaussian in all three dimensions and given by the quantum nature of the synchrotron radiation emission. The stochastic excitation of particle oscillations by synchrotron radiation increases with the fourth power of particle energy, while an integral damping effect increases with the third power. Both together are responsible for the Gaussian shape of the bunches, which in all dimensions increase linearly with energy (height, width, and relative energy spread). For different machines, but with the same energy, the energy spread is inversely proportional to the square root of the bending magnet radius (9). The relative energy spread at maximum energy for each listed storage ring is of the order of 10^{-3} to 10^{-4}. Its narrowness is ideally suited for analyzing narrow particle resonances. But at low energy this Gaussian energy distribution can be disturbed by the turbulent *bunch lengthening*, an effect that increases the energy spread in low-energy storage rings by factors up to

4 or 5 (Section 3) and decreases particle production rates in very narrow resonances by the same factor.

1.4 *Background*

Interaction regions in electron-positron storage rings have an extremely small volume because of the short bunch length and the very small transverse beam dimensions. The number of reactions from e^+e^- collisions originating from this small volume exceeds the number of reactions with the residual gas and cosmic ray events by a very large factor. In general, discrimination of one type of these events against the others does not present any difficulty. A greater problem is the general radiation background in the detectors. This background originates from particles lost near the detector region due to bremsstrahlung or large-angle scattering. The problem is aggravated by the experimenter's desire for small-diameter beam pipes at the interaction point in order to improve the resolution of vertex detectors. Particle tracking programs are now used to understand better the source of the high-energy background and to design countermeasures (10). Another serious source of background can be the high-energy tail of the synchrotron radiation spectrum. Making the bending magnets nearest the interaction point very weak, the SPEAR group decreased this kind of background significantly (11), a technique now used in all electron-positron storage rings. Monte Carlo programs are used to calculate synchrotron radiation background and to design suitable absorbers (12, 13).

1.5 *Polarization*

Electron and positron beams can be transversely polarized. This is because, for the two possible spin states in the magnetic guide field, the synchrotron radiation processes associated with a spin flip have different probabilities. Hence, a polarization of the initially unpolarized beams may gradually build up. The mechanism was predicted by Sokolov & Ternov (14) and has since been observed in several storage rings. Apart from the physics aspects of such a polarization and the additional handle it gives on interpreting colliding-beam events, there are also important machine physics aspects. Measurement of depolarizing resonance frequencies makes it possible to determine the energies of stored beams with unparalleled accuracy ($\Delta E/E < 10^{-6}$). Rotation of the transverse spin direction into the direction of motion at the collision points will be of great importance in very high-energy machines in order to investigate weak interaction effects. Unfortunately, polarization is very sensitive to the smallest machine imperfections. Polarized beams are more difficult to obtain in the very high-energy machines for which polarization aspects become most interesting. In Section 5 we summarize the present state of the art.

In the following an attempt is made to describe some of the progress made in the performance of e^+e^- storage rings and some of the more fundamental limits as seen today. While the energy that can be reached with a new storage ring project depends mostly on the available budget, luminosities are limited for a number of reasons. Some of the problems can be defined in an analytical way, but much of our present "understanding" comes from extended simulation and beam tracking programs made possible by modern computer technology. At the end of this article a glossary defines some of the symbols and special terms (marked with an asterisk) used by us. For the reader who wants to learn more about the physics of electron-positron storage rings we recommend the excellent introduction written by M. Sands (9).

2. ELECTRON AND POSITRON ACCUMULATION

In electron-positron storage rings, particles must be accumulated in a few small bunches. Radiation damping* permits particles to be injected off-axis. The injection repetition frequency is typically between 4 and 100 Hz, corresponding to damping times of the order of 100 to 4 ms. The effect of radiation damping together with the stochastic excitation of betatron* and phase oscillations* through synchrotron radiation eliminates all injection peculiarities and concentrates all particles eventually in the same small phase volume. Electron-positron storage rings are quite different in this respect from proton storage rings in which radiation excitation and damping effects at energies presently achieved have been quite negligible so far owing to the large proton mass. Injection problems arise in conjunction with positron production and accumulation.

Since positron injection rates can be very small, they usually determine the time it takes to refill the storage ring and thereby the average luminosity. Positrons are produced in a double conversion process by bombarding tungsten targets with high-energy electrons. Depending on the energy of the electrons and the details of the converter systems, positron yields may vary between 0.1 and 15%. There are two limitations to the positron accumulation rate: The short electron bursts required to produce the short positron bunches are limited in their peak current, and the repetition frequency of the injector has to match the injection rate into the storage ring as given by radiation damping times. Two new systems have been developed during the last few years to overcome these limitations.

2.1 *Bunch Coalescing Through Vernier Phase Space Compression*

In the CESR project (15) a linear accelerator (linac) produces a train of 60 positron bunches per injection pulse. Since in the electron part of the linac the peak current is limited, one now gets 60 times as many positrons as in a single bunch. These positrons are postaccelerated in a second linac to an energy of 200 MeV and in a synchrotron to 8 GeV. At that energy the storage ring has sufficiently strong radiation damping to permit injection and accumulation at a rate of 60 Hz in the storage ring. After the desired number of positrons have been accumulated, the 60 bunches are put on top of each other by a novel vernier phase space compression scheme: The storage ring CESR is arranged concentric to the synchrotron and has a 2.4% larger circumference. By ejecting one bunch at a time and injecting it back into the synchrotron and allowing it to catch up for a few turns with the beam in the storage ring, one can, after transfer back into the storage ring, put each of the original 60 positron bunches into the same bunch location. The practical overall gain factor for this coalescing scheme compared to a direct injection of single bunches into the storage ring is about 15. With this the positron accumulation time is 15 minutes; otherwise it would have been 3.5 hours.

2.2 *Positron Accumulator Rings*

Injection energy at the PETRA storage ring (16) is limited to 7 GeV because of the available injector synchrotron. At that energy, radiation damping in PETRA limits injection rates to 8 Hz. To overcome this handicap and at the same time the problem of peak current limitation in the positron-producing electron linac, an accumulating ring PIA (positron intensity accumulator) (Figure 2) was built (17). This small storage ring with a circumference of 28.8 m employs very strong magnets with a bending radius of 1 m, such that radiation damping at the operating energy of 450 MeV permits injection and accumulation at 50 Hz. The linac injects a bunch train of positrons with a length of 30 m. Radiation damping in the accumulator ring collects these positrons in one 1-m long bunch (the accelerating rf system works on the circumferential frequency of the particles). After 10 linac pulses have been accumulated, this bunch is further compressed by a higher harmonic rf acceleration system. It is then transferred to the injector synchrotron where it is accelerated as a single bunch to the injection energy for PETRA. Such positron accumulator rings make the positron accumulation rate in e^+e^- storage rings independent of the radiation damping times at injection; they become particularly useful for very large rings where the injection energy usually is much lower than the maximum energy. The PIA ring produces eight positron bunches per second, each containing 2×10^9 positrons. With this, injection times for PETRA are reduced to the order of one minute.

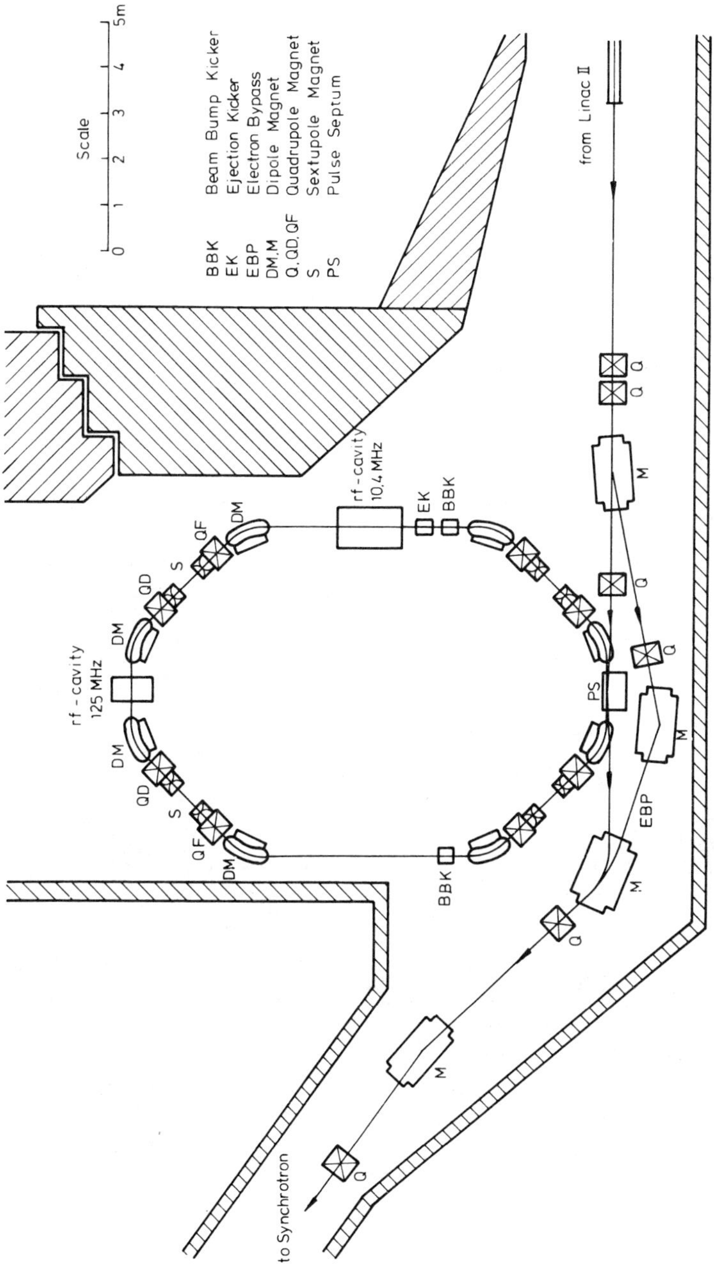

Figure 2 Positron intensity accumulator (PIA).

3. SINGLE-BEAM LIMITATIONS

One of the most severe problems of particle accumulation in e^+e^- storage rings is the limitation of single-beam currents due to intensity-dependent phenomena. Since the circulating particles are focussed in the longitudinal and transverse directions, the beams form an oscillatory system in all three space directions. At high intensities of the stored beam, the circulating particle bunches generate strong electromagnetic fields that can excite electromagnetic resonances in the elements of the vacuum chamber (bellows, kicker, tanks, etc) and particularly in the accelerating rf resonators. The electromagnetic forces induced by these interactions can produce various instabilities of the oscillating particles, so that the intensity of the stored beam will be limited. The interaction of the stored beam with the "environment" is most dangerous for very high-energy machines in which the injection energy is relatively small. These machines need a large number of accelerating resonators for operation at high energies. But these resonators already interact with the beam at low energies where instabilities easily arise because of the lower stiffness of the beam. Much work has gone into the investigation of these current-dependent phenomena.

3.1 *Satellite Resonances*

Among the fundamental perturbations of the stability of particle motion are the current-dependent resonances.

3.1.1 EXCITATION OF SATELLITES BY SPURIOUS DISPERSION IN THE CAVITIES When a particle bunch passes through an rf resonator the individual particle experiences an energy change depending on its longitudinal position within the bunch. In the presence of dispersion* in the resonator, the energy change leads to a displacement of the equilibrium closed orbit*. For particles off this new orbit, this leads to an excitation of betatron oscillations*. Since the energy change depends on the longitudinal position within the bunch and this position changes according to synchrotron oscillations*, the excitation of betatron oscillations is modulated with the synchrotron frequency. As a consequence new "stop bands"* appear as side bands (satellites) of the integer stop bands. With increasing intensities in the beam, the bunches also excite highly nonlinear longitudinal fields in the resonators, such that the longitudinal motion of the individual particles becomes highly nonlinear. This generates higher harmonics of the longitudinal oscillation and produces higher-order satellite stop bands that are current dependent and spaced around the integer stop bands by multiples of the synchrotron frequency. The width of the stop bands increases with current and may considerably reduce the range of possible machine tunes (18–21).

3.1.2 EXCITATION OF SATELLITES BY TRANSVERSE FIELDS If a particle bunch passes through an rf resonator off-axis, it excites transversely acting fields. These fields contain high-frequency components such that the field strength varies within the bunch. With this the transverse forces on the individual particles depend on the longitudinal position, which changes with time. As in the case of spurious dispersion (i.e. unavoidable small values of dispersion due to imperfections of the storage ring), the modulated transverse forces produce current-dependent satellite stop bands around the integer stop bands. Both types of satellite-producing mechanisms have been observed, and although satellite resonances have been investigated with great effort (22–24), they still severely limit single-bunch currents. These satellite effects can be reduced by the following procedures, which lead to a better performance of the storage rings:

1. precise control of transverse and longitudinal oscillation frequencies,
2. minimization of horizontal and vertical dispersion in the rf resonators, and
3. precise control of closed orbit position.

3.1.3 EXCITATION OF SATELLITE RESONANCES AT NONVANISHING CHROMATICITY If the chromaticity* is not completely compensated, the betatron frequency depends on the particle's energy deviation from the ideal synchronous particle. But this deviation changes for each particle with the synchrotron motion. The result is a frequency modulation of the betatron oscillation leading to satellite side bands to resonances. The higher-order satellites become current dependent if the synchrotron motion becomes nonlinear due to the excitation of nonlinear longitudinal electric fields (25, 26).

Satellite resonances belong to the class of incoherent phenomena: Although the higher-order satellites are induced by collective action of all particles in the bunch, the particles become unstable individually because of resonance conditions. Another important example of this class of incoherent current-dependent effects is connected with the higher-order mode losses.

3.2 *Higher-Order Mode Losses*

When a high-intensity particle bunch passes through an rf resonator (or through a cavity-like object in the ring), it excites an infinite number of resonator modes that differ in frequency and field configuration. A fraction of the energy gained by passing through the accelerating cavity is immediately lost through excitation of higher cavity modes by the bunch (27–36). This energy loss increases with the bunch current and is equivalent to a reduction of the accelerating voltage, such that individual particles may leave the stable phase focussing regime. This in turn may lead to a reduction of beam lifetime. At high storage ring energies, which normally are limited by the available rf power, higher-order mode losses cause an intensity limitation. The simplest way to avoid this limitation is to provide sufficient rf power. However, recent progress in the application of computational methods using modern computer technology allows the design of rf resonators for optimum conditions. In existing machines (e.g. PEP, PETRA) great effort was made to keep the vacuum chambers as smooth as possible in order to reduce the higher-order mode losses.

3.3 *Coherent Single-Beam Instabilities*

While satellite resonances and the reduction of lifetime due to higher-order mode excitation are examples of incoherent phenomena, the coherent instabilities represent another class of severe current limitations. Coherent instabilities are accompanied by an unstable coherent motion of particle groups in the beam.

3.3.1 COUPLED-BUNCH INSTABILITIES In the simplest case of coherent instabilities, the bunches of the beam oscillate as a whole. For a given number of bunches B there are B eigenmodes possible differing in the phase between adjacent bunches. An unstable growth of the oscillation amplitude can occur if the oscillating bunches excite parasitic resonator modes with a decay time longer than the time between bunch passages. Because of this long decay time all bunches are coupled and a phase shift can arise between the oscillating bunches and the forces acting back on the beam. This phase shift then can lead to a coupled-bunch instability. Since the bunches are coupled, the currents of all individual bunches contribute. Thus the total current is limited by this kind of instability, which can occur in the longitudinal or transverse direction depending on the field configuration of the excited resonator modes (37–39). One of the basic problems in the case of coupled-bunch instabilities is the identification of the excited resonator modes. Once the offending modes have been identified, "absorbing systems" can be built to match the frequency and the field configuration of the mode. These absorbing systems reduce the "coupling impedance" so that the instability be cured. In the multibunch machine DORIS I, the rf spectrum of the unstably oscillating beam was observed and analyzed. This led to a mode identification. Accordingly constructed and installed absorbers improved considerably the DORIS performance (40). For machines with a small number of bunches where the time-of-flight interval between the bunch passage is sufficiently long (0.5 μs), a "feed-back" system is the most efficient cure for coupled-bunch instabilities. Transverse and longitudinal feed-back systems have been constructed for several machines

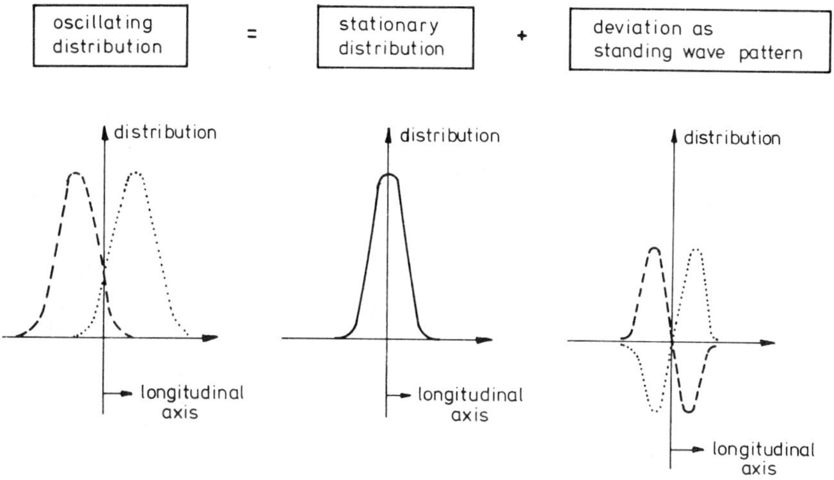

Figure 3 Classification of longitudinal modes.

and are successfully operated to improve the storage ring performance (41–44).

The principal mechanism of a feed-back system is simple: An electrical signal generated from the oscillating bunches will be detected by a pick-up station that distinguishes between individual bunches. This signal passes an amplifier that provides sufficient gain and controls a deflecting device acting on the individual bunches. With an appropriate choice of the phase advance between pick-up and feed-back station, the bunch system can be strongly damped.

3.3.2 SINGLE-BUNCH INSTABILITIES Besides satellite resonances, coherent single-bunch instabilities are the source of the severest limitations of storage ring performance. This kind of instability is characterized by the fact that the bunches of the beam become unstable individually. A single bunch is described by a three-dimensional particle distribution, and in such a distribution an infinite number of internal longitudinal and transverse oscillation modes can exist. Those modes leading to a center-of-mass motion of the bunch form only one special configuration.

For the longitudinal motion, the time-dependent particle distribution is quantitatively described by its deviation from the stationary distribution. This deviation is analyzed in terms of standing wave patterns for the classification of modes. Figure 3 shows the case where the bunch oscillates as a whole.

The transverse configurations so far observed in e^+e^- storage rings are characterized by the fact that particles at a fixed longitudinal position within the bunch oscillate with the same phase. Between particles at different longitudinal positions, however, any phase shift may exist. Accordingly, the transverse modes are classified by analyzing the transverse deformation of the bunch along the longitudinal axis in terms of standing wave patterns (Figure 4).

In principle any longitudinal or transverse mode can become unstable. However, modes with a high mode number can be driven only by fields varying strongly within the bunch. These fields contain high-frequency components that normally are less strongly excited. The excitation of electromagnetic fields by the beam is described in terms of two fundamental quantities, the longitudinal and transverse impedance function*. If these functions are known for all objects in the ring, the strength of single-bunch instabilities can be calculated in the framework of a modern formalism developed by F. Sacherer (45–48). The theoretical determination of the impedance functions is one of the basic problems in this field. Recently it has become possible to find numerical solutions for the impedance functions using new computational methods and large computers (49).

3.3.3 BUNCH LENGTHENING From a general point of view the concept of single modes is only a mathematical tool. For weak interactions (low influences), the single modes have a physical meaning because they are eigenstates of the dynamic system. If the intensity increases, the isolation of single modes becomes physically meaningless. The coupling of longitudinal modes at high intensities leads to a severe single-bunch instability known as "turbulent bunch lengthening" (50–58).

Because of the unstable motion of particles within the bunch, the energy spread of the bunch increases and therefore the bunch length also increases. As a consequence of an increased energy spread, the lifetime can be considerably reduced. Bunch lengthening with increasing single-bunch current was observed in nearly all existing e^+e^- storage rings (59, 60). The

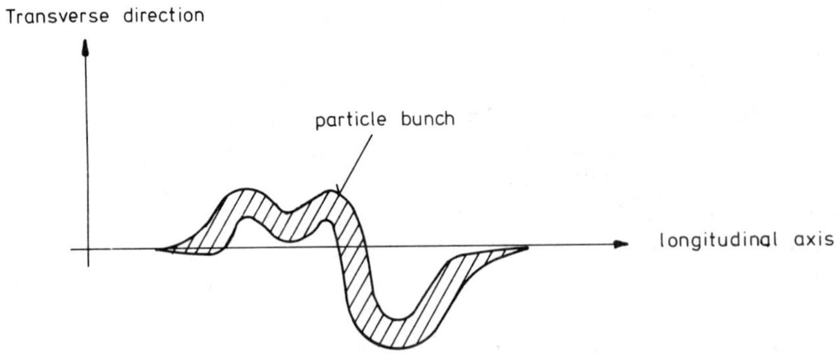

Figure 4 Transverse deformation of particle bunch due to internal transverse oscillation.

investigations of this effect were performed with great effort because of its importance for the large storage rings under construction. Attempts were made to derive "scaling laws" that would allow one to scale bunch lengthening from one machine to another without the explicit knowledge of the longitudinal impedance function. The progress in the numerical computation of resonator fields as part of longitudinal tracking programs led to a better understanding of the observed effects (61).

3.3.4 THE HEAD-TAIL EFFECT The head-tail effect is the classical transverse single-bunch instability that has been observed in nearly all storage rings and was first analyzed for the ADONE ring (62, 63). The mechanism of this effect can be easily demonstrated if the particle distribution of a bunch is approximated by two rigid particle "collectives." If one of the collectives of the circulating bunch is leading and the other is trailing, the leading collective having a small transverse displacement leaves a wake field behind that acts transversely on the trailing collective. Since particles (collectives) differing from the equilibrium longitudinal position perform synchrotron oscillations, the local constellation of two collectives changes with time, and after half a synchrotron period the two collectives have exchanged their longitudinal positions. The two collectives drive each other via this exchange.

The synchrotron oscillation is accompanied by a change of the energy, so that the two collectives have different energies during the exchange of their positions. If the chromaticity of the machine is not compensated, different energies effect different betatron frequencies. Because of this a phase shift between the collectives driving each other causes the head-tail instability. If the chromaticity is compensated, the instability can be avoided. Therefore the head-tail effect does not cause a severe current limitation in e^+e^- storage rings, provided the chromaticity can be compensated. However, the investigation of this effect is of great importance. Since transverse instabilities are governed by the transverse impedance function, the observation of the head-tail instability as a function of various machine parameters (current, chromaticity, bunch length, etc) gives detailed experimental information about this fundamental quantity. Besides an instability the head-tail mechanism causes an observable current-dependent shift of the betatron frequency. The experimental and theoretical investigations of both the head-tail instability and the coherent frequency shift have contributed to a better understanding of transverse effects.

3.3.5 TRANSVERSE "TURBULENCE" After PETRA began to operate the first high-current experiments showed that the single-bunch currents were limited by a vertical instability occurring even at compensated chromaticity. This single-bunch instability seemed to be in disagreement with all mechanisms of known transverse instabilities. The effect turned out to be a severe limitation of the PETRA performance when the number of accelerating resonators was increased. Theoretical and experimental studies showed that this instability could be explained—in analogy to the longitudinal case—by transverse mode coupling (transverse "turbulence"). The theory predicted an increase of the maximum single-bunch current with increasing bunch length, with increasing synchrotron frequency, and with a reduced amplitude function in the rf sections (64, 65). These predictions were confirmed experimentally (66). For the luminosity performance of PETRA the instability can be avoided by an artificial increase of the bunch length during injection and by using a special injection optics with reduced amplitude functions in the rf sections. In the meantime the new instability has also been observed at PEP (67, 68). Since this instability so severely limits single-bunch intensities in large storage rings, further effort is needed to find new concepts for controlling it in future large e^+e^- storage rings such as LEP (69).

A most effective system that can cure various types of instabilities was proposed a long time ago (70): the higher harmonic rf system. The idea of applying such an additional rf system is based on the fact that the gradient of the accelerating voltage can be compensated at the bunch center without affecting those parts of the rf voltage that determine the beam lifetime. This compensation effects a considerable reduction of the small amplitude synchrotron frequency without loss of beam lifetime (71, 72). As a consequence of the reduced synchrotron frequency, the bunch length increases without a change in energy spread.

Since the restoring force of the synchrotron motion becomes strongly nonlinear, the higher harmonic system contributes sufficient Landau damping* (73) to cure longitudinal coupled-bunch instabilities. The reduction of the small amplitude synchrotron frequency shifts the current-limiting satellites out of the range of normal machine tunes.

A higher harmonic rf system was installed in SPEAR (74), and the satellite resonances were completely suppressed. In PETRA the influence of a higher harmonic rf system operating on the second harmonic was studied in detail at the injection energy. In agreement with the observations at SPEAR, the satellite resonances disappeared. In addition the turbulent transverse instability was suppressed up to single-bunch currents far above the values needed for good luminosity performance.

As a result of these intensive investigations of various current-dependent phenomena, the current limitations in all existing e^+e^- storage rings are understood at least qualitatively. However, we are left with the problem of quantitative predictions. These predictions are of great importance for the proper design of the new machines. The problems may be solved by an extended application of computational methods utilizing modern computer technology.

4. SPACE CHARGE LIMITATIONS

4.1 *General Description*

The most important luminosity limitation in storage rings is given by incoherent beam-beam interaction at the collision points. Often this effect is also referred to as space charge limitation. Space charge forces are proportional to $1/\gamma^2$ (γ = energy divided by rest energy). In a single beam, they are totally negligible in high-energy electron-positron storage rings. Space charge forces between the beam and its image currents in the resistive wall of the vacuum chamber can also be neglected, but the electric and magnetic fields of an opposing beam add up in their effects. If electrons and positrons travel on the same orbit (as is almost always the case in a single electron-positron storage ring and in the absence of transverse electric fields), the effect of these space charge forces on particles close to the axis is that of a focussing lens. This focussing lens increases the horizontal and vertical betatron wave numbers* Q_x and Q_z (betatron oscillations per turn) by ΔQ_x and ΔQ_z (75), (ΔQ is the wave number change per interaction region):

$$\Delta Q_x = \frac{r_e N_B \beta_{xip}}{2\pi\gamma\sigma_x(\sigma_x + \sigma_z)} \qquad 1.$$

$$\Delta Q_z = \frac{r_e N_B \beta_{zip}}{2\pi\gamma\sigma_z(\sigma_x + \sigma_z)} \qquad 2.$$

where r_e is the classical electron radius; N_B is the number of particles in the opposing bunch; $\beta_{x,zip}$ are the beta functions* at the interaction point; and $\sigma_{x,zip}$ are the beam dimensions at the interaction point (standard deviation).

Such a change of the number of betatron oscillations per turn can move the operating point in the Q_x-Q_z diagram onto a resonance* and cause beam loss. More important though is the fact that this focussing lens is linear, i.e. independent of the transverse position, only in the area where the current density of the opposing beam is constant. But since particle densities have Gaussian distributions, the space charge lens as given by the opposing beam is highly nonlinear, creating on its own a large number of new nonlinear resonances. The strength of these resonances increases with the opposing current. Many of these resonances are still too weak or may be avoided by the right choice of Q values at small currents. As the currents increase, one reaches a point where stable storage is no longer possible. This point is characterized by a maximum linear tune shift ΔQ for particles close to the axis. But it must be stressed that this linear tune shift is not the cause of the problem. Its effect could be compensated by a readjustment of storage ring parameters. The ΔQ serves only as a convenient way to describe the strength of the nonlinear effects, which are proportional to the linear effects and are the real cause of current limitations.

4.2 *Observations*

The most recent summary of observations on incoherent beam-beam interaction in electron-positron storage rings and its effect on luminosity was given in a special session on beam-beam effects at the International Accelerator Conference 1980 in Geneva (76). The observations at different storage rings are more or less in agreement and can be summarized in the following way:

1. As the currents in the opposing beams are increased, the vertical beam size, which normally is small as compared to the horizontal beam size, increases. This reduces the specific luminosity L_{sp} ($L_{sp} = L/i^+i^-$).
2. As the currents are further increased, the luminosity itself may reach a maximum and drop again. Eventually the beam lifetime will become very short. In some cases the point of short beam lifetime is reached before the luminosity has gone through a maximum.
3. The point of highest luminosity is characterized by vertical ΔQ values of usually between 0.025 and 0.06 (e.g. 77). This space charge limit ΔQ increases somewhat with the radiation damping of betatron oscillations, i.e. in a given storage ring with the beam energy. Proton-antiproton storage rings with no radiation damping have reached ΔQ values of 0.0035 (78).
4. If there is more than one bunch of each kind in the ring, the space charge limit per interaction region is usually lower, but there are also cases where the limit, with up to three bunches in the beam, was not smaller than that with only one bunch of each kind. With the larger bunch number, a luminosity three times as large could be obtained (79).
5. The space charge limits are higher if the vertical betatron frequency is slightly above an integral number (77). But only a few storage rings can be operated in this region.
6. If the opposing bunches have different currents, the stronger one is much less affected by beam-beam forces. Unequal beam currents result in unequal vertical blow-ups and in smaller specific luminosities. It is therefore important to make both currents as equal as possible. Sometimes it is difficult to prevent an unequal increase of vertical beam size even with identical currents [flip-flop effect (80)].

4.3 *Theory and Computer Tracking Results*

The theoretical understanding of this space charge effect has proven to be exceedingly difficult and has resulted only in qualitatively correct predictions. According to Chirikov (81) the width of nonlinear resonances created by the space charge forces is proportional to ΔQ. If there were only one degree of freedom for particle motion, stability would exist as long as the widths of the resonances were smaller than their spacing (Chirikov criterion). If resonances overlap, stochastic or chaotic motion will result with subsequent beam blow-up. But even in the case of narrow resonances there are stochastic regions that in the case of multidimensional motion (for instance, two degrees of betatron motion and one degree of synchrotron motion) may be connected such that a gradual blow-up may result (Arnold diffusion).

Quantitative agreement between experiment and the Chirikov theory is not very good. One problem may be that small imperfections in the storage ring optics (slightly different beam size at the different interaction points, small differences in the betatron phase advance* between interaction points, differences in dispersion or small spurious vertical dispersion at the interaction points, and many others) may have a profound effect on the strength of certain nonlinear resonances but are very difficult to measure.

During the last few years it has become possible to simulate beam-beam interactions with large computers (82, 83). Such simulations take all conceivable effects into account. The kick a particle receives through the electromagnetic fields of the opposing bunch will lead to a slightly different betatron motion in each turn. Simulations assume the proper charge distribution at the interaction points (already affected by beam-beam interactions) and include also phase motion of the particles. The stochastic excitation of particle oscillations due to synchrotron emission as well as the radiation damping are included. Assumptions about machine errors of the kind mentioned above are found to be very important. Simulations allow the separation of the different effects to such an extent that the effectiveness of each machine error in lowering the space charge limit can be judged. A comparison of computer predictions with observations indicates that, with reasonable assumptions of machine errors, good agreement can be reached. Computer simulations seem to describe all the observed features of incoherent beam-beam interactions correctly: The time dependence of the blow-up when both beams are brought to collision, the difference between strong-weak beam and strong-strong beam effects and the energy dependence of the maximum ΔQ shifts (84).

4.4 *Ways to Improve the Space Charge Limited Luminosity*

The incoherent space charge interaction limits the currents that can be brought to collision and thereby the luminosity. For Gaussian current distributions one can derive the equation for the luminosity to be

$$L = \frac{N_1 N_2 f_c}{4\pi\sigma_x\sigma_z B}. \qquad 3.$$

Here $N_{1,2}$ are the numbers of electrons (positrons) in each beam, B is the number of bunches in each beam, f_c is the circumferential frequency, and $\sigma_{x,z}$ are the standard deviations of the beam dimensions at the interaction points. If one eliminates from Equation 3 the quantities N_1 and N_2 by using Equations 1 and 2, one gets an expression for the space charge limited luminosity:

$$L = \frac{\pi B f_c \gamma^2}{r_e^2} \frac{(1+K)^2 \varepsilon_x \Delta Q^2}{\beta_{zip}}. \qquad 4.$$

In this expression ε_x is the horizontal emittance* of the beam and K is the "coupling factor" describing the vertical emittance ($\varepsilon_z = K\varepsilon_x$). The beam dimensions* have been expressed by $\sigma = \sqrt{\varepsilon\beta}$. It is assumed that the limiting $\Delta Q_z = \Delta Q_x$ and that both ΔQ have been made the same by the right choice of β_{xip} at the interaction points ($\beta_{xip} = \beta_{zip}/K$).

Equation 4 indicates all possibilities for improving the space charge limited luminosity in electron-positron storage rings. For given ΔQ and ε_x the luminosity increases quadratically with γ, i.e. with energy. But since ε_x itself increases quadratically with energy (for a given storage ring optics), the luminosity normally would increase with the fourth power of energy. Since ΔQ also may have a small energy dependence (85), the overall energy dependence of the space charge limited luminosity is usually stronger than the fourth power. The luminosity is proportional to the number of bunches B in each beam, provided a larger B does not decrease the space charge limit ΔQ (see above). The coupling factor K is not a free parameter; K normally is determined by betatron coupling in imperfectly aligned quadrupoles and by vertical betatron excitation from synchrotron radiation in conjunction with spurious vertical dispersion (also due to machine imperfections). In a well-aligned machine with small currents, K may be as small as a few percent. The first indication of beam-beam interaction is an increase of K, which subsequently allows for larger currents and thereby larger luminosities. If the vertical beam size in both beams would increase by the same amount, the luminosity would increase until the vertical beam dimensions reach a limiting aperture. At that moment the lifetime would be drastically reduced. The only truly free parameters that can be adjusted for high luminosity are the horizontal emittance ε_x, the vertical beta function β_{zip}, and the bunch number B.

4.5 *Variation of the Number of Bunches in Each Beam*

In a single ring the minimum number of bunches B in each beam is determined by the desired number of beam-beam collision points, which is $2B$. An increase in B will only lead to higher luminosities if the space charge limit per interaction point Q is not drastically reduced with the now larger number of interaction points. To overcome this problem one could build two separate rings for the electron and positron beams such that they intersect each other only at a small number of interaction points. In such a setup a very large number of bunches in each beam should be permissible, since each bunch would only meet an opposing bunch a few times per turn. This was the rationale for the original design of the DORIS I storage rings: In two independent rings up to 480 bunches could be accumulated. The two rings intersected each other at two points. One would assume that the space charge limited luminosity is now 480 times as large as that of a single ring with only one bunch per beam. At low energies DORIS indeed had larger luminosities than comparable single rings. But the factor 480 was never realized. It turned out that the space charge limits for crossing-beam geometries are an order of magnitude lower than those for head-on collisions (86). Further, at higher energies the total current was limited by multibunch instabilities. A luminosity advantage of such a two-ring arrangement as compared to a single ring existed only for beam energies smaller than 2 GeV. In single rings a way of increasing the number of bunches B in each beam is by avoiding unwanted beam-beam collisions with electrostatic separator fields. This technique is planned for LEP (87), where in the so-called Phase I four electron bunches are to collide with four positron bunches at four interaction points. The other four potential beam-beam collisions are avoided by vertical beam separation. Similar techniques are being prepared for the storage ring CESR.

4.6 *Adjustment of the Horizontal Emittance*

According to Equation 4, the space charge limited luminosity is proportional to the horizontal emittance. For a given storage ring optics, i.e. for a given lattice configuration and for given focus strengths of the quadrupole magnets in the ring, the emittance is proportional to γ^2. At a certain maximum energy the circulating beam will use up the acceptance of the machine as determined by vacuum chamber cross sections and orbit distortions or by the effect of nonlinear resonances. At energies below that limiting energy, the luminosity could be increased by an artificial increase of beam emittance and a simultaneous increase of beam current such as to stay at the space charge limit. Such emittance control can be achieved in three different ways.

4.6.1 EMITTANCE CONTROL THROUGH WIGGLER MAGNETS Wiggler magnets (88) consist of a series of dipole magnets with alternating polarity such that the overall deflection of the beam is zero. They are arranged in straight sections of the storage ring. If the magnetic field strength in the wigglers dominates that in the rest of the machine, stochastic excitation of betatron and phase oscillations together with an increased integral damping effect due to the larger average synchrotron radiation energy losses will result in larger beam emittances. The result is a set of beam parameters similar to those of the machine at higher energy. The disadvantage of emittance control through wigglers is the unavoidable increase in energy spread and the increased overall synchrotron radiation losses.

4.6.2 EMITTANCE CONTROL THROUGH VARIABLE OPTICS The beam emittance depends on the storage ring optics. Lattices with weaker focussing result in larger beam emittances. By choosing the optics that will maximize beam emittance at a given energy, the space charge limited luminosity can be maximized. The energy spread is not changed with this change in optics because it depends only on the beam energy and the radius of curvature in the magnetic guide field. A drawback of this method is that it does not permit a continuous emittance variation. By varying the focussing strength, one also changes the frequency of the betatron oscillations* and would soon lose the beams on a resonance.

4.6.3 EMITTANCE CONTROL THROUGH VARIATION OF DAMPING PARAMETERS The emittance is determined by the equilibrium between radiation damping and stochastic excitation of betatron and phase oscillations through the emission of synchrotron radiation. According to a theorem formulated by Robinson (89), the sum of the three damping rates for horizontal and vertical betatron oscillations and phase oscillations is a constant, dependent only on machine radius and beam energy. But the partition of total damping rate onto the three modes of oscillation depends on the lattice and can be changed. Horizontal betatron oscillations can be less strongly damped by transferring some of that damping to phase oscillations; this will increase the horizontal emittance. In large storage rings such a change of damping rates is extremely simple. The partition of damping rates depends critically on the average beam position in the quadrupole elements. A small change in orbit radius as it can be easily facilitated by a small change of frequency of the accelerating cavities can change the damping rate for horizontal betatron oscillations to the point where they are antidamped (Figure 5). The corresponding change in orbit radius is, for example, only 2.9 mm for the PETRA storage ring.

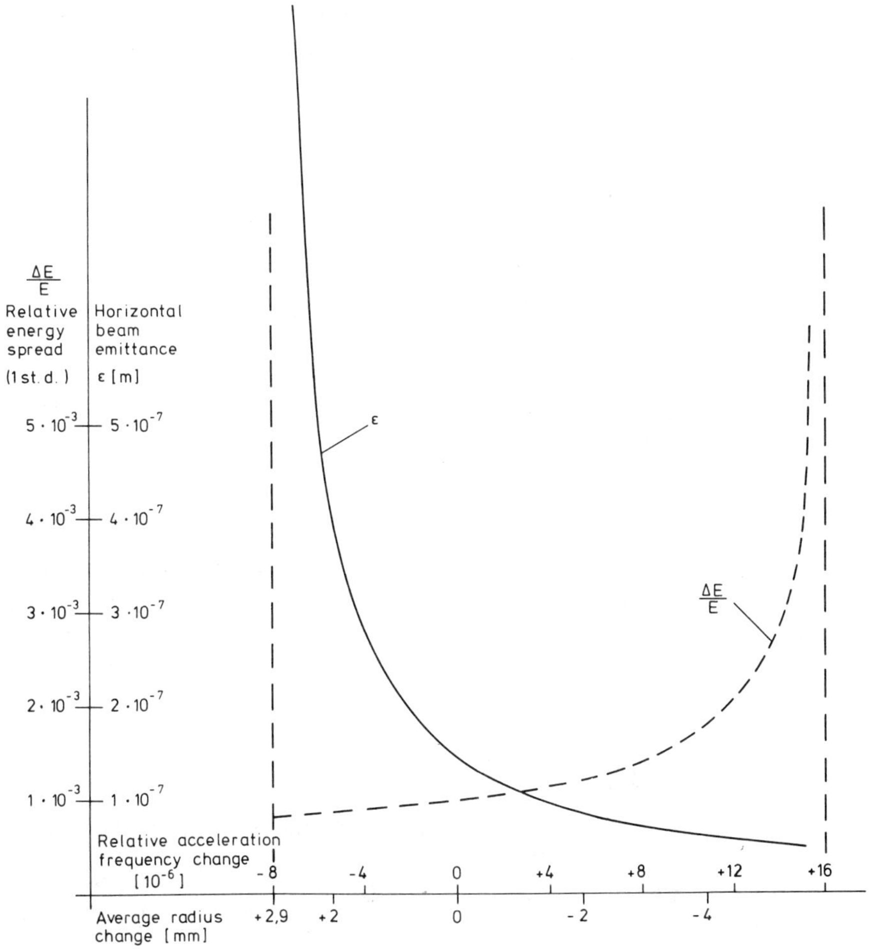

Figure 5 Variation of damping parameters by variation of orbit radius in PETRA.

Wiggler magnets are presently in use at the PEP facility (79). Variable optics and the damping rate adjustment technique are being used in the PETRA storage ring (16), and all three techniques are planned for LEP (90).

4.7 *Low-beta Interaction Regions*

The most obvious and promising way of increasing the space charge limited luminosity is by decreasing the value of the vertical beta function at the interaction points (see Equation 4). This technique was first proposed for the CEA-bypass project (91) and also tried out on that machine (92). By arranging quadrupole magnets near the interaction point one could focus the beam to an extremely small cross section (0.5 mm wide, 0.07 mm high). The corresponding beta functions were $\beta_x = 7$ cm, $\beta_z = 22$ cm. This "low-beta" interaction region increased the luminosity by about two orders of magnitude as compared to an interaction point in the normal storage ring lattice. Low-beta interaction regions have since been incorporated in most storage rings. The limit to how small beta can be made is given by chromatic effects and the problems of how to correct them (next section).

4.8 *Chromaticity Problems*

For a given beam emittance* the beam divergence at the interaction point is inversely proportional to the beam size. Very small beam cross sections at the interaction points imply large angular divergences and consequently large beam cross sections at the adjacent quadrupoles. The beta function at the first lens is given by

$$\beta(s) = \beta_{\mathrm{ip}} + \frac{s^2}{\beta_{\mathrm{ip}}}, \qquad 5.$$

with s being the distance to the interaction point. Large beta functions at these lenses cause large chromatic errors: Particles with energies deviating from the central energy will be focussed differently. This in turn will lead to a different betatron wave number. The energy dependence of the betatron wave number is called chromaticity:

$$\xi = \gamma \frac{\mathrm{d}Q}{\mathrm{d}\gamma}. \qquad 6.$$

The chromaticity contribution of a single quadrupole magnet to the total chromaticity is

$$\Delta\xi = -\frac{\beta}{4\pi f}, \qquad 7.$$

where f is the focal length of the quadrupole magnet. Large beta functions at the quadrupole cause large chromaticity contributions. The overall natural chromaticity of a storage ring is negative. Low-beta interaction regions contribute a large portion to the overall chromaticity. This chromaticity must be corrected to values close to zero for several reasons:

1. The strongest coherent single-beam instability is the so-called head-tail instability (see Section 3). For the most important mode of this instability, the strength is directly proportional to the negative chromaticity. By correcting ξ to zero this instability is suppressed.
2. The dependence of the betatron wave number on energy will cause satellite resonances (see Section 3). These resonances are side bands to

the main linear and nonlinear resonances and are caused by the frequency modulation of betatron oscillations by phase oscillations. The strength of these resonances is directly proportional to the chromaticity.
3. The sensitivity of the storage ring to orbit errors is much larger for a chromatically uncorrected machine (93).
4. The acceptance for off-momentum particles is greatly reduced if the chromaticity does not equal zero.

For these reasons it is desirable, even necessary, to correct the machine chromaticity. Such a correction can only be done with sextupole magnets at locations of beam dispersion, i.e. where particles of different momentum travel on different orbits. For horizontal beam displacements sextupoles act like quadrupoles with a strength depending on orbit position. To particles travelling on different orbits due to different energy deviation, they can give the necessary extra focussing (or defocussing) strength to make the wave number independent of energy. These new elements produce, on the other hand, very bad side effects: For particles with vertical displacement they act like quadrupoles rotated by 45°, giving rise to a coupling of horizontal to vertical betatron oscillations. They also give rise to nonlinear betatron resonances for particles with large betatron amplitudes. The result is a limitation of the possible beam emittance that can be stably stored. Large chromaticities need strong sextupole corrections and consequently lead to small machine acceptances. Many investigations have been made to find the best chromatic corrections which preserves the largest machine acceptance (see, for example, 94–96). It is obvious that the desire for a very small beta function at the interaction points has to be balanced against the difficulty of correcting the chromaticity. In order to reach the largest space charge limited luminosity, one must make a compromise between these conflicting demands. This then determines the smallest value of β in a low-beta interaction region. (Other limitations may be given by the bunch length, see Section 4.10.)

4.9 *Mini-beta Interaction Regions*

The focal length of the quadrupoles near the interaction region is proportional to the distance between quadrupole and interaction point. The chromaticity contribution (Equations 5 and 7) of these quadrupoles can therefore be rewritten as

$$\Delta\xi \approx \frac{s^2/\beta_{\mathrm{ip}} + \beta_{\mathrm{ip}}}{4\pi s} \approx \frac{s}{\beta_{\mathrm{ip}}} \qquad \text{for } s \gg \beta_{\mathrm{ip}}. \tag{8.}$$

If one wants smaller beta values without increasing the chromaticity, s should be made as small as possible. An optical arrangement in the storage

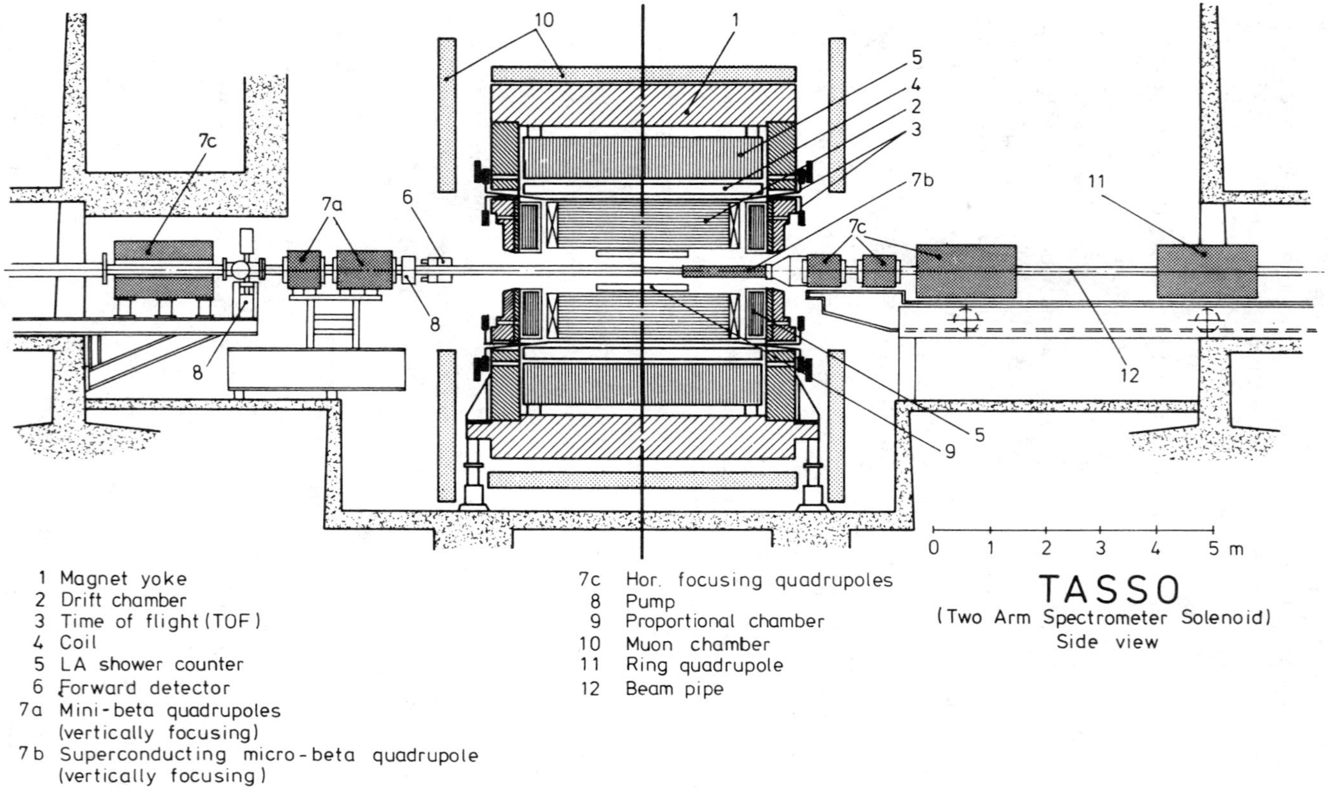

Figure 6 (*Left*) Mini-beta arrangement with TASSO detector (in operation). (*Right*) Micro-beta arrangement with TASSO detector (as planned).

ring PETRA, in which the quadrupoles were moved as close to the detector as possible, increased luminosity by a factor of 3 (97, 98). This arrangement was dubbed "mini-beta interaction region." Similar reductions of the quadrupole distance from the interaction point were subsequently introduced in the PEP and CESR storage rings. Figure 6 (*left*) shows a PETRA "mini-beta interaction region."

4.10 *Micro-beta Interaction Regions*

The limit of beta in a "mini-beta" region is reached when the quadrupoles forming the final focus are just outside the detector. If one wants to further decrease β_{ip}, quadrupoles would have to move into the detector. Most of the detectors use longitudinal magnetic fields that are incompatible with a steel quadrupole magnet. This, and the fact that when the focussing quadrupole moves ever closer to the interaction point the focal strength must get larger and larger, led to the proposal of air core superconducting quadrupoles inside the magnetic detector very close to the interaction point (99). An arrangement as shown in Figure 6 (*right*) is proposed for the PETRA storage ring. It is expected that beta functions as small as $\beta_{xip} = 70$ cm and $\beta_{zip} = 4$ cm can be reached, which would further increase luminosity by a factor of 2 as compared to the "mini-beta" solution. A similar arrangement is planned for the LEP storage ring at CERN.

What now constitutes the limit of beta reduction in a "micro-beta" region? In storage rings with very high energy it is the maximum gradient in superconducting quadrupoles. For LEP a minimum distance of 3.5 m between interaction point and front face of the first quadrupole seems to be a practical limit if one wants to stay within tolerable levels of chromaticity. This then results in a β_{zip} of 10 cm. For storage rings with smaller energies it was shown by computer tracking (83) that the space charge limit ΔQ is independent of beta only if beta is large compared with the bunch length. The maximum luminosity is reached when $\beta_{zip} = 2\sigma_L$ (σ_L = one standard deviation of the bunch length).

4.11 *Space Charge Compensation*

An interesting attempt to overcome the space charge limitation is being made with the DCI storage ring at Orsay (100). Space charge forces are made to vanish with a second bunch of the same shape and intensity but with opposite charge travelling in the same direction. Beam collisions now take place between four beams, an electron and a positron beam coming from both sides (Figure 7). So far space charge compensation with four beams has not resulted in larger luminosities. An explanation may be that small differences in the cross sections of the four beams create nonlinear fields strong enough to cancel the benefits of any partial compensation.

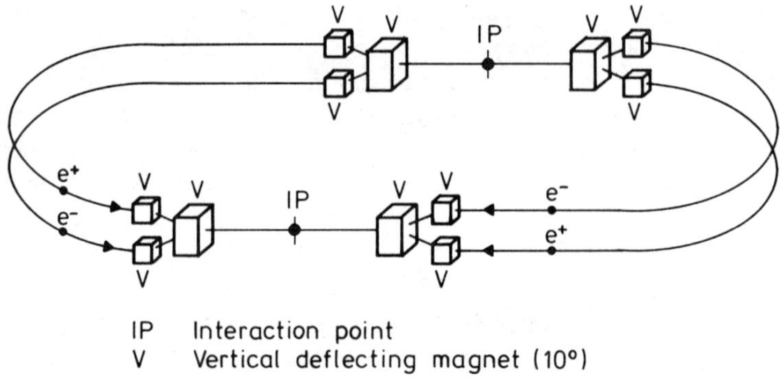

Figure 7 The four-beam space charge compensation scheme (schematic). Each of the two rings contains one electron and one positron bunch.

There also seem to be coherent beam-beam instabilities between the four beams.

5. POLARIZATION

5.1 *History and Observations*

Observations of spin polarization in electron storage rings were first made in 1971 independently by the Orsay (101) and Novosibirsk (102) groups. Polarization was measured by the spin-dependent intrabeam elastic scattering (Touschek scattering). The observed build-up times and the observed degree of polarization agreed well with the Sokolov-Ternov predictions, and evidence for various types of depolarizing resonances was found. The main resonances occur when the spin precession f_{sp} is an exact multiple n of the circumferential frequency f_c:

$$f_{sp} = a\gamma f_c = nf_c$$

with

$$a = \frac{g-2}{2} = 1.16 \times 10^{-3},$$

where g is the gyromagnetic moment of the electron.

This means that the main imperfection resonances occur at energy intervals of 440 MeV. By applying longitudinal or transverse magnetic fields with a frequency matched to f_{sp}, the beams can be depolarized at a given energy γ. This made it possible for the Novosibirsk group to determine the exact masses of the ϕ, $K^{\pm}$, K^0, ψ, and ψ' mesons and recently the Y resonance with an accuracy of 3×10^{-5} (103). At PETRA an accurate measurement of the frequency that depolarizes the beam allowed an absolute energy measurement at an energy of 16 GeV with an accuracy of 3 $\times 10^{-6}$ (104). In 1975 the SPEAR group made use of the polarization: By measuring the distribution of the azimuthal angle for the jet production in e^+e^- collision experiments it was possible to determine the spin of the initial state quarks to be $-1/2$ (120). Also, a new method for measuring the polarization was developed at SPEAR: Circularly polarized laser photons were Compton-backscattered from the stored beam. Since the vertical distribution of the high-energy backscattered photons depends on the spin direction of the electrons, this is an effective way for measuring the degree of polarization (105). This method is now generally used and is also applicable to new large storage rings such as HERA or LEP (106, 108). Shátunov (107) applied longitudinal magnetic fields on the spin precession resonance in such a way as to produce a complete spin flip of the electrons without appreciable depolarization. This method may have great significance for very large storage rings in which individual bunches could now have different spin states. This together with a spin rotation into the longitudinal direction at the interaction points would be an excellent means of investigating weak interaction effects (108).

5.2 *Polarization and Depolarization*

The time constant τ for the synchrotron radiation induced build-up of transverse polarization is, according to Sokolov & Ternov (14),

$$\tau = 98.7 \frac{\rho^3}{E^5} \frac{R}{\rho},$$

where ρ is the radius of curvature in the guide field measured in meters, E is the particle energy in GeV, R is the average radius of the machine, and τ is measured in seconds.

The theoretical maximum polarization that can be reached in the absence of depolarizing effects is 92%. The principal mechanism of depolarization is that of spin diffusion: In an ideal plane machine the equilibrium spin will always be perpendicular to the plane of motion. Particles not 100% polarized will, classically speaking, precess about this equilibrium spin direction with the frequency $f_{sp} = a\gamma f_c$. If the machine is disturbed by imperfections or by special inserts, a new equilibrium spin direction can be defined: a spin motion that closes upon itself after one turn. This alone would reduce polarization because as soon as the equilibrium spin no longer points in the direction of the magnetic guide field the polarizing mechanism is lessened. In such a disturbed machine, the spins of particles with different momenta do not necessarily have the same direction at each point along the circumference. The change of this vector with momentum is called spin dispersion in close analogy to orbital motion. But particles with large betatron oscillations will also have a different spin motion. At the instant when a particle emits a photon through synchrotron radiation, the off-momentum closed orbit and betatron amplitude change abruptly. The spin has not changed its direction (except for the rare case of a spin flip), but with respect to the new closed orbit and betatron oscillation amplitude the spin direction no longer corresponds to that of a fully polarized particle. After the excited betatron and phase oscillations have damped down (which happens in a short time as compared to the polarization time) the spin will have changed as compared to the time just before the photon emission. This change is a vector, which may be designated **d** and which should be called a "spin orbit coupling vector." Its magnitude determines the rate of spin diffusion that sets in because of the uncorrelated photon emissions. If **d** is known everywhere in the machine, the degree of polarization may be calculated using an equation derived by Derbenev & Kondratenko (109):

$$P = 0.92\frac{\langle \rho^{-3}\mathbf{B}(\mathbf{n}-\mathbf{d})\rangle}{\langle \rho^{-3}[1-\frac{2}{9}(\mathbf{n}\boldsymbol{\beta})^2+\frac{11}{18}(\mathbf{d})^2]\rangle},$$

where ρ is the radius of curvature, **B** is the magnetic field direction, **n** is the equilibrium spin direction, and $\boldsymbol{\beta}$ is the particle direction.

If the machine distortions are known, **d** and thereby the polarization can be numerically calculated, e.g. with the SLIM program developed by A. W. Chao (110). SLIM accurately predicts the polarization, as long as one is not right on a resonance, where linear approximations are not valid. The main imperfection resonances ($f_{sp} = k_0 f_c$) and the betatron side-band resonances [$f_{sp} = (k_0 \pm Q)f_c$] together with synchrotron side-band resonances [$f_{sp} = (k_0 \pm Q_s)f_c$] are accurately described, but not the effect of nonlinear resonances. The only way in which nonlinear resonances excited by beam-beam interaction or by imperfect magnetic fields may be evaluated in their effect on polarization is through tracking programs (111). Figure 8 shows a typical prediction for the polarization in the HERA storage ring. Because machine imperfections and beam-beam effects turn the spin by an amount proportional to $a\gamma$, their effectiveness grows linearly with energy. Special efforts are necessary to maintain polarization in large machines.

One way of eliminating the effect of resonances is to make the machine "spin transparent." Originally this method was invented by Chao & Yokoya (112) to cancel the depolarizing effects of spin rotators. For such an insert, ten integral conditions can be defined to assure that the spin orbit coupling **d** remains zero. Steffen (113) extended these spin transparency conditions to improve the performance of imperfect machines. Rossmanith & Schmidt (114) showed through SLIM calcu-

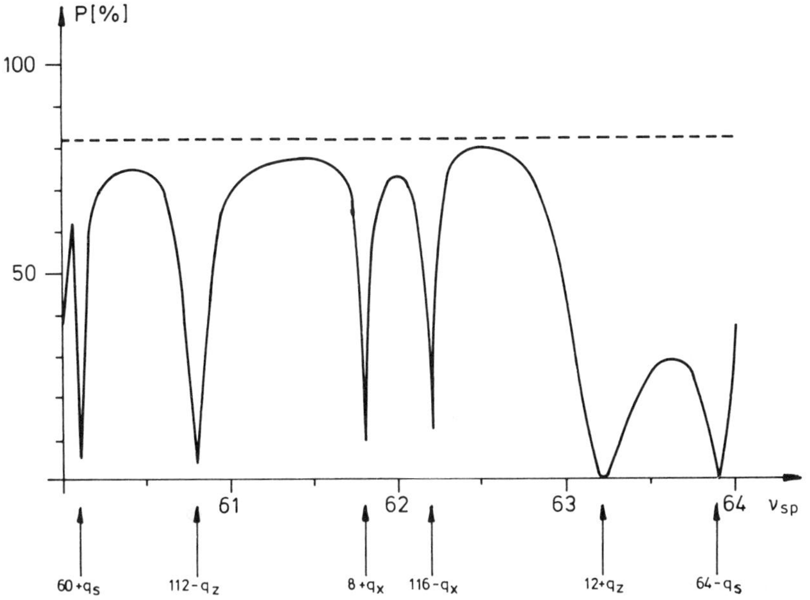

Figure 8 Polarization in a storage ring with an antisymmetric spin rotator (predictions by the SLIM program).

lations that, by cancelling the driving forces of those resonances closest to the working point, depolarization should be greatly reduced. Such cancellation is relatively easily done with the help of a few vertical steering coils. The experiment performed in PETRA (115) showed indeed a dramatic improvement of polarization, from 20 to 80%, when such empirical corrections were applied.

The increasing destructiveness of machine errors with increasing energy is not the only reason why high-energy electron storage rings have difficulties with polarization: The absolute spread in energy gets quite large in these rings and this means that the spread in spin precession frequencies grows to the point where it can no longer fit between the main imperfection resonances, which are spaced at intervals of 440 MeV. The solution to this problem may be the so-called Siberian Snakes (109): spin rotators that turn the spin such that in one half of the ring it is opposite to that in the other half. In this way the spin precession frequency is made to be 1/2, independent of particle energy. The spread in spin precession frequency is reduced to zero, and staying away from harmful resonances is no longer a problem. But at the same time the Sokolov-Ternov mechanism of polarization is reduced to zero. Additional wigglers with unequal field strength in the alternating poles may restore some of the polarization.

Even if polarization can be maintained for single beams under such adverse conditions it will be of no use if beam-beam effects depolarize the beams. Polarization has to be maintained under operating conditions with luminosity. The analytical treatment of this problem is extremely difficult. Estimates (116) make it likely that depolarization through beam-beam interaction may become a strong effect when the space charge forces begin to blow up the beams, i.e. when the specific luminosity begins to drop. In this connection, measurements on PETRA (117) that seem to confirm these estimates are of great interest. With colliding beams characterized by a space charge interaction of $\Delta Q_z = 0.023$, a polarization of 80% was measured; with more intense beams, polarization dropped to 20%. The best values were 80% polarization at a luminosity of 4×10^{30} $\text{cm}^{-2}\ \text{s}^{-1}$.

5.3 *Spin Rotators*

Most of the interest in polarization with very high-energy storage rings is connected with weak interaction physics. For this, longitudinal polarization at the interaction points with both helicities is essential. That way it may be possible, for example, to discover effects of right-handed neutral currents, should they exist. For this, different kinds of spin rotators have been designed (e.g. 118), but are not built yet. Figure 9 shows a spin rotator for the proposed e-p storage ring HERA (119). Most rotators have the drawback that they only work for a particular energy unless they permit some flexibility in their beam geometry, which is quite limited even in the best cases. Designing a rotator that maintains a high degree of polarization is a formidable task. Optics requirements and spin transparency form a set of twelve or more matching conditions that need to be simultaneously fulfilled.

Polarization represents a new dimension of storage ring technology, as yet not fully exploited. There are no reasons known why polarized electron-positron beam collisions should not be possible even at the highest storage ring energies, provided the necessary precautions are taken. And such beam-beam collisions should even be possible with longitudinally polarized beams. But all this requires a high degree of sophistication and complete mastering of the art of accelerator technology. At this moment one is only beginning to get into this field.

6. CONCLUSION AND OUTLOOK

During the last 15 years electron-positron storage rings have evolved into one of the most important tools in experimental high-energy physics. Their energy has increased from 2 × 500 MeV in the ACO storage ring to 2 × 59 GeV for the LEP (Phase I) storage ring presently under construction at

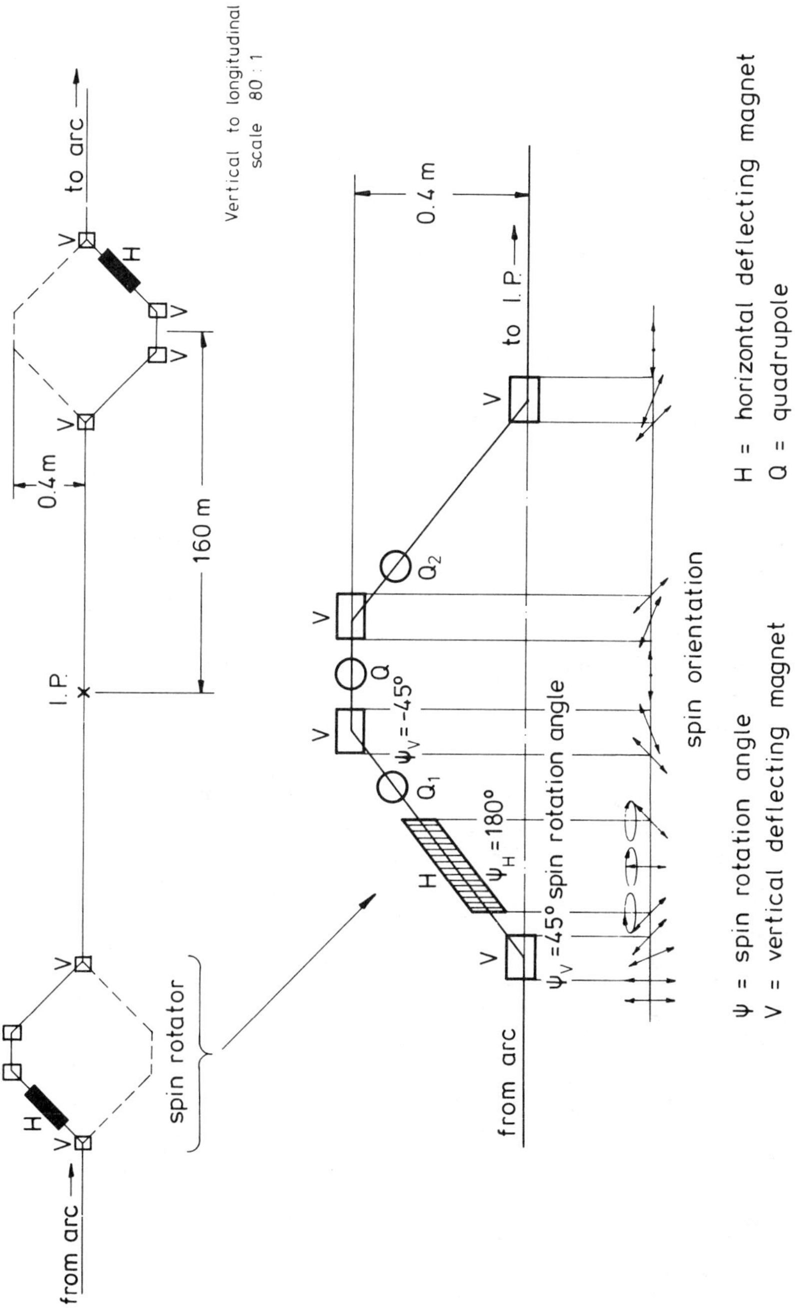

Figure 9 Spin rotator for the HERA storage ring (side view) (schematic).

CERN. At the same time the luminosity of these devices has increased by 3 to 4 orders of magnitude to values in the 10^{31} cm^{-2} s^{-1} range.

This remarkable development was made possible through a major machine physics effort reflected in a very large number of publications in this new field. The main directions of research were in the areas of single-beam stability and beam-beam interaction. Understanding of the effects of the "environment" (vacuum chambers, radiofrequency cavities, etc) on the stability of high-density current bunches has matured to the point where predictions for machines not yet built can be made with confidence. Although the analytical treatment of these problems is very difficult, quantitative descriptions of the relevant effects have become possible through numerical methods for calculating the electromagnetic fields excited by the circulating beams together with tracking programs to determine the interplay between fields and charge distribution in the beams. Similar methods have been very successful in describing the effects of beam-beam interaction. From all this the importance of certain machine features has become evident: storage rings must be built with smooth vacuum chambers to avoid the excitation of fields that can act back on the beams, and with strong focussing at the places of radiofrequency cavities to minimize the effect of beam induced fields on the beam. For high luminosity an extremely strong focussing at the interaction points is essential.

Storage ring technology has improved to the point where major breakthroughs can perhaps no longer be expected. There are only two areas in which we can look forward to major developments. (*a*) Superconducting rf systems may make storage rings cheaper (by perhaps as much as 50%?) and improve their performance somewhat. Because of the higher gradients in superconducting cavities, their required number will be smaller and this will result in a smaller impedance as seen by the beams. (*b*) Polarization of the beams, and particularly longitudinal polarization at the interaction points, is a distinct possibility, although much work is still necessary to make this a practical tool.

Acknowledgments

We would like to thank our colleagues from DESY and other laboratories for helping us with the material for this article. We also wish to thank Drs. D. Degèle, K. Steffen, and Professor L. Hand for helpful comments regarding the manuscript.

Glossary of Some of the Special Technical Terms

Betatron oscillations
Beta function or amplitude function

Betatron wave number
Betatron phase advance: Owing to the transverse focussing, particles deviating from the closed orbit perform transverse oscillations about the closed orbit. If s is the coordinate along the closed orbit and y the deviation from the closed orbit, the betatron oscillation can be described as an amplitude- and phase-modulated harmonic oscillation:

$$y = A\sqrt{\beta(s)}\cos[\phi(s)+\phi_0].$$

The term $\beta(s)$ is called the *amplitude function* or *beta function*, $\phi(s)$ is the *phase advance*, and ϕ_0 and A are constants, where A is the invariant amplitude of the *betatron oscillations*. If C is the circumference of the closed orbit one has

$$\phi(s+C) = \phi(s)+2\pi Q.$$

Here Q is the *betatron wave number* (number of betatron oscillations per revolution). Typical values for large e^+e^- storage rings are $15 < Q < 70$.

Chromaticity: Change of betatron wave number with energy deviation

$$\xi = \frac{\Delta Q}{\Delta E/E}.$$

Closed Orbit: That particular particle trajectory which, at equilibrium energy, closes upon itself after one revolution.

Dispersion: Deviation of a particle with relative energy deviation $\Delta E/E_0 = 1$ from the closed orbit.

Emittance
Beam dimensions: Emittance is the area in phase space filled with particles. It is customary to define

ε_x as the area in the horizontal (x, dx/ds) plane
ε_z as the area in the vertical (z, dz/ds) plane.

The emittance ε is independent of s, and the *transverse beam dimensions* along the closed orbit are described by $\sqrt{\varepsilon\beta(s)}$.

Impedance function: The action of the e.m. forces on the particles is measured in terms of electric potentials. These potentials are proportional to the beam current. The factor of proportionality has the dimension of an impedance and depends on the frequency spectrum of the current. It is called the impedance function.

Landau damping: If the single particles of a particle distribution have different (longitudinal or transverse) frequencies, the individual particles excited by a common perturbation run out of phase after a time that can be characterized by the Landau damping time. This causes a damping of collective motion.

Radiation damping: Synchrotron radiation in electron-positron storage rings damps the betatron and synchrotron oscillations of individual particles with damping rates that are of the order of the energy loss per turn divided by particle energy. The radiation damping rates increase with the third power of the beam energy. Typical damping times τ in large e^+e^- storage rings are $1 < \tau < 200$ ms.

Resonances

Stop bands: Since any imperfection in the magnetic guiding fields of the ring causes a periodic perturbation for the circulating particles, certain values of the wave number Q lead to resonant conditions so that the oscillations of the particles become unstable. Such resonant conditions can occur for the horizontal (Q_x) and vertical (Q_z) direction; they effect in the Q_x, Q_z plane forbidden resonance lines (*stop bands*), on which the particle motion becomes unstable.

Synchrotron oscillations or
Phase oscillations: Particles deviating from the equilibrium energy E_0 by an amount ΔE perform *synchrotron* or *phase oscillations*. Because of these oscillations the energy changes periodically within the interval $\pm\Delta E$. The energy oscillations are accompanied by a periodic change of the time-of-arrival (phase) at the accelerating rf resonators. The time-of-arrival oscillations and the energy oscillations are out of phase by 90°. The number of synchrotron oscillations Q_s along one revolution is the *synchrotron wave number* Q_s (typical values for large e^+e^- machines are $0.03 < Q_s < 0.1$).

Literature Cited[1]

1. Bernardini, C., et al. 1962. *Nuovo Cimento* 23:202
2. Bayer, V. N. 1963. In *Proc. Int. Conf. High-Energy Accel.*, ed. A. A. Kolomenski, p. 274. Moscow: Dubna
3. Barber, W. C., et al. 1966. In *Proc. Int. Symp. on Electron Positron Storage Rings*, ed. E. Cremien-Alcan. Paris: Saclay
4. Budger, G. I. 1966. See Ref. 3
5. Marin, P. 1967. In *Proc. Symp. on Electron Photon Interactions at High Energies*, p. 376. Stanford: AEC (Atomic Energy Commission)
6. Richter, B. 1976. *Nucl. Instrum. Methods* 136:473
7. Voss, G.-A. 1982. In *Proc. of ECFA-RAL Conf.* Geneva: CERN/ECFA
8. Kouptsidis, J. S. 1977. In *Proc. 7th Int. Vacuum Congress and 3rd Int. Conf. on Solid. Surfaces*, p. 93, ed. R. Dobrozemsky. Vienna
9. Sands, M. 1971. In *Proc. Int. Sch. Phys. "Enrico Fermi", Course XL VI 1969*, ed. B. Touschek. New York, London: Academic
10. Bartel, W., Wrulich, A. 1981. *DESY F11-81/03*. Hamburg: DESY
11. *SLAC Annual Report*. 1975. Stanford: SLAC
12. Scholz, K.-U. 1978. *DESY PET-78/03*. Hamburg: DESY

[1] Much of the material has been published in laboratory reports. The best way of obtaining those reports may be by requesting them from the corresponding laboratory libraries.

13. Potter, K. 1981. Gen. Meet. on LEP, *ECFA 81/54*. Geneva: ECFA
14. Sokolov, A. A., Ternov, I. M. 1964. *Sov. Phys. Dokl.* 8 1203:91
15. Tigner, M. 1977. *IEEE Trans. Nucl. Sci.* NS24:1849
16. PETRA Storage Ring Group. 1980. In *Proc. 11th Int. Conf. on High Energy Accel.*, p. 16. Basel: Birkhäuser
17. Febel, A., Hemmie, G. 1980. *DESY M-80/17*. Hamburg: DESY
18. Crowley-Milling, M. C., Rabinowitz, I. I. 1971. *IEEE Trans. Nucl. Sci.* NS18:1052
19. Donald, M. H. R. 1973. *RHEL/M/Nim 18*. Chilton: Rutherford High Energy Lab.
20. Piwinski, A., Wrulich, A. 1976. *DESY 76/07*. Hamburg: DESY
21. Chao, A. W., Piwinski, A. 1977. *DESY PET-77/09*. Hamburg: DESY
22. Vinokurov, N. A., et al. 1977. In *Proc. 10th Int. Conf. on High-Energy Accel.*, p. 254, ed. Yu. M. Ado. Protvino: IFWE
23. Sundelin, R. M. 1979. *IEEE Trans. Nucl. Sci.* NS26:3604
24. Piwinski, A. 1980. See Ref. 16, p. 638
25. Orlov, Yu. F. 1957. *Sov. Phys. JETP* 5:45
26. Robinson, K. W. 1958. *CEA 54*. Cambridge: CEA
27. Liboff, R. L. 1970. *J. Math. Phys.* 11:1295
28. Keil, E. 1972. *Nucl. Instrum. Methods* 100:419
29. Keil, E., et al. 1975. *Nucl. Instrum. Methods* 127:475
30. Chao, A. W. 1975. *PEP 105/SPEAR 198*. Stanford: SLAC
31. Chao, A. W. 1975. *PEP 119*. Stanford: SLAC
32. Zotter, B. 1979. *PEP-310*. Stanford: SLAC
33. Bassetti, M., et al. 1979. *DESY 79/07*. Hamburg: DESY
34. Halbach, K. 1976. *Part. Accel.* 7:213
35. Wilson, P. B. 1977. *SLAC PUB 1908/PEP 240*. Stanford: SLAC
36. Weiland, T. 1980. *CERN ISR-TH/80-07*. Geneva: CERN
37. Pellegrini, C. 1979. *Nuovo Cimento* 64A:477
38. Robinson, K. W. 1969. *CEATM-183*. Cambridge: CEA
39. Karliner, M. M. 1970. *IYa F Report 45-70*. Novosibirsk: Institut Yadernoi fisiki SOAN SSSR
40. Kohaupt, R. D. 1975. *IEEE Trans. Nucl. Sci.* NS22:1456
41. Pruett, C. H., Otte, R. A., Mills, F. E. 1965. In *Proc. 5th Int. Conf. on High-Energy Accel.*, ed. M. Guilli, p. 343. Rome: Comitato Nazionale Energia Nucleare
42. Pellegrin, J. L. 1975. *IEEE Trans. Nucl. Sci.* NS22:1500
43. Heins, D., et al. 1979. *DESY PET-79/06*. Hamburg: DESY
44. Wille, K. 1979. *DESY PET-79/01*. Hamburg: DESY
45. Sacherer, F. J. 1972. *CERN/SI,BR/72.5*. Geneva: CERN
46. Sacherer, F. J. 1973. *IEEE Trans. Nucl. Sci.* NS20:825
47. Sacherer, F. J. 1974. In *Proc. 9th Int. Conf. on High Energy Accel.*, p. 346. Stanford: SLAC
48. Sacherer, F. J. 1977. *IEEE Trans. Nucl. Sci.* NS24:1393
49. Weiland, T. 1982. *DESY 82-015*. Hamburg: DESY
50. Keil, E., Schnell, W. 1969. *CERN-TH-RF/69-48*. Geneva: CERN
51. Sessler, A. M. 1973. *PEP Note-28*. Stanford: SLAC
52. Channell, P. J., Sessler, A. M., 1976. *Nucl. Instrum. Methods* 136:473
53. Messerschmid, E., Month, M. 1976. *BNL-22411*. Brookhaven: BNL
54. Renieri, A. 1976. *LNF-76/11(R)*. Rome: Comitato Nazionale Energia Nucleare
55. Chao, A. W., Gareyte, J. 1976. *SPEAR-197/PEP-224*. Stanford: SLAC
56. Sacherer, F. J. 1977. *CERN/PS/BR/77-5*. Geneva: CERN
57. Ruggiero, A. 1977. *IEEE Trans. Nucl. Sci.* NS24:1205
58. Hardt, W., Kohaupt, R. D. 1977. *DESY 77/20*. Hamburg: DESY
59. Wilson, P. B., et al. 1977. *IEEE Trans. Nucl. Sci.* NS24:1211
60. Kohaupt, R. D. 1977. *DESY 77/66* Hamburg: DESY
61. Weiland, T. 1981. *DESY 81-088*. Hamburg: DESY
62. Pellegrini, C. 1969. *Nuovo Cimento* 64A:447
63. Sands, M. 1969. *SLAC-TN-69-8*. Stanford: SLAC
64. Kohaupt, R. D. 1980. *DESY 80/22*. Hamburg: DESY
65. Kohaupt, R. D. 1980. *DESY M-80/19*. Hamburg: DESY
66. Kohaupt, R. D. 1980. See Ref. 16, p. 582
67. Satoh, K., Chin, Y. 1982. *KEK 82-2*. Tsukuba: KEK
68. Chao, A. W., et al. 1982. To be published. Stanford: SLAC
69. Zotter, B. 1982. *LEP-Note 402*. Geneva: CERN
70. Averill, R., et al. 1973. *IEEE Trans. Nucl. Sci.* NS20:3
71. Gram, R., Morton, P. 1967. *SLAC TN-67-30*. Stanford: SLAC
72. Bramham, P., et al. 1977. *IEEE Trans. Nucl. Sci.* NS24:1490

73. Voss, G.-A. 1976. In *Proc. Int. Sch. Part. Accel. Erice*. Geneva: CERN
74. Wiedemann, H. 1977. See Ref. 22, p. 430.
75. Amman, F., Ritson, D. 1961. In *Proc. Int. Conf. on High-Energy Accel.* Brookhaven: BNL
76. Panel Discussion on Beam Beam Effects. 1980. See Ref. 16, p. 741
77. Amman, F. 1971. In *Proc. 8th Int. Conf. on High-Energy Accel.*, p. 63. Geneva: CERN
78. Hofmann, A., Vos, L., Zotter, B. 1980. See Ref. 16, p. 713
79. Paterson, J. M. 1980. See Ref. 16
80. Donald, M. H. R., Paterson, J. M. 1979. *IEEE Trans. Nucl. Sci.* NS26: 3580
81. Chirikov, B. V. 1979. *Phys. Rep.* 52: 263
82. Piwinski, A. 1980. *DESY 80/131*. Hamburg: DESY
83. Meyers, S. 1981. *CERN-ISR-RF/81-08*. Geneva: CERN
84. Piwinski, A. 1981. *IEEE Trans. Nucl. Sci.* NS28: 2440
85. Piwinski, A. 1981. *DESY 81-066*. Hamburg: DESY
86. DORIS Storage Ring Group. 1979. *IEEE Trans. Nucl. Sci.* NS26: 3135
87. Meyers, S. 1982. *LEP-Note*. Geneva: CERN
88. Paterson, J. M., et al. 1975. *PEP-125*. Stanford: SLAC
89. Robinson, K. W. 1948. *Phys. Rev.* 75: 1912
90. LEP Study Group. 1979. *CERN/ISR-LEP/79-33*. Geneva: CERN
91. Robinson, K. W., Voss, G.-A. 1966. See Ref. 3
92. Averill, R., et al. 1971. See Ref. 77, p. 163
93. Piwinski, A. 1976. *PETRA Note NR.79*. Hamburg: DESY
94. Autin, B. 1976. *CERN/ISR-LTD/76-37*. Geneva: CERN
95. Wiedemann, H. 1976. *PEP-220*. Stanford: SLAC
96. Zyngier, H. 1977. *LAL 77/35*. Paris: LAL
97. Steffen, K. G. 1979. *DESY PET-79/07*. Hamburg: DESY
98. PETRA Storage Ring Group. 1981. *IEEE Trans. Nucl. Sci.* NS28: 2025
99. Steffen, K. G., Voss, G.-A., Wolf, G. 1981. *DESY M-81/20*. Hamburg: DESY
100. Marin, P. 1974. See Ref. 47
101. Serednyakov, S. I., et al. 1976. *Sov. Phys. JETP* 44: 1063
102. LeDuff, J., et al. 1973. *Orsay 4-73*. Orsay: LAL
103. Artanonov, A. S., et al. 1982. *Novosibirsk Preprint 82-94*. Novosibirsk: Institut Yadernoi fisiki SOAN SSSR
104. Neumann, R., Rossmanith. R. 1982. *Nucl. Instrum. Methods* 204: 29
105. Gustavson, D. B., et al. 1979. *Nucl. Instrum. Methods* 165: 177
106. Montague, B. W. 1982. *DESY-82/09*. Hamburg: DESY
107. Shátunov, Ju. 1982. *DESY M-82/09*. Hamburg: DESY
108. Schwitters, R. 1982. *DESY M-82/09*. Hamburg: DESY
109. Derbenev, Ya. S., Kondratenko, A. M. 1973. *Sov. Phys. JETP* 37: 968
110. Chao, A. W. 1981. *Nucl. Instrum. Methods* 180: 29
111. Kewisch, J. 1982. *DESY M-82/09*. Hamburg: DESY
112. Chao, A. W., Yokoya, K. 1981. *KEK 81-7*. Tsukuba: KEK
113. Steffen, K. G. 1982. *DESY HERA 82-02*. Hamburg: DESY
114. Rossmanith, R., Schmidt, R. 1982. *DESY M-82/09*. Hamburg: DESY
115. Rossmanith, R., Schmidt, R. 1982. *DESY 82-026*. Hamburg: DESY
116. Courant, E. 1982. *DESY M-82/09*. Hamburg: DESY
117. Bremer, H. D., et al. 1982. In *Proc. of High-Energy Spin Phys.* Brookhaven: BNL
118. Buon, J. 1982. *SLAC Rep. 250*. Stanford: SLAC
119. HERA Proposal. 1981. *DESY HERA 81/10*. Hamburg: DESY
120. Schwitters, R. F., et al. 1975. *Phys. Rev. Lett.* 35: 1609

Appendix

Reprinted from Reviews of Modern Physics, Vol. 56, No. 2, Part II (April 1984)

Stable Particle Table

Quantities in italics are new or have changed by more than one (old) standard deviation since April 1982.

Particle	$I^G(J^P)C$	Mass (MeV)	Mean life (sec) $c\tau$ (cm)	Partial decay mode: Mode	Fraction		p or p_{max} (MeV/c)
				LEPTONS			
ν_τ	$J=\frac{1}{2}$	< 164					
τ	$J=\frac{1}{2}$	1784.2 ±3.2	$(3.4\pm0.5)\times10^{-13}$ $c\tau=0.010$	$\tau^-\rightarrow$ (or $\tau^+\rightarrow$chg. conj.)			
				$\mu^-\bar{\nu}\nu$	(18.5 ± 1.1)%		889
				$e^-\bar{\nu}\nu$	(16.5 ± 0.9)%		892
				hadron$^-$ neutrals	(*48.1* ± *2.0*)%	S=1.1*	
				3(hadron$^\pm$) neutrals	(*17.0* ± *1.3*)%	S=1.2*	
				5(hadron$^\pm$) neutrals	(<1.4)%		
				†[3(hadron$^\pm$)ν	(5 ± 4)%		
				3(hadron$^\pm$)$\nu(\geq1\gamma)$	(12 ± 4)%]		
				†[$\pi^-\nu$	(10.3 ± 1.2)%		887
				$\rho^-\nu$	(22.1 ± 2.4)%		726
				$K^-\nu$	(*1.3* ± *0.5*)%		824
				K^- neutrals	(*small*)%]		

(continued next page)

Stable Particle Table *(cont'd)*

Particle	$I^G(J^P)C$	Mass (MeV)	Mean life (sec) $c\tau$ (cm)	Partial decay mode: Mode	Fraction	p or p_{max} (MeV/c)
				$\tau^- \rightarrow$ (or $\tau^+ \rightarrow$ chg. conj.)		
$\boldsymbol{\tau}$ (continued)				†[$K^{*-}(892)\nu$	(1.7 ± 0.7)%	669
				$K^{*-}(1430)\nu$	(<0.9)%	323
				$\pi^-\rho^0\nu$	(5.4 ± 1.7)%]	718
				e^- chgd.parts. + μ^- chgd.parts.	(<4)%	
				$\mu^-\gamma$	(<5.5)$\times10^{-4}$	889
				$e^-\gamma$	(<6.4)$\times10^{-4}$	892
				$\mu^-\mu^+\mu^-$	(<4.9)$\times10^{-4}$	876
				$e^-\mu^+\mu^-$	(<3.3)$\times10^{-4}$	886
				$\mu^-e^+e^-$	(<4.4)$\times10^{-4}$	889
				$e^-e^+e^-$	(<4.0)$\times10^{-4}$	892
				$\mu^-\pi^0$	(<8.2)$\times10^{-4}$	884
				$e^-\pi^0$	(<2.1)$\times10^{-3}$	887
				μ^-K^0	(<1.0)$\times10^{-3}$	819
				e^-K^0	(<1.3)$\times10^{-3}$	823
				$\mu^-\rho^0$	(<4.4)$\times10^{-4}$	722
				$e^-\rho^0$	(<3.7)$\times10^{-4}$	726

CHARMED NONSTRANGE MESONS

$D^{\pm}$	$\frac{1}{2}(0^-)$	1869.4 ± 0.6	$(9.2^{+1.7}_{-1.2})\times 10^{-13}$ $c\tau = 0.028$	$D^+ \to$ (or $D^- \to$ chg. conj.)				
				e^+ anything	$(19\ ^{+4}_{-3})\%$			
				K^- anything	$(16 \pm 4)\%$			
				$\overline{K}^0$ any + K^0 any	$(48 \pm 15)\%$			
				K^+ anything	$(6.0 \pm 3.3)\%$			
	$m_{D^\pm} - m_{D^0} = 4.7$ ± 0.3			η anything	$^k(< 13)\%$			
				$\mu^+\nu$	$(<2)\%$			932
				†[$K^-\pi^+\pi^+$	$(4.6 \pm 1.1)\%$		S=1.3*	845
				$K^-\pi^+\pi^+\pi^0$	$(2.6\ ^{+3.1}_{-1.0})\%$			816
				$K^-\pi^+\pi^+\pi^+\pi^-$	$(<4)\%$			772
				$\overline{K}^0\pi^+$	$(1.8 \pm 0.5)\%$			862
				$\overline{K}^0\pi^+\pi^0$	$(13 \pm 8)\%$			845
				$\overline{K}^0\pi^+\pi^+\pi^-$	$(8.4 \pm 3.5)\%$			814
				$\overline{K}^0K^+$	$(0.45 \pm 0.30)\%$			792
				$K^+K^-\pi^+$	$(<0.6)\%$			744
				$K^+\pi^+\pi^-$	$(<0.23)\%$			845
				$\pi^+\pi^0$	$(<0.5)\%$			925
				$\pi^+\pi^+\pi^-$	$(<0.4)\%$]			908
				†[$\overline{K}^{*0}\pi^+$	$(<3.7)\%$]			714

(continued next page)

Stable Particle Table *(cont'd)*

Particle	$I^G(J^P)$	Mass (MeV)	Mean life (sec) $c\tau$ (cm)	Partial decay mode: Mode	Fraction		p or p_{max} (MeV/c)
				$D^0 \to$ (or $\overline{D}^0 \to$ chg. conj.)			
D^0 $\overline{D}^0$	$\frac{1}{2}(0^-)$	1864.7 ± 0.6	$(4.4\ ^{+0.8}_{-0.6}) \times 10^{-13}$ $c\tau = 0.013$	e^+ anything	$(5.3\ ^{+2.9}_{-1.3})\%$		
				K^- anything	$(44 \pm 10)\%$	S=1.3*	
				$\overline{K}^0$ any + K^0 any	$(33 \pm 10)\%$		
	$\lvert m_{D_1^0} - m_{D_2^0} \rvert$	$< 6.5 \times 10^{-10}$ MeV $^{\ell}$		K^+ anything	$(8 \pm 3)\%$		
				η anything	$^{k}(< 13)\%$		
				†[$K^-\pi^+$	$(2.4 \pm 0.4)\%$		861
				$K^-\pi^+\pi^0$	$(9.3 \pm 2.8)\%$		844
	$\dfrac{\lvert \tau_{D_1^0} - \tau_{D_2^0} \rvert}{\text{average}}$	$< 0.55^{\ell}$		$K^-\pi^+\pi^+\pi^-$	$(4.6 \pm 1.4)\%$	S=1.2*	812
				$K^-\pi^+\pi^0\pi^0$	(seen)		815
				$\overline{K}^0\pi^0$	$(2.2 \pm 1.1)\%$		860
				$\overline{K}^0\pi^+\pi^-$	$(4.2 \pm 0.8)\%$		842
	$\dfrac{\Gamma(D^0 \to \overline{D}^0 \to K^+\pi^-)}{\Gamma(D^0 \to K\pi)}$	< 0.16		$\pi^+\pi^-$	$(7.9 \pm 3.8) \times 10^{-4}$		922
				$\pi^+\pi^+\pi^-\pi^-$	$(< 1.0)\%$		880
				K^+K^-	$(2.7 \pm 0.8) \times 10^{-3}$]		791
				†[$K^{*-}\pi^+$	$(3.4 \pm 1.4)\%$		711
	$\dfrac{\Gamma(D^0 \to \overline{D}^0 \to \mu^- \text{ anything})}{\Gamma(D^0 \to \mu^\pm \text{ anything})}$	< 0.044		$\overline{K}^{*0}\pi^0$	$(1.4\ ^{+2.3}_{-1.4})\%$		711
				$K^-\rho^+$	$(7.2\ ^{+3.0}_{-3.1})\%$		679
				$\overline{K}^0\rho^0$	$(0.1\ ^{+0.6}_{-0.1})\%$		677
				$\overline{K}^{*0}\rho^0$	$(0.7\ ^{+0.8}_{-0.7})\%$		423
				$K^-\pi^+\rho^0$	$(3.9\ ^{+1.3}_{-1.6})\%$		613
				$\overline{K}^{*0}\pi^+\pi^-$	$(< 2.3)\%$		685
				$K^-A_2^+$	$(< 0.8)\%$]		198

			CHARMED STRANGE MESON			
				$F^+ \rightarrow$ (or $F^- \rightarrow$ chg. conj.)		
$F^{\pm}$	$0(0^-)$	1971 ± 6	$(1.9^{+1.3}_{-0.7})\times 10^{-13}$ $c\tau=0.006$	$\phi\pi^+$	(*seen*)	713
				$\eta\pi^+$	(*possibly seen*)	903
				$\eta\pi^+\pi^+\pi^-$	(*possibly seen*)	857
				$\eta'\pi^+\pi^+\pi^-$	(*possibly seen*)	679
				$\phi\rho^+$	(*possibly seen*)	411
			BOTTOM MESONS			
				$B^+ \rightarrow$ (or $B^- \rightarrow$ chg. conj.)		
$B^{\pm}$	$\frac{1}{2}(0^-)$	5270.8 ± 3.0		$\overline{D}^0\pi^+$	(4.2 ± 4.2)%	2303
				$D^{*-}\pi^+\pi^+$	(4.8 ± 3.0)%	2243
				$B^0 \rightarrow$ (or $\overline{B}^0 \rightarrow$ chg. conj.)		
B^0 $\overline{B}^0$	$\frac{1}{2}(0^-)$	5274.2 ± 2.8		$\overline{D}^0\pi^+\pi^-$	(13 ± 9)%	2298
				$D^{*-}\pi^+$	(2.6 ± 1.9)%	2253
$B^{\pm}$, B^0, $\overline{B}^0$ (not separated)			$(14\pm 4)\times 10^{-13}$ $c\tau=0.042$	$e^{\pm}\nu$ hadrons	(13.0 ± 1.3)%	
				$\mu^{\pm}\nu$ hadrons	(12.4 ± 3.5)%	
				D^0 anything	(80 ± 28)%	
				K anything	(seen)	
				p anything	(> 3.6)%	
				Λ anything	(> 2.2)%	
				e^+e^- anything	(< 0.8)%	
				$\mu^+\mu^-$ anything	(< 0.7)%	

(continued next page)

Stable Particle Table *(cont'd)*

Particle	$I^G(J^P)$	Mass (MeV)	Mean life (sec) $c\tau$ (cm)	Partial decay mode: Mode	Fraction	p or p_{max} (MeV/c)
NONSTRANGE CHARMED BARYON						
Λ_c^+	$0(\frac{1}{2}^+)$	2282.0	$(2.3^{+1.0}_{-0.6})\times 10^{-13}$	$pK^-\pi^+$	(2.2 ± 1.0)%	820
		±3.1	$c\tau$=0.007	$p\overline{K}^0$	(1.1 ± 0.7)%	870
		S=1.8*		$p\overline{K}^0\pi^+\pi^-$	(<4, seen)%	751
				Λ anything	(33 ± 29)%	
				†[$\Lambda\pi^+$	(0.6 ± 0.5)%	861
				$\Lambda\pi^+\pi^+\pi^-$	(< 3.1, seen)%	804
				$\Sigma^0\pi^+$	(seen)]	822
				†[pK^{*0}	(0.48 ± 0.30)%	681
				$\Delta^{++}K^-$	(0.45 ± 0.27)%	706
				$pK^{*-}\pi^+$	(seen)]	575
				e^+ anything	(4.5 ± 1.7)%	
				†[pe^+ anything	(1.8 ± 0.9)%	
				Λe^+ anything	(1.1 ± 0.8)%]	

Reprinted from Reviews of Modern Physics, Vol. 56, No. 2, Part II (April 1984)

Quantities in italics are new or have changed by more than one (old) standard deviation since April 1982.

Meson Table

	$I^G(J^P)C_n$ — *estab.*	Mass M (MeV)	Full Width Γ (MeV)	Partial decay mode: Mode	Fraction(%) Upper limits (%) are 90% CL	p or p_{max} (MeV/c)
$\eta_c(2980)$	$\underline{0^+}(0^-)\underline{+}$	2981 ±6	< 20	$\eta\pi^+\pi^-$	seen	1426
				$2(\pi^+\pi^-)$	seen	1458
				$K^+K^-\pi^+\pi^-$	seen	1343
				$p\bar{p}$	seen	1158

(continued next page)

Meson Table *(cont'd)*

	$I^G(J^P)C_n$ ___ *estab.*	Mass M (MeV)	Full Width Γ (MeV)	Partial decay mode: Mode	Fraction(%) Upper limits (%) are 90% CL	p or p_{max} (MeV/c)
J/ψ(3100)	$\underline{0^-(1^-)-}$	3096.9 ±0.1	0.063 ±0.009	e^+e^-	7.4 ± 1.2	1548
				$\mu^+\mu^-$	7.4 ± 1.2	1545
				hadrons + radiative	85 ± 2	

Decay modes into stable hadrons

Mode	Fraction(%)	p or p_{max} (MeV/c)
†[$2(\pi^+\pi^-)\pi^0$	3.7 ± 0.5	1496
$3(\pi^+\pi^-)\pi^0$	2.9 ± 0.7	1433
$\pi^+\pi^-\pi^0K^+K^-$	1.2 ± 0.3	1368
$4(\pi^+\pi^-)\pi^0$	0.9 ± 0.3	1345
$\pi^+\pi^-K^+K^-$	0.72 ± 0.23	1407
$p\bar{p}\pi^+\pi^-$	0.53 ± 0.06	1107
$2(\pi^+\pi^-)$	0.4 ± 0.1	1517
$3(\pi^+\pi^-)$	0.4 ± 0.2	1466
$n\bar{n}\pi^+\pi^-$	0.38 ± 0.36	1106
$\Xi\bar{\Xi}$	0.32 ± 0.08	818
$2(\pi^+\pi^-)K^+K^-$	0.31 ± 0.13	1320
$K_S^0K^\pm\pi^\mp$	0.26 ± 0.07	1440
$\Sigma^+\bar{\Sigma}^-$	0.24 ± 0.26	988
$p\bar{p}\eta$	0.23 ± 0.04	948
$p\bar{p}$	0.22 ± 0.02	1232
$p\bar{n}\pi^-$ or $\bar{p}n\pi^+$	0.21 ± 0.02	1174
$n\bar{n}$	0.18 ± 0.09	1231

Decay modes into hadronic resonances

Mode	Fraction(%)		p or p_{max} (MeV/c)
†[$\rho\pi$	1.22 ± 0.12		1449
$\omega 2\pi^+2\pi^-$	0.85 ± 0.34		1392
ρA_2	0.84 ± 0.45		1126
$\omega\pi\pi$	0.68 ± 0.19		1435
$K^{*0}(892)\bar{K}^{*0}(1430)$+c.c.	0.67 ± 0.26		1009
$K^\pm K^{*\mp}(892)$	0.34 ± 0.05		1373
$B^\pm(1235)\pi^\mp$	0.29 ± 0.07		1298
$K^0\bar{K}^{*0}(892)$+c.c.	0.27 ± 0.06		1370
ωf	0.23 ± 0.08	S=1.2*	1143
$\phi\pi^+\pi^-$	0.21 ± 0.09		1365
$\eta' p\bar{p}$	0.18 ± 0.06		596
$\phi K\bar{K}$	0.18 ± 0.08		1176
$\omega p\bar{p}$	0.16 ± 0.03		768
$\omega K\bar{K}$	0.16 ± 0.10		1265
$\phi\eta$	0.10 ± 0.06		1320
$\phi f'(1525)$	0.037 ± 0.013		871
$\phi S(975)$	*0.026 ± 0.006*		1184

$p\bar{p}\pi^+\pi^-\pi^0$	0.16 ± 0.06	1033
$\Sigma^0\bar{\Sigma}^0$	0.13 ± 0.04	988
$\Lambda\bar{\Lambda}$	0.11 ± 0.02	1074
$p\bar{p}\pi^0$	0.11 ± 0.01	1176
$2(K^+K^-)$	0.07 ± 0.03	1131
K^+K^-	0.022 ± 0.008	1468
$\pi^+\pi^-$	0.011 ± 0.005	1542
$\Lambda\bar{\Sigma}$	< 0.015	1032
$K^0_S K^0_L$	< 0.009]	1466

$\pi^\pm A_2^\mp$	< 0.43	1263
$K^{*0}(1430)\bar{K}^{*0}(1430)$	< 0.29	606
$K^0\bar{K}^{*0}(1430)$+c.c.	< 0.2	1158
$K^\pm K^{*\mp}(1430)$	< 0.2	1159
$\phi 2\pi^+ 2\pi^-$	< 0.15	1318
$\phi\eta'$	< 0.13	1192
$K^{*0}(892)\bar{K}^{*0}(892)$	< 0.05	1261
ϕf	< 0.037	1037
$\omega f'(1525)$	< 0.016]	1003

Radiative decay modes

†[$\gamma 2(\pi^+\pi^-)$	*0.49 ± 0.17*	1517
$\gamma\rho\rho$	*seen*	1344
$\gamma\iota(1440) \to \gamma K\bar{K}\pi$	0.42 ± 0.12	1214
$\gamma\eta'$	0.36 ± 0.05	1400
γf	0.15 ± 0.04	1286
$\gamma\eta$	0.086 ± 0.009	1500
$\gamma\pi^0$	0.007 ± 0.005	1546

Radiative decay modes (cont'd)

$\gamma\eta_c(2980)$	*seen*	114
$\gamma\theta(1690)$	*seen*	1087
$\gamma\eta\pi\pi$	*seen*	1487
$\gamma D(1285)$	< 0.6	1283
2γ	< 0.05	1548
$\gamma f'(1525)$	< 0.03	1173
$\gamma p\bar{p}$	< 0.01	1232
3γ	< 0.006]	1548

$\chi(3415)$	$\underline{0^+(0^+)+}$	3415.0 ± 1.0	$2(\pi^+\pi^-)$ (incl. $\pi\pi\rho$)	4.3 ± 0.9	1679
			$\pi^+\pi^-K^+K^-$ (incl. $\pi K\bar{K}^*$)	3.4 ± 0.9	1580
			$3(\pi^+\pi^-)$	1.7 ± 0.6	1633
			$\pi^+\pi^-$	0.9 ± 0.2	1702
			$\gamma J/\psi(3100)$	0.8 ± 0.3	303
			K^+K^-	0.8 ± 0.2	1635
			$p\bar{p}\pi^+\pi^-$	0.6 ± 0.2	1320
			For upper limits, see footnote *o*		

(continued next page)

Meson Table *(cont'd)*

	$I^G(J^P)C_n$ — *estab.*	Mass M (MeV)	Full Width Γ (MeV)	Partial decay mode: Mode	Fraction(%) [Upper limits (%) are 90% CL]	p or p_{max} (MeV/c)
χ(3510)	$0^+(1^+)+$	3510.0 ±0.6		$\gamma J/\psi(3100)$	28 ± 3	389
				$3(\pi^+\pi^-)$	2.4 ± 0.9	1683
				$2(\pi^+\pi^-)$ (incl. $\pi\pi\rho$)	1.8 ± 0.5	1727
				$\pi^+\pi^-K^+K^-$ (incl. $\pi K\bar{K}^*$)	1.0 ± 0.4	1632
				$\pi^+\pi^-p\bar{p}$	0.15 ± 0.10	1381
χ(3555)	$0^+(2^+)+$	3555.8 ±0.6		$\gamma J/\psi(3100)$	15.5 ± 1.8	429
				$2(\pi^+\pi^-)$ (incl. $\pi\pi\rho$)	2.3 ± 0.5	1750
				$\pi^+\pi^-K^+K^-$ (incl. $\pi K\bar{K}^*$)	2.0 ± 0.5	1656
				$3(\pi^+\pi^-)$	1.2 ± 0.8	1706
				$\pi^+\pi^-p\bar{p}$	0.35 ± 0.14	1410
				$\pi^+\pi^-$	0.20 ± 0.11	1772
				K^+K^-	0.16 ± 0.12	1708
→ ψ(3685)	$0^-(1^-)-$	3686.0 ±0.1	0.215 ±0.040	e^+e^-	0.9 ± 0.1	1843
				$\mu^+\mu^-$	0.8 ± 0.2	1840
				hadrons + radiative	98.1 ± 0.3	

$m_{\psi(3685)} - m_{\psi(3100)} = 589.06 \pm 0.13$

Particle	$I^G(J^{PC})$	Mass	Width	Radiative decay modes	Fraction	p	Decay modes into hadrons	Fraction	p
				†[$\gamma\chi(3415)$	8.2 ± 1.4	261	†[$J/\psi\pi^+\pi^-$	33 ± 2	477
				$\gamma\chi(3510)$	8.0 ± 1.3	172	$J/\psi\pi^0\pi^0$	17 ± 2	481
				$\gamma\chi(3555)$	7.4 ± 1.3	128	$J/\psi\eta$	2.8 ± 0.6	196
				$\gamma\eta_c(2980)$	0.43 ± 0.26	638	$2(\pi^+\pi^-)\pi^0$	0.35 ± 0.15	1799
				$\gamma\eta_c'(3590)$	0.2 to 1.3	91	$\pi^+\pi^-K^+K^-$	0.16 ± 0.04	1726
				$\gamma\pi^0$	< 0.5 (CL=95%)	1841	$J/\psi\pi^0$	0.10 ± 0.03	528
				$\gamma\eta$	< 0.02	1802	$p\bar{p}\pi^+\pi^-$	0.08 ± 0.02	1491
				$\gamma\eta'$	< 0.02	1719	$K^{*0}(892)K^-\pi^+$ + cc.	0.067 ± 0.025	1674
				$\gamma\iota(1440) \to \gamma K\bar{K}\pi$	$< 0.012^n$]	1562	$2(\pi^+\pi^-)$	0.05 ± 0.01	1817
							$\rho^0\pi^+\pi^-$	0.042 ± 0.015	1751
							$p\bar{p}$	0.019 ± 0.005	1586
							$3(\pi^+\pi^-)$	0.015 ± 0.010	1774
							K^+K^-	0.010 ± 0.007	1776
							$\pi^+\pi^-$	0.008 ± 0.005	1838
							$\rho\pi$	< 0.1	1760
							$\Lambda\bar{\Lambda}$	< 0.04]	1467
$\psi(3770)$	<u>$(1^-)-$</u>	3770	25				e^+e^-	0.0011 ± 0.0002	1885
		± 3	± 3				$D\bar{D}$	dominant	242
	$m_{\psi(3770)} - m_{\psi(3685)} = 83.9 \pm 2.4$ S=1.8*								
$\psi(4030)$	<u>$(1^-)-$</u>	4030	52				e^+e^-	0.0014 ± 0.0004	2015
		± 5	± 10				hadrons	dominant	
							†[$D\bar{D}$	*seen*	752
							$D\bar{D}^* + D^*\bar{D}$	*seen*	559
							$D^*\bar{D}^*$	*seen*]	177
$\psi(4160)$	<u>$(1^-)-$</u>	4159	78				e^+e^-	0.0010 ± 0.0004	2079
		± 20	± 20				hadrons	dominant	
$\psi(4415)$	<u>$(1^-)-$</u>	4415	43				e^+e^-	0.0010 ± 0.0003 S=1.4*	2207
		± 6	± 20				hadrons	dominant	

(continued next page)

Meson Table *(cont'd)*

Particle	$I^G(J^P)$ — *estab.*	Mass M (MeV)	Full Width Γ (MeV)	Partial decay mode: Mode	Fraction(%) [Upper limits (%) are 90% CL]	p or p_{max} (MeV/c)
Υ(9460) or Υ(1S)	$\underline{(1^-)-}$	9460.0 ±0.3 S=1.6*	0.0443 ±0.0066	$\mu^+\mu^-$	2.9±0.5	4729
				e^+e^-	2.5±0.5	4730
				$\tau^+\tau^-$	*3.4±0.8*	4381
$\chi_b(9875)$ or $\chi_b(1\,^3P_0)$	( )+	*9872.9 ±5.8*		γΥ(9460)	*seen*	404
$\chi_b(9895)$ or $\chi_b(1\,^3P_1)$	( )+	*9894.5 ±3.5*		γΥ(9460)	*43±11*	425
$\chi_b(9915)$ or $\chi_b(1\,^3P_2)$	( )+	*9914.6 ±2.4*		γΥ(9460)	*20.0±4.4*	444
Υ(10025) or Υ(2S)	$\underline{(1^-)-}$	10023.4 ±0.3	0.0296 ±0.0047	$\mu^+\mu^-$	*1.9±1.8*	5011
				e^+e^-	1.6±0.3	5012
				Υ(9460)ππ	*19.5±1.7*	476
	$m_{\Upsilon(10025)} - m_{\Upsilon(9460)}$ = *563.3±0.4*			$\gamma\chi_b(9875)$	*3.5±1.4*	149
				$\gamma\chi_b(9895)$	*5.9±1.4*	128
				$\gamma\chi_b(9915)$	*6.1±1.4*	108
→ $\chi_b(10255)$ or $\chi_b(2\,^3P_1)$	( )+	*10253.7 ±3.4*		γΥ(9460)	*seen*	763
				γΥ(10025)	*seen*	228
$\chi_b(10270)$ or $\chi_b(2\,^3P_2)$	( )+	*10271.0 ±2.4*		γΥ(9460)	*seen*	779
				γΥ(10025)	*seen*	245

$\Upsilon(10355)$ or $\Upsilon(3S)$	$\underline{(1^-)-}$	10355.5 ±0.5	*0.0177* *±0.0051*	e^+e^-	*2.0 ± 0.7*	5178
				$\mu^+\mu^-$	*3.3 ± 2.0*	5177
				$\Upsilon(9460)\pi^+\pi^-$	*5.1 ± 1.1*	814
	$m_{\Upsilon(10355)} - m_{\Upsilon(9460)}$ = *895.5 ± .6*			$\Upsilon(10025)\pi^+\pi^-$	*3 ± 3*	177
				$\gamma\chi_b(10235)$	*7.6 ± 3.5*	122
				$\gamma\chi_b(10255)$	*15.6 ± 4.2*	101
				$\gamma\chi_b(10270)$	*12.7 ± 4.1*	84
$\Upsilon(10575)$ or $\Upsilon(4S)$	$\underline{(1^-)-}$	10573 ±4	14 ±5	e^+e^-	0.0017 ± 0.0007	5286
	$m_{\Upsilon(10575)} - m_{\Upsilon(9460)}$ = 1113 ± 4					

CHARMED, NONSTRANGE MESONS

D^+	$\underline{1/2(0^-)}$	1869.4		See Stable Particle Table		
D^0		1864.7				
$D^{*+}(2010)$	$1/2(1^-)$	2010.1 ±0.7	< 2.0	$D^0\pi^+$	64 ± 11	39
				$D^+\pi^0$	28 ± 9	38
				$D^+\gamma$	8 ± 7	136
	$m_{D^{*+}} - m_{D^0}$ = 145.4 ± 0.2 MeV					
$D^{*0}(2010)$	$1/2(1^-)$	2007.2 ±2.1	< 5	$D^0\pi^0$	55 ± 15	44
				$D^0\gamma$	45 ± 15	137

SUBJECT INDEX

D

E

F

G

T

U

V

W